Applications of Polyhedral Oligomeric Silsesquioxanes

ADVANCES IN SILICON SCIENCE

VOLUME 3

Series Editor:

JANIS MATISONS
School of Chemistry, Physics and Earth Sciences, Flinders University, South Australia.

Advances in Silicon Science is a book series which presents reviews of the present and future trends in silicon science and will benefit those in chemistry, physics, biomedical engineering, and materials science. It is aimed at all scientists at universities and in industry who wish to keep abreast of advances in the topics covered.

Series Editor

Professor Janis Matisons,
Nanomaterials Group,
Chair of Nanotechnology,
School of Chemistry, Physics and Earth Sciences,
Flinders University,
GPO Box 2100,
Adelaide 5001
SOUTH AUSTRALIA

Volume 3

Applications of Polyhedral Oligomeric Silsesquioxanes

Volume Editor

Claire Hartmann-Thompson
Michigan Molecular Institute
Midland, MI
USA

For other titles published in this series, go to http://www.springer.com/series/7926

Claire Hartmann-Thompson
Editor

Applications of Polyhedral Oligomeric Silsesquioxanes

Springer

Editor
Claire Hartmann-Thompson
Michigan Molecular Institute
1910 W. Saint Andrews Rd.
Midland Michigan 48640
USA
thompson@mmi.org

ISBN 978-90-481-3786-2 e-ISBN 978-90-481-3787-9
DOI 10.1007/978-90-481-3787-9
Springer Dordrecht Heidelberg London New York

Printed on acid-free paper

Springer is part of Springer Science+Business Media (www.springer.com)

Foreword: The Re-Birth of Polyhedral Oligosilsesquioxane Chemistry

Frank J. Feher

Fundamental research is a wonderful example of chaos. Each researcher makes individual choices about which fields to pursue, which experiments to do and how to do them, how to interpret their data and the data of others, where to seek funding, who to consult or collaborate with, and where to present or publish their work. The end results from fundamental research are deterministic because they are ultimately governed by universal Laws of Nature, but the path to discovery is impossible to predict and the order in which Nature's secrets are unlocked depends greatly on seemingly trivial decisions. In many ways, fundamental research seems analogous to the situation described by Lorenz [1], where the flapping of a butterfly's wings in Brazil can set off a tornado in Texas.

When I was asked to write the foreword to this book, I was struck by the enormous progress made with polyhedral oligosilsesquioxanes (POS) over the past two decades. Numerous researchers have made important contributions to the field, and it is truly remarkable what has been accomplished using these versatile and now ubiquitous organosilicon compounds. To a newcomer, the contents of this book will probably create an impression that the current state-of-the-art was preordained to evolve over the past 25 years. Maybe it was, but as one of the earlier researchers responsible for the renaissance of POS chemistry, I have often wondered where the field would be now if some of the early butterflies had not flapped their wings at just the right moment. From my vantage point, POS chemistry as we know it today is the result of many unexpected twists and turns that uniquely shaped how the field developed. I am grateful to the editor of this book (Claire Hartmann-Thompson) for the opportunity to share my personal perspective about the early days of this renaissance and some of reasons for the success enjoyed by its participants.

The discovery of polyhedral oligosilsesquioxanes (POS) was inevitable because many $[RSiO_{3/2}]_{2n}$ frameworks form rapidly via hydrolytic condensation of $RSiX_3$, and spontaneously crystallize in pure form from common organic solvents. In fact, silsesquioxanes were undoubtedly synthesized (by accident) shortly after the first trifunctional organosilanes (i.e., $RSiX_3$) were prepared, and there are reports consistent with silsesquioxane formation back as far as the 1870s [2]. Not surprisingly, much of the early work in the field was done by researchers at the major silicone producers. Scott of General Electric (GE) described the formation of $Me_8Si_8O_{12}$ in 1946 [3]. Although his analytical data was only good enough to establish that the molecule was an oligomer with the formula $[MeSiO_{3/2}]_{2n}$, he correctly realized that the molecule had a highly symmetric structure because of its poor solubility and tendency to sublime without melting. Several years later, Barry and Gilkey (Dow Corning) [4,5] convincingly established that the molecule indeed possessed a cube-octameric structure, and that a variety of crystalline POS frameworks with the formulae $R_6Si_6O_9$, $R_8Si_8O_{12}$, and $R_{12}Si_{12}O_{18}$ could be prepared from readily available organyltrichlorosilanes ($RSiCl_3$ with R = Me, Et, Pr, Bu, Cy, Ph).

The early pioneers in POS chemistry faced truly formidable challenges because most of the analytical tools we now take for granted were not yet available. Quite often, the only data available for making structural assignments were from elemental analysis (including silicon) and molecular weight measurements in solution. This was generally sufficient to identify pure compounds with the formula $[RSiO_{3/2}]_{2n}$, but it was a major obstacle to making timely progress in the broader field. This situation began to change by the 1960s, when IR spectroscopy became widely available. While it is likely that many POS researchers during this time felt blessed to have IR spectroscopy available to characterize their products and mixtures, it was still extremely difficult to make unambiguous structural assignments for many POS compounds, especially compounds with low symmetry, mixed organyl groups, or multiple stereoisomeric possibilities.

By the 1970s, analytical methods for characterizing organic compounds were advancing rapidly, a sizable body of early knowledge about POS was being accumulated, and efforts to identify commercially viable applications were underway. At the same time, related studies of silicones were moving forward with dramatic success; in comparison, development of POS and POS applications was much less fruitful. Despite three decades of work, the availability of tank-car quantities of several trifunctional silanes (e.g., $PhSiCl_3$, $MeSiCl_3$ and vinyl-$SiCl_3$), and the ability to make some POS compounds in practically quantitative yield, no one had discovered a large-scale commercial application for POS. Time seemed to run out because continuation of this work was difficult to justify during the economic climate of the late 1970s and early 1980s. So as the early pioneers retired or moved on to other fields, the first generation of POS research came to an end.

Completely oblivious to *anything* related to POS, I started graduate school at the University of Rochester in 1980 to pursue a Ph.D. in Chemistry. Although I initially planned to specialize in natural product (i.e., drug) synthesis, I eventually teamed up with a starting assistant professor (William D. Jones) to explore whether homogeneous (i.e., soluble) organometallic complexes could be used to catalytically activate C-H bonds for functionalization by small organic molecules (e.g., CO, CO_2, C_2H_4). In the course of my studies, I became familiar with a related body of work, seeking to understand and mimic the chemistry of metal surfaces using soluble model compounds containing small metal clusters (e.g., $Os_3(CO)_{12}$). These clusters exhibited a rich reaction chemistry with hydrocarbons, and provided valuable insights into possible mechanisms for the reactions of hydrocarbons on metal surfaces. However, none of these cluster models truly behaved like the metal surfaces they were intended to model. This sparked my interest in developing soluble clusters that could better mimic the chemistry of catalyst surfaces.

Towards the end of my graduate career, a fellow graduate student (Prudence Bradley) from another research group introduced me to POS through a presentation she made at a late-night group meeting. It was Prudence's turn to give a review of any "special topic" of her choosing, and by chance she saw a recent review article by Voronkov [6] and decided to focus on $[RSiO_{3/2}]_{2n}$ polyhedra for her presentation. Because the people in attendance were transition-metal chemists with little interest in the chemistry of organosilicon clusters, the presentation did not generate much enthusiasm. Most attendees simply drank more beer than usual and revisited familiar topics of discussion. At one point near the end of our meeting, the group began a familiar debate about the value of using small molecular clusters to model surface chemistry. This time the debate took an interesting turn as I stared at a hand-drawn structure of $R_8Si_8O_{12}$ on an adjacent chalkboard. Rather than questioning the value of cluster models for surfaces, it occurred to me that these models might work well for cases where the surface could be reasonably approximated as an insulating solid with a localized electronic structure. I then suggested that molecules similar to POS might be great models for silica surfaces, if methods could be developed for synthesizing structurally well-defined silsesquioxane frameworks containing reactive Si-OH groups. One of my mentors (Richard Eisenberg) reacted favorably to the idea, but quickly asserted that I would "never be able to make useful quantities of these frameworks." As our meeting adjourned, I took that as a challenge and promptly headed off to the library in search of a silsesquioxane framework that might be suitable as a cluster model for silica, or as a ligand in cluster models for silica-supported catalysts. By the morning, I had found what I was looking for, was reasonably convinced that the idea could be made to work, and was well on my way to creating an independent research program for this purpose.

Upon reading Voronkov's review [6] and diving into the early silsesquioxane literature, it became apparent that a wide variety of structurally well-defined silsesquioxane frameworks could be obtained in synthetically useful quantities, even though synthetic methods, product yields, procedures for separation and purification, characterization data, structural assignments, and mechanistic explanations for product formation often left a lot to be desired. Frameworks with reactive Si-OH groups were much less common and generally less well characterized than fully condensed $[RSiO_{3/2}]_{2n}$ frameworks, but they still seemed to be accessible. John F. Brown Jr.'s work at GE appeared to offer the best starting point, and his work with cyclohexyl-substituted silsesquioxanes [7] provided the most compelling evidence for the formation of structurally well-defined, incompletely-condensed silsesquioxane frameworks. Trisilanol **1**, in particular, seemed to be an ideal candidate for modeling both the chemistry of silica and the chemistry of monometallic silica-supported catalysts. This eventually proved to be the case and provided the foundation for many of my research group's contributions to the re-birth of POS chemistry.

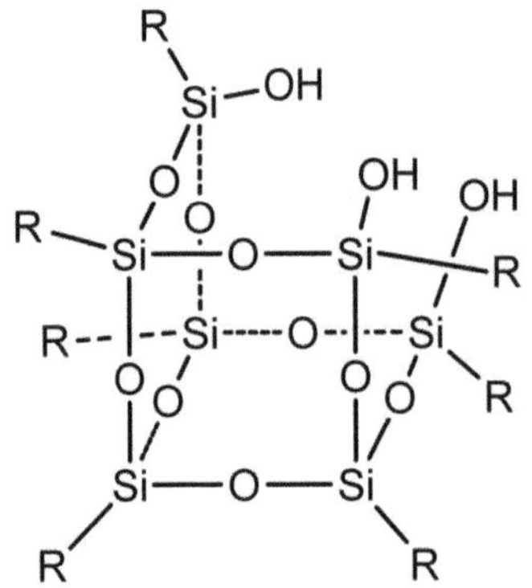

Compound **1**

R = c-C_6H_{11}

Starting in 1985 with trisilanol **1** and a goal to develop realistic solution-state cluster models for silica and silica-supported catalysts, my group at the University of California, Irvine (UCI) began a journey that ultimately produced a large number of new silsesquioxane frameworks and metal-containing frameworks. Some of these frameworks indeed appeared to be good models for silica surface sites [8, 9], as well as excellent ligands for catalytically active cluster models for silica-supported catalysts [10]. At the same time, an increasing number of researchers were starting to discover practical routes to POS frameworks with synthetically useful groups attached to Si (e.g., H, $CH_2CH_2CH_2Z$ with Z = Cl, SH and NH_2). The pool of known POS frameworks expanded rapidly over a very short period of time as general and highly efficient methodology was developed for synthetically manipulating both the Si/O framework and the pendant groups attached to silicon.

Efforts to develop practical applications for discrete POS frameworks gained a major champion in the early 1990's when Joseph D. Lichtenhan initiated a research program at Edwards Air Force Base (California, USA) to use POS-containing polymers as precursors to hybrid inorganic/organic materials. Joe was a former UCI student who learned quickly how to scale up the synthetic methods developed in my laboratory. His team quickly identified a number of promising applications for POS-containing polymers in military applications, and along the way they became so skilled at making POS that they were able to supply samples to Aldrich Chemical Company for sale to other researchers who were interested in working with POS. Eventually, Joe and several colleagues left Edwards Air Force in the late 1990s to form the company now known as Hybrid Plastics, Inc., which is recognized as the commercial leader for POS production and applications.

Much of the new generation of researchers' success with POS was a direct consequence of advances in routine analytical instrumentation for characterizing complex organic molecules, especially multi-nuclear NMR spectrometers, mass spectrometers, and single-crystal X-ray diffraction systems. The first generation of POS researchers simply did not have the necessary tools for the quick and unambiguous assignment of molecular structures, or for the analyses of complex mixtures of products. This was highlighted to me during a visit to GE around 2000, where I had the pleasure of meeting John F. Brown Jr., who was about to start his sixth decade with the company. Although he had not worked with silsesquioxanes since the mid-1960's, he still remembered an impressive amount of information about his earlier work and was enthusiastic about discussing the topic. At one point during our discussion, John volunteered a few seemingly minor details that undoubtedly played an important role in his successes. In particular, John told me that his practice of allowing POS compounds to crystallize slowly from resinous mixtures of products over days, weeks, months or sometimes even years stemmed from his graduate research with sugars, which can be notoriously slow to crystallize from syrupy mixtures. Without this patient approach to synthesis, which was a unique characteristic of his early work with silsesquioxanes, it might have been very difficult to create conditions conducive to the formation of trisilanol **1.** Isolation of pure **1** from other polycondensation products might also have been very difficult if a more expedient synthetic approach had been pursued. Upon hearing this, and realizing that my entry into POS chemistry was largely motivated by John's successful synthesis of trisilanol **1**, I could not resist wondering whether a butterfly had been flapping its wings in Schenectady, New York.

Better analytical tools were only partly responsible for the re-birth of POS chemistry and applications. Two other important reasons for the successes observed after 1985 were an increased focus on mechanistic studies involving POS frameworks, and the deliberate use of POS as precursors to more complex Si/O and Si/O/M frameworks (as well as materials containing these frameworks). In addition to attracting broad interest, expertise, and generous funding from outside the traditional organosilicon community, this emphasis shifted the focus of POS research away from identification and isolation of compounds that spontaneously formed during hydrolytic condensation of $RSiX_3$, and towards the rational design and synthesis of specific compounds for specific purposes. This approach eventually paid big and unexpected dividends when studies aimed at developing methods for selective cleavage of Si-O-Si linkages in POS frameworks led to the discovery that the $R_7Si_7O_9(OH)_3$ structure shown for trisilanol **1** is actually the thermodynamic product when many different trifunctional organosilanes are subjected to hydrolytic condensation in the presence of alkali metal ions. This discovery opened the door to a broad range of options for producing new POS derivatives, and ultimately allowed Hybrid Plastics, Inc. to develop practical methods for large-scale manufacturing of many potentially useful materials. Between the compounds derived from trisilanols such as **1**, and other compounds derived directly or indirectly from hydrolytic condensation of readily available $RSiX_3$ feedstocks, it is not hard to imagine a day in the near future when it might be possible to make POS compounds and derivatives with almost any chemically reasonable structure.

Looking forward it seems likely that future successes with POS will be just as likely to emerge from chaos as they will from deliberate efforts to develop commercially viable materials and practical applications with compelling value in the marketplace. This is the nature of fundamental research. What should be kept in mind as you read the contributions to this book is that regardless of how we arrived at the current state-of-the-art, many of the interesting things accomplished with POS over the past 25 years were initially thought to be beyond reach. This bodes well for the continued development of POS technology; the field is still advancing rapidly, seems destined to offer real materials as solutions for a broad range of applications, and presents tantalizing opportunities for discovering new, as yet unimaginable, applications that will undoubtedly emerge from efforts to tailor the properties and performance of these materials.

References

1. Lorenz EN (1996) The Essence of Chaos. University of Washington Press, Seattle, pp. 181.

2. Landenburg A (1873) Ber 6:379.

3. Scott DW (1946) J Am Chem Soc 68:356.

4. Barry AJ, Gilkey JW (1949) US Pat. 2465188.

5. Barry AJ, Daudt WH, Domicone JJ, Gilkey JW (1955) J Am Chem Soc 77:4248.

6. Voronkov MG, Lavret'yev V (1982) Top Curr Chem 102:199.

7. Brown Jr JF, Vogt LH (1965) J Am Chem Soc 87:4313.

8. Feher FJ, Newman DA, Walzer JF (1989) J Am Chem Soc 111:1741.

9. Feher FJ, Budzichowski TA, Rahimian K, Ziller JW (1992) J Am Chem Soc 114:3859.

10. Feher FJ, Blanski RL (1993) Makromol Chem Macromol Symp 66:95.

Preface

Claire Hartmann-Thompson

It has been said that the polymer was the defining material of the twentieth century, and some predict that the defining material platform of the twenty-first century could very well be the hybrid, where two or more components are combined in a single material to give new and previously unattainable combinations of useful properties. However, hybrid materials require building blocks, and many exciting ones are available from the much-hyped field of 'nanotechnology' including dendritic polymers, carbon nanotubes, graphenes, buckminsterfullerenes, and polyhedral oligomeric silsesquioxanes- the latter being subject of this book.

Polyhedral oligomeric silsesquioxanes may be considered as a nano form of silica, and in their most common form ($Si_8O_{12}R_8$) are comprised of a cubic cage of eight silicon corner atoms and twelve oxygen edge atoms, where each of the eight silicon atoms may carry one of an extremely wide range of functional groups (R), generating hundreds of possible compounds, many of which may now be manufactured on a large scale and purchased at competitive prices. In their properties, polyhedral oligomeric silsesquioxanes (comprised of SiO_3R units) occupy a fascinating middle ground between silica (comprised of SiO_4 units) and polysiloxanes (comprised of SiO_2R_2 units), and combine many of the benefits of silica (thermal, chemical and radiation stability, and optical transparency) with those of siloxane polymers (solubility, ease of manipulation using conventional chemical techniques, ease of processing, low toxicity, and potential to tailor the structure by varying R to achieve the desired properties). In addition, polyhedral oligomeric silsesquioxanes are extremely versatile, and may be chemically bonded or physically blended into a companion material, resulting in a hybrid nanomaterial having the combined benefits of polyhedral oligomeric silsesquioxanes (above) and the companion material. This is what we expect of hybrid materials technology at its best, and it should not be confused with simply combining two new components for the sake of it, and characterizing the resulting materials, but achieving no real advantage.

The commercial availability of polyhedral oligomeric silsesquioxanes has opened up the field to everybody. It is no longer limited to sol-gel and organoelement specialists, and has expanded to include anyone who wishes to exploit the unique properties of polyhedral oligomeric silsesquioxanes for their own materials science and chemical applications. I am one of the comparative newcomers, but I was transfixed by the possibilities of what could be achieved using this particular molecular play set after I attended the 'POSS® Nanotechnology' conference in Huntington Beach, California in September 2002, and where I first met many of the contributors to this book. We begin with a foreword by Frank Feher, a true pioneer in this field. In the first chapter, Matisons, Markovic and Constantopolous give a chronological overview of the development of polyhedral oligomeric silsesquioxane chemistry, and in the second chapter Lickiss and Cordes give a comprehensive review of the preparation of the vast range of known polyhedral oligomeric silsesquioxane compounds and their chemical and physical characterization. Examples of hybrid materials in useful applications are then covered in the succeeding chapters, including Maschmeyer, Ward and Masters on metallasilsesquioxanes for catalysis (where a metal replaces a silicon atom, generating a system chemically similar to a silica-supported metal catalyst), Coughlin, Williams and Gido on polymer materials carrying covalently bound polyhedral oligomeric silsesquioxanes, DeArmitt on polymer materials carrying physically blended polyhedral oligomeric silsesquioxanes, Mabry, Smith, Peloquin and Iacono on superhydrophobic polymer nanocomposites carrying fluorinated polyhedral oligosilsesquioxanes, my own chapter on polyhedral oligomeric silsesquioxanes in electronics and energy applications (covering liquid crystal and electroluminescent materials, lithographic resists, sensors, fuel cell membranes, battery electrolytes and lubricants), Brandhorst on polyhedral oligomeric silsesquioxanes in space applications, and Seifalian, Ghanbari, Marashi, Rafiei and Chaloupka on polyhedral oligomeric silsesquioxanes in biomedical applications. Hence in the initial chapters, the polyhedral cage is synthesized and characterized, and in the remaining chapters, either the cage itself is modified by introducing metal atoms at silicon positions, or else other materials are themselves modified by physically blended or chemically bonded polyhedral cages. In some of these cases the presence of the polyhedron improves the physical properties (e.g., thermal or mechanical), in other cases the polyhedron is used as a scaffold to carry a useful group (e.g., superhydrophobic, sensing or proton conducting), and in other cases the polyhedron imposes a useful geometry upon the system (e.g., when it dictates liquid crystalline phase transitions, or when it prevents π-stacking, excimer formation and quenching in electroluminescent or fluorescent materials).

This book is truly interdisciplinary, and should be of interest to chemists, materials scientists, polymer physicists, plastics engineers, surface scientists, and anybody with a commercial or academic interest in plastics, composite materials, space materials, dental materials, tissue engineering, drug delivery, lithographic fabrication, fuel cells, batteries, or LC, LED, sensor, photovoltaic or biomedical devices. I hope that this book proves a useful and comprehensive source of reference, both to those familiar with the subject matter and to those looking for something new, diverting and above all, fun.

Finally I would like to thank my publisher Springer, my publishing editor Dr. Sonia Ojo, the series editor Prof. Janis Matisons, my current and former colleagues Dr. Petar Dvornic and Dr. Mike Owen (who went this way before me with Volume 2, 'Silicon-Containing Dendritic Polymers'), Rusty Govitz, Judy Eastland, Dr. James Plonka, Dr. Robert Nowak and Dr. Steven Keinath, and my family and particular friends for their patience, encouragement and interest in this undertaking.

Claire Hartmann-Thompson
Michigan Molecular Institute, Midland, Michigan, USA
August, 2010

Biographical Note

Dr. Claire Hartmann-Thompson

Dr. Claire Hartmann-Thompson was born and educated in England, and received an M.A. in chemistry from the University of Oxford in 1993, and a Ph.D. in physical organic chemistry and high energy materials from the University of Exeter in 1996. She is currently a senior scientist at Michigan Molecular Institute in Midland, Michigan, and moved to the USA in 2001 after several years in industrial polymer research and development with Raychem, Ltd and with Dow Corning Corporation in Great Britain. Her research interests cover polyhedral oligomeric silsesquioxanes, other organosilicon compounds and polymers, dendritic polymers, and the use of these building blocks to create hybrid nanomaterials for sensor, fuel cell and space applications.

Contents

Contributors

Henry W. Brandhorst, Jr.
Director, Space Research Institute, Auburn University, AL, USA

Karla Chaloupka
University College London Division of Surgery and Interventional Science
Royal Free Hampstead NHS Trust Hospital, Hampstead Campus, London, UK

Kristina Constantopolous
School of Chemistry and Physics, Flinders University, Adelaide, South Australia

David B. Cordes
Department of Chemistry, Imperial College London, South Kensington, London, UK

Edward Bryan Coughlin
Department of Polymer Science and Engineering, University of Massachusetts
Amherst, Amherst, MA, USA

Chris DeArmitt
Phantom Plastics, Hattiesburg, MS, USA

Frank J. Feher
The Goodyear Tire and Rubber Company, Akron, OH, USA

Hossein Ghanbari
University College London Division of Surgery and Interventional Science
Royal Free Hampstead NHS Trust Hospital, Hampstead Campus, London, UK

Samuel Paul Gido
Department of Polymer Science and Engineering, University of Massachusetts
Amherst, Amherst, MA, USA

Scott T. Iacono
U.S. Air Force Academy, Department of Chemistry, USAF Academy, Colorado Springs, CO, USA

Paul D. Lickiss
Department of Chemistry, Imperial College London, South Kensington, London, UK

Joseph M. Mabry
Air Force Research Laboratory, Space & Missile Propulsion Division, Edwards AFB, CA, USA

Sayed Mahdi Marashi
University College London Division of Surgery and Interventional Science
Royal Free Hampstead NHS Trust Hospital, Hampstead Campus, London, UK

Elda Markovic
School of Chemistry and Physics, Flinders University, Adelaide, South Australia

Thomas Maschmeyer
Laboratory of Advanced Catalysis for Sustainability, School of Chemistry F11, University of Sydney, New South Wales, Australia

Anthony F. Masters
Laboratory of Advanced Catalysis for Sustainability, School of Chemistry F11, University of Sydney, New South Wales, Australia

Janis G. Matisons
School of Chemistry and Physics, Flinders University, Adelaide, South Australia

Andrew J. Peloquin
U.S. Air Force, Patrick AFB, FL, USA

Yasmin Rafiei
University College London Division of Surgery and Interventional Science
Royal Free Hampstead NHS Trust Hospital, Hampstead Campus, London, UK

Alexander M. Seifalian
University College London Division of Surgery and Interventional Science
Royal Free Hampstead NHS Trust Hospital, Hampstead Campus, London, UK

Dennis W. Smith, Jr.
Department of Chemistry & Center for Optical Materials Science & Engineering Technologies, Clemson University, Clemson, SC, USA

Claire Hartmann-Thompson
Michigan Molecular Institute, Midland, MI, USA

Antony J. Ward
Laboratory of Advanced Catalysis for Sustainability, School of Chemistry F11, University of Sydney, New South Wales, Australia

Katherine Grace Williams
Department of Polymer Science and Engineering, University of Massachusetts Amherst, Amherst, MA, USA

Chapter 1
Polyhedral Oligomeric Silsesquioxanes: From Early and Strategic Development through to Materials Application

Elda Markovic, Kristina Constantopolous and Janis G. Matisons

1.1 Introduction

The chemistry of organo-functionalized silsesquioxanes has emerged as a fascinating new field of modern nanotechnology [1]. Nanostructured polyhedral oligomeric silsesquioxanes (POS) compounds have been used to design novel hybrid nanocomposites now used in a variety of applications [2-5]. Polyhedral oligomeric silsesquioxanes remain well-defined, three-dimensional nanobuilding blocks that can create unique hybrid materials, where a precise control of nanostructure and properties is needed [6-10].

The term sil-ses-quioxane indicates that each silicon is connected to three oxygen atoms, specifically referring to a class of silicon compounds that have the general formula $(RSiO_{1.5})_{2n}$, [where n = 4 and R = H, alkyl, aryl, halogen, etc., which are then called hydridosilsesquioxanes, alkylsilsesquioxanes, arylsilsesquioxanes, and halosilsesquioxanes, respectively]. The architecture of these compounds varies, with random structures, cage structures and partial cage structures known, see Fig. 1.1 [11].

Harrison [1] describes polyhedral silsesquioxanes as compounds with structures based on several Si-O linkages forming a cage having a silicon atom at each vertex. Their substituents are coordinated around the silicon vertices tetrahedrally, and these substituents are important as they determine many of the physical properties of the cage molecule. Silicate cages, also known as polyhedral oligosilsesquioxanes (POS), have structures that involve several rings connected in a confined, three-dimensional molecular skeleton. The prefix oligo- denotes the existence of a

E. Markovic, K. Constantopolous and J.G. Matisons
School of Chemistry and Physics, Flinders University
GPO Box 2100, Adelaide 5001, South Australia
Email: Janis.Matisons@Flinders.Edu.Au

C. Hartmann-Thompson (ed.), *Applications of Polyhedral Oligomeric Silsesquioxanes*, Advances in Silicon Science 3, DOI 10.1007/978-90-481-3787-9_1,

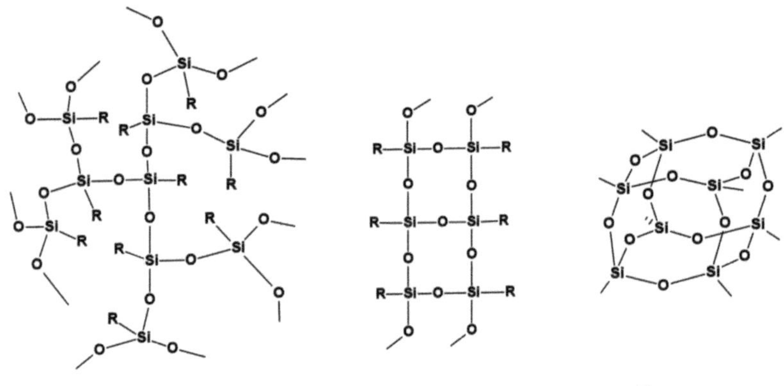

Fig. 1.1 Random, ladder and cage structure conformations of silsesquioxanes (reprinted with permission from reference 11)

small number of silsesquioxane units present in the compound. The prefix 'octa' specifically indicates the number of links, viz. $(XSiO_{1.5})_8$. The nomenclature most frequently used can be understood as follows:

- *sil* – silicon;
- *sesqui* – each Si atom is bound to an average of <u>one and a half</u> oxygens;
- *ane* – Si atom bound to one hydrocarbon group;
- the silicon-oxygen cluster forms a polyhedron (cage) not a cube as the silicon-oxygen-silicon and oxygen-silicon-oxygen bonds are at angles $\neq 90°$.

Some other prefixes are commonly used to show the number of repeat units such as hexa-, octa-, deca- with octahedral silsesquioxane being the most commonly used.

Silsesquioxanes are also referred to as spherosiloxanes because their polyhedral structures are topologically equivalent to a sphere. Other common names include: hydrosilsesquioxanes; hydridosilsesquioxanes; oligosilsesquioxanes; organylsilsesquioxanes; and POSS®.

Two methods are used to name silicon compounds such as polyhedral oligomeric silsesquioxanes: the formal IUPAC and the General Electric (GE) terminology. The terminology and nomenclature that follows the IUPAC system is cumbersome, and so has lead to the widespread use of the shorthand terminology developed by GE. This terminology is dependent on the substitution pattern of oxygen atoms attached to silicon. If the silicon atom is attached to one oxygen atom, it is given the symbol **M** (mono-substituted), to two oxygen atoms **D** (di-substituted), three oxygen atoms **T** (tri-substituted), and to four oxygen atoms **Q** (quaternary or tetra-substituted) (see Fig. 1.2)[10]. The symbols **M, D, T,** and **Q** then represent such empirical structures as $R_3SiO_{0.5}$, R_2SiO, and $RSiO_{1.5}$, and SiO_2 respectively (where R = alkyl group). Generally octahedral silsesquioxanes are then designated as T_8R_8. The R group here may represent hydrogen, alkyl, alkenyl, aryl, arylene, or their functionalized derivatives.

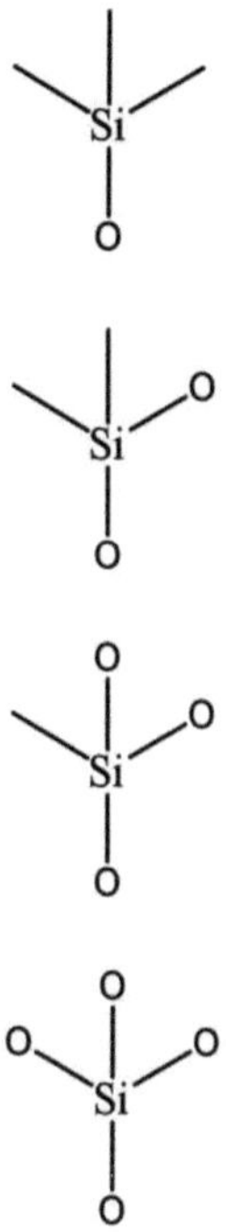

M $(CH_3)_3SiO_{1/2}$ monofunctional moiety

D $(CH_3)_2SiO$ difunctional moiety

T $(CH_3)SiO_{3/2}$ trifunctional moiety

Q $SiO_{4/2}$ quaternary-functional moiety

♦ *Superscript letter represents substituent other than methyl group:*

$HSiO_{1.5}$ ⇒ T^H
$(CH_2{=}CH)SiO_{1.5}$ ⇒ T^{Vi}

♦ *Silsesquioxane structures represented as R-T or H-T:*

MQ e.g. M_8Q_8 or alternatively Q_8M_8; similarly T_8H_8

Fig. 1.2 **M**, **D**, **T**, and **Q** silicone units

Polyhedral oligosilsesquioxanes are often referred to as the smallest particles of silica possible, however, unlike silica, each molecule contains an *exo*-cage of organic substituents which can be specially designed to be non-reactive or reactive. The precise three-dimensional structure of nanoscopic size has an Si-Si distance equal to 0.5 nm and an approximate R-R distance of 1.5 nm (see Fig. 1.3).

Polyhedral oligosilsesquioxanes are useful building blocks, with hydridosilsesquioxanes (Fig. 1.3; R = H) being of interest as cores for dendritic molecules via reaction of their eight reactive hydrogens [11]. However, the chemistry of silsesquioxanes has remained underdeveloped for decades because of

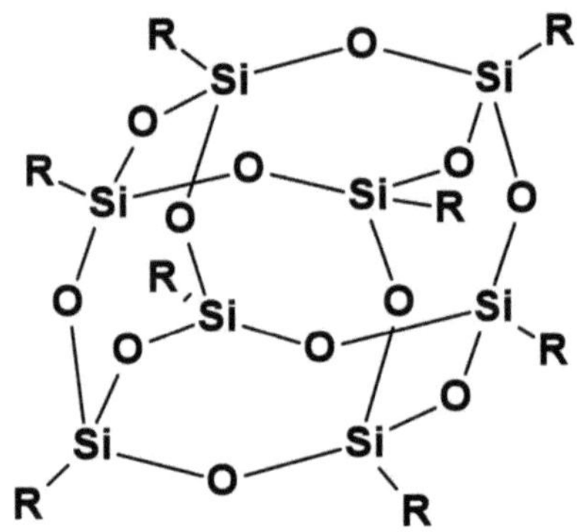

Fig. 1.3 The general structure of octahedral silsesquioxane

the lack of straightforward methods for their synthesis (see Chapter 2).

The most common process for making POS compounds is the hydrolytic condensation of functional silane trialkoxymonomers $XSiY_3$, where X is a chemically stable organic substituent, and Y is a highly reactive substituent such as methoxy, ethoxy or even Cl [10].

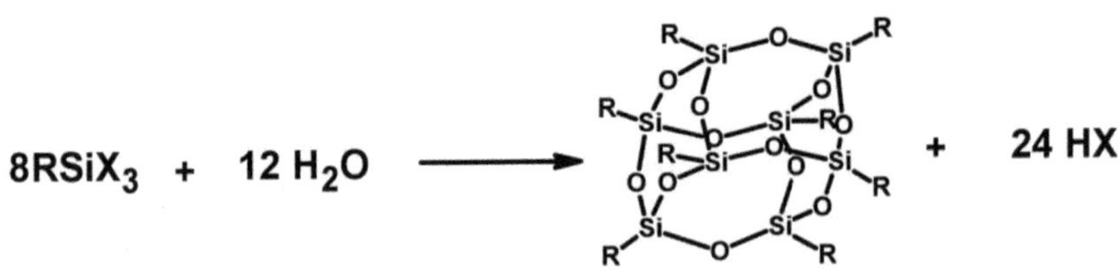

Fig. 1.4 Synthesis of POS by hydrolytic condensation (reprinted from [10] with permission from Springer)

1.2 Early Synthesis of Polyhedral Oligosilsesquioxanes (POS)

Andrianov [12,13] in the 1930's likely formed silsesquioxane structures, by hydrolyzing alkyltriethoxysilanes in water, but a combination of non-English publication, Western distrust of Eastern scientific data and lack of suitable characterisation techniques has seen this early pioneering work not generally acknowledged. The hydrolysis of phenyltrichlorosilane [14] to yield brittle white solids with high silicon content is most likely the first isolation of a phenyl silsesquioxane, as this corresponds to similar later syntheses of this material. Around this time, Scott [15] also discovered high-temperature polymer products in his hydrolysis of a mixture of methyltrichlorosilane and silicon tetrachloride, and ascribed them cyclic structures.

Barry and Gilkey [16] in an early patent gave empirical formulae to the low molecular weight, polycyclic hydrolysis products $(n\text{-}C_3H_7SiO_{1.5})_8$, $(C_2H_5SiO_{1.5})_8$ and $(n\text{-}C_4H_9SiO_{1.5})_8$ produced by heating the corresponding alkyltrichlorosilane with powdered alkali salts. A few relatively simple cyclic products have been isolated from hydrolysis products of trifunctional silicon compounds, and these polycyclic structures usually have not been characterized, although it was evident that many must exist.

A silsesquioxane was really first formed and identified as such, though in very low yield, in 1946 by Scott, as the product $(CH_3SiO_{1.5})_n$ from the thermolysis of the polymeric products of methyltrichlorosilane hydrolysis [15]. His two-step synthetic process started with the hydrolysis of methyltrichlorosilane to form the trisilanol derivative, and was then followed by condensation to form a silsesquioxane oligomer. Although, this method was initially limited to thermally-stable organofunctional groups on the trichlorosilane, such as alkyl and aryl groups, the yield could be increased by employing alkali hydroxides or triethyl amine as a catalyst [17]. In 1955 Sprung and Guenther used the same method to isolate a product with the structure $(CH_3SiO_{1.5})_8OSiCH_3(OC_4H_9)$, that initiated development of this area of silicon chemistry [18].

On the other side of the iron curtain, Russian researchers discovered a noteworthy route to POS structures through the condensation of oligoorganylcyclosiloxanes [19]. Their synthesis involved a base-catalyzed dehydrocondensation of two cyclic siloxanes: 1,3,5,7-tetramethylcyclotetrasiloxane and 1,3,5,7-tetraphenylcyclotetra-1,3,5,7-siloxanol.

1.3 Hydrolysis and Condensation in Making Oligosilsesquioxanes

The common method that emerges from this early work and one that is still used in the synthesis of polyhedral oligomeric silsesquioxanes, is the hydrolytic condensation of trifunctional monomers $RSiX_3$, where R is a chemically stable organic substituent, and X is a highly reactive substituent such as Cl, alkoxy or ethoxy (i.e., X = Cl, OR, OAc in Eq. 1.1 to 1.3). In such a synthesis, the silane monomer, possessing three hydrolyzable groups and one organyl group is hydrolysed (Eq. 1.1), affording a trisilanol monomer that subsequently undergoes self-condensation to produce the Si-O-Si bonds of the silsesquioxane cage [10].

Hydrolysis: $RSiX_3 + 3\,H_2O \rightarrow RSi(OH)_3 + 3\,HX$ (Eq. 1.1)

Condensation: $3\,RSi(OH)_3 \rightarrow 3\,RSiO_{1.5} + 1.5\,H_2O$ (Eq. 1.2)

Condensation: $RSi(OH)_3 + RSiX_3 \rightarrow RSiO_{1.5} + 3\,HX$ (Eq. 1.3)

The structures and yields of the products are very much dependent upon the processing parameters. To summarize the reaction conditions involved in hydrolytic polycondensation and subsequent formation of polyhedral oligosilsesquioxanes is difficult because each reaction is uniquely sensitive to an interdependent combination of experimental factors. An extensive review by Voronkov [10] published in 1982 covers the methods of synthesizing POS compounds up to that time.

The general scheme of the silsesquioxane synthesis is shown in Fig. 1.4. The first step of the hydrolytic condensation reaction is always the hydrolysis of the monosilane to give the corresponding trisilanol with the rate of the reaction decreasing in the following functional group order: Cl > OH > OCOR > OR [10]. Kudo and Gordon [20] performed a theoretical study of the mechanism for the hydrolysis of a trichlorosilane ($HSiCl_3$) to form the trihydroxysilane ($HSi(OH)_3$) (Eq. 1.4). Using ab initio quantum mechanical methods, they determined that the reactions take place in a stepwise manner (Eq 1.4 to 1.7). The energy barrier for the first step (Eq. 1.4) was predicted to be the highest and also found to be higher than the barriers for the subsequent condensation steps [21], suggesting that reaction (Eq. 1.4) is the rate-determining step. In a later study [22], for X=OMe, the correlated levels of theory predicted essentially no change in the barrier from step to step. For X=Cl, it was suggested that intramolecular hydrogen bonding stabilizes the transition state in the second and third steps; however, this was not the case for X=OMe.

Intramolecular condensations are favored over intermolecular condensations when the systems are dilute, i.e. the concentration of the precursor ($RSiX_3$) is kept low. Voronkov et al. [10] suggested the following:

(a) alkyltrichlorosilanes having lower alkyl substituents – molar concentration range 0.1-0.2M;

(b) alkyltrichlorosilanes bearing higher alkyl substituents – higher molar concentration of approximately 2.2M;

$$HSiCl_3 + H_2O \rightarrow HSiCl_2(OH) + HCl \quad \text{(Eq. 1.4)}$$
$$HSiCl_2(OH) + H_2O \rightarrow HSiCl(OH)_2 + HCl \quad \text{(Eq. 1.5)}$$
$$HSiCl(OH)_2 + H_2O \rightarrow HSi(OH)_3 + HCl \quad \text{(Eq. 1.6)}$$
$$HSiCl_3 + 3H_2O \rightarrow HSi(OH)_3 + 3HCl \quad \text{(Eq. 1.7)}$$

(c) alkoxytrichlorosilanes – molar concentration range 0.3-0.5M.

The precise mechanism that occurs during the synthesis of polyhedral oligosilsesquioxanes is complex and is still being investigated. A number of factors are deemed responsible for the successful synthesis of POS molecules:

- the type and concentration of the initial monomer
- the nature of the solvent
- the nature of the substituents R and X in the initial monomer

- the character of catalyst used
- temperature
- the rate and quantity of water addition

Although, the strongly interactive and synergistic effects of the above factors compound the complexity of the condensation process [1], it is possible to observe some general trends. For example, the nature of the R group of starting silane $RSiX_3$ determines whether a fully condensed or incompletely condensed silsesquioxane can be isolated. If silanes containing bulky R groups such as *tert*-butyl, phenyl, and cyclohexyl are used, then incompletely condensed silsesquioxanes can be first isolated, while the silanes with smaller groups like methyl or hydrogen yield completely condensed silsesquioxane cages as the only isolated products [23-25]. The nature of the hydrolyzable group X is also important in the overall condensation process. For example, Cl groups hydrolyze faster than either OEt or OMe groups; the latter two are then able to control the condensation to a greater extent. If X = Cl then the HCl formed as the hydrolysis product also enhances subsequent further hydrolysis and condensation [1].

The rate and quantity of water addition was reported to influence the extent of silane hydrolysis, which modifies the concentration of silanol groups present in the mixture [1]. High dilution favors silsesquioxane cage formation (*vide infra*), whereas oligomeric and ladder species occur at higher concentrations. A theoretical study on the mechanism of hydridosilsesquioxane synthesis further reports that the addition of only one water molecule to the reaction increases rates of either hydrolysis or condensation in the process, by reducing energy barriers [20].

As solvent molecules interact with the reaction species present in solution, the solvent plays a strong role in influencing the synthesis of silsesquioxanes [26]. Polar solvents have the ability to stabilize incompletely-condensed silsesquioxane compounds by interacting with OH groups through hydrogen bonding. Similarly, solvents with higher dielectric constant and dipole moment are able to stabilize silsesquioxanes containing an OH functionality [26]. Organyltrichlorosilanes and organyltrialkoxysilanes are highly reactive species, therefore the synthesis of their oligomers is carried out in organic solvents. In addition, many alkoxysilanes are immiscible with water so here alcohols are used to homogenize the reaction mixture. Octaphenylsilsesquioxane is readily formed in benzene, nitrobenzene and pyridine [17].

The hydrolysis and condensation of silsesquioxanes is catalysed by either acidic or basic media. Base-catalyzed hydrolytic condensation of bulky R substituents affords higher yields than acid-catalyzed reactions [17,35]. However, octamers with C_2H_5 and CH_2=CH are formed in alcoholic media without the addition of a catalyst [10]. The most commonly employed acidic catalysts are HCl, $FeCl_3$, $FeCl_2$ and $AlCl_3$. There are reports on a variety of POS compounds synthesized by acidic catalysis, with simple functional groups such as $H_8Si_8O_{12}$ [27-29] $Me_8Si_8O_{12}$ [30] and $(CH_2{=}CHCH_2)_8Si_8O_{12}$ [31]. In addition, POS functionalized with larger peripheral groups, such as $(n\text{-Bu})_8Si_8O_{12}$ [32] and $(m\text{-MeC}_6H_4)_8Si_8O_{12}$ [7] can also be synthesised using acidic catalysts. POS compounds can also be synthesized using base-catalysed methods, despite the fact that the reaction products usually also contain mixtures of polymeric materials. With precise control of reaction conditions, it is still possible to isolate the POS products having functional groups, such as $Me_8Si_8O_{12}$ [33], $Ph_8Si_8O_{12}$ [17], and $(o\text{-}Me_2NC_6H_4)_8Si_8O_{12}$ [34].

The rates of hydrolysis and condensation are both influenced by the temperature at which the reaction is carried out. Reactions performed at a lower temperature afford polyhedral oligosilsesquioxanes, while higher temperatures often lead to silsesquioxane polymers as the only product [1].

More recently the successful treatment of a range of triethoxysilanes in a THF solution of tetrabutylammonium fluoride provided a mild method of hydrolysis and lead to yields of T_8 cages up to 95% [35]. It was found the yield of T_8 cage depended upon the nature of the carbon adjacent to the silicon, where R = cyclopentyl gave 95%, cyclohexyl gave 84%, octyl gave 65%, 2-bicycloheptyl gave 56%, phenyl gave 49%, hexyl gave 44%, allyl gave 3%, vinyl gave 1% and methyl gave 0%. As the steric effect of the alkyl groups decreased the yield of T_8 cage also decreased. A few representative examples of polycondensations which lead to synthetically useful quantities of various POS frameworks under either acidic or basic conditions are summarized in Table 1.

Table 1.1 POS Synthesis via Hydrolytic Condensation of $RSiX_3$

$RSiX_3$	Product	Yield (%)	Catalyst	Ref.
$HSi(OMe)_3$	$H_8Si_8O_{12}$	13	Acid	[37]
$HSiCl_3$	$H_8Si_8O_{12}$	23	Acid	[29]
$MeSi(OEt)_3$	$Me_8Si_8O_{12}$	89	Base	[33]
$CH_2{=}CHSiCl_3$	$(CH_2{=}CH)_8Si_8O_{12}$	20	--	[76]
$C_6H_5SiCl_3$	$(C_6H_5)_8Si_8O_{12}$	88-89	Base	[17]
$C_6H_5Si(OEt)_3$	$(C_6H_5)_8Si_8O_{12}$	49	Base	[35]

1.4 Synthesis of Hydridooctasilsesquioxane, $H_8Si_8O_{12}$ (T_8H_8) and Octakis-(Hydridodimethylsiloxy)Octasilsesquioxane, $[H(CH_3)_2SiO]_8Si_8O_{12}$ ($Q_8M_8H_8$)

The first hydridosilsesquioxane was isolated in a yield of less than 1%, and characterized by Müller et al. in 1959 [36]. Later in 1970, Frye and Collins were able to increase the yield to 3% by modifying Müller's initial method [37], through the use of trimethoxysilane in place of trichlorosilane, and using high dilution in an emulsion polymerization. Owing to the difficulties of preparing trimethoxysilane $[HSi(OCH_3)_3]$ in those days, and the requirement for large quantities of reagents and solvents (more than 4 L for several grams of product) this method proved problematic for large-scale production of this very useful material.

In 1987 Agaskar et al. developed a much improved synthesis, leading to the production of pure T_8^H in yields up to 18% [27,38]. The modified procedure involved a two-phase reaction mixture: a polar phase containing a concentrated solution of ferric chloride $FeCl_3$ in HCl/MeOH, and a toluene/hexane mixture for the non-polar phase. Using the partially hydrated $FeCl_3$ as the source of water for the hydrolysis of $HSiCl_3$, T_8^H is formed along with the related hydridosilsesquioxanes T_{10}^H, T_{12}^H, and T_{14}^H, which are also all isolated. The formation of the hydridosilsesquioxane (T_8^H) can be represented by Eq. 1.8 below, where sufficient water of hydration associated with the hydrated $FeCl_3$ must be present to produce the required stoichiometry.

$$8HSiCl_3 + 12H_2O \rightarrow (HSiO_{1.5})_8 + 24HCl \qquad \text{(Eq. 1.8)}$$

Both Bassindale and Gentle in 1993 [28] as well as Crivello and Malik in 1997 [29] improved Agaskar's method by increasing the dilution of the trichlorosilane present in the mixture with the resulting yields now up to 23%. The increase in yield was thought to be due to the reduced trichlorosilane concentration in the reaction. The separation of several higher order silsesquioxanes (T_n^H), where n = 10, 12, 14, 16, 18 was still effective.

As various members of my group have sought to reproduce the Agaskar, the Bassindale and the Crivello syntheses, we have never been able to attain the quoted yields of 23% in the latter two cases, or of 18% in Agaskar's case. For the Bassindale and the Crivello syntheses our yields varied from 2% to 17%, with most syntheses resulting in less than 10% yield. For Agaskar's procedure we recorded yields between 7% and 15% with most syntheses resulting in greater than 10% yield. Sol-gel reactions are notoriously fickle with respect to a host of reaction conditions, and the reproducibility of the chemistry also depends very much on the skill and experience of the practitioner, so variations in yield are not unexpected in such cases. Nevertheless, the earlier Agaskar approach remains the favored approach within my group for making T_8^H.

Synthesis of another frequently utilized SiH-terminated POS compound was optimized by Hasegawa's group [39,40]. The synthesis of $Q_8M_8^H$ involves a two-step process: synthesis of a silicate octa-anion, which then undergoes silylation with hydrochlorosilane to afford octakis(hydridodimethylsiloxy)octasilsesquioxane, $Q_8M_8^H$. The synthesis of silicate anion is easily achieved using tetra-alkyl ammonium hydroxide and tetraethoxysilicate. However, extensive study of the reaction conditions which direct the condensation of tetraethoxysilicate to a specific polyhedral structure showed a size dependence associated with the alkyl chains in tetraalkyl ammonium hydroxide [41-44]. Only tetramethyl ammonium hydroxide as starting reagent favored the condensation of the octahedral silsesquioxane core, whilst longer alkyl chains such as tetrabutyl ammonium hydroxide led to a decahedral silsesquioxane core. In this way different core sizes could now be accessed in high yield. However, a recent, revised, study of the synthesis of $Q_8M_8^H$

and its derivatives demonstrated that the yield does depend on the amount of solvent employed, particularly when scaling up the reaction to larger quantities [45]. Overall the high yields obtained in both the syntheses of $Q_8M_8^H$ and its organofunctional derivatives make these compounds good precursors to other POS hybrid compounds.

Mono- and oligo-functional polyhedral silsesquioxanes can be synthesized from any of the hydridosilsesquioxanes by hydrosilylation. Most POS polymer precursors are developed by the hydrosilylation reactions of simple T_8^H or $Q_8M_8^H$ cages with various unsaturated alkene or alkyne groups.

1.5 Hydrosilylation

The hydrosilylation reaction is a method used for the synthesis of organosilicon compounds and organosilicon derivatives, in which generally the addition of Si-H to unsaturated alkene or alkyne groups takes place in the presence of a transition metal catalyst (most commonly platinum) [46]. Hydrosilylation reactions can occur across double or triple bonds in two possible ways: α-addition (Markovnikov) or β-addition (anti-Markovnikov) (Fig. 1.5). The β (anti-Markovnikov) addition product is sterically favored and is thus the major product. In some hydrosilylations a small amount (0 – 20 %) of the α (Markovnikov) product is also obtained.

In the absence of a catalyst, the reaction is believed to proceed via a free radical mechanism. The most active catalyst used in the hydrosilylation reaction is platinum. There are two classes of platinum catalysts; those of the Pt^0 oxidation state, e.g., bis(divinyltetramethyldisiloxane)Pt^0 or $Pt_2(H_2C{=}CHSiMe_2OSiMe_2CH{=}CH_2)_n$ (commonly called Karstedt's catalyst) [47], and those at higher oxidation states (e.g., Pt^{IV}, Speier's catalyst, H_2PtCl_6) [48]. The hydrosilylation reactions that employ platinum catalysts can be performed in a variety of solvents, including tetrahydrofuran, chloroform and other chlorinated solvents, benzene, toluene and silicones, although solvents are not obligatory [49].

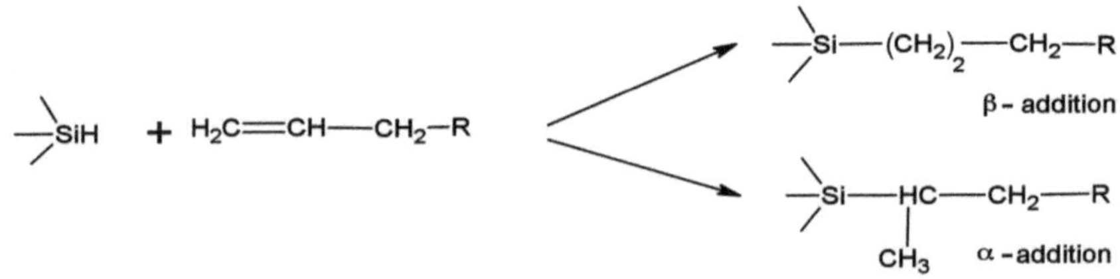

Fig. 1.5 The formation of α- and β-isomers during hydrosilylation

1.6 Octa-Functionalized POS Macromonomers

POS compounds can be converted into various octafunctionalized macromonomers by hydrosilylation between the terminal SiH groups on a POS molecule and any unsaturated C=C bond in the presence of a platinum catalyst [46,50]. Octafunctionalized POS compounds are used as precursors in the synthesis of hybrid nanocomposites (see Chapter 5); examples include alkyl [51], allyl alcohol, 2-allyloxy ethanol [52], epoxy [53-55], styryl [56,57] norbornyl [58], mesogen [59], methacrylate [60], amine [61,62], amino and bromo-phenyls [63,64].

By controlling stoichiometry, hydrosilylation also provides a good way for making mono-substituted silsesquioxanes [65]. However, the end-capping of incompletely condensed silsesquioxane precursors, developed by Feher [66-69], offers an alternative route into monofunctional silsesquioxane chemistry.

1.6.1 *Macromonomers Derived by the Hydrosilylation of Octahydridosilsesquioxane ($H_8Si_8O_{12}$; T_8^H)*

The first synthesis of products derived from octahydridosilsesquioxane (T_8^H) were reported by Day et al. in 1985 [70]. This work included the photochlorination of $H_8Si_8O_{12}$ to obtain $Cl_8Si_8O_{12}$ and its subsequent methoxylation with methylnitrite, (CH_3ONO), which resulted in the synthesis of $(CH_3O)_8Si_8O_{12}$. The single-crystal X-ray diffraction study confirmed that after methoxylation the stereochemical configuration of the Si_8O_{12} core was retained. The methoxylation reaction was also attempted with ionic reagents such as CH_3OLi or CH_3ONa, which resulted in the degradation of the Si_8O_{12} core. In 1990 the palladium-catalyzed deuterium exchange of $H_8Si_8O_{12}$ afforded $D_8Si_8O_{12}$, which was the second reported example using this octa-substituted silsesquioxane [71]. The assumption that these reactions proceed through a radical-type mechanism initiated studies on the ability of $H_8Si_8O_{12}$ to form Si-C bonds via hydrosilylation, through a radical-type reaction.

The hydrosilylation of hex-1-ene and methylenecyclohexane in the presence of Speier's catalyst was the first report on the formation of octa-substituted derivatives from $H_8Si_8O_{12}$ [72]. The octa-substituted silsesquioxane products from both hydrosilylations were obtained in excellent yields of 90% (though it was not discussed whether the β-isomer was the only product). Subsequently, platinum-catalyzed hydrosilylation reactions of $H_8Si_8O_{12}$ have provided access to a wide variety of functionalized silsesquioxanes, though in most cases the products were mixtures of both α- and β-isomers.

Furthermore, Matisons et al. [73] first established the crystal structure of the $Q_8M_8^H$ compound, and then reported the hydrosilylation of $H_8Si_8O_{12}$ with allyl bromide and showed that no α-isomer was formed.

Bassindale and Gentle [74] reported the hydrosilylation of $H_8Si_8O_{12}$ with a series of alkenes of different chain lengths. Here ^{29}Si NMR studies confirmed octa-substituted β-adducts as the only hydrosilylation products. The same paper reported the hydrosilylation of vinyl- and allyl-siloxanes onto $H_8Si_8O_{12}$. The NMR spectra for the reactions of vinyl-siloxanes gave evidence of the occurrence of both α- and β-adducts, but surprisingly when the allyl-siloxanes were hydrosilylated, ^{29}Si NMR and ^{13}C NMR spectroscopy confirmed that α-addition had occurred.

The homologous series of octaalkyl-substituted octasilsesquioxanes had been prepared by hydrosilylation of alkenes with $H_8Si_8O_{12}$ in the presence of Speier's catalyst in separate reactions [75]. The ^{13}C NMR spectra for these products proved that no Markovnikov addition took place, and only β-adducts formed. The melting points of the alkyl-substituted octasilsesquioxanes were studied by DSC, and higher values were obtained for POS compounds containing shorter alkyl chains. Relatively low melting points with an odd-even carbon number pattern occurred for alkyl chains longer than four methylene units.

Hydrosilylation of $H_8Si_8O_{12}$ with epoxy and 1-propenoxy functional groups were reported by Crivello and Malik [29]. The ^{29}Si NMR spectrum of the hydrosilylation of 1-hexene to $H_8Si_8O_{12}$ showed both α and β isomers present, with the major peak assigned to the β adduct. Clearly, steric hindrance of the organic substituents plays an important role in the hydrosilylation reactions of $H_8Si_8O_{12}$.

Dittmar et al. [77] made a series of octa-(propylsilsesquioxanes) $[(3\text{-}XC_3H_6)_8(Si_8O_{12})]$ by the hydrosilylation of $H_8Si_8O_{12}$ with allyl compounds with X = *p*-$(CH_3)OC_6H_4$, C_6H_5, OC_4H_9, OCH_2-CH-O-CH_2, $OCOCH_3$, CN, $OSi(CH_3)_3$, C_6F_5, OC_6H_5, $SO_2C_6H_5$ and $Si(CH_3)_3$. Some hydrosilylations, for example where X was an ether, ester or cyanide group, gave both α- and β-adducts. However, the α-isomer yields could be lowered by the addition of very small amounts of platinum catalyst, and by using a large excess of the allyl compound.

A hydroxyl functionalized T_8 silsesquioxane was synthesized by hydrosilylation of $H_8Si_8O_{12}$ and pentenol in the presence of Speier's catalyst, H_2PtCl_6 [78]. When the reaction temperature was kept at 90 °C there were no observable side reactions of SiH with OH groups. Additionally, a branched siloxane derivative of $H_8Si_8O_{12}$ with 16 Me_3Si groups was synthesised in a two-step reaction sequence: first the OH group on pentenol was replaced by $Si(Me)(OSiMe_3)_2$, and then the resulting molecule was hydrosilylated onto $H_8Si_8O_{12}$ in the presence of Karstedt's catalyst to give an almost quantitative yield of branched octa-substituted $[(CH_2)_5OSi(Me)(OSiMe_3)_2)]_8Si_8O_{12}$. Similarly, platinum-catalyzed hydrosilylation of $H_8Si_8O_{12}$ with excess 3-methyl-but-1-en-3-ol ($H_2C{=}CHCMe_2OH$) yielded compound $(HOCMe_2CH_2CH_2)_8Si_8O_{12}$ as the only product [79]. This demonstrated that increasing the steric bulk at the α position, by using 3,3-substituted β-alkene, suppressed the α-addition to give the high yield of $(HOCMe_2CH_2CH_2)_8Si_8O_{12}$.

Dare et al. [80] carried out the hydrosilylation of $H_8Si_8O_{12}$ with 2-chloroethylvinyl ether (catalyzed by H_2PtCl_6) quantitatively affording the octa-substituted β-adduct $Si_8O_{12}(CH_2CH_2OCH_2CH_2CH_2Cl)_8$ as the only product. The formation of any α-adduct was suppressed by using excess 2-chloroethylvinyl ether. The product was identified crystallographically, and its thermal stability was determined by thermogravimetric analysis (TGA).

Hydrosilylation of 4-bromostyrene onto $H_8Si_8O_{12}$, reported by Lin et al. [81] led to the synthesis of octa-functional precursors that could be further reacted with conjugated polymers by Suzuki coupling, Heck reaction, or Yamamoto coupling (see Section 7.2.3). The β-adduct (84%) again formed in higher yield than the α-adduct (16%) because it is sterically less hindered.

Allyl glycidyl ether was hydrosilylated onto $H_8Si_8O_{12}$ in the presence of Karstedt's catalyst. ^{1}H NMR and ^{13}C NMR showed that the methyl signal due to α-addition was too weak to detect [82]. However, ^{29}Si NMR showed two resonances, the major signal at – 65.2 ppm assigned to β-addition and the small signal at – 67.6 ppm is assigned to α-addition, indicating that just because no ^{1}H NMR and ^{13}C NMR signal is seen for the α adduct does not necessarily affirm it is not formed in some minor amount.

1.6.2 Macromonomers Derived by the Hydrosilylation of Octa(Hydridodimethylsiloxy)Octasilsesquioxane [(HSiMe$_2$O)$_8$Si$_8$O$_{12}$; (Q$_8$M$_8$H$_8$)]

Octa(hydridodimethylsiloxy)octasilsesquioxane [$(HSiMe_2O)_8Si_8O_{12}$, ($Q_8M_8^H$)] is used as a suitable starting material for the hydrosilylation of various polymerizable groups. The first reported synthesis of functionalized derivatives of $Q_8M_8^H$ was the hydrosilylation of vinylferrocene to this particular POS cage using Karstedt's catalyst [83]. This work also indicated that hydrosilylation of substituents bearing electronegative groups minimises any α-addition.

Sellinger and Laine [59,60] investigated the functionalization of T_8^H and $Q_8M_8^H$ with both allyl glycidyl ether and propargyl methacrylate groups. When four equivalents of allyl glycidyl ether were used, the products ranged from mono- to septa-substituted derivatives. When eight equivalents were used, $Q_8M_8^H$ underwent complete substitution while T_8^H only gave the septa-substituted product. ^{1}H NMR, ^{13}C NMR, and ^{29}Si NMR of both tetra-substituted and octa-substituted $Q_8M_8^H$ indicated that exclusive β addition took place, without the side reactions usually seen in the hydrosilylation of allyloxy moieties [60,84]. On the other hand, when four equivalents of propargyl methacrylate were hydrosilylated onto T_8^H and $Q_8M_8^H$ tetra-substituted products were obtained in both cases. Such reactions demonstrate the different reactivity of two very similar SiH-functionalized POS molecules towards hydrosilylation. Higher reaction temperatures were required for T_8^H (80 °C), while $Q_8M_8^H$ easily hydrosilylated at 40 °C. The choice of catalyst also played an important role in the hydrosilylation of propargyl methacrylate. If platinum dicyclopentadienyl dichloride, Pt(dcp), is used, hydrosilylation proceeded successfully, whereas hydrosilylation catalyzed with platinum divinyltetramethyldisiloxane, (Pt(dvs), Karstedt's catalyst) only resulted in crosslinked gels. ^{1}H NMR indicated that various isomer distributions occurred, as both α- and β-adduct shifts were observed under all reaction conditions.

The hydrosilylation reaction of $Q_8M_8^H$ with allyl alcohol and 2-allyloxyethanol resulted in the synthesis of compounds that are potential precursors to novel hybrid organic-inorganic materials [85]. As mentioned previously [79] unsaturated alcohols hydrosilylate easily to T_8^H ($H_8Si_8O_{12}$) with no observable side reactions. In general however, hydrosilylation of alcohols gives a mixture of products where O-silylation competes successfully with C-silylation, so that the hydroxyl group of the alcohol must be protected to avoid O-silylation (Fig. 1.6) [86-88]. However, in the hydrosilylation of allyl alcohol to $Q_8M_8^H$, the O-silylation was avoided simply by using either toluene or heptane as solvent. If either THF or CH_2Cl_2 were used, then both C- and O- silylations were observed. Similarly, using Karstedt's catalyst also resulted in both C- and O- silylations.

$Q_8M_8^H$ is readily hydrosilylated with allyl chloride using Speier's catalyst

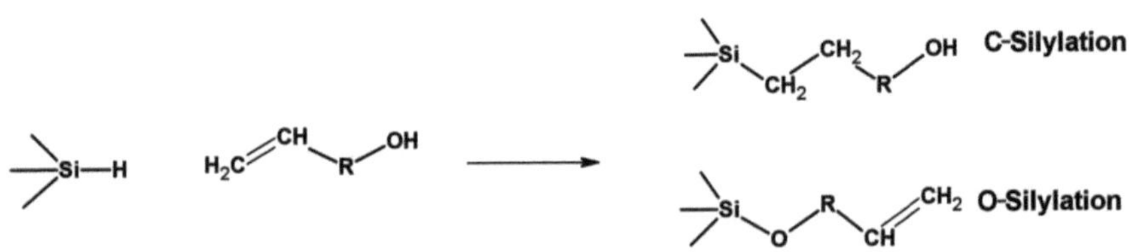

Fig. 1.6 Possible C- and O-silylation reaction products

(H_2PtCl_6) [89] to afford a mixture of products. Here the ^{29}Si NMR spectrum surprisingly showed signals attributed to both Q and D type silicon units; the latter resulting from the hydrolysis of the terminal Si-H groups together with dimerization of the cubic precursor.

In more recent times, water-soluble POS derivatives have been made by attaching water soluble oligo- or polymeric groups to $Q_8M_8^H$ using hydrosilylation. Oligomeric poly(ethylene oxide)-functionalized silsesquioxanes were made by hydrosilylation using Karstedt's catalyst, and the thermal behaviour of the PEO chains was affected by their links to the POS core, as shown by the increase of PEO glass transition temperature, T_g. Though the melting temperature, T_m, increased slightly, a notable suppression of crystallization was indicated by the decrease of ΔH_m when studied by TGA and DSC [90]. Even allyl-terminated monomethyl poly(ethylene oxide) with a number average molecular weight of 2000 could be hydrosilylated onto $Q_8M_8^H$, but ^{1}H NMR showed that only six poly(ethylene oxide) chains on average were attached to $Q_8M_8^H$ no doubt due to the steric hindrance that would develop around the POS core if additional high molecular weight PEO chains were attached [91]. Hydrosilylation of $Q_8M_8^H$ with lower molecular weight of poly(ethylene oxide), (average molecular weight 200), did afford octa-functional POS-PEO materials. Here the NMR spectra of these compounds confirmed that the hydrosilylation occurred exclusively through β-addition, without any side reactions [92].

1.7 Organic-Inorganic Hybrid Materials Prepared from POS: Octasilsesquioxane-containing Polymers

POS materials having nano-dimensions, inorganic cores and multiple reactive peripheral organic groups are ideal compounds for incorporation into polymers by either grafting or copolymerization, thus generating organic-inorganic hybrids [93]. If all eight functional groups of the cubic POS cage are modified by a polymerizable organic group, then the POS molecules can become either a cross-linking agent or a multiarm starlike filler. Multifunctional POS cages usually form several bonds from the POS cage into the matrix. These organic-inorganic hybrid materials employ POS cages at the centre of a local cross-linked network. The integration of the POS nanoclusters into a polymer matrix can result in significant improvements in polymer properties, which include, but are not limited to, temperature and oxidation resistance, surface hardening, improved mechanical properties, reduced flammability, diminished exothermicity on combustion and reduced viscosity during processing [94] (see also Chapters 4 and 5).

POS polymers are materials that contain nanoscale POS segments directly bound to the polymer chains. Such nanocomposite materials have shown synergistic properties when compared against properties predicted by the rule of mixtures of the component materials [32,33]. Both mono- and octa-functional POS macromonomers can be incorporated into most existing polymers through blending, grafting or copolymerization, resulting in significant improvements to several polymer properties [34].

Copolymerization can be achieved by covalently reacting a single polymerizable functional group on a POS cage with suitably miscible monomers that allow formation of copolymers having POS molecules as pendants within a linear copolymeric backbone. When such mono functional POS monomers are copolymerized with other crosslinking monomers, it is possible to decrease the resulting resin's crosslink density, since the mono-substituted cage cannot in this case contribute to the crosslinking [93]. These materials will be discussed in more detail in the next Section.

1.7.1 Hybrid Organic-Inorganic Crosslinked Materials Containing POS

A number of polyhedral octasilsesquioxane macromonomers bearing more than one polymerizable organic group can be copolymerized and so made to function as nano-crosslinkers. Such POS derivatives are usually synthesized by hydrosilylation of reactive SiH functional groups with unsaturated compounds that either possess (or can be further modified into) polymerizable groups. Alternatively, the synthesis of such polymerizable octafunctional POS derivatives can be accomplished by the hydrolytic condensation of trifunctional monomers $RSiX_3$, where R is the functional polymerizable group (such as the vinyl groups shown in Table 1), and X is a highly reactive substituent such as Cl or alkoxy [10,76].

Poly(methyl methacrylate) hybrid materials containing POS macromonomers (PMMA-POS) have been synthesised by the polymerization of the methacryloyl functionalized POS monomers [61,85] to achieve hard, transparent, crosslinked polymers. We subsequently reported a similar synthesis of highly crosslinked PMMA-POS materials [95]. Our modified approach however used methacrylic acid anhydride instead of methacryloyl chloride [85] to prepare the methacryloyl functional POS, so eliminating a minor back-biting side reaction inherent in the earlier work, and producing a product of significantly higher purity. Highly crosslinked polymers were then obtained by free radical polymerization of the methacryloyl-functionalized POS with methyl methacrylate using benzoyl peroxide. By varying the concentration of octa-functionalized POS, a very significant increase in both decomposition and glass transition temperatures could be achieved with small amounts of the POS crosslinkers.

Similarly, a variety of synthetic approaches were employed using functionalized POS cages as crosslinkers in making epoxy nanocomposites, which had significantly improved thermomechanical properties [53-55]. Vinyloctasilsesquioxane (T_8^{Vi}) and octakis(vinyldimethylsiloxy)silsesquioxane ($Q_8M_8^{Vi}$) were reacted with *m*-chloro-peroxybenzoic acid to generate epoxide-functionalized POS, which was then polymerized using $ZnCl_2$, BF_3OEt_2 or $H_2NCH_2NH_2$ [53]. Alternatively, the hydrosilylation of allylglycidylether onto T_8^H and $Q_8M_8^H$ cages gave the epoxide-functionalized silsesquioxanes, which when subsequently polymerized by UV light using triaryl sulfonium hexafluorophosphate catalyst, yielded highly crosslinked materials [54]. In addition, 4,4'-diaminodiphenylmethane (DDM) has also been used as a rather short organic tether between several octa-epoxy functionalized POS cages [55,96,97]. Differential scanning calorimetry (DSC) revealed that hybrid crosslinked polymers did show an increase in their glass transition temperatures when compared to their linear counterparts (synthesized by reacting DDM with the diglycidylether of bisphenol A).

Hybrid epoxy-based thermosets can be prepared by reacting octaglycidyl epoxy POS cages with 4,4'-diaminodiphenyl sulfone (DDS), a common high peformance hardener [98]. Nanometer-scale deformation processes associated with void formation are responsible for the toughening seen in these compounds, established by a combination of transmission electron microscopy (TEM) and thermal analysis. Dynamic mechanical analysis (DMA) revealed a significant dependence of the cure mechanism and kinetics upon the DDS content, since an increase in the DDS concentration (i.e., an increase in crosslink density) within these hybrids increased the dynamic storage modulus. Microshear yielding between voids prevailed, conferring a balance of stiffness, strength and toughness in the thermoset nanocomposite.

Octa-functional cubic silsesquioxanes can also be used as additives to improve

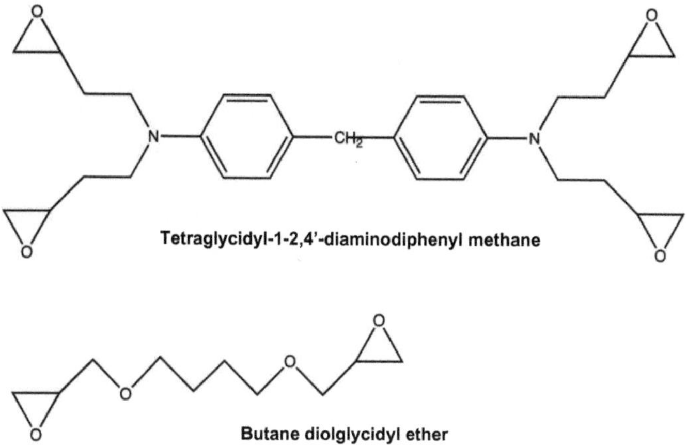

Fig. 1.7 Components of an aerospace grade epoxy resin

the thermomechanical properties and the processability of aerospace-grade epoxy resin, specifically by using a mixture of tetraepoxide (tetraglycidyl-4,4'-diaminodiphenyl methane) and diepoxide (butane diolglycidyl ether); see Fig. 1.7 [99]. Incorporating epoxy-functionalized POS macromers significantly improves the dimensional thermal stability of the resultant aerospace epoxy resin. The storage modulus of the hybrids also increases, especially at the elevated temperatures of importance in aerospace applications.

Crosslinked POS-containing polymers can be uniquely synthesised by the hydrosilylation of vinyl functionalized POS $[(CH_2{=}CH)(CH_3)Si]_8Si_8O_{20}$ ($Q_8M_8^{Vi}$) onto the SiH groups of the hydrido POS, $Q_8M_8^{H}$, in the molar ratio of 1:1 [100]. The rigid POS molecules do not allow for accessibility of all reactive functional groups beyond the gelation point of the polymer product, so the resulting rigid and brittle polymeric materials contain some unreacted vinyl and hydrido silane groups. These hybrid materials were microporous crosslinked polymers with a specific surface area of 300 m^2g^{-1}. The hydrosilylations of $Q_8M_8^{Vi}$ with $[(CH_3)_3SiO_2]_2CH_3SiH$ and $(C_2H_5)_3SiH$ did not follow the Markovnikov rule and gave the products $[(CH_3)_3SiO_2CH_3Si(CH_2)_2Si(CH_3)]_8Si_8O_{20}$ and $[(C_2H_5)_3Si(CH_3)_2]_8Si_8O_{20}$ respectively. In addition, a series of hybrid polymers were synthesized with Q_8M_8 linked together by chain-like, cyclic, and branched tethers, which showed very low specific surface areas of some 1-8 m^2g^{-1} [101]. Cross-linked polymers synthesized by the hydrosilylation of T_8^H and the vinyl functional Q_8M_8 ($Q_8M_8^{Vi}$), and $Q_8M_8^H$ with the allyl functional Q_8M_8 ($Q_8M_8^A$) also produced materials with low specific surface areas (2-3 m^2g^{-1}). Interestingly, the polymers obtained by the hydrosilylation of $Q_8M_8^{Vi}$ with $Q_8M_8^H$, and $Q_8M_8^A$ with $Q_8M_8^H$, showed higher specific surface areas (200-300 m^2g^{-1}) [102].

Similar work by Harrison and Kannengiesser indicated that if T_8^{Vi} is hydrosilylated with either T_8^H or $Q_8M_8^H$ then polymers with high specific surface areas could be synthesized [103]. Equally, the hydrosilylation of vinyl and silyl POS by Zhang et al. [104] suggested that the synthesis of crosslinked POS containing polymers made through the copolymerization of T_8^H or $Q_8M_8^H$ with $Q_8M_8^{Vi}$ can also lead to materials with very high specific surface areas (~380-530 m^2g^{-1}). All polymers had excellent thermal stability with the onset of decomposition in nitrogen starting above 300 °C. The condensation polymerisation increased yields from 43% to 81% as the chain lengths on the POS units increased. POS compounds with longer spacer groups between cages were more reactive than those with no spacer. Solid-state ^{13}C NMR and ^{29}Si CP/MAS-NMR indicated that the copolymerization proceeded predominantly via the β-hydrosilylation pathway.

The incorporation of POS into polymeric materials often results in dramatic improvements in polymer properties. The use of molecular dispersions of multi-functional POS units, such as the tetra-epoxidized $(C_6H_5CHCHO)_4(Si_8O_{12})(CH{=}CHC_6H_5)_4$ POS crosslinker in epoxy resins significantly improved both the thermal and mechanical properties [105]. For example, if 5-25% of this crosslinker is copolymerized with the low viscosity aliphatic epoxides, the storage modulus of the resulting hybrids increases substantially. Similarly, if this tetra-epoxidised POS crosslinker is incorporated into phenolic resin at only 5 wt %, the T_g increases by 11-14 °C.

Similarly, another epoxy functionalized POS molecule, octa(propylglycidyl ether) polyhedral oligomeric silsesquioxane, can be crosslinked into poly(4-vinylpyridine) (P4VP) [106]. The formation of a crosslinked hybrid structure is attributed to the reaction that occurs between the pyridine rings and the epoxide groups. Here too the thermal stability and the T_g's of the resulting crosslinked hybrids increased with increasing POS concentration in the nanocomposites; a result arising from both the formation of a crosslinked structure between P4VP and the octa-epoxy-POS, and the insulating effect of the inorganic POS cages homogeneously distributed throughout the P4VP.

POS-containing polyurethane hybrid networks can likewise be synthesized using octa-aminophenyl polyhedral oligomeric silsesquioxane as a crosslinking agent together with 4,4'-methylenebis-(2-chloroaniline) [107]. The POS-containing hybrid polyurethane networks showed increased glass transition temperatures, substantially increased storage modulii for the glassy state, as well as an increased rubber plateau when compared to that of the POS-free control polyurethane. Here even very low concentrations of POS integrated into the hybrid network produced very large property changes.

1.7.2 Star-Shaped Hybrid Organic-Inorganic Materials Containing POS as a Macroinitiator

Where the silsesquioxane cage has all its eight corners modified by an organic functionality, the silsesquioxane may become either a crosslinking monomer (described above) or a multiarm starlike initiator. Such starlike polymers form three-dimensional compounds, such that the numbers of arms branching out from the central POS core may become integrated within a complimentary branched polymer matrix, since different topologies can arise from differing associations within the polymer [108,109]. Well-defined star polymers are especially interesting compounds, and can show different thermal and rheological properties when compared to those of linear polymers of the same molecular weight. These materials may be incorporated as nanocomposite fillers within thermoplastics. In addition, since the cores can also be used as initiation sites, such branched POS fillers can be dispersed to a greater extent within the thermoplastic polymer.

Two different methods are used to synthesize these starlike polymers: the "arm-first" method [110-112], in which the pre-formed linear chains are linked to a suitable POS core, and the "core-first" method [113,114] where an active multifunctional POS core is used to initiate the growth of polymer chains.

Using the second "core-first" approach, a hybrid organic-inorganic star polymer with POS core has been synthesized from an hydroxyl functional (Q_8M_8), where the multifunctional core was then used as a polymerization initiator. The arms of the star polymer were initiated from the POS core by atom transfer radical polymerization (ATRP). The initial hydroxyl functional Q_8M_8 was synthesized by the hydrosilylation of $Q_8M_8^H$ with allyl alcohol, which was then further reacted with 2-bromo-2-methylpropionyl bromide to attach an ATRP initiator species to all eight corners of the POS cube. This octakis(2-bromo-2-methylpropionoxypropyldimethyl siloxy) octasilsesquioxane (OBPS) initiator was then polymerized with methyl methacrylate monomer under ATRP conditions using CuCl as the catalyst [115].

Similarly, POS spherosilicate cages of the form $[Si_8O_{20}]^{8-}$ were used as inorganic cores for the preparation of macroinitiators for thermal and photoinduced free radical and controlled radical polymerizations [116]. Two different approaches to the synthesis of these macroinitiators were investigated: (i) the direct modification of the Q_8M_8 octa-anion with chlorosilane-functionalized initiators, and (ii) the hydrosilylation of SiH-substituted cages. The synthesis of the macroinitiators using the latter hydrosilylation method gave better-defined reaction products. Here, various allyl-modified initiators were first made and then reacted with $Q_8M_8^H$ by hydrosilylation. The resulting macroinitiators were then used in free radical polymerization of styrene and methyl methacrylate, both by UV light and ATRP. The UV light free radical polymerization afforded polymers with broad polydispersities, while the copper-catalyzed ATRP showed that good control of polymerization and polydispersity could be achieved.

Again the "core-first" approach was used in the ring opening polymerization of 2-methyl-2-oxazoline initiated with various octa-functional POS molecules to prepare organic-inorganic star-shaped polyoxazoline hybrids [117]. Changing the ratio of POS to 2-methyl-2-oxazoline enabled the properties of the resulting hybrids to be reasonably controlled [117]. The NMR and GPC studies showed that generally hybrid polyoxazolines with an average of four or five arms emanating from the POS core were synthesized, but the very slow rate of polymerization led to polymers with differing arm lengths emanating from the core. However, these star-shaped polymers did display greater thermal stability than their linear polyoxazoline analogues.

Similarly, a series of hybrid star polycaprolactones (PCLs) were synthesised through ring-opening polymerization of ε-caprolactone using an octa-hydroxyl functional POS initiator [118]. The star PCLs were then reacted with α- and γ-cyclodextrins (CD) to form inclusion complexes (ICs), characterised by XRD, solid-state ^{13}C CP/MAS NMR spectroscopy, ^{1}H NMR spectroscopy and FTIR spectroscopy. The XRD and solid state ^{13}C CP/MAS NMR showed that the star polymers lost their original crystalline properties and formed columnar structures when confined within the CD channels. Supramolecular inclusions, such as these CD's, direct the topological interactions that normally occur within polymeric associations, by a combination of primary structure and non-covalent directional and non-directional interactions, leading to unexpected nanostructural forms, and so are very important in extending the range of possible structure-property relationships within hybrid polymer materials.

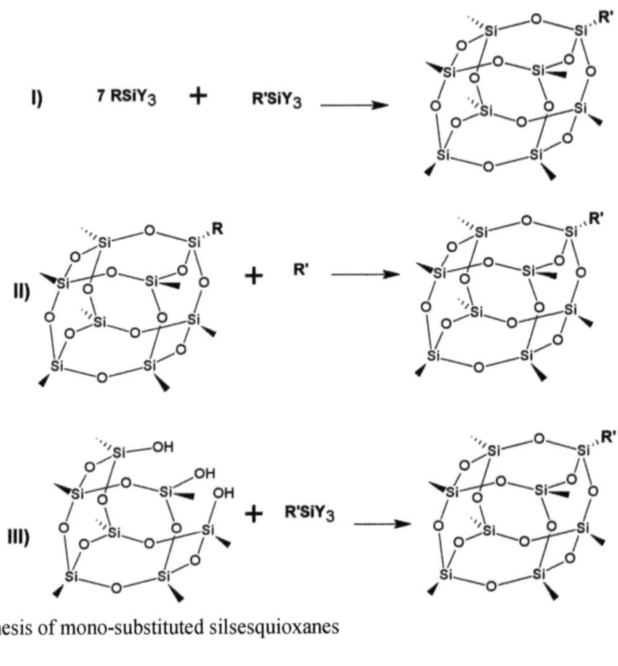

Fig. 1.8 Synthesis of mono-substituted silsesquioxanes

1.8 Mono-Substituted Polyhedral Oligomeric Silsesquioxane Macromonomers

Mono-substituted octasilsesquioxanes with the general formula $R'R_7Si_8O_{12}$ are molecules of particular interest as they are regarded as a first step in the synthesis of heterogeneously functionalized silsesquioxanes. These molecules have one distinct (and usually reactive) R' organic group providing the opportunity for further functionalization, whilst keeping the other R functionalities inert but useful in aiding the homogeneous dispersion of the POS core within various media. Hybrids using these singular functional compounds have found utility in surface modification [1]. Mono-substituted octasilsesquioxanes are generally synthesized by three general methods (see Fig. 1.8) [119], and much of the work in this field is a direct result of the pioneering studies undertaken by the Feher group [120].

Route I represents the standard synthetic process, where the polycondensation of monomers is employed in the synthesis of mono-substituted silsesquioxanes. However, a mixture of hetero-substituted compounds, including the desired mono-substituted species, is generally obtained if the reaction is carried out in the presence of monomers containing different R groups.

Route II represents a more practical approach to the synthesis of mono-substituted silsesquioxanes, where R = H or $OSi(CH_3)_2H$. In most cases the R' group is then introduced by stoichiometrically controlled hydrosilylation of silsesquioxanes. Still a mixture of products generally results, together with both Markovnikov and anti-Markovnikov additions occurring through the hydrosilylation.

Feher [120] developed the best practical synthesis of mono-substituted silsesquioxanes by the pathway shown in route III. Although, this route leads selectively to the mono-substituted silsesquioxanes as the only products, in the early days it was limited to POS products containing large organic groups as the R substituent.

1.8.1 Synthesis of Mono-Substituted Silsesquioxanes by Hydrolysis of Trifunctional Silanes

The first reported synthesis of mono-substituted silsesquioxane by the co-hydrolysis of $RSiY_3$ and $R'SiY_3$ silane monomers utilised compounds with R = organic group and R' = H [10]. However, the products obtained by this route were mixtures of silsesquioxanes containing all possible ratios of R/R'. Martynova and Chupakhina [121] investigated the significance of reaction conditions on the hydrolysis of (i) chlorosilanes: CH_3SiCl_3 and $C_2H_3SiCl_3$ and (ii) ethoxysilanes: $CH_3Si(OCOCH_3)_3$ and $C_2H_3Si(OCOCH_3)_3$ in separate reactions. They found that the reaction course and the yield of the silsesquioxane products were greatly affected by the concentrations of both alkali and water present in the reaction mixture. The maximum amount of mono-substituted products was obtained when the silane monomers $RSiY_3$ and $R'SiY_3$ were in the ratio 1:7. Corriu [122-124] described the syntheses of several mono-substituted octahedral heptahydrosilsesquioxanes. $RH_7Si_8O_{12}$ silsesquioxanes with R' = CH_2-CHR'', CH=CHR'', or $Co(CO)_4$ bonds were prepared and their crystal structures were determined. Later Calzaferri [125] synthesized $PhH_7Si_8O_{12}$ by co-hydrolysis of $HSiCl_3$ and $PhSiCl_3$, which at that time was the first reported synthesis of a mono-substituted silsesquioxane containing the Si-Ph bond. The structure of this POS compound was determined by X-ray crystallography, and revealed that the mono-substitution of a silsesquioxane caused a reduction in the POS symmetry.

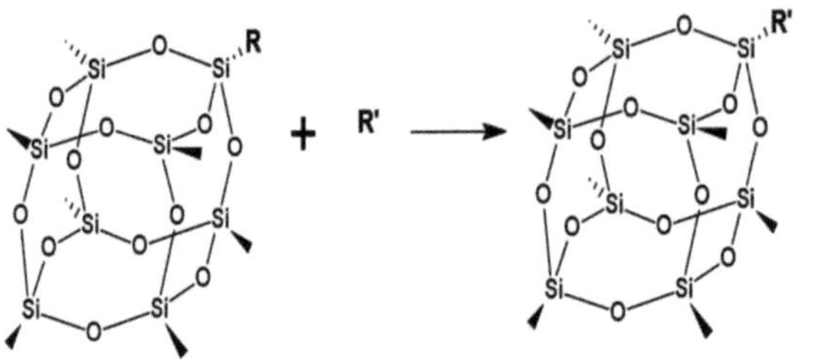

Fig. 1.9 Various mono-substituted silsesquioxanes (reprinted from [123] with permission from Wiley) [123,126,127]

1.8.2 Synthesis of Mono-Substituted Silsesquioxanes by Hydrosilylation

Despite the fact that the efficient synthesis of compounds with only one substituent is difficult, it appears that the best method is the reaction of one SiH group of the silsesquioxane cage with substituents containing unsaturated bonds. The addition of a double bond to an SiH group is achieved by hydrosilylation in the presence of a platinum catalyst; either Speier's, H_2PtCl_6, or Karstedt's are generally employed [64,65], see Section 1.5. There are a small number of reports on successful hydrosilylation where only the mono-substituted silsesquioxane was obtained using Speier's catalyst (H_2PtCl_6). Some examples are shown in Fig. 1.9 [123,126,127].

A similar reaction was described by Morán [128] where (η-C_5H_5)Fe(η-$C_5H_4CH = CH_2$) was hydrosilylated onto $[H(CH_3)_2SiO]_8Si_8O_{12}$ using Karstedt's catalyst. This synthetic route provided access to the first example of a polymer backbone structure comprised of ferrocene and silsesquioxane moieties. The reaction shown in Fig. 1.9 produced the first example of an organometallic-substituted silsesquioxane [124]. The SiH group can cleave metal-metal bonds. The first row transition metals have the weakest metal-metal bonds amongst transition metals. The reaction between hydridosilsesquioxane $H_8Si_8O_{12}$ and $Co_2(CO)_8$ in a 1:1 molar ratio led to the first organometallic silsesquioxane containing a silicon-metal bond (Fig. 1.10). This unique molecule contains no C-H bonds yet is readily soluble in very non-polar hydrocarbon solvents such as cyclohexane.

Frey [129,130] reported the synthesis of two mono-functional silsesquioxane monomers. The first example was the preparation of the monovinyl-spherosiloxane (Fig. 1.11), obtained through the initial hydrosilylation of octahydridosilsesquioxane with dibrominated decadiene, which was followed by a second hydrosilylation of

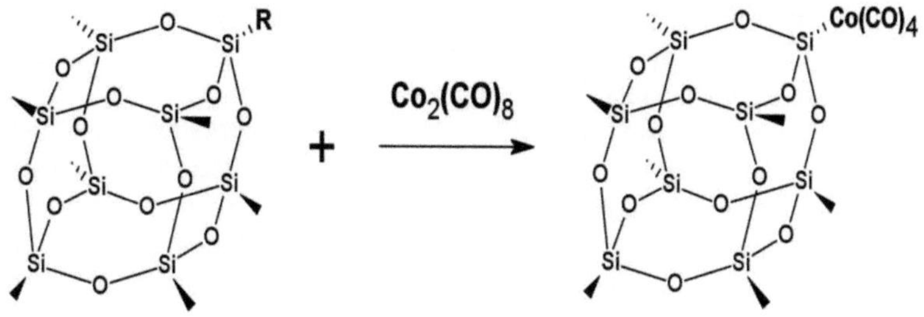

R=H

Fig. 1.10 An organometallic silsesquioxane containing a Si-Co bond (reprinted from [124] with permission from the Royal Society of Chemistry)

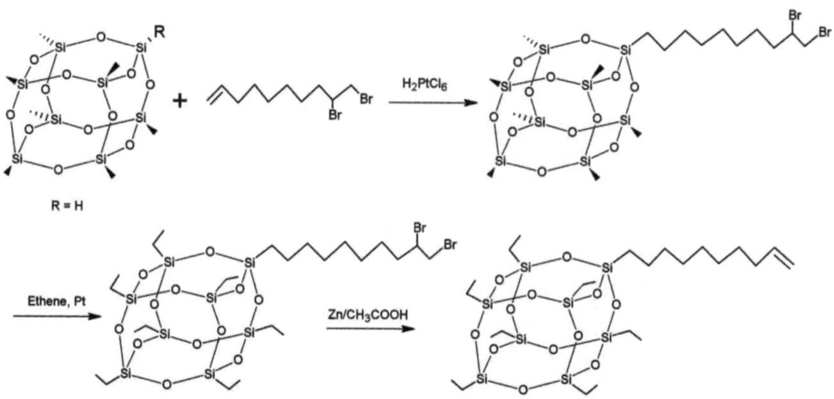

Fig. 1.11 Monovinyl-functional silsesquioxane (reprinted from [129] with permission from the American Chemical Society)

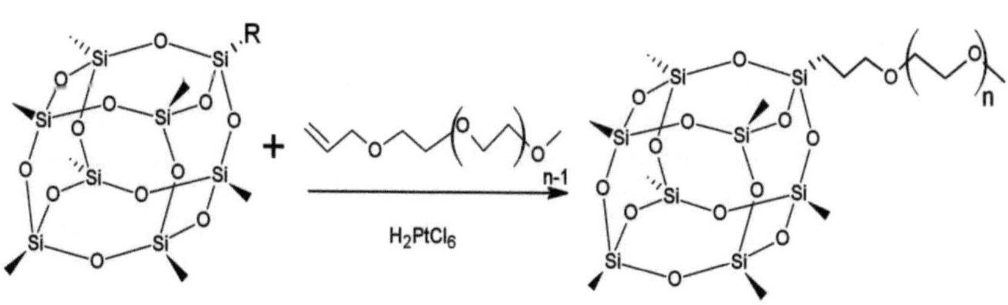

R = H

Fig. 1.12 Mono-oligo(ethylene oxide) hydridosilsesquioxane (reprinted from [130] with permission from the American Chemical Society)

ethylene onto the remaining unreacted SiH groups. Subsequent debromination and chromatographic purification yielded the unsaturated product. This mono-functional silsesquioxane was then homo- and co-polymerized with ethylene and propylene respectively, to generate high molecular weight polymers containing pendant POS cubes. The thermal analysis of these novel organic inorganic-hybrid compounds demonstrated that incorporation of such silsesquioxanes into the polyethylene polymer greatly improved its thermal stability in air.

Frey then went on to synthesize amphiphilic spherosilsesquioxanes using a similar approach (Fig. 1.12) [130]. The mono-substituted POS derivative was synthesized by hydrosilylating a hydridosilsesquioxane with an allyl-terminated oligo(ethylene oxide) (M_n = 750 g mol^{-1}). This mono-functional silsesquioxane possesses very interesting amphiphilic properties, in that a relatively hydrophobic spherosiloxane core is associated with a very hydrophilic oligo(ethylene oxide) chain. Aggregation of the uncondensed amphiphile demonstrated the potential for the preparation of novel core/shell type silicate nanoparticles.

1.8.3 Synthesis of Mono-Substituted Silsesquioxanes by Corner-Capping Reactions

Synthesis of mono-substituted silsesquioxanes by a method of corner-capping reactions has been developed by Feher, and starts from his pioneering discovery of the synthesis of an incompletely condensed silsesquioxane that contains three reactive OH groups [$R_7Si_7O_9(OH)_3$] [66-69,131]. The three silanol groups of the incompletely condensed silsesquioxane are very reactive, readily reacting with virtually any organotrichlorosilane (R'$SiCl_3$) to yield the fully condensed cage (see Fig. 1.8 part III). A variety of mono-functional silsesquioxanes (as the sole reaction products) can then be easily synthesized by simply varying the R' group of the organotrichlorosilane. Furthermore, subsequent modifications of this R' group have since led to the synthesis of various mono-functional monomers that could be polymerized to make organic-inorganic hybrid materials.

One example of a mono-functional POS derivative obtained by the Feher synthetic route is [$(c\text{-}C_6H_{11})_7Si_8O_{12}(CHPR_3)$] (where R = Me or Ph). These POS molecules are synthesized from $(c\text{-}C_6H_{11})_7Si_8O_{12}Cl$ or $(c\text{-}C_6H_{11})_7Si_8O_{12}H$ by reaction with Me_3PCH_2 and Ph_3PCH_2, respectively [17]. These compounds are surprisingly able to react with a variety of aldehydes as Wittig reagents, which is certainly not the case if analogous reactions of $H_8Si_8O_{12}$ and $Cl_8Si_8O_{12}$ with phosphoranes are performed. Incompletely condensed silsesquioxanes have also found applications as precursors for the synthesis of hetero- and metallasilsesquioxanes, where a hetero main-group or transition-metal is incorporated into the silsesquioxane cage [14,16,131-133] (see also Chapter 3).

1.9 Chemistry of Incompletely Condensed Silsesquioxanes

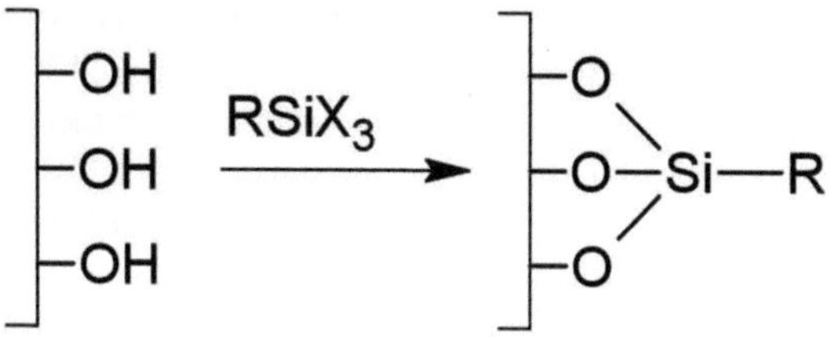

Fig. 1.13 Silylation of a silica surface with $RSiX_3$

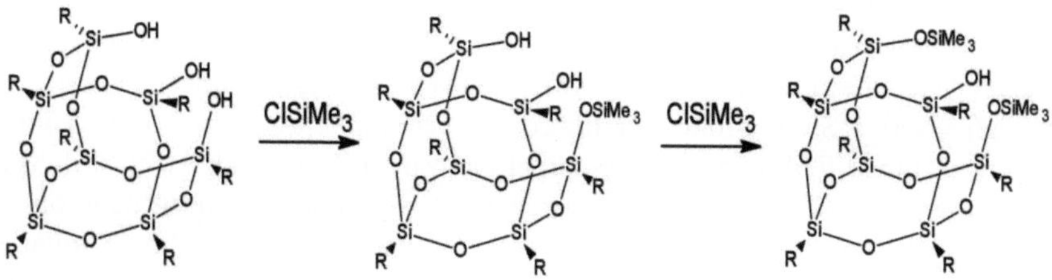

Fig. 1.14 Silylated incompletely condensed silsesquioxanes (reprinted from [120] with permission from the American Chemical Society) [120,143].

Incompletely condensed silsesquioxanes [$R_7Si_7O_9(OH)_3$] contain Si-OH groups, similar to those associated with silica and silicate surfaces, which make such POS derivatives suitable as soluble model compounds for investigating the surface chemistry of silica and silicate based materials [134,135]. Furthermore, such incompletely condensed silsesquioxanes with their spacious silicon-oxygen framework should resemble functionalized silicas in their electronic properties [136]. Surface-modified silica materials have applications as stationary phases in chromatography [137,138], in ion collection [139], heterogeneous hydrogenation catalysts [140], inorganic polymer fillers [141], and drug delivery agents [142]. The properties of such surface-modified silicas can be tailored through the chemical

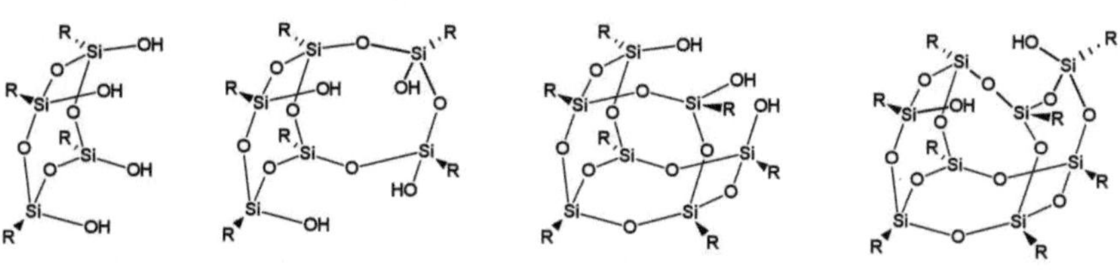

$R_4D_4(OH)_4$ $R_8T_2D_4(OH)_4$ $R_7T_4D_3(OH)_3$ $R_8T_6D_2(OH)_2$

Fig. 1.15 Examples of incompletely condensed POS systems (reprinted from [10] with permission from Springer)

attachment of organic groups to the surface silanol groups, using silane coupling agents such as the organotrichlorosilane used in the previous section (see also Fig. 1.13) [1,141]. Although, such surface silylation on POS is a very simple process, it appears to be much more complex when employed in silicas.

Fig. 1.14 shows some examples of the variations possible in reacting chlorosilanes onto incompletely condensed silsesquioxanes [120,143].

The chemistry of POS compounds containing incompletely condensed frameworks (Fig. 1.15) has mostly focused on reactions of the silanol groups. While many incompletely condensed POS structures have been isolated and characterized (Fig. 1.15), only the $R_7T_4D_3(OH)_3$ structure in Fig. 1.15 is formed in high yield and is readily scalable to multi-kg quantities. The main goals of such research into incompletely condensed POS nanostructures is to model structures, to generate a better understanding of silica surfaces, and to decrease the polysilanol functionality of these systems [144].

POS macromonomers bearing one or two reactive functionalities are of interest, since they can be used for grafting onto existing macromolecular structures through reactive SiOH groups, or for copolymerization with reactive macromers leading to linear POS-containing polymers and hybrid materials [145,146]. Most of the chemistry of incompletely condensed POS systems has been done on the $R_7T_4D_3(OH)_3$ compound (Fig. 1.15) because of its ready availability. Similar to completely condensed POS systems, incompletely condensed silsesquioxanes are usually synthesized by hydrolytic condensation of $RSiCl_3$ precursors. Work on incompletely condensed POS systems, however, was first reported by Brown [17,147]. At present this area of research using POS compounds is recognized in the pioneering work performed by the Feher group [120, 131-133,136,148-156].

1.9.1 Synthesis of Incompletely Condensed Silsesquioxanes

Synthetic methodologies for incompletely condensed silsesquioxanes have largely resulted from the studies of the Feher group [120, 131-133,136,148-156]. The most common route for synthesis of incompletely condensed silsesquioxanes is hydrolytic condensation of trifunctional silanes. In addition, acid- or base-catalyzed cleavage of fully condensed silsesquioxanes has been reported as a successful alternative method for the preparation of these compounds [153-155]. Incompletely condensed silsesquioxanes can be synthesized by the slow hydrolytic condensation of the corresponding silane $RSiCl_3$ (R = organic group; X = Cl, OMe, OEt; see Fig. 1:16) [1].

The first synthesis of an incompletely condensed silsesquioxane with R = cyclohexyl was reported by Brown and Vogt [147]. This compound was synthesized by the hydrolytic condensation of trichlorosilane (c-C_6H_{11})$SiCl_3$ in aqueous acetone. This first-reported synthesis of an incompletely condensed silsesquioxane cage was very slow, and it took three years to achieve a final yield of 60-70 %. However, the

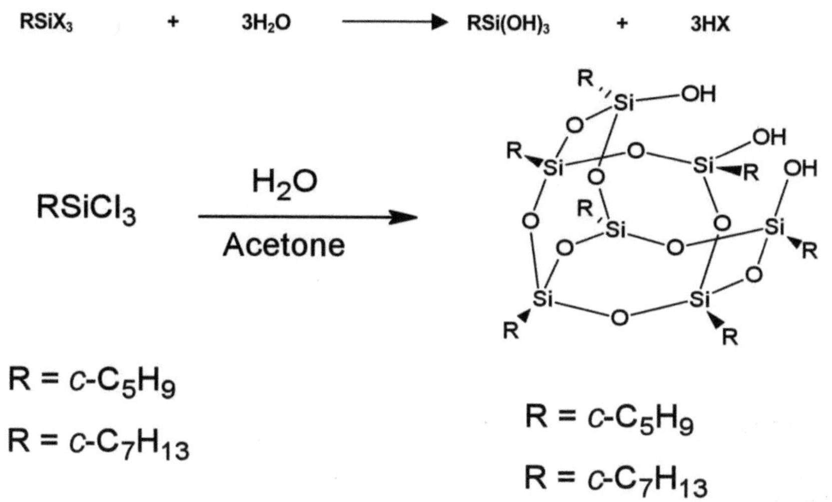

Fig. 1.16 Synthesis of incompletely condensed silsesquioxanes by hydrolytic condensation (reprinted from [120] with permission from the American Chemical Society)

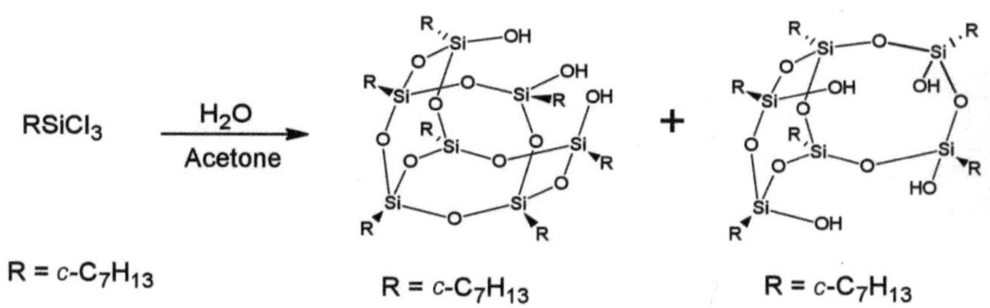

Fig. 1.17 Synthesis of mixtures of incompletely condensed silsesquioxanes (reprinted from [148] with permission from the American Chemical Society)

importance of incompletely condensed silsesquioxanes, as well as their potential applicability in silicon chemistry, convinced Feher to revisit and study this synthesis in more detail. First products reported by Feher's group were $[(c\text{-}C_5H_9)_7Si_7O_9(OH)_3]$ and $[(c\text{-}C_7H_{13})_7Si_7O_9(OH)_3]$ (see Fig. 1.16). The synthesis involved hydrolytic condensation of $c\text{-}C_5H_9SiCl_3$ and $c\text{-}C_7H_{13}SiCl_3$ in refluxing aqueous solution [148]. In the case when the R group of the starting silane was $c\text{-}C_7H_{13}$ the tetrasilanol $[(c\text{-}C_7H_{13})_6Si_6O_9(OH)_4$ was also afforded in 7% yield. (Fig. 1.17).

The synthesis of these compounds was straightforward but it took six weeks to obtain products in reasonable yield. However, the condensation of $c\text{-}C_5H_9SiCl_3$ and $c\text{-}C_7H_{13}SiCl_3$ was improved by increasing the temperature of the condensation of the silane from room temperature to reflux, though the condensation of $c\text{-}C_6H_{11}SiCl_3$ remained slow even at reflux temperatures [134]. Furthermore, the yield of $[(c\text{-}C_5H_9)_7Si_7O_9(OH)_3$ increased to 64% over 18 hours by using acetonitrile instead of acetone as a solvent. In addition, such improved syntheses afforded

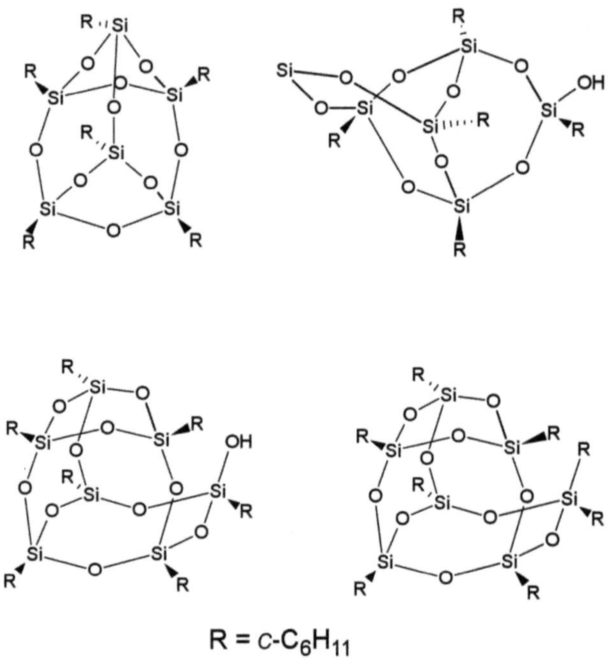

Fig. 1.18 Silsesquioxane isomers formed in the synthesis of incompletely condensed POS with acetonitrile (R= c-C_6H_{11}) (reprinted from [135] with permission from Wiley)

compounds where a wider variety of R groups such as phenyl, methyl, *n*-octyl and allyl could be made. More recently the dual role of acetonitrile in accelerating the formation of [(c-C_5H_9)$_7$$Si_7O_9(OH)_3$] as the only product, and increasing its yield was affirmed [135]. The synthesis of cyclohexylsilsesquioxanes produced four isomeric structures (Fig. 1.18). Here the higher solubility of the cyclohexylsilsesquioxane relative to that of cyclopentylsilsesquioxanes was thought to be responsible for the lower selectivity.

A number of factors are important in the synthesis of incompletely condensed silsesquioxanes (Section 1.3 vide supra). Those factors and their interactivity are essential in determining whether the synthesis will afford fully or incompletely condensed POS cages. For example, the nature of the R group of starting silane $RSiX_3$ determines whether the condensed or incompletely condensed silsesquioxane will be formed. If the silanes containing bulky R groups such as *t*-butyl, phenyl, or cyclohexyl are used in a hydrolytic condensation reaction, then the synthesis products are incompletely condensed silsesquioxanes, while silanes with smaller groups like methyl or hydrogen yield completely condensed cages [157-159]. Pescarmona [26,160] studied the importance of the solvent in the synthesis of incompletely condensed POS. Polar solvents were able to stabilize incompletely condensed silsesquioxane compounds by interacting with their OH groups through hydrogen bonding. In addition, solvents with higher dielectric constant and dipole moment were similarly able to stabilize silsesquioxane compounds containing

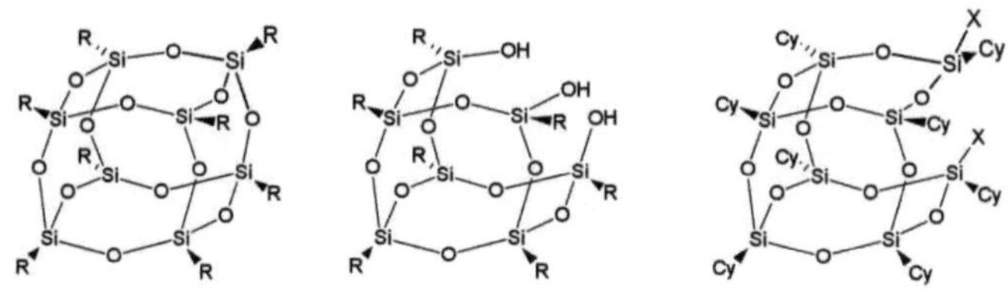

1 R = $OSiR_1R_2R_3$
2 alkyl or aryl
3 c-C_6H_{11}
4 c-C_6H_5

5 R = Cy
6 c-C_5H_9

7 X = F
8 $OSnMe_3$
9 OH
10 OTf

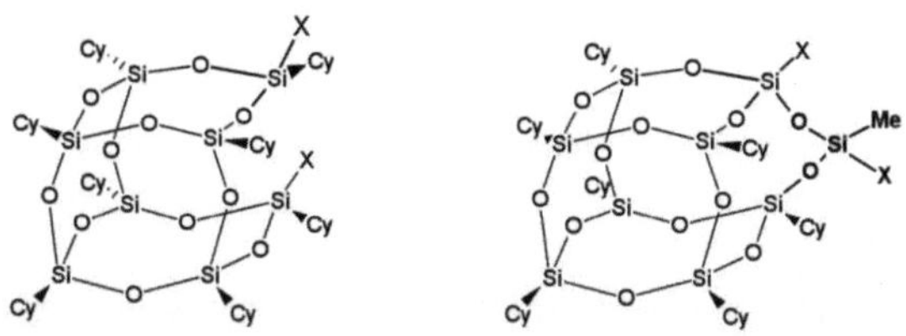

11 X = OTf
12 OH

13 X = H
14 $(CH)_2CMe_2CH_2CO_2Me$

Fig. 1.19 Synthesis of incompletely condensed silsesquioxanes by controlled cleavage of a completely condensed cage (reprinted from [152] with permission from the Royal Society of Chemistry)

OH functionalities, by favoring their synthesis over that of less polar, completely condensed silsesquioxane species.

The synthesis of incompletely condensed silsesquioxanes by the controlled cleavage of the silsesquioxane framework marked a substantial breakthrough by the Feher group [152]. Here controlled cleavage by either acid or base of fully condensed cube-octameric polyhedral silsesquioxanes ($R_8Si_8O_{12}$) now afforded a radical new way for manufacturing an incompletely condensed silsesquioxane framework [152-154,156].

In an initial paper on the controlled cleavage of $R_8Si_8O_{12}$, Feher [152] outlined a strategy for preparing the incompletely condensed silsesquioxanes from their

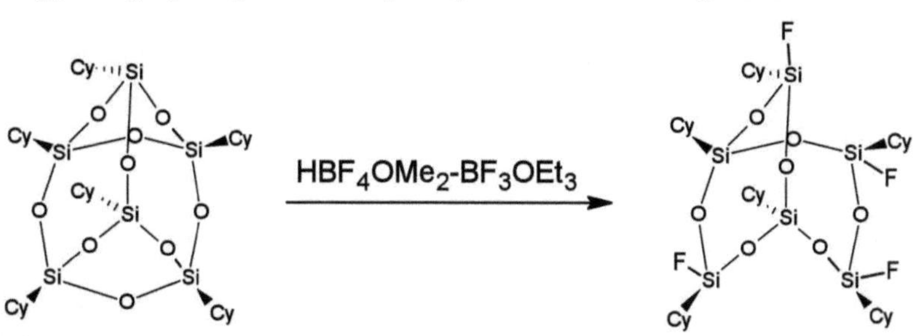

Fig. 1.20 Acid-mediated cleavage and rearrangement of (c-C_6H_{11})$_6Si_6O_9$ (reprinted from [153] with permission from the Royal Society of Chemistry)

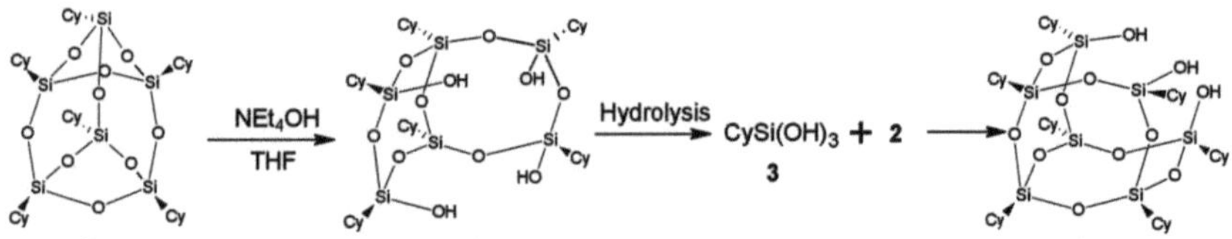

1 2 4

Fig. 1.21 Synthesis of incompletely condensed silsesquioxanes by base-catalyzed controlled cleavage of a fully condensed cage. Reprinted from [156] with permission from the Royal Society of Chemistry.

completely condensed silsesquioxane precursors ($R_8Si_8O_{12}$, where R = c-C_6H_{11}; compound **3** in Fig. 1.19) by the addition of an excess of $HBF_4.OMe_2$ and $BF_3.OEt_2$ to a solution of compound **3** in C_6D_6 or $CDCl_3$. Further studies with other strong acids showed that it is not the formation of the Si-F bond but rather the acidity of HBF_4 that drives the cleavage of the Si-O-Si framework. Compound **3** (Fig. 1.19) in excess triflic acid (CF_3SO_3H) produces the ditriflate **11**, having no Si-F bond. Though the difluoride 7 does not react with water, the ditriflate **11**, easily hydrolyses to give the disilanol **12**.

The efficiency of acid-catalyzed cleavage of completely condensed systems was affirmed when $(c\text{-}C_6H_{11})_6Si_6O_9$ was rearranged by either triflic acid or methanesulfonic acid ($MeSO_3H$) to produce $(c\text{-}C_6H_{11})_6Si_6O_8X_2$ (X = OTf, OMs,) in good yield, which then underwent further rearrangement to give a variety of incompletely condensed silsesquioxanes $(c\text{-}C_6H_{11})_6Si_6O_8X_2$ (X = OH, $OSiMe_2H$, F,

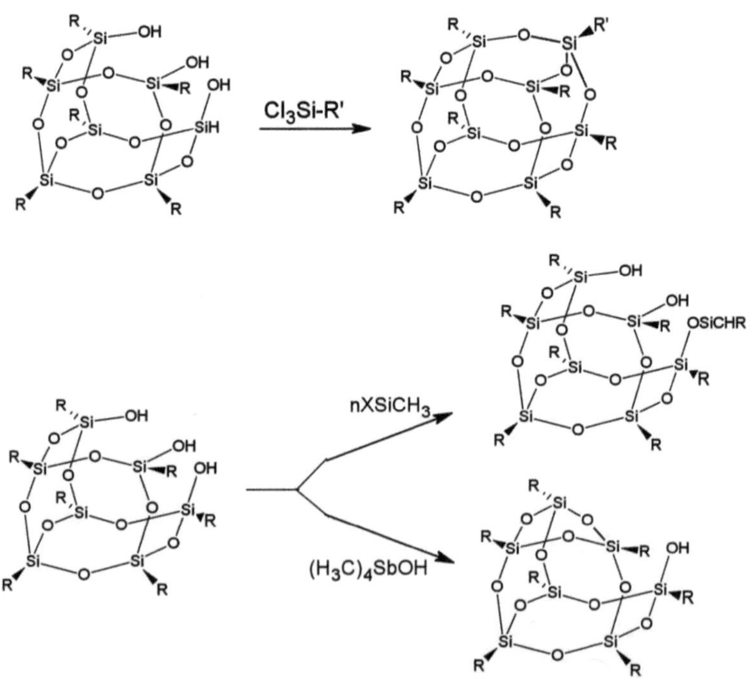

Fig. 1.22 Reactions of incompletely condensed POS. Reprinted from [131] with permission from the American Chemical Society.

NMe_2 and CCPh [153]. If the acid used is $HBF_4.OMe_2$-$BF_3.OEt_2$, the product is a C_2-symmetric tetrafluoride (Fig. 1.20), obtained by the irreversible cleavage of the Si-O framework.

Similar to acid-mediated cleavage of completely condensed silsesquioxanes, it is possible to synthesize incompletely condensed silsesquioxanes by the base-catalyzed controlled cleavage of fully condensed systems [156]. $Cy_6Si_6O_9$ readily reacts with aqueous NEt_4OH in THF to afford endo-C_{2h}-$Cy_6Si_6O_7(OH)_4$, that then hydrolyzes to produce $CySi(OH)_3$ fragments which further react with the intermediate endo-$C_{2h}Cy_6Si_6O_7(OH)_4$ to give the endo-C_3-$C_7Si_7O_9(OH)_3$ product. (Fig. 1.21)

Both acid- and base-catalyzed controlled cleavages of completely condensed silsesquioxanes are important in silsesquioxane chemistry, since the two approaches use different starting materials. Large-scale syntheses of most silsesquioxane materials have been developed by Hybrid Plastics under the guidance of Joe Lichtenhan, the first researcher in the Feher group to work on POS materials. The reactivity of the three silanol groups within incompletely condensed POS cages has resulted in the use of such materials as precursors in various syntheses of both monofunctional POS monomers and polymers bearing monofunctional silsesquioxanes as pendant groups.

1.9.2 Chemistry of Incompletely Condensed Silsesquioxanes

Incompletely condensed silsesquioxanes can be used as versatile precursors to a variety of Si/O or Si/O/M silsesquioxane frameworks, because the reactions of OH groups can lead to the synthesis of a variety of new siloxanes and heterosiloxanes [136,161-164].

Firstly, the trisilanol functionality of the incompletely condensed silsesesquioxane molecule can be partially or fully reduced by reaction with silanes such as $XSi(CH_3)_3$, where X = Cl, $N(CH_3)_2$ and n = 1 – 3 (Fig. 1.22) [131]. Secondly, the vacant corner in the starting incompletely condensed silsesquioxane may be occupied by a transition metal resulting in a tridentate complex. However, there are also the examples where the silsesquioxane has been singly or doubly coordinated to such metal centres, and here it proved useful to react one or two of the precursor's hydroxyl groups with chlorotrimethylsilane $(CH_3)_3SiCl$, before introducing the transition metal [161]. Recently, a number of reactions have been reported involving incompletely condensed silsesquioxanes and their silylated derivatives with a metal centre. Some of these metallasilsesquioxanes find application as model compounds for metal centres grafted onto the surface of silicas and homogenous catalysts [57].

Studies on the hydrosilylation of incompletely condensed silsesquioxanes also demonstrated the feasibility of selective silylation of just one of the three silanols (though the di- and tri-silylated products were also produced), particularly if the reaction was performed in the presence of an amine base [143].

1.9.3 Hybrid Organic-Inorganic Materials Derived from Mono-Substituted POS Monomers

POS macromers possessing seven inert groups (e.g., cyclopentyl, cyclohexyl or isobutyl to enhance solubility) and one polymerizable group can be covalently incorporated into polymer backbones. These hybrid organic-inorganic materials incorporate POS as a pendant group within a linear backbone, which decreases polymer chain mobility and so increases both the glass transition and degradation temperatures, whilst reducing viscosity and improving oxygen permeability. There are various reports on attempts to synthesize polymers that contain POS macromers as pendants within a linear polymer backbone.

It is useful at this point to remember that only 15-19% of appropriately surface-functionalized silica can be blended into polystryrene without loss of mechanical or thermal properties. However, Hybrid Plastics has shown that the appropriately functionalized POS cages can be successfully blended into polystryrene with concomitant improvements in both mechanical and thermal properties. It is with this in mind that we can turn our attention to copolymers involving mono-substituted POS cages.

For example, styryl-substituted POS, $R_7(Si_8O_{12})(CH_2C_6H_4CH{=}CH_2)$ (where R = cyclohexyl or cyclopentyl) can be copolymerized together with styrene monomer by carrying out a conventional AIBN-initiated free-radical polymerization. Similarly, copolymerization with 4-methylstyrene gave alternative copolymers [56]. Simply polymerizing the styryl POS on its own produced a homopolymer (R = cyclohexyl) that had greater solubility in THF, and both a higher glass transition temperature T_g and decomposition temperature T_d, than the cyclopentyl-substituted styryl-POS/styrene copolymer. An increase in the mole percent of POS resulted in a corresponding increase in both T_g and T_d. Rheological studies of the styryl POS/4-methyl styrene copolymers showed that copolymers with low POS mole fraction retained melting and solid state properties similar to those of poly(4-methyl styrene). As the POS mole fraction increased in the hybrids, modified viscoelastic behaviour was observed, resulting in rubber-like features at higher temperatures [57].

Likewise, a series of hybrid poly(acetoxystyrene-*co*-isobutylstyryl-POS) were synthesized and characterized, where the POS content was controlled by varying the monomer feed ratio [165]. Thermal analysis again demonstrated that inclusion of POS can efficiently increase the T_g of the hybrid compounds, particularly at relatively high POS concentrations, and afford the final hybrid copolymer with a narrower molecular weight distribution.

POS-styrene macromers, $R_7Si_8O_{12}$styrene, (where R = isobutyl, cyclopentyl, cyclohexyl, or phenyl) have been synthesized and free radically polymerized together with styrene to produce copolymers that show a variation in modulus above the glass transition temperature [166]. Here the importance of the R group on the POS cage with respect to bulk polymer properties was assessed. The variation in modulus is much more significant for copolymers containing 30 wt % cycloalkyl POS than that observed for isobutyl POS. Furthermore, the packing of the POS cages within the resulting glassy matrix was strongly affected by the type of group on the POS cage.

Hemi-telechelic POS-polystyrene copolymers have been synthesized by reacting POS cages terminated with isocyanate groups with a series of hydroxyl-terminated polystyrenes having various molecular weights. The reaction occurs by urethane bond formation catalyzed by dibutyltin dilaurate. WAXS analysis shows that an increase in the aggregate size of POS domains occurs with an increase in the weight percent of POS [167].

Recently a novel approach described the synthesis of POS-polystyrene by polymerizing Al-containing POS cages with styrene [168]. This polymerization involved the grafting of varying amounts of aluminosilsesquioxane onto a pre-synthesized random copolymer of styrene and vinyl-diphenylphosphine oxide, where coordination between the aluminum and phosphine oxide occurs. Different molecular weight poly(styrene-diphenylphosphine) polymers were synthesized with the same molar ratio of styryl and vinyldiphenylphosphine, using free-radical polymerization. The rheology studies of the resulting copolymers indicated that the interchain POS-POS interactions dominated the viscoelastic dynamics of these copolymers.

Ring opening metathesis polymerization (ROMP) is relatively new to polymer chemistry. Since its discovery, this type of polymerization has developed fast, and has recently become one of the most studied in the polymer chemistry [181]. The large choice of monomer and catalyst systems offers synthetic routes to a wide range of polyolefin products that have broad industrial applications. The most suitable catalyst systems employed in ROMP are derived from transition-metal complexes and co-catalysts such as BuLi, $EtAlCl_2$, R_3Al, and R_4Sn (where R is Me, Bu, Ph, etc) [181]. The most prominent transition-metal catalyst complexes are those of molybdenum, tungsten, and ruthenium (Schrock- and Grubbs-type catalysts) [182-189].

There are a number of monomers that have been successfully polymerized using ROMP [190]. Typical examples of monomers include norbornenes, norbornadienes, and 7-oxonorbornenes [186,191-193], but also other strained cyclic olefins such as azanorbornenes [187,194] cyclobutenes [195], cyclooctenes [196], cyclooctadienes [199] and others. ROMP can be carried out in bulk, in solution or heterogeneously (e.g., emulsion, suspension) [198-200]. A number of solvents can be used such as benzene, toluene, dichloromethane, acetone, alcohols, water, and many others. Solvent mixtures are sometimes used. For the synthesis of functional polymers and tailor-made complexes, ROMP is emerging as a particularly suitable method.

A few papers have been reported on preparation of POS-containing copolymers using ROMP [8,58]. Poly (norbornyl-POS) copolymers have been synthesized, where the influence of both the weight fraction of POS-norbornyl monomer and its corner group composition upon the physical properties of copolymers were examined [8,58]. Semicrystalline polyethylene hybrid copolymers containing POS (PE-POS) have been prepared using ROMP [201]. This synthesis involved the ROMP of cyclooctane and the POS monomer.

Ring opening metathesis polymerisation (ROMP) using a catalyst, such as $Mo(C_{10}H_{12})(C_{12}H_{17}N)(OC_4H_9)$, offers a new polymerization method that can incorporate POS cages. Heptacyclohexyl or heptacyclopentyl monofunctional norbornyl POS monomers were polymerized with norbornene by ROMP, and the effect on the glass transition temperature of POS inclusion into the resulting norbornene copolymers was monitored [58]. Dynamic mechanical thermal analysis (DMTA) clearly showed that the α-relaxation temperature T_α (corresponding to T_g) increased as the POS content increased. Furthermore, shape memory effects were also studied by DMTA, by measuring the recovered strain while re-heating the samples back above the T_g. This revealed that the incorporation of POS monomers caused a small decrease in the percentage recovery of such copolymers, but improved thermal stability drastically [8]. Atomistic molecular dynamics simulations were used to study the effect of introducing norbornene-substituted pendant POS monomers onto polynorbornene chains [169]. Such studies indicated that the main polynorbornene chains were sensitive to the presence of POS monomers; the simulations predicted that the tensile, bulk and shear modulii would be enhanced, and change less significantly with increasing temperature, in comparison to the same properties of polynorbornene homopolymers having similar molecular weights.

It is not only thermoplastic copolymers that show dramatic property improvements; such effects are also possible in thermoset polymers. Hybrid thermoset copolymers prepared by ROMP with dicyclopentadiene, norbornenylethyl polyhedral oligomeric silsesquioxane (1NB-POS) and tris(norbornenylethyl)-POS (3NB-POS) [170], showed that while the incorporation of the mono-substituted monomer into the polycyclopentadiene polymer decreased the crosslink density, the 3NB-POS increased the crosslink density. Here the glass transition temperatures (T_g) of the resulting hybrid materials are only slightly affected by the incorporation of POS. The hybrid containing 20 wt % of 1NB-POS results in a T_g of 114 °C (decreased from the 128 °C T_g value seen for polycyclopentadiene).

Vinyl–POS copolymers have been synthesized by the polymerization of heptaethy-9-enyl POS monomer with ethylene and propylene [129]. The monovinyl-functional spherosiloxane was first synthesized by the controlled hydrosilylation of octahydridosilsesquioxane ($H_8Si_8O_{12}$) with dibrominated decadiene, followed by reaction of the seven unreacted SiH groups with ethylene and propylene. The homopolymerization and copolymerization of this mono-functional POS monomer with either ethylene or propylene using methylaluminoxane-activated catalyst gave high mass polymers bearing POS pendant to a polymer backbone. The thermal stability studied by thermal gravimetric analysis (TGA) indicated an increase in the decomposition temperatures in air when compared to the corresponding homopolymers [171]. Wide angle X-ray scattering (WAXS) of these hybrid materials further indicated that they aggregate and crystallize as nanocrystals, since the crystallites of both the polyethylene and the pendant POS coexisted within the copolymers [172]. This demonstrates a "mutually dependant" and unique microstructure, where none of the individual components dominates, yet a mutually synergistic hybrid structure is achieved having unique properties. The existence of POS monomers in such nanocomposites lowers the crystallinity of the polyethylene component, causing smaller, and disordered, polyethylene crystallites.

Similarly, mono-substituted propyl methacrylate POS monomers have been made that possess only one polymerizable group together with seven inert cyclohexyl or cyclopentyl groups. The methacryl POS bearing cyclohexyl groups was twice as soluble as the cyclopentyl-substituted POS monomer in THF, toluene, and benzene [173]. X-ray powder diffraction studies showed that both macromers were crystalline. The homopolymers of these POS monomers can undergo free radical polymerizion using AIBN as the initiator, though the resulting molecular weights are not particularly high. The incorporation of methacrylate-POS monomers as pendant materials in methacrylate copolymers, however, significantly changes the properties. For instance, the DTMA of copolymers with only 10 mol % cyclopentyl-substituted-POS-methacrylate in butyl methacrylate copolymers,shows that the copolymer has a T_g of 150 °C, which is 130 °C higher than the T_g (20 °C) of atactic poly(butyl methacrylate). Similarly, the POS-methyl methacrylate hybrids were synthesized from cyclopentyl-substituted POS-methyl methacrylate monomer by atomic transfer radical polymerisation (ATRP) [174,175]. The gas permeability of the resulting hybrids was enhanced by the incorporation of the POS monomer into its backbone, and the oxygen permeability of the 10 wt % cyclopentyl-substituted POS-methyl methacrylate hybrid was nine times higher than that of PMMA.

Matyjaszewski's group developed the capability to introduce methacrylate-functional POS monomers into polyacrylate materials using atom transfer radical polymerisation (ATRP) [174,175,202-205]. Varieties of well-defined star diblock and ABA triblock copolymers have been created by ATRP, where the POS moieties are attached to the copolymer backbone as pendant side-chains.. In the synthesis of ABA triblock copolymers possessing a soft middle polyacrylate segment and POS-containing outer blocks via ATRP, Matyjaszewski showed that this method affords microphase-separated structures by modifying the composition and molar mass of each of the polymer segments [205].

Inorganic polymers backbones may also readily incorporate POS. Siloxane-POS copolymers have been synthesized by Lichtenhan [145] using two methods. The first method involved reacting a silanol-functionalised POS macromer with a bis(dimethyl amine)-substituted silane. The second method used the hydrosilylation of allyl-substituted POS macromers, $R_7Si_8O_{12}(CH_2\text{-}CH{=}CH_2)$, where R = cyclohexyl or cyclopentyl, onto poly(hydridomethyl-*co*-dimethylsiloxane) to generate pendant POS functionality along a siloxane polymer chain. These POS-siloxane copolymers exhibit quite complex depolymerisation-decomposition processes at very high temperatures. The decomposition of evolved cyclic dimethyl siloxanes occurs at 400 °C, cyclohexyl hydrocarbons from 450-550 °C and hydrogen evolution from 700-1000 °C. The decomposition of the POS cage takes place concomitantly from

450-650 °C [176].

POS-containing homopolymers and copolymers can also be used in the synthesis of polyurethanes. Polyurethane elastomers were synthesized by incorporating a POS containing two hydroxy groups and a POS molecule containing 3-(allylbisphenol-A) as a chain extender, together with diisocyanates. The resulting hybrid properties were then compared with those of the neat polyurethane containing no POS [177]. Wide-angle X-ray powder diffraction (WAXD) demonstrated that POS molecules formed nanoscale crystals in the hard segment domains of polyurethanes. Interesting POS-polyurethanes that are amphiphilic hybrid materials have been prepared by forming direct urethane linkages between different molecular weight poly(ethylene glycols) and the isocyanate group of a mono-functional POS monomer [178]. Thermal analysis of the hybrids demonstrated that the crystallinity significantly decreased with increased POS content, with the hybrid materials becoming completely amorphous at POS contents above 50 mol %.

Finally, epoxy resins have also been prepared using mono-functional POS-epoxy macromers [179]. Mono-functional epoxy POS was used with low melting point aromatic diamines to synthesize precursors for incorporation into an epoxy network [180]. Gel permeation chromatography (GPC) was employed to follow the distribution of the species within the hybrid network. Differential-scanning calorimetry (DSC) then demonstrated that an increase in the glass transition temperature took place without increasing the extent of cross-linking within the hybrid network. Such behaviour arises from the nanoscopic size of the POS cages, which hinders any motion of the molecular chains and network crosslinks.

1.10 Summary

Organically modified POS finds increasing attention in the design of modern hybrid materials. Improvements in the syntheses of organofunctional POS compounds has resulted in an exponential increase in research including these materials. The attraction of this class of cage compounds lies in the construction of hybrid materials with new and potentially useful properties arising from their rigid, silica core structure and their polyfunctionality, all on the 'nano' scale. Modified POS molecules then remain as readily synthesized discrete units, and their unique architecture makes them promising scaffolds for advanced functional nanomaterials. In principle, modified POS molecules can be structurally likened to dendrimers in that both types of compounds can be prepared as multi-armed units containing precisely placed, peripheral moieties. Unlike dendrimers, however, the chemical modification of POS is immediately achievable at up to eight corners and this degree of functionality can be realized in earlier generations than for dendrimers, where two or three generations are required to achieve octa-functionality.

Researchers have used POS molecules as building blocks to prepare highly conjugated moieties suitable for utility in organic and molecular electronics (see also Chapter 7). Their high stability and unique architecture also makes these compounds ideal candidates for functionalization with a variety of functional and/or polymerizable groups. The key to the appeal of the use of polyhedral octasilsesquioxanes as building blocks is the ease with which their organic functionality can be derivatized in a controlled and high yielding manner. Complete conversion is a desired prerequisite for the construction of structurally discrete molecules. Silicon chemistry offers several quantitative reactions suitable for functionalizing POS with attractive functional groups. The most prevalent of these reactions is hydrosilylation. Hydrosilylation can be used in the synthesis of linear, branched and hyperbranched polymer systems, and generally results in monodisperse materials that have symmetrical molecular structures.

POS is and will remain an effective scaffold for preparing both linear and multi-arm macromolecules and organic-inorganic hybrid materials, where POS is used in the side chain of many polymeric backbones. POS derivatives have been prepared with reactive functionalities suitable for copolymerization, grafting or blending. These nanosized building blocks are utilized for preparation of organic-inorganic hybrid nanocomposite materials as they offer precise control and definition of the organic architecture between the dispersed rigid silica cores. The creation of nanocomposite materials based on POS building blocks has generated active research and tremendous excitement due to the simplicity in processing and the comprehensive improvement in the properties of this class of organic-inorganic hybrid material.

1.11 References

1. Harrison, PG (1997) J Organomet Chem 542:141.
2. Constable GS, Lesser AJ, Coughlin EB (2004) Macromolecules 37:1276.
3. Xu H, Kuo SW, Lee JL, Chang FC (2002) Macromolecules 35:8788.
4. Leu CM, Chang YT, Wei KH (2003) Chem Mater 15:3721.
5. Fu BX, Lee A, Haddad TS (2004) Macromolecules 37:5211.
6. Sanches C, Soller-Ilia GJ, Ribot F, Lalot T, Mayer CR, Cabuil V (2001) Chem Mater 13:3061.
7. Feher FJ, Budzichowski TA (1989) J Organomet Chem 379:33.
8. Jeon HG, Mather PT, Haddad TS (2000) Polym Int 49:453.

9. Gilman JW, Schlitzere DS, Lichtenhan JD (1996) J Appl Polym Sci 60:591.

10. Voronkov MG, Lavrent'yev VI (1982) Topics Curr Chem 102:199.

11. Baney RH, Itoh M, Sakikabara A, Suzuki T (1995) Chem Rev 95:1409.

12. Andrianov KA (1938) J Gen Chem (USSR) 8:1255.

13. Andrianov KA (1946) J Gen Chem (USSR) 16:633.

14. Andrianov KA, Breitmann BM (1947) J Gen Chem (USSR) 17:1522.

15. Scott DW (1946) J Am Chem Soc 68:356.

16. Barry AJ, Gilkey GW (1949) US Pat. 2465188.

17. Brown Jr JF, Vogt Jr LH, Prescott PI (1964) J Am Chem Soc 86:1120.

18. Sprung MM, Guenther FO (1955) J Am Chem Soc 77:3996.

19. Andrianov KA, Tikhonov VS, Makhneva GP (1973) Izv. Akad Nauk SSSR Ser Khim:956.

20. Kudo T, Gordon MS (1998) J Am Chem Soc 120:11432.

21. Kudo T, Gordon MS (2000) J Phys Chem A 104:4058.

22. Kudo T, Gordon MS (2002) J Phys Chem A 106:11347.

23. Winkhofer N, Roeski HW, Noltemeyer M, Robinson WT (1992) Angew Chem Int Ed 31:599.

24. Takiguchi T (1959) J Am Chem Soc 81:2359.

25. Ishida H, Koenig JL (1979) J Polym Sci Polym Phys Ed 17:1807.

26. Pescarmona PP, van der Waal JC, Maxwell IE, Maschmeyer T (2001) Angew Chem Int Ed 40:740.

27. Agaskar PA (1991) Inorg Chem 30:2707.

28. Bassindale AR, Gentle TE (1993) J Mater Chem 3:1319.

29. Crivello JV, Malik R (1997) J Polym Sci Part A Polym Chem 35:407.

30. Agaskar PA, Klemperer WG (1995) Inorg Chim Acta 229:355.

31. Martynova TN, Korchkov VP, Semyannikov PP (1983) J Organomet Chem 258:277.

32. Olsson K (1958) Ark Kemi 13:367.

33. Vogt Jr LH, Brown Jr JF (1963) Inorg Chem 2:189.

34. Tacke R, Lopez-Mras A, Sheldrick WS, Sebald A (1993) Z Anorg Allg Chem 619:346.

35. Bassindale AR, Liu Z, MacKinnon IA, Taylor PG, Yang Y, Light ME, Horton PN, Hursthouse MB (2003) Dalton Trans:2945.

36. Müller R, Kohne R, Sliwinski SJ (1959) Prakt Chem 9:71.

37. Frye CJ, Collins WT (1970) J Am Chem Soc 92:5586.

38. Agaskar PA, Day VW, Klemperer WG (1987) J Am Chem Soc 109:5545.

39. Hasegawa I, Sakka S, Sugahara Y, Kuroda K, Kato C (1989) Chem. Commun:208.

40. Hasegawa I, Motojima S (1992) J Organomet Chem 441:373.

41. Hoebbel D, Wieker W (1971) Z Anorg Allg Chem 384:45.

42. Hoebbel D, Wieker W, Franke P, Otto A (1975) Z Anorg Allg Chem 418:35.

43. Hoebbel D, Vargha A, Fahlke B, Engelhardt G (1985) Z Anorg Allg Chem 521:61.

44. Hoebbel D, Pitsch I, Reiher W, Hiller H, Jancke H, Müller Z (1989) Z Anorg Allg Chem 576:160.

45. Hasegawa I, Kadzuko I, Ohnishi H (2003) Appl Organomet Chem 17:287.

46. Marciniec B (1992) Comprehensive Handbook on Hydrosilylation, Pergamon: New York.

47. Karstedt BD (1973) US Pat. 3775452.

48. Speier JL, Webster JA, Barnes GH (1957) J Am Chem Soc 79:974.

49. Brook MA (2000) Silicon in Organometallic and Polymer Chemistry, John Wiley & Sons, Inc., New York.

50. Lewis LN, Lewis N (1986) J Am Chem Soc 108:7228.

51. Bolln C, Tsuchida A, Frey H, Műlhaupt R (1997) Chem Mater 9:1475.

52. Zhang C, Laine RM, J Am Chem Soc 122:6979.

53. Zhang C, Laine RM (1996) J Organomet Chem 521:199.

54. Sellinger A, Laine RM (1996) Chem Mater 8:1592.

55. Laine RM, Choi J, Lee I (2001) Adv Mater 13:800.

56. Haddad TS, Lichtenhan JD (1996) Macromolecules 29:7302.

57. Romo-Uribe A, Mather PT, Haddad TS, Lichtenhan JD (1998) J Polym Sci Part B Polym Phys 36:1857.

58. Mather PT, Jeon HG, Romo-Uribe A, Haddad TS, Lichtenhan JD (1999) Macromolecules 32:1194.

59. Sellinger A, Laine RM, Chou V, Viney C (1994) J Polym Sci Part A Polym Chem 32:3069.

60. Sellinger A, Laine RM (1996) Macromolecules 29:2327.

61. Gravel MC, Zhang C, Dinderman M, Laine RM (1999) Appl Organomet Chem 13:329.

62. Feher FJ, Windham KD (1998) Chem Commun:3223.

63. Tamaki R, Choi J, Laine RM (2003) Chem Mater 15:793.

64. Lin W-J, Che W-C, Wu W-C, Niu, Y-H, Jen K-Y (2004) Macromolecules 37:2335.

65. Calzaferri G, Herren D, Imhof R (1990) Helv Chim Acta 73:698.

66. Feher FJ (1986) J Am Chem Soc 108:3850.

67. Feher FJ, Weller KJ (1990) Organometallics 9:2638.

68. Feher FJ, Weller, KJ (1991) Inorg Chem 30:880.

69. Feher FJ, Weller KJ, Schwab JJ (1995) Organometallics 14:2009.

70. Day VW, Klemperer WG, Mainz VV, Millar DM (1985) J Am Chem Soc 107:8262.

71. Bürgy H, Calzaferri G (1990) Helv Chim Acta 73:698.

72. Herren D, Bürgy H, Calzaferri G (1991) Helv Chim Acta 74:24.

73. Provatas A, Luft M, Mu JC, White AH, Matisons JG, Skelton BW (1998) J Organomet Chem 565:159.

74. Bassindale AR, Gentle TE (1993) J Mater Chem 3:1319.

75. Bolln C, Tsuchida A, Frey H, Mülhaupt R (1997) Chem Mater 9:1475.

76. Harrison PG, Hall C (1997) Main Group Metal Chemistry 20:515.

77. Dittmar U, Hendan BJ, Flörke U, Marsmann HC (1995) J Organomet Chem 489:185.

78. Bassindale AR, Gentle T (1996) J Organomet Chem 521:391.

79. Said MA, Roesky HW (1999) Angew Chem Int Ed 38:661.

80. Dare EO, Olatunji GA, Ogunniyi DS (2004) J Appl Polym Sci 93:907.

81. Lin WJ, Chen WC, Wu WC, Niu Y-H, Jen AKJ (2004) Macromolecules 37:2335.

82. Liu Y, Zheng S (2006) J Polym Sci Part A Polym Chem 44:1168.

83. Morán M, Casado CM, Cuadrado I (1993) Organometallics 12:4327.

84. Laine RM, Zhang C, Sellinger A, Viculis L (1998) Appl Organomet Chem 12:715.

85. Zhang C, Laine RM (2000) J Am Chem Soc 122:6979.

86. Ghose BN (1979) J Organomet Chem.164:11.

87. Braun F, Willner L, Hess M, Kosfeld R (1987) J Organomet Chem 332:63.

88. Braun F, Willner L, Hess M, Kosfeld R (1989) J Organomet Chem 366:53.

89. Carmo DR, Dias Filho NL, Stradiotto NR (2004) Mater Res 7:499.

90. Maitra P, Wunder S (2002) Chem Mater 14:4494.

91. Mya KY, Li X, Chen L, Ni X, Li J, He C (2005) J Phys Chem B 109:9455.

92. Lee YJ, Huang JM, Kuo SW, Chang FC (2005) Polymer 46:10056.

93. Li G, Wang L, Pittman Jr CU (2001) J Inorg Organomet Polym 11(3):123.

94. Pittman Jr CU, Li GZ, Ni H (2003) Macromol Symp 196:301.

95. Toepfer O, Neuman D, Choudhury NR, Whittaker A, Matisons J (2005) Chem Mater 17:1027.

96. Laine RM, Choi J, Costa ROR (2000) Polym Prepr (Am Chem Soc Div Polym Chem) 41:524.

97. Choi J, Harcup J, Yee AF, Zhu Q, Laine RM (2001) J Am Chem Soc 123:11420.

98. Kim GM, Qin H, Fang F, Mather PT (2003) J Polym Sci Part B Polym Phys 41:3299.

99. Mya KY, He C, Huang J, Xiao Y, Dai J, Show Y-P (2004) J Polym Sci Part A Polym Chem 42:3490.

100. Hoebel D, Pitsch I, Heidemann D (1990) Z Anorg Allg Chem 538:133.

101. Hoebel D, Pitsch I, Heidemann D (1991) Z Anorg Allg Chem 592:207.

102. Hoebel D, Endres K, Reinert T, Pitsch I (1994) J Non-Cryst Solids 176:179.

103. Harrison PG, Kannengiesser R (1996) Chem Commun:415.

104. Zhang C, Babonneau F, Bonhomme C, Laine RM, Soles CL, Hristov HA, Yee AF (1998) J Am Chem Soc 120:8380.

105. Li GZ, Wang L, Toghiani H, Daulton TI, Koyama K, Pittman Jr CU (2001) Macromolecules 35:8686.

106. Liu Y, Meng F, Zheng S (2005) Macromol Rapid Commun 26:920.

107. Liu H, Zheng S (2005) Macromol Rapid Commun 26:196.

108. Kamigaito M, Ando T, Sawamoto M (2001) Chem Rev 101:3689.

109. Angot S, Taton D, Gannou Y (2000) Macromolecules 33:5418.

110. David BA, Kinning DJ, Thomas EL, Fetters LJ (1986) Macromolecules 19:215.

111. Hadjichristidis N, Guyot AN, Fetters LJ (1978) Macromolecules 11:668.

112. Morton M, Helminiak TE, Gakary SD, Bueche F (1962) J Polym Sci 57:471.

113. Jacob S, Mojoros I, Kennedy JP (1996) Macromolecules 29:8631.

114. Schappacher M, Deffleur A (1992) Macromolecules 25:6744.

115. Costa ROR, Vasconcelos WL, Tamaki R, Laine R (2001) Macromolecules 34:5398

116. Holzinger D, Kickelbick G (2002) J Polym Sci Part A Polym Chem 40:3858.

117. Chujo Y, Ouchi Y, Kim K-M (2003) Polym Bull 49:341.

118. Chan S-C, Kuo S-W, Chang F-C (2005) Macromolecules 38:3099.

119. Marcolli C, Calzaferri G (1999) Appl Organomet Chem 13:213.

120. Feher FJ, Newman DA, Walzer JF (1989) J Am Chem Soc 111:1741.

121. Martynova TN, Chupakhina TI (1988) J Organomet Chem 345:11.

122. Corriu RJP, Jutzi P (1996) Tailormade Silicon Oxygen Compounds, Vieweg, Braunscheweig.

123. Calzaferri G, Herren D, Imhof R (1991) Helv Chim Acta 74:1278.

124. Calzaferri G, Imhof R, Törnroos KW (1993) J Chem Soc Dalton Trans: 3741.

125. Calzaferri G, Marcolli C, Imhof R, Törnroos KW (1996) Dalton Trans:3313,

126. Calzaferri G, Imhof R (1992) Dalton Trans:3391.

127. Marcolli C, Calzaferri G, Imhof RT (1997) Microchim Acta 14:493.

128. Morán M, Casado CM, Cuadrado I, Losada J (1993) Organometallics 12:4327.

129. Tsuchida A, Bolln C, Serntez FG, Frey H, Mülhaupt R (1997) Macromolecules 30:2818.

130. Knischka R, Dietsche F, Hanselmann R, Frey H, Mülhaupt R (1999) Langmuir 15:4752.

131. Feher FJ, Budzichowski TA, Weller KJ (1989) J Am Chem Soc 111:7288.

132. Feher FJ, Budzichowski TA (1991) Organometallics 10:812,

133. Feher FJ, Walzer JF (1991) Inorg Chem 30:1689.

134. Krjnen S, Harmsen RJ, Abbenhuis HCL, van Hooff JCH, van Santen RA (1999) Chem Commun:501.

135. Abbenhuis HCL (2000) Chem Eur J 6:25.

136. Feher FJ, Budzichowski TA (1995) Polyhedron 14:3239.

137. Unger KK, Porous Silica, Elsevier, New York, 1979.

138. Horvath C (1988) High Performance Liquid Chromatography, Academic Press, New York.

139. Leyden DE, Luttrell GH, Sloan AE, De Angelis NJ (1976) Anal Chim Acta 84:97.

140. Iwasawa Y, Tailored Metal Catalysts, Reidel, Boston, MA, 1986.

141. Plueddemann EP, Silane Coupling Agents, Plenum Press, New York, 1992.

142. Unger KK, Rupprecht H, Valentin B, Kirhcner W (1983) Drug Dev Ind Pharm 9:69.

143. Feher FJ, Newman DA (1990) J Am Chem Soc 112:1931.

144. Salamone JC (1996) Polymeric Materials Encyclopedia, CRC Press, Boca Raton.

145. Lichtenhan JD, Vu NQ, Carter JA, Gilman JW, Feher FJ (1993) Macromolecules 26:2141.

146. Haddad TS, Lichtenhan JD (1995) J Inorg Organomet Polym 5:237.

147. Brown JF, Vogt LH (1965) J Am Chem Soc 87:4313.

148. Feher FJ, Budzinowski TA, Blanski RL, Weller KJ, Ziller JW (1991) Organometallics 10:2526.

149. Feher FJ, Phillips SH, Ziller JW (1997) Chem Commun:829.

150. Feher FJ, Phillips SH, Ziller JW (1997) J Am Chem Soc 119:3397.

151. Feher FJ, Soulivong D, Lewis GT (1997) J Am Chem Soc 119:11323.

152. Feher FJ, Soulivong D, Eklund AG (1998) Chem Commun:399.

153. Feher FJ, Soulivong D, Nguyen F (1998) Chem Commun:1279.

154. Feher FJ, Nguyen F, Soulivong D, Ziller JW (1999) Chem Commun:1705.

155. Feher FJ, Terroba R, Ziller JW (1999) Chem Commun:2153.

156. Feher FJ, Terroba R, Ziller J (1999) Chem Commun:2309.

157. Winkhofer N, Roeski HW, Noltemeyer M, Robinson WT (1992) Angew Chem Int Ed 31:599,

158. Takiguchi T (1959) J Am Chem Soc 81:2359.

159. Ishida H, Koenig JL (1979) J Polym Sci Polym Phys Ed 17:1807.

160. Pescarmona PP, van der Waal JC, Maschmeyer T (2004) Eur J Inorg Chem:978.

161. Herman WA, Anwander R, Dufaud V, Scherer W (1994) Angew Chem Int Ed 33:1285.

162. Herman WA, Anwander R, Dufaud V, Scherer W (1994) Angew Chem

106:1338.

163. Crocker M, Herold RHM, Orpen AG (1997) Chem Commun:2411.

164. Murugavel R, Chandrasekhar V, Roesky HW (1996) Acc Chem Res 29:183.

165. Xu H, Kuo S-W, Lee J-S, Chang F-C (2002) Macromolecules 35:8788.

166. Haddad TS, Viers BD, Phillips SH (2002) J Inorg Organomet Polym 11:155.

167. Cardoen G, Coughlin EB (2004) Macromolecules 37:5123.

168. Lee A, Xiao J, Feher FJ (2005) Macromolecules 38:438.

169. Bhardwaj BK, Berry RJ, Farmer BL (2000) Polymer 41:7209.

170. Constable GS, Lesser AJ, Coughlin EB (2004) Macromolecules 37:1276.

171. Zheng L, Farris LJ, Coughlin EB (2001) Macromolecules 34:8034.

172. Zhang L, Waddon AJ, Farris RJ, Coughlin EB (2002) Macromolecules 35: 2375.

173. Lichtenhan JD, Otonary YA, Carr MJ (1995) Macromolecules 28:8435.

174. Pyun J, Matyjaszewski K (2000) Macromolecules 33:217.

175. Mather PT, Chun SB, Pyun J, Matyjaszewski K (2000) Polym Prepr (Am Chem Soc Div Polym Chem) 51:582.

176. Mantz RA, Jones PF, Cheffee KP, Lichtenhan JD, Gilman JW, Ismail IMK, Burneister MJ (1996) Chem Mater 8:1250.

177. Fu BX, Hsiao BS, Pagola S, Stephens P, White H, Rafailovich M, Sokolov J, Mather PT, Jeon HG, Phillips S, Lichtenhan J, Schwab J (2001) Polymer 42(2):599.

178. Kim B-S, Mather PT (2002) Macromolecules 35:8378.

179. Lee A, Lichtenhan JD (1998) Macromolecules 31:4970.

180. Ramirez C, Vilariño JML, Abad MJ, Barral L, Bouza R, Cano J, Diez FJ, García-Garabal S, López J (2004) J Appl Polym Sci 92:1576.

181. Ivin KJ, Saegussa T (1984) Ring-Opening Polymerization, Elsevier Applied Science, p 1.

182. Schrock RR (2002) Chem Rev 102:145.

183. Schrock RR, Hoveyda AH (2003) Angew Chem Int Ed 42:4592.

184. Schrock RR (2004) J Mol Catal A Chem 213:21.

185. Trnka TM, Grubbs RH (2001) Acc Chem Res 34:18.

186. Schitter RME, Steinhäusler T, Stelzer F (1997) J Mol Catal A Chem 115:11.

187. Schitter RME, Jocham D, Saf R, Mirtl C, Stelzer F, Hummel K (1998) J Mol Catal A Chem 133:75.

188. Grubbs RH (2004) Tetrahedron 60:7117.

189. Grubbs RH (2003) Handbook of Metathesis, Wiley, VCH, Weinheim.

190. Lapinte V, Broose JC, Fontaine I (2004) Macromol Chem Phys 205:824.

191. Pollino JM, Stubbs LP (2003) Macromolecules 36:223

192. Delaude L, Demoncceau A, Noels F (2003) Macromolecules 36:1446.

193. Rule JD, Moore JS (2002) Macromolecules 35:7878.

194. Charvet R, Novak BM (2001) Macromolecules 34:7680.

195. Breitenkamp K, Emrick T (2003) J Am Chem Soc 125:12070.

196. Bielawski CW, Benitez D, Grubbs RH (2003) J Am Chem Soc 125:8424.

197. Lee BS, Mahajan S, Clapham B, Janda KD (2004) J Org Chem 69:3319.

198. Chemtop A, Héroguez V, Gannou Y (2002) Macromolecules 35:9262.

199. Mecking S, Held A, Bauers FM (2002) Angew Chem 114:564.

200. Demel S, Schoefberger W, Slugovic C, Stelzer F (2003) J Mol Catal A Chem 200:11.

201. Zheng L, Farris RJ, Coughlin EB (2001) J Polym Sci Part A Polym Chem 39:2920

202. Matyjaszewski K, Pyun J, Gaynor SG (1998) Macrom Rapid Commun 19:665.

203. Wang J-S, Matyjaszewski K (1995) J Am Chem Soc 117:5614.

204. Patten TE, Matyjaszewski K (1999) Acc Chem Res 32:895.

205. Pyun J, Matyjaszewski K, Wu J, Kim G.-M, Chun SB (2003) Polymer 44:2739

Chapter 2
Preparation and Characterization of Polyhedral Oligosilsesquioxanes

David B. Cordes and Paul D. Lickiss

2.1 General Comments

Polyhedral OligoSilSesquioxanes are often abbreviated by the acronym POS or POSS® and are compounds of the general formula $Si_nO_{3n/2}R_n$ (often denoted as T_nR_n). The commonest of these compounds are the cubic T_8R_8 POS species since many hydrolysis/condensation reactions of simple $RSiX_3$ compounds (X = Cl, OMe or OEt) give the T_8R_8 cage as a major product, usually along with small amounts of other T_nR_n cages and polymeric by-products. Few T_4R_4 POS have been prepared as they seem to require very bulky R-groups to stabilize the strained POS system and to prevent formation of larger, more stable POS cages. T_6 derivatives are more common than those containing T_4 cages, but again, only a limited number of compounds have been isolated, and ring strain is probably also a problem in these compounds. $T_{10}R_{10}$ POS are much more common than T_4 and T_6 derivatives owing to reduced ring strain, however, even larger POS are also rare and are often found as components of mixtures containing mainly T_8R_8, or polymeric materials and other T_nR_n species derived from $RSiX_3$ hydrolyses. The isolation of individual pure forms of the higher silsesquioxanes is further complicated by the existence of multiple isomers. For example, in the case of the T_nH_n POS, identification of isomers by high-resolution single-ion-detection GC has shown that although for n = 12, only one isomer was formed, four were found for n = 14, nine for n = 16, and ten or twelve isomers for n = 18 [1]. NMR studies of mixtures of isomers have allowed the structures of some to be determined: the D_{2d} form for $T_{12}H_{12}$, D_{3h} (major isomer) and C_{2v} for $T_{14}H_{14}$, and a D_{4d} structure for $T_{16}H_{16}$ [1]. The POS products that have been

D.B. Cordes and P.D. Lickiss
Department of Chemistry, Imperial College London, South Kensington, London SW7 2AZ, U.K.

C. Hartmann-Thompson (ed.), *Applications of Polyhedral Oligomeric Silsesquioxanes*, Advances in Silicon Science 3, DOI 10.1007/978-90-481-3787-9_2,

isolated include those with n = 12, 14, and 16, as well as smaller POS derivatives; however, larger POS molecules ($n \geq 18$) have been identified using either SEC [2], or GC [1], as minor components from $RSiX_3$ hydrolyses.

2.2 Synthesis of T_nR_n Compounds where R = H, Alkyl or Alkenyl

2.2.1 *Hydrolysis*

2.2.1.1 T_4 and T_6 Compounds

Only two alkyl T_4 POS, $T_4(i\text{-Pr})_4$, and $T_4(t\text{-Bu})_4$, have been prepared by hydrolysis; using acidic conditions *i*-$PrSiCl_3$ and *t*-$BuSiCl_3$ afford the T_4 derivatives with yields of 55% and 95%, respectively [3]. Hydrolysis of other alkyl $RSiCl_3$ compounds, e.g. R = Me or Et, leads to the formation of larger T_nR_n compounds or polymeric siloxanes, see below.

The synthesis of a small number of T_6 alkyl-POS has been reported including early work on compounds identified as T_6R_6, obtained as sublimable solids in low yield, by the distillation of products from the hydrolysis of $RSi(OEt)_3$ (R = Me, Et) [4, 5]. The properties of these materials were noted to be different to those of the previously known T_8 compounds, and as a result were identified as T_6R_6 species.

The attempted hydrolysis in organic solvents of alkyltrichlorosilanes with alkyl substituents the size of a cyclohexane ring or larger, has a tendency to produce incompletely condensed products, such as $T_7Cy_7(OH)_3$ [6]. If, however, the reaction uses solely aqueous HCl as the hydrolysis medium, then the products consist of mixtures of fully condensed species, including T_6R_6 derivatives, for the cases when R = *n*-C_7H_{15} [7], *n*-C_8H_{17}, or *i*-C_9H_{19} [8]. Vacuum distillation of the *n*-octyl and *i*-nonyl products was found to lead to thermal rearrangement of the compounds and the isolation of pure T_6R_6 derivatives. In a similar fashion T_6Cy_6 may be prepared by the aqueous hydrolysis/condensation of $CySiCl_3$, giving the pure product in 10% yield [9]. For trichlorosilanes with alkyl groups in most cases smaller than cyclohexane, the presence of non-aqueous, or scarce-water, hydrolysis media can give rise to the desired T_6R_6 products. For trichlorosilanes with R = *i*-Bu, *c*-C_5H_9, Cy, Ph, *n*-Oct, $(CH_2)_3C_6H_4$-4-OMe, and $(CH_2)_2CMe_2CH_2CO_2Me$, hydrolysis in DMSO led to T_6R_6 in moderate yields [10].

2.2.1.2 T_8 Compounds

A range of T_8 POS substituted with alkyl and related derivatives can be prepared by hydrolysis and condensation reactions. These include substituents from the very simplest, such as T_8H_8 and T_8Me_8 to more complicated functional groups, for example $T_8[(CH_2)_3Cl]_8$. Details for the preparation of these POS derivatives are presented in Table 2.1. It may be seen that the reaction products show a very wide range of yields, this being one of the biggest problem areas in the use of hydrolysis reactions to prepare POS compounds. Optimization of reaction conditions can lead to yields being improved, significantly in some cases, and the use of the basic tetrabutylammonium fluoride as a catalyst [11] has allowed synthesis of previously inaccessible, simple T_8 POS, and led to significantly improved yields of others. Several of these compounds are useful precursors to a wide range of other POS derivatives. For example, T_8H_8 yields alkylated products via hydrosilylation, $T_8(CH{=}CH_2)_8$ undergoes metathesis reactions, additions, and Heck reaction chemistry, and $\{T_8[(CH_2)_3NH_3]_8\}Cl_8$ and its deprotonated derivative $T_8[(CH_2)_3NH_2]_8$, are precursors to a range of amides.

The synthetic route to T_8H_8 has improved significantly, from the initially reported yield of less than 1% for the acid-hydrolysis of $HSiCl_3$ [12], to the current yield of 18% for the scarce-water acid hydrolysis of $HSiCl_3$ [13]. Similarly, the yield of $T_8(CH{=}CH_2)_8$ has been increased from 6% [14], to a best yield of 41%, when the hydrolysis is carried out in the presence of acidic Amberlite (IR-120 PLUS) ion exchange resin [15]. A contemporary preparation for $T_8(CH{=}CH_2)_8$ shows a significantly higher yield (80%), from a phase-transfer-catalysed hydrolysis of $CH_2{=}CHSi(OEt)_3$ [16]. In contrast to these two compounds, the synthetic yields of $\{T_8[(CH_2)_3NH_3]_8\}Cl_8$ have stayed constant at about 35% despite a number of minor variations to synthetic conditions. This salt can be deprotonated using basic Amberlite IRA-400 ion exchange resin to give the unstable free amine in quantitative yield [17]. However, there have been recent reports claiming the synthesis of the free amine directly from the starting silane in yields ranging up to 93% [18]. These direct preparations of the free amine are reported to lead to a much more stable material.

Table 2.1 Hydrolytic routes to T_8R_8 compounds, R = H, alkyl or alkenyl

R or T_8R_8	Starting materials	Yield (%)	Refs.
-H	$HSiCl_3$ + $FeCl_3$, HCl	18	[13]
-Me	$MeSi(OEt)_3$ + poly(2-hydroxyethyl methacrylate), NH_3 or HCl	23	[19]
-Me	$MeSi(OEt)_3$ + NEt_3, HCl, AcOH	88	[20]
$-CH{=}CH_2$	$CH_2{=}CHSiCl_3$ + acid Amberlite,	41	[15]
$-CH{=}CH_2$	$CH_2{=}CHSi(OEt)_3$ + NMe_4OH	80	[16]
-*i*-Bu	*i*-$BuSi(OEt)_3$ + NBu_4F	26	[11]
-*i*-Bu	*i*-$BuSi(OMe)_3$ + KOH	96	[21]

R or T_8R_8	Starting materials	Yield (%)	Refs.
$-(CH_2)_2(CF_2)_3CF_3$	$CF_3(CF_2)_3(CH_2)_2Si(OEt)_3 + KOH$	99	[22]
$-(CH_2)_3NH_2$	$H_2N(CH_2)_3Si(OEt)_3 + NMe_4OH$	93	[18]
$\{T_8[(CH_2)_3NH_3]_8\}Cl_8$	$H_2N(CH_2)_3Si(OEt)_3 + HCl$	35	[17]
$-(CH_2)_3SH$	$HS(CH_2)_3Si(OMe)_3 + HCl$	17	[23]
$-(CH_2)_3Cl$	$Cl(CH_2)_3Si(OEt)_3 + HCl$	37	[24]
$-(CH_2)_3Br$	$Br(CH_2)_3SiCl_3 + FeCl_3$, HCl	3	[25]
-*n*-C_6H_{13}	*n*-$C_6H_{13}Si(OEt)_3 + NBu_4F$	44	[11]
-*c*-C_5H_9	*c*-$C_5H_9Si(OEt)_3 + NBu_4F$	95	[11]
-Cy	$CySi(OEt)_3 + NBu_4F$	94	[11]

The mechanism by which the T_8 POS cage is formed during the hydrolytic reactions of simple $HSiX_3$ compounds, via $HSi(OH)_3$ condensation, to give T_8H_8 has been the subject of a theoretical study showing that the overall formation of T_8H_8 is exothermic by 48.3 kJ mol^{-1}, and that H-bonding is a significant factor in determining the nature of the intermediates and the mechanistic route [26].

The capping of partially condensed silsesquioxanes, $R_7Si_7O_9(OH)_3$, by reaction with chloro- or alkoxysilanes to give T_8R_7R' derivatives follows the same general reaction scheme (Eq. 2.1). The partially condensed trisilanol species have been known since 1965 [27], and the recent interest in the synthesis of T_8R_7R' derivatives derives from their utility in incorporating POS cage substituents into other materials via a single connection point. The most useful compounds for this are thus T_8R_7R' species where the R groups are inert and the lone R' substituent contains a functional group.

$$R_7Si_7O_9(OH)_3 + R'SiX_3 \xrightarrow{\text{base}} T_8R_7R' \qquad \text{(Eq. 2.1)}$$

$$X = Cl, O\text{-alkyl}$$

In T_8R_7R' compounds with a sufficiently bulky R group, an R' group that would normally be too reactive to give a stable $T_8R'_8$ compound may be incorporated, as for example in the preparation of $T_8R_7(CH_2)_3NH_2$ (R = *i*-Bu, *c*-C_5H_9 or Ph), T_8R_7Cl (R = *i*-Bu, *c*-C_5H_9 or Cy), and $T_8Cy_7OSiCl_3$. The synthetic routes to a selection of representative T_8R_7R' compounds are presented in Table 2.2.

Table 2.2 Alkyl-T_8R_7R' compounds prepared by corner capping of $R_7Si_7O_9(OH)_3$

Substituents		Starting materials	Yield (%)	Refs.
R	R'			
-*i*-Bu	-H	$(i\text{-Bu})_7Si_7O_9(OH)_3 + HSiCl_3$, NEt_3	67	[28]
-*i*-Bu	$-(CH_2)_3NH_2$	$(i\text{-Bu})_7Si_7O_9(OH)_3 + NH_2(CH_2)_3Si(OEt)_3$, NEt_3	-	[29]

Substituents R	R'	Starting materials	Yield (%)	Refs.
-*i*-Bu	$-(CH_2)_3Cl$	$(i\text{-Bu})_7Si_7O_9(OH)_3 + Cl(CH_2)_3SiCl_3$, NEt_3	73	[30]
$-(CH_2)_2CF_3$	-H	$Na_3\{[CF_3(CH_2)_2]_7Si_7O_9(O)_3\} + HSiCl_3$, NEt_3	76	[31]
$-(CH_2)_2CF_3$	-Me	$Na_3\{[CF_3(CH_2)_2]_7Si_7O_9(O)_3\} + MeSiCl_3$, NEt_3	72	[32]
$-c\text{-}C_5H_9$	$-CH_2Cl$	$(c\text{-}C_5H_9)_7Si_7O_9(OH)_3 + ClCH_2SiCl_3$, NEt_3	78	[33]
$-c\text{-}C_5H_9$	$-(CH_2)_3Cl$	$(c\text{-}C_5H_9)_7Si_7O_9(OH)_3 + Cl(CH_2)_3SiCl_3$	82	[34]
$-c\text{-}C_5H_9$	$-(CH_2)_3NH_2$	$(c\text{-}C_5H_9)_7Si_7O_9(OH)_3 + NH_2(CH_2)_3Si(OMe)_3$	78	[35]
-Cy	-H	$T_7Cy_7(OH)_3 + HSiCl_3$, NEt_3	88	[36]
-Ph	$-(CH_2)_3NH_2$	$Ph_7Si_7O_9(OH)_3 + NH_2(CH_2)_3Si(OMe)_3$, toluene	68	[37]

2.2.1.3 T_{10}, T_{12} and Larger Compounds

A small number of alkyl-POS compounds larger than T_8 have been prepared by hydrolysis of their parent trichloro- or trialkoxysilanes. Among the most common syntheses are the scarce-water hydrolyses, for example, $HSiCl_3$ affords $T_{10}H_{10}$ in less than 4% yield, as a by-product of the formation of T_8H_8 [13]. An alternative hydrolysis route leads to the formation of $T_{12}H_{12}$, again as a by-product, in similar yield [1]. Another scarce-water hydrolysis, in this case using tetrabutylammonium fluoride and $c\text{-}C_5H_9SiCl_3$, has been reported to produce $T_{10}(c\text{-}C_5H_9)_{10}$ in 12% yield [38].

A more conventional aqueous hydrolysis of $CF_3(CH_2)_2SiCl_3$ gives $T_{12}[(CH_2)_2CF_3]_{12}$, although no yield was reported for this reaction [39]. A study of the acidic hydrolysis and condensation of vinyltrimethoxysilane showed that a mixture of products was formed. This was reported to include the large vinyl-POS $T_n(CH{=}CH_2)_n$ for n=14 and 16, although neither compound was isolated from the product mixture [40].

2.2.2 *Substitution*

Substitution reactions at the corner silicon atoms of alkyl POS compounds can produce a very wide range of products. The simplest of these reactions is that of preparing T_8D_8 from T_8H_8 using deuterium gas, catalysed by palladium on carbon [41]. The most common reaction involving substitution at silicon is hydrosilylation, a simple platinum-catalyzed reaction allowing the addition of a Si-H to an unsaturated compound such as an alkene or alkyne. The general hydrosilylation reaction can be summarized in Eq. 2.2, and the syntheses of a range of T_8R_8 derivatives are presented in Table 2.3. The potential problems of hydrosilylation reactions are two-fold; first that the product can be either the α- or the β-isomer, and second that reaction may take place at functional groups other than alkene or alkyne within the added compound, most often at oxygen.

The first problem may be overcome by the strong directing effect of silicon, which favours the α-product, leaving the β-product generally as an impurity to be removed. The occurrence of reaction at other functionalities can often be reduced during optimization of a reaction by careful choice of the platinum catalyst, solvent and reaction conditions. A wide variety of functional groups may be introduced into POS systems by hydrosilylation including those, such as alcohols, that could not be introduced via the hydrolysis of $RSiX_3$ monomer synthetic route. Similar reactions are also seen for the T_8R_7H compounds, giving rise to T_8R_7R' species.

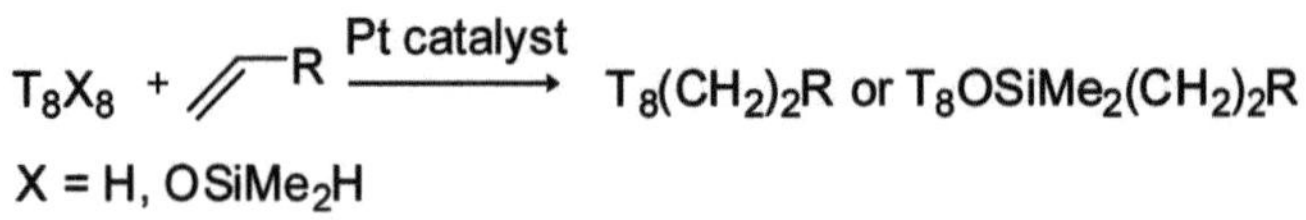

(Eq. 2.2)

Table 2.3 Alkyl-T_8R_8 compounds prepared by hydrosilylation

R	Starting materials	Yield (%)	Refs.
$-C_nH_{2n+1}$ (n = 2-10)	$T_8H_8 + CH_2{=}CH_2$, or $CH_2{=}CH_2C_{n-2}H_{2n-3}$, H_2PtCl_6	76-92	[42]
$-(CH_2)_2CMe_2OH$	$T_8H_8 + CH_2{=}CHCMe_2OH$, H_2PtCl_6	62	[43]
$-(CH_2)_2C_6H_4$-4-Br	$T_8H_8 + CH_2{=}CHC_6H_4$-4-Br, Pt(dvs)	81	[44]
$-(CH_2)_2C_6H_4$-4-CH_2Cl and $-CHMeC_6H_4$-4-CH_2Cl	$T_8H_8 + CH_2{=}CHC_6H_4$-4-CH_2Cl, H_2PtCl_6, $Cl(CH_2)_2Cl$	85[a]	[45]
$-(CH_2)_3Cy$	$T_8H_8 + CH_2{=}CHCH_2Cy$, H_2PtCl_6, $Cl(CH_2)_2Cl$, $MeO(CH_2)_2OMe$	94	[46]
$-(CH_2)_3Ph$	$T_8H_8 + CH_2{=}CHCH_2Ph$, H_2PtCl_6, $Cl(CH_2)_2Cl$, $MeO(CH_2)_2OMe$	69	[46]
$-(CH_2)_3CN$	$T_8H_8 + CH_2{=}CHCH_2CN$, H_2PtCl_6	-	[47]
$-(CH_2)_5OH$	$T_8H_8 + CH_2{=}CH(CH_2)_3OH$, H_2PtCl_6	-	[48]
$-(CH_2)_3OCH_2CH(O)CH_2$	$T_8H_8 + CH_2{=}CHCH_2OCH_2CH(O)CH_2$, $Pt(COD)Cl_2$	24	[23]
$-(CH_2)_3SiMe_3$	$T_8H_8 + CH_2{=}CHCH_2SiMe_3$, H_2PtCl_6	93	[23]
$-(CH_2)_3SiCl_3$	$T_8H_8 + CH_2{=}CHCH_2SiCl_3$, H_2PtCl_6	-	[49]
$-(CH_2)_4CH(O)CH_2$	$T_8H_8 + CH_2{=}CH(CH_2)CH(O)CH_2$, Pt(dvs)	99	[50]
$-(CH_2)_nBr$ (n = 4, 5, 7, 8)	$T_8H_8 + CH_2{=}CH(CH_2)_{n-2}Br$, H_2PtCl_6	88-99	[51]

R	Starting materials	Yield (%)	Refs.
-CH=CHCMe$_2$OH and –OCMe$_2$C≡CH	T_8H_8 + CH≡CCMe$_2$OH, H_2PtCl_6	-[a]	[49]
-CH=CH(CH$_2$)$_4$Cl and –C(=CH$_2$)(CH$_2$)$_4$Cl	T_8H_8 + CH≡C(CH$_2$)$_4$Cl, H_2PtCl_6	79[a]	[51]

[a] Consists of a mixture of these products.

A similar method can be used to prepare unusual T_8 compounds derived from monohydrosilylation of T_8H_8, to give T_8H_7R products. Thus, treatment of T_8H_8 with 9,10-dibromo-1-decene in the presence of H_2PtCl_6 gives the mono-substituted compound $T_8H_7[(CH_2)_8CHBrCH_2Br]$ [52] and further substitution was realized by reaction with ethylene, again using H_2PtCl_6 catalyst, to give $T_8Et_7[(CH_2)_8CHBrCH_2Br]$. Likewise T_8H_8 has been treated with three long-chain alkenes using H_2PtCl_6 to give the $T_8H_7(CH_2)_nMe$ (n = 18, 20, 22) products [53]. The substitution patterns resulting from the reaction of T_8H_8 with varying molar ratios of either phenol and undec-1-ene or phenol and 3,3-dimethylpent-4-enoate have been studied [54]. When complicated mixtures of products were clearly present, patterns observed in the ^{29}Si NMR spectra of the product mixtures suggest that identification of particular substitutional isomers may still be possible. The unusual hydrosilylation reaction of T_8H_8 with 1,3-divinyl-1,1,3,3-tetramethyldisiloxane proceeds with an α : β product ratio of 3.4 : 1, but it also shows signs of creating some linkages between adjacent T_8 cages [55]. Similar reactions, but using more complicated divinyl derivatives, and Pt(dcp) as catalyst, allow preparation of di-POS compounds when using a divinyl compound to POS ratio of 1:2, polymeric POS systems were seen at other reagent ratios [56].

2.2.3 *Cage Rearrangement*

Fragmentation and rearrangement of existing POS cages can be an effective method for the preparation of new larger POS cage materials. The expansions of a series of T_8 alkyl-POS functionalized with either linear alkyl chains, or a variety of 3-substituted propyl chains, $(CH_2)_3R$ (R = Cl, Br, I, NCS, C_6F_5) have been prepared by the action of basic catalysts such as K_2CO_3, NaOH, or Na_2SiF_6 on the corresponding T_8 compounds [57, 58]. The resulting products comprised mixtures of the corresponding T_8, T_{10}, and sometimes T_{12}, POS compounds with the T_8 species predominating, although the individual cage forms were not isolated. Another T_{10} derivative that can be prepared by fragmentation and rearrangement reactions is $T_{10}[(CH_2)_3N_3]_{10}$ which was discovered serendipitously when attempting to prepare $T_8[(CH_2)_3N_3]_8$ from $T_8[(CH_2)_3Cl]_8$ and NaN_3 [30]. A mixture of products is again produced by this reaction; however, in this case, the major product is the T_{10} POS, with T_8 and T_{12} as minor products.

2.2.4 *Modification of R*

2.2.4.1 T_8 Compounds

One of the largest categories of reactions producing alkyl POS is that of modification of R groups using a variety of reactions such as substitutions, reactions involving acid derivatives and reactions of unsaturated systems. There has been a wide range of reported substitution reactions for alkyl-POS compounds. These range from relatively conventional substitutions, taking place close to the POS core, to more complicated reactions, often taking place at a greater remove. These more complicated reactions have included Sonogashira couplings, oxidations, and reactions with lithium reagents. An unusual alkyl-POS substitution reaction involving an enzymatically catalyzed substitution of $T_8[(CH_2)_3NH_2]_8$ by catechin was reported in 2005 [59]. The product was reported to be disubstituted, with catechin bound to two of the amine groups of the POS, either as an imide or as a secondary amine. Unsurprisingly, given the nature of the reaction, the substitution pattern on the POS cube is unknown. Of the simpler reactions, a large proportion of these have involved substitution of the chlorine in $T_8[(CH_2)_3Cl]_8$ by a wide range of nucleophiles. A selection of these substitution reactions is presented in Table 2.4.

Table 2.4 Alkyl-T_8R_8 compounds prepared by substitution reactions of functional groups

R	Starting materials	Yield (%)	Refs.
$-CH{=}CHC_6H_4\text{-}4\text{-}C{\equiv}CSiMe_3$	$T_8(CH{=}CHC_6H_4\text{-}4\text{-}Br)_8 + HC{\equiv}CSiMe_3$, $[Pd(PPh_3)_2Cl_2]$, CuI, PPh_3, NEt_3	37	[60]
$-(CH_2)_2P({=}O)(OEt)_2$	$T_8(CH_2)_2Br + P(OEt)_3$	-	[61]
$-(CH_2)_2Si(CH_2PMe_2)_3$	$T_8[(CH_2)_2SiCl_3]_8 + LiCH_2PMe_2$	97	[62]
$-(CH_2)_2Si(CH_2PPh_2)_3$	$T_8[(CH_2)_2SiCl_3]_8 + LiCH_2PPh_2$	93	[63]
$-(CH_2)_2SiMe_2H$	$T_8[(CH_2)_2SiMe_2Cl]_8 + LiAlH_4$	61	[64]
$-(CH_2)_2SiMe_2OH$	$T_8[(CH_2)_2SiMe_2H]_8 + H_2O$, Pd	65	[64]
$-(CH_2)_2SiMe[(CH_2)_2SiMe_2OH]_2$	$T_8\{(CH_2)_2SiMe[(CH_2)_2SiMe_2H]_2\}_8 + H_2O$, Pd	96	[64]
$-(CH_2)_2SiMe_n[C_6H_4\text{-}3\text{-}P(NMe_2)_2]_{3-n}$ (n =0, 1, 2)	$T_8[(CH_2)_2SiMe_nCl_{3-n}]_8 + BrC_6H_4\text{-}3\text{-}P(NMe_2)_2$, *n*-BuLi	-	[65]
$-(CH_2)_3NHCH_2PPh_2$	$T_8[(CH_2)_3NH_2]_8 + H_2CO/Ph_2PH$	37	[17]
$-(CH_2)_3N_3$	$\{T_8[(CH_2)_3NH_3]_8\}Cl_8 + CF_3(CF_2)_3SO_2N_3$, $CuSO_4\cdot 5H_2O$, $NaHCO_3$	67	[66]
$-(CH_2)_3OH$	$T_8[(CH_2)_3Cl]_8 + Ag_2O$	91	[67]
$-(CH_2)_3ONO_2$	$T_8[(CH_2)_3I]_8 + AgNO_3$	99	[17]
$-(CH_2)_3PPh_2$	$T_8[(CH_2)_3Cl]_8 + KPPh_2$	83	[23]

R	Starting materials	Yield (%)	Refs.
$-(CH_2)_3SMe$	$T_8[(CH_2)_3Cl]_8 + NaSMe$	55	[23]
$-(CH_2)_3SCN$	$T_8[(CH_2)_3Cl]_8 + NaSCN$	90	[23]
$-(CH_2)_3I$	$T_8[(CH_2)_3Cl]_8 + NaI$	97	[23]

Similar reactions to those used on the T_8R_8 systems have also been applied to T_8R_7R' compounds to produce an equally wide variety of products with varying substitutions. Formation of bis-T_8 derivatives can be achieved in the unusual reaction between $T_8(i\text{-Bu})_7(CH_2)_3SH$ and the *ortho*-chloro substituted benzene derivative 1,2-$(CN)_2$-C_6H_2-4,5-Cl_2 in the presence of a base, leading to the formation of 1,2-$(CN)_2C_6H_2$-4,5-$[S(CH_2)_3T_8(i\text{-Bu})_7]_2$ in 94% yield [68]. Similarly, treatment of $T_8(c\text{-}C_5H_9)_7(CH_2)_3I$ with the *para*-hydroxy substituted compound HOC_6H_2-2,5-Br_2-4-OH, again in the presence of base, led to the formation of the bis-T_8 derivative C_6H_2-2,5-Br_2-1,4-$[O(CH_2)_3T_8(c\text{-}C_5H_9)_7]_2$, although in only 15% yield [69].

Those reactions involving formation of a carboxylic acid derivative break down fairly evenly between those forming amides and esters; however hydrolysis of esters being used as protecting groups can also be used to give CO_2H groups. Perhaps somewhat surprisingly, most reactions involving the formation of an amide or ester do not involve a POS compound containing a CO_2H group, this probably being associated with the difficulty of introducing unprotected acid functionalities into POS species. A selection of reactions involving acid derivatives is presented in Table 2.5. Although some of these reactions occur without a catalyst, most do require one, even when using reactive acid derivatives. This may be due to the steric interactions arising when attempting to fully functionalize all eight substituents in an alkylated T_8R_8 compound. Many of the hydrolyses of alkyl-T_8 derivatives involve the deprotection of an alcohol but there are some where hydrolysis of an ester leads to the formation of carboxylic acid.

Table 2.5 Preparation of alkyl-T_8R_8 compounds involving carboxylic acid derivatives

R	Starting materials[a]	Yield (%)	Refs.
$-(CH_2)_3NH$(L-Lys) TFA salt	$T_8[(CH_2)_3NH_3]_8Cl_8$ + BLDCH, HBTU, HOBT, $N(i\text{-Pr})_2Et$, citric acid, CF_3CO_2H	47	[70]
$-(CH_2)_3N[(CH_2)_2CO_2H]_2$	$T_8\{(CH_2)_3N[(CH_2)_2CO_2\text{-}t\text{-Bu}]_2\}_8 + HCO_2H$,	99	[71]
$-(CH_2)_3N[(CH_2)_2C(=O)NH(CH_2)_2NH_2]_2$	$T_8\{(CH_2)_3N[(CH_2)_2CO_2Me]_2\}_8$ + $H_2N(CH_2)_2NH_2$	92	[72]
$-(CH_2)_3NHC(O)Ph$	$\{T_8[(CH_2)_3NH_3]_8\}Cl_8 + PhC(O)Cl$	49	[17]
$-(CH_2)_3NAc_2$	$T_8[(CH_2)_3NH_3]_8Cl_8 + Ac_2O$	59	[73]
$-(CH_2)_3NCO$	$T_8[(CH_2)_3NH_2]_8 + C(O)Cl_2$	11-47	[17]

[a] BLDCH - $N_\alpha N_\varepsilon$-di(Boc)-L-Lys dicyclohexylammonium salt; HBTU - 2-(1*H*-benzotriazole-1-yl)-1,1,3,3-tetramethyluronium hexafluorophosphate; HOBT - 1-hydroxybenzotriazole hydrate

Some interesting differences in product have been observed upon varying the reaction conditions when treating $T_8[(CH_2)_3NH_2]_8$ with a series of anhydrides. At room temperature, succinic, maleic [17], and acetic [74] anhydrides all react to form primary amides; however, at 180° C, the reaction with succinic anhydride affords a cyclic secondary amide [74].

Various biologically-relevant acid derivatives of alkyl-POS compounds have been prepared, usually derived from the octa-amine $T_8[(CH_2)_3NH_2]_8$. Amino acids and oligopeptides have been introduced in their *N*-protected forms by reaction with $\{T_8[(CH_2)_3NH_3]_8\}Cl_8$ to give the amides in yields ranging from 44% to 94% [75]. It has also proven possible to synthesize oligopeptide POS derivatives using protection/deprotection methods, as well as by reaction with a pre-prepared oligopeptide [75]. More recently, this idea has been applied to making biologically-based dendrimers by repeatedly functionalising $T_8[(CH_2)_3NH_2]_8$ with Boc-protected lysine salts. Four sequential reactions led to the formation of the lysine-dendrimers $T_8[(CH_2)_3NH(\text{L-Lys})]_8$, $T_8[(CH_2)_3NH(\text{L-Lys})(\text{L-Lys})_2]_8$, $T_8[(CH_2)_3NH(\text{L-Lys})(\text{L-Lys})_2(\text{L-Lys})_4]_8$ and $T_8[(CH_2)_3NH(\text{L-Lys})(\text{L-Lys})_2(\text{L-Lys})_4(\text{L-Lys})_8]_8$ as their TFA salts [70]. $T_8[(CH_2)_3NH_2]_8$ has also been coupled with the lactones δ-lactonolactone, ε-caprolactone, δ-maltonolactone, and δ-gluconolactone, to give octa-amide products, reactions proceeding in 23% to 53% yield [17].

Some of the first work on the synthesis of vinyl POS derivatives was carried out as a study of cross-metathesis reactions of $T_8(CH{=}CH_2)_8$ with a selection of simple and substituted alkenes [76]. Two different catalysts were compared; the first generation Grubbs catalyst, $[RuCl_2({=}CHPh)(PCy_3)_2]$, and Schrock's catalyst, $\{Mo[{=}NC_6H_3\text{-}2,6\text{-}(i\text{-}Pr)_2]({=}CHCMe_2Ph)[OCMe(CF_3)_2]_2\}$, and the results from the more successful syntheses are presented in Table 2.6 [76].

Table 2.6 Alkynyl-T_8R_8 compounds prepared by alkene metathesis

R	Starting materials[a]	Yield (%)
$-CH{=}CHPh$	$T_8[CH{=}CH_2]_8 + CH_2{=}CHPh$, [Mo]	81
$-CH{=}CHCH_2Si(OMe)_3$	$T_8[CH{=}CH_2]_8 + CH_2{=}CHCH_2Si(OMe)_3$, [Mo]	99
$-CH{=}CH(CH_2)_2CH_3$	$T_8[CH{=}CH_2]_8 + CH_2{=}CHC_3H_7$, [Mo]	99
$-CH{=}CH(CH_2)_3Br$	$T_8[CH{=}CH_2]_8 + CH_2{=}CH(CH_2)_3Br$, [Ru]	76
$-CH{=}CH(CH_2)_3OH$	$T_8[CH{=}CH_2]_8 + CH_2{=}CH(CH_2)_3OH$, [Ru]	<10
$-CH{=}CH(CH_2)_6Si(OMe)_3$	$T_8[CH{=}CH_2]_8 + CH_2{=}CH(CH_2)_6Si(OMe)_3$, [Mo]	99
$-CH{=}CH(CH_2)_8CO_2Et$	$T_8[CH{=}CH_2]_8 + CH_2{=}CH(CH_2)_8CO_2Et$, [Mo]	99

[a] [Mo] - $\{Mo[{=}NC_6H_3\text{-}2,6\text{-}(i\text{-}Pr)_2]({=}CHCMe_2Ph)[OCMe(CF_3)_2]_2\}$, [Ru] - $[RuCl_2({=}CHPh)(PCy_3)_2]$

This study was followed by another, looking at the preparation of other T_8-vinyl derivatives by either cross-metathesis, using the first or second generation Grubbs catalysts, or silylative coupling, using $[RuHCl(CO)(PCy_3)_2]$ [77]. It was found that the choice of either metathesis or silylative coupling catalysts does not affect the product yield for 1-hexene, allyltrimethylsilane and styrene, but that the silylative coupling reactions for 1-hexene and allyltrimethylsilane produce a mixture of *cis*/*trans* isomers. In contrast, however, for the reactions of vinyltrimethylsilane, 1-vinyl-2-pyrrolidinone, butylvinyl ether, *tert*-butyl vinyl ether, and trimethylsilyl vinyl ether, only the silylative coupling catalyst led to the production of product [77]. No reaction was observed for *tert*-butyl vinyl sulfide using either the first generation Grubbs catalyst or the silylative coupling catalyst; however when the second generation Grubbs catalyst was tried, the reaction proceeded. A related cross-metathesis study involved contrasting the couplings carried out using the first generation Grubbs catalyst, and silylative coupling catalyst $[RuHCl(CO)(PCy_3)_2]$, of $T_8(i\text{-Bu})_7CH{=}CH_2$, with $CH_2{=}CHR$ (R = Ph, C_6H_4-4-Me, C_6H_4-4-OMe, C_6H_4-4-Cl, C_6H_4-4-Br, CH_2Ph and CH_2SiMe_3) to produce monovinyl-substituted products [78]. Further studies on metathesis reactions of T_8 POS derivatives have been more focused on the synthesis of specific products, using the first generation Grubbs catalyst, for example, the formation of dendrimers derived from $T_8(CH{=}CH_2)_8$, by coupling the POS core to functionalized styrenyl derivatives [60, 79, 80], and of photoluminescent compounds, by decorating it with vinylbiphenyl derivatives [81]. In addition to this, $T_8[CH{=}CHC_6H_4C_6H_3\text{-}3,5\text{-}(CHO)_2]_8$ has been prepared by the cross-metathesis of octavinyl POS with 4'-vinylbiphenyl-3,5-dicarbaldehyde [81]. Upon the subsequent reduction of the aldehyde to the alcohol, using either $NaBH_4$ or $LiAlH_4$, the internal carbon-carbon double bond was found to be unaffected.

As well as metathetic reactions, a range of addition reactions to unsaturated compounds involving POS species to give alkyl-POS compounds is known. Most of these reactions take place between an unsaturated POS derivative, whether simple systems like $T_8(CH{=}CH_2)_8$ or more complex, and an appropriate reagent; however, there are a small number of reactions known between a T_8 POS derivative and an unsaturated compound. A selection of these addition reactions are presented in Table2.7.

Table 2.7 Alkyl-T_8R_8 compounds prepared by addition reactions of alkenes

R	Starting materials	Yield (%)	Refs.
-Et	$T_8(CH{=}CH_2)_8 + H_2$, Pd/C	89	[82]
$\text{-CHDCH}_{1.14}D_{1.86}$	$T_8(CH{=}CH_2)_8 + D_2$, Pd/C	-	[82]
$\text{-(CH}_2)_2CHO$	$T_8(CH{=}CH_2)_8$ + CO/H_2, $PtCl_2$(sixantphos), $SnCl_2$	-	[21]
$\text{-(CH}_2)_2CO_2Me$	$T_8(CH{=}CH_2)_8$ + CO + MeOH, $MeSO_2OH$, Pd_2dba_3, C_6H_4-1,2-$[CH_2P\text{-}(t\text{-Bu}_2)]_2$	43	[83]
$\text{-(CH}_2)_2Ph$	$T_8(CH{=}CH_2)_8 + C_6H_6 + AlCl_3$	73	[84]
$\text{-(CH}_2)_2PEt_2$	$T_8(CH{=}CH_2)_8 + HPEt_2$, AIBN	> 90	[63]
$\text{-(CH}_2)_2S(CH_2)_2(CF_2)_nCF_3$ (n = 5, 7)	$T_8(CH{=}CH_2)_8 + HS(CH_2)_2(CF_2)_nCF_3$, AIBN	90, 91	[85]
$\text{-(CH}_2)_2SiMe_nCl_{3-n}$ (n = 2, 1, 0)	$T_8[CH{=}CH_2]_8 + HSiMe_nCl_{3-n}$, H_2PtCl_6	93-99	[86]

R	Starting materials	Yield (%)	Refs.
$-(CH_2)_2SiMe_n(CH{=}CH_2)_{3-n}$ (n = 2, 1, 0)	$T_8[(CH_2)_2SiMe_nCl_{3-n}]_8 + CH_2{=}CHMgBr$	52-81	[86]
$-(CH_2)_2SiMe_2C_6H_4$-4-Br	$T_8(CH{=}CH_2)_8 + HSiMe_2C_6H_4$-4-Br, Pt(dvs)	67	[60]
$-(CH_2)_2Br$	$T_8(CH{=}CH_2)_8$ + HBr, $(BzO)_2$	65	[83]
-CH=CHPh	$T_8(CH{=}CH_2)_8$ + PhI, $[Pd(OAc)_2]$, PPh_3, NEt_3	51	[60]
$-(CH_2)_3NHC(O)NHCH_2CH{=}CH_2$	$T_8[(CH_2)_3NH_2]_8 + CH_2{=}CHCH_2NCO$	90	[17]
$-(CH_2)_3N[(CH_2)_2CO_2Me]_2$	$T_8[(CH_2)_3NH_2]_8 + CH_2{=}CHCO_2CH_3$	73	[87]
$-CH_2CH(O)CH_2$	$T_8(CH_2CH{=}CH_2)_8$ + MCPBA	75	[88]

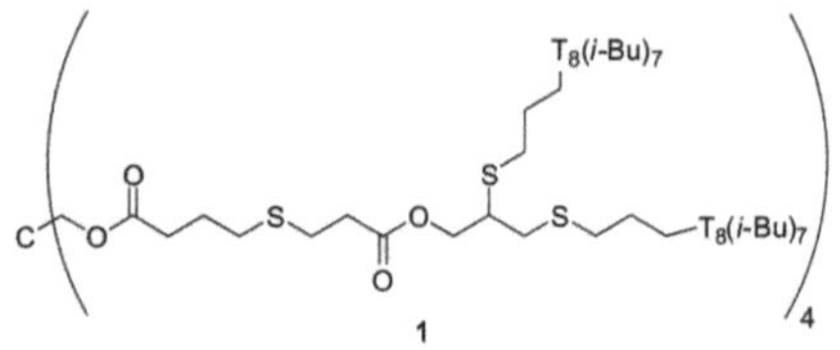

1

Phosphorus-containing derivatives of alkyl-POS have been prepared by UV-promoted photoaddition of phosphines to carbon-carbon double bonds. Both $T_8(CH{=}CH_2)_8$ and $T_8(CH_2CH{=}CH_2)_8$ were converted to the respective octa-substituted dimethylphosphino derivative by UV irradiation in the presence of Me_2PH [89]. The ethyldimethylphosphine derivative was further functionalized by reaction with sulphur to give $T_8[(CH_2)_2P({=}S)Me_2]_8$, and then complexed to a series of metal carbonyl fragments.

An octa-POS compound has recently been prepared by the addition reaction of the mono-thiol POS $T_8(i\text{-Bu})_7(CH_2)_3SH$ with the tetra-alkyne $C[CH_2OC({=}O)(CH_2)_3S(CH_2)_2CO_2CH_2C{\equiv}CH]_4$ [90]. Spectroscopic analysis indicated the reaction had proceeded to give the octa-POS compound **1**.

Two related addition reactions of dichlorocarbene (derived from chloroform) to $T_8(CH{=}CH_2)_8$ in the presence of base have been seen to give rise to either the octa- or the tetra-substituted derivatives, $T_8(2,2\text{-}Cl_2\text{-}c\text{-}C_3H_3)_8$ and $T_8(CH{=}CH_2)_4(2,2\text{-}Cl_2\text{-}c\text{-}C_3H_3)_4$ in about 40% yield [84]. Several other cyclization reactions are known for unsaturated POS derivatives. Two octa-substituted cyclohexane-epoxides, differing in the tether to the POS core, have been prepared from the cyclohexene compounds using Oxone® or MCPBA [50]; and the partial epoxidation of $T_8(CH{=}CH_2)_8$ by MCPBA resulted in the formation of $T_8(CH{=}CH_2)_6[CH(O)CH_2]_4$ [91].

Other alkyl-POS compounds containing more than one type of substituent can be prepared by addition reactions. On treatment of $T_8(CH{=}CH_2)_8$ with HOTf, the monoaddition product $T_8(CH{=}CH_2)_7(CH_2)_2OTf$, resulted. This product can be hydrolyzed to give the monoalcohol (which can be reduced to $T_8Et_7(CH_2)_2OH$) or treated with 2-mercaptopyridine to give the monopyridyl derivative [92].

A limited number of cycloaddition reactions involving alkyl-POS compounds have been carried out, the majority of which were aimed at making T_8R_7R' products [93-95]. However, there has very recently been a series of nine T_8 POS-octatriazole compounds prepared using "click" reactions. These syntheses involved the treatment of $T_8[(CH_2)_3N_3]_8$ with a series of alkynes, ranging from simple alkynes like phenylacetylene, to complicated alkyne derivatives involving protected systems, and a pyrazolylborate derivative. Various different catalyst and solvent systems were tested for these alkynes, and optimization of conditions led to generally high yields (63% to 90%) [66].

Of the studies of cycloaddition reactions leading to T_8R_7R' products, the most simple of these involved the reaction of a POS-azide derivative with phenylacetylene to form a 1,2,3-triazole derivative by a "click" reaction [93, 94]. An in-depth investigation of a series of cycloadditions has compared the efficacies of microwave-assisted and conventional reaction conditions for the synthesis of a series of POS derivatives containing isoxazolidine or isoxazoline rings [95]. The microwave-assisted reactions often performed better, resulting in higher yields, and product mixtures that could be separated to give individual isomers.

2.2.4.2 T_{10} and T_{12} Compounds

There have been two recent reports of reactions modifying the functional groups of T_{10} alkyl-POS compounds, both starting from $T_{10}(CH{=}CH_2)_{10}$. Starting from a mixture of T_{10} and T_{12} cages, comprising one or two phenyl substituents, the rest being vinyl, or vice versa, a number of reactions were carried out at the vinyl groups of these compounds [96]. These included metathesis reactions with 4-bromostyrene, and Heck couplings; both with vinyltriethoxysilane, and with the pre-metathesis product, to form oligomeric species. Starting from a simpler vinyl-POS mixture comprising T_8, T_{10} and T_{12} POS compounds, thiol-ene reactions have been attempted with large alkyl-POS compounds [97]. Following free-radical-initiated reaction of the POS mixture with $HS(CH_2)_2CO_2Bu$, the products $T_{10}[(CH_2)_2S(CH_2)_2CO_2Bu]_{10}$ and $T_{10}[(CH_2)_2S(CH_2)_2CO_2Bu]_4(CH{=}CH_2)_6$ were identified, depending on the ratio of the starting materials.

2.2.5 *Other Synthetic Methods*

2.2.5.1 T_6 Compounds

With sufficiently large R groups, hydrolysis of $RSiCl_3$ can lead to stable, isolable silanetriols, $RSi(OH)_3$, or 1,1,3,3-tetrahydroxydisiloxanes, $[RSi(OH)_2]_2O$. Following isolation of such silanols, it has been found possible to dehydratively couple them using dicyclohexylcarbodiimide (DCC), giving rise to T_6R_6 species in yields ranging from 20 %, where R is 2,4,6-*i*-$Pr_3C_6H_2$ [98], to 26% or 41%, where R is *t*-Bu or CMe_2-*i*-Pr, respectively [99]. Similarly, 1,3-diisopropyl-1,1,3,3-tetrahydroxydisiloxane can be converted to the cyclotetrasiloxane, which, when treated with $(i\text{-PrSiCl}_2)_2O$, gives $T_6(i\text{-Pr})_6$ in 25% yield [100]. Another unusual route to $T_6(t\text{-Bu})_6$, involves reaction of *t*-$BuSi(OH)_3$ with NaH to give a mixture of silsesquioxane products from which the T_6 compound can be isolated in 35% yield [101].

2.2.5.2 T_8 Compounds

Two reactions related to the conventional hydrolysis and condensation reactions have also provided unusual routes to T_8 alkyl-POS. One is the cleavage, hydrolysis and condensation reaction that takes place by the action of potassium hydroxide on a mixture of tetraethylcyclotetrasiloxane and pentaethylcyclopentasiloxane resulting in the formation of T_8Et_8 in 37% yield [102]. The second involves the reaction between tetrabutylammonium fluoride and $[Me(MeO)SiO]_4$, producing T_8Me_8 in 52% yield, which is one of the better yields for this compound [103]. Other reactions have been used to prepare alkyl-POS from siloxanes; with DCC effecting the dehydration of the cyclotetrasiloxane $[i\text{-Pr(OH)OSi}]_4$ to give $T_8(i\text{-Pr})_8$ [104], and of the disiloxane $[Cy(OH)_2Si]_2O$ to give T_8Cy_8 [105]. Another unusual synthetic route to alkyl-POS arises from the depolymerization of a combination of polymethylsilsesquioxane and polyvinylsilsesquioxane, using tetrabutylammonium fluoride, NBu_4F. This results in the formation of a mixture of T_n cages (n = 8, 10, 12) in reasonable yield, but the only characterization of the cages formed and their pattern of substitution was obtained using MALDI-ToF mass spectrometry [106].

2.2.5.3 T_{10} and T_{12} Compounds

Two more unusual routes to T_{10} and T_{12} alkyl-POS compounds have been reported. Firstly, the base-catalyzed reformation of mixtures of methycyclosiloxanes, which resulted in a sublimable product, $T_{10}Me_{10}$, suitable for single crystal X-ray diffraction [107]; and secondly, the preparation of $T_{10}(CH{=}CH_2)_{10}$ and $T_{12}(CH{=}CH_2)_{12}$ in moderate yields by hydrolysis of $[CH_2{=}CHSi(OEt)_2]_2O$ in the presence of NBu_4F [38].

2.3 Synthesis of T_nR_n Compounds where R = Aryl

2.3.1 *Hydrolysis*

2.3.1.1 T_8 Compounds

There are surprisingly few aryl- or heteroaryl-POS compounds that have been prepared by the hydrolysis of chloro- or alkoxysilane precursors, especially as, until recently, it has proven difficult to functionalize T_8Ph_8 specifically without the formation of mixtures of isomers of aryl-POS. The most common hydrolytic syntheses are presented in Table 2.8, the most important being that of T_8Ph_8 [108], as it is a useful precursor to substituted aryl-T_8 derivatives. This compound was originally synthesized by the acidic hydrolysis of $PhSiCl_3$ [109], with significant improvements coming in the use of tetrabutylammonium fluoride for hydrolysis [11], and in the use of an initial hydrolysis followed by base-catalysed rearrangements of the resulting siloxane mixture [110].

Table 2.8 Examples of aryl-T_8R_8 compounds prepared by hydrolysis

R	Starting materials	Yield (%)	Refs.
-Ph	$PhSiCl_3$ + $PhCH_2NMe_3OH$	98	[108]
-C_6H_4-4-Me	4-Me-$C_6H_4SiCl_3$	19	[111]
-C_6H_4-3-Me	3-Me-$C_6H_4SiCl_3$	25	[112]
-C_6H_4-4-CH_2Cl	4-$ClCH_2C_6H_4SiCl_3$	15	[113]
-C_6H_3-3,5-Me_2	3,5-$Me_2C_6H_4SiCl_3$	22	[112]
-C_6H_4-2-NMe_2	2-$NMe_2C_6H_4Si(OMe)_3$ + HCl	84	[114]

A recent study on the synthesis of substituted-phenyl POS derivatives has led to the preparation of $T_8(C_6H_4\text{-}4\text{-}R)_8$ POS in 9% to 21% yields by hydrolysis of the arylchlorosilanes RC_6H_4-4-$SiCl_3$ (R = Me, *i*-Pr, $SiMe_3$) [115]. The same hydrolyses carried out in the presence of NBu_4F, rather than HCl, for the methyl and trimethylsilyl species also formed the T_8 POS, albeit in lower yield. However, these reactions also produced larger POS species $T_n(C_6H_4\text{-}4\text{-}R)_n$ (n = 10, R = Me, $SiMe_3$; n = 12, R = $SiMe_3$), in up to 41% yield for n = 10, and 9% yield for n = 12. Two other, little studied, aryl-POS species, the octakis(1-naphthyl) [111] and octakis(2-thienyl) [116] POS have been prepared by hydrolysis of 1-naphthyltrimethoxysilane and 2-thienyltrimethoxysilane respectively. A small number of T_8R_7R' derivatives where R is aryl have also been prepared. These include the capping of $R_7Si_7O_9(OH)_3$ (R = *c*-C_5H_9, [117] or Cy [118]) with $ClCH_2C_6H_4$-4-$SiCl_3$ to give $T_8R_7C_6H_4$-4-CH_2Cl in good yield.

2.3.1.2 T_{10} and T_{12} Compounds

Only three larger aryl-POS compounds, other than the three $T_8(C_6H_4\text{-4-R})_8$ compounds described above, seem to have been prepared via simple hydrolyses. Thus, T_nPh_n, n = 10, 12, are isolable following base-treatment of the initial hydrolysis products resulting from $PhSiCl_3$ [110], and $T_{12}(\text{2-thienyl})_{12}$ has been prepared by the basic hydrolysis of (2-thienyl)Si(OMe)$_3$ [116].

2.3.2 *Modification of R*

2.3.2.1 T_8 Compounds

A significant amount of effort has been devoted to exploring the possibility of achieving a uniform substitution of the phenyl rings via nitration, halogenation, and recently, sulfonation [119, 120], but, with a few exceptions, most of these attempts have given rise to mixtures. A selection of these reactions is presented in Table 2.9. The nitration of T_8Ph_8 to give $T_8(C_6H_4NO_2)_8$ is important as the nitro derivative is the precursor to $T_8(C_6H_4NH_2)_8$, which is widely used as an additive in polymers [120]. This reaction was first observed on the dissolution of T_8Ph_8 in HNO_3 [109], and more recently was reinvestigated and optimized [121]. It was observed that $T_8(C_6H_4NO_2)_8$ did not form regioselectively, as a mixture of predominantly monosubstituted *meta-* and *para*-isomers are formed. In contrast, the di-nitration of T_8Ph_8, although reported to give a mixture of nitration products, [122] can, under the appropriate conditions, afford predominantly the 2,4-dinitrated product [123].

Bromination of T_8Ph_8 gives an apparently monobrominated T_8 product comprising a mixture of 3- and 4-substituted phenyl rings [124], but the product containing, on average, 5.3 bromine atoms, $T_8Ph_8Br_{5.3}$, was chosen as the most suitable for further reaction due to its comprising a mixture of compounds with the fewest multiply-brominated rings, and the highest proportion of 4-brominated rings [125]. As a result of the incomplete substitution, reactions of this product result in complicated product mixtures. Compounds with higher or lower proportions of bromination can be produced by this method, including products with an average of 15.7 bromines per T_8 core [125]. While it has not been possible to exclusively brominate all eight rings of T_8Ph_8 in the 4-position, an important recent report has shown that $T_8(C_6H_4\text{-4-I})_8$ can be prepared which will be a useful precursor to a range of novel $T_8(C_6H_4\text{-4-R})_8$ species via a variety of synthetic routes [126]. The use of chlorosulfonic acid to introduce SO_3H has been reported to give $T_8(C_6H_4\text{-3-}SO_3H)_8$ in 74% yield [127].

Other studies report that the product has only an average of 40% of the phenyl rings bearing a sulfonic acid group after milder sulfonating conditions were used [128], or that no preference for one substitutional isomer over another was observed in the product [129] (see also discussion in Section 7.6, Chapter 7). Friedel Crafts alkylation of T_8Ph_8 was also reported [127]. The unusual multiple bromination of

T_8(2-thienyl)$_8$ to give the tribromothiophene product in excellent yield has also been reported [116]. Four different aryl-T_8 compounds, T_8Ph_8, $T_8(C_6H_4$-n-Me$)_8$ (n = 3 or 4), and $T_8(C_6H_3$-3,5-Me$_2)_8$, have been hydrogenated in yields ranging from 88 to 95 % using palladium on charcoal as catalyst to give the saturated analogues [112].

Table 2.9 Aryl-T_8R_8 compounds prepared by direct substitution at the aromatic ring

R or T_8 compound	Starting materials	Yield (%)	Refs.
-$C_6H_4NO_2$	T_8Ph_8 + HNO_3	90[a]	[121]
-$C_6H_3(NO_2)_2$	T_8Ph_8 + HNO_3, H_2SO_4	78[a]	[122]
-C_6H_3-2,4-$(NO_2)_2$	T_8Ph_8 + HNO_3, H_2SO_4	82	[123]
-C_6H_4-3-SO_3H	T_8Ph_8 + $ClSO_2OH$	74	[127]
-C_6H_4Br	T_8Ph_8 + Br_2, $FeCl_3$	60[a]	[124]
-C_6H_4-4-I	T_8Ph_8 + ICl	90	[126]
$T_8Ph_8Br_{5.3}$	T_8Ph_8 + Br_2, Fe	87[a]	[125]

[a] Contains a mixture of substitutions.

A wide range of reactions has been successfully carried out at substituents on aryl-POS compounds without cleaving the T_8 cage. These include substitutions at benzylic carbons by a variety of nucleophiles, and reaction at substituted benzene rings by lithiations, Grignard reactions, and catalyzed reactions such as Heck, Sonogashira and Suzuki chemistries. A selection of these reactions is presented in Table 2.10. One unusual substitution reaction involving T_8 aryl-POS is the recent preparation of an iminofullerene-POS compound, prepared in 36% yield by the reaction of T_8(*i*-Bu)$_7C_6H_4$-4-CH_2N_3 with C_{60} [130]. Another unusual reaction is that of $T_8(C_6H_4Br)_8$ with the Grignard reagent formed from 2-bromo-3-(2'-ethylhexyl) thiophene to give the predominantly octa-substituted product [131], although some of the analytical data may suggest that the Grignard reagent has partially degraded the silsesquioxane core.

Table 2.10 Aryl-T_8R_8 compounds prepared by substitution reactions at functional groups

R	Starting materials	Yield (%)	Refs.
-C_6H_4-4-CH_2I	$T_8(C_6H_4$-4-$CH_2Cl)_8$ + NaI	99	[113]
-C_6H_4-4-CH_2OH	$T_8(C_6H_4$-4-$CH_2I)_8$ + $AgClO_4$	95	[113]
-C_6H_4-4-CH_2ONO_2	$T_8(C_6H_4$-4-$CH_2I)_8$ + $AgNO_3$	93	[113]
-C_6H_4-4-CH_2OPPh_2	$T_8(C_6H_4$-4-$CH_2OH)_8$ + Ph_2PCl	70	[132]
-C_6H_4-4-$CH_2P(O)Ph_2$	$T_8(C_6H_4$-4-$CH_2I)_8$ + $Ph_2P(O)Et$	83	[132]
-C_6H_4-4-CH_2PPh_2	$T_8[C_6H_4$-4-$CH_2P(O)Ph_2]_8$ + AlH_3	64	[132]
-C_6H_4-4-CH=CH-4-C_5H_4N	$T_8(C_6H_4$-4-I$)_8$ + CH_2=CH-4-C_5H_4N, Pd_2dba_3, tri-*o*-tolylphosphine, H_2CuSn, NEt_3	48	[126]

R	Starting materials	Yield (%)	Refs.
$-C_6H_4$-4-CH=CHC_6H_4-4-Me	$T_8(C_6H_4\text{-4-I})_8$ +CH_2=CHC_6H_4-4-Me, Pd_2dba_3, tri-*o*-tolylphosphine, Cu, NCy_2Me	87	[126]
$-C_6H_4$-4-C≡CPh	$T_8(C_6H_4\text{-4-I})_8$ + HC≡CPh, CuI, $Pd(PPh_3)_4$, NEt_3	90	[126]
$-C_6H_4$-4-Ph	$T_8(C_6H_4\text{-4-I})_8$ + $PhB(OH)_2$, Pd_2dba_3, tri-o-tolyl-phosphine, Ag_2O	83	[126]
$-C_6H_4NH_2$	$T_8(C_6H_4NO_2)_8$ + HCO_2H, Pd/C, NEt_3	82[a]	[121]
$-C_6H_4N_3$	$T_8C_6H_4NH_2$ + H_2SO_4, $NaNO_2$, NaN_3	-[a]	[133]
$-C_6H_3(NH_2)_2$	$T_8[C_6H_3(NO_2)_2]_8$ + HCO_2H, Pd/C, NEt_3	70[a]	[122]
Methylstyrene derivatives of $T_8Ph_8Br_{5.3}$	$T_8Ph_8Br_{5.3}$ + CH_2=CHC_6H_4-4-Me, Pd[P(*t*-Bu)$_3$]$_2$, Pd_2dba_3, NCy_2Me	78[a]	[125]
Phenylacetylene derivatives of $T_8Ph_8Br_{5.3}$	$T_8Ph_8Br_{5.3}$ + CH≡CPh, CuI, $Pd_2(dba)_3$, Pd[P(*t*-Bu)$_3$]$_2$, NEt_3	85[a]	[134]
Phosphonic phenylester derivatives of $T_8Ph_8Br_{5.3}$	$T_8Ph_8Br_{5.3}$ + P(=O)(OPh)$_2$Cl, *n*-BuLi	78[a]	[127]

[a] Contains a mixture of substitutions.

One of the most frequently used aryl-POS derivatives, $T_8(C_6H_4NH_2)_8$, can be prepared from $T_8(C_6H_4NO_2)_8$ in high yield by reduction using formic acid in the presence of Pd/C, to give a product suitable for incorporation into a wide variety of polymers [121]. Similarly, mixed isomer hexadecanitro POS, $T_8[C_6H_3(NO_2)_2]_8$, can be converted to $T_8[C_6H_3(NH_2)_2]_8$, the hexadecamine compound [122]; however, the preparation of the isomer-specific 2,4-dinitrate has not been reported. Further to this, $[T_8(C_6H_4NH_3)_8]Cl_8$, has recently been prepared from $T_8(C_6H_4NH_2)_8$ for use in DNA-detection experiments [135]. Recently, reaction of HNO_3 with $T_8(C_6H_4\text{-4-}SiMe_3)_8$ has been shown to give exclusively the *para*-substituted isomer $T_8(C_6H_4\text{-4-}NO_2)_8$ in 94% yield [115], and the related nitro-derivative $T_8(C_6H_3\text{-3-}NO_2\text{-4-Me})_8$ can be prepared by the nitration of $T_8(C_6H_4\text{-4-Me})_8$ using either nitric acid or a copper (II) nitrate, acetic anhydride mixture [115]. These pure compounds should open up routes to a range of new precursors for materials chemistry, as will the recent synthesis of the specifically iodinated $T_8(C_6H_4\text{-4-I})_8$, which can be used to prepare many novel compounds via Heck, Suzuki and Sonogashira reactions [126, 134].

Carboxylate derivatives of T_8 aryl-POS compounds have been prepared. For example, esters can be prepared from $T_8(C_6H_4\text{-4-}CH_2OH)_8$ by reaction with the reactive acid derivatives acetic anhydride and 4-nitrobenzoyl chloride, and by reaction with terephthaloyl chloride and methanol to give $T_8[C_6H_4\text{-4-}CH_2OC(O)Me]_8$, $T_8[C_6H_4\text{-4-}CH_2OC(O)C_6H_4\text{-4-}NO_2]_8$, and $T_8[C_6H_4\text{-4-}CH_2OC(O)C_6H_4\text{-4-}CO_2Me]_8$ respectively in near quantitative yield [113]. Similarly, several amino acid-derived POS compounds were prepared by the reaction of $T_8(C_6H_4\text{-4-}CH_2OH)_8$ with N-protected amino acids and oligopeptides [75]. Hydrolysis of the ester $T_8[C_6H_4N(CH_2CO_2Me)_2]_8$ can be used to give the acid functionalized POS

$T_8[C_6H_4N(CH_2CO_2H)_2]_8$ [136].

2.3.2.2 T_{10} and T_{12} Compounds

Reactions of functional groups in T_{10} and T_{12} aryl-POS compounds are usually similar to those observed for the T_8 analogues. Thus, $T_{12}Ph_{12}$ may be nitrated using HNO_3, giving $T_{12}(C_6H_4NO_2)_{12}$ in 95% yield [122], although as is the case for the T_8 systems, this is not a regiospecific reaction, with predominantly *meta*, some *para* and a trace of *ortho* substitution. These nitro groups can be converted to amines using Pd/C, in the presence of formic acid, in 70% yield [122]. Similarly, the specifically substituted 4-nitrophenyl T_n POS (n = 10 and 12) has been prepared in 80% and 33% yield, respectively, by treating $T_n(C_6H_4\text{-}4\text{-}SiMe_3)_n$ with HNO_3 [115]. Hydrogenation of $T_{12}Ph_{12}$ using palladium on charcoal proceeds with an excellent 93% yield, but the equivalent reaction for $T_{10}Ph_{10}$ has not been reported [112]. Again in parallel to what has been done for the T_8 compounds, T_{12}(2-thienyl)$_{12}$ has been triply brominated to give the dodeca(tribromothiophene) derivative in 90% yield [116].

2.3.3 *Other Synthetic Methods*

The potentially highly useful regiospecifically brominated T_8Ph_8 derivative $T_8(C_6H_4\text{-}4\text{-}Br)_8$ has been prepared by acid polycondensation of the cyclic sodium silanolate Na_4-*c*-$[BrC_6H_4\text{-}4\text{-}SiO(O)]_4$, followed by base-catalyzed rearrangement to give the product in moderate 58% yield [137]. While this synthetic route may prove highly useful in a limited number of cases, given the necessity of preparing the starting cyclotetrasiloxane sodium silanolate, it is unlikely to be widely used, especially given the relative ease of preparation of $T_8(C_6H_4\text{-}4\text{-}I)_8$, see above.

The ethynyl derivative $T_8(C_6H_4C{\equiv}CSiMe_3)_8$ has been prepared, along with the related T_4 and T_6 species (detectable by MALDI-TOF mass spectrometry in the crude T_8 product), by the hydrolysis of an unusual starting material, the ionic λ^5-substituted spirosilicate, pyridinium bis(2,3-naphthalendiolato)-4'-(trimethylsilylethynyl) phenylsilicate [138]. However, although the hydrolysis proceeded in 88% yield, this methodology is unlikely to become widespread, since the starting spirosilicate requires a seven-step synthesis.

2.4 Synthesis of T_nR_n Compounds where R=Alkoxy

There have been only a very limited number of alkoxy-POS compounds prepared, all of which are $T_8(OR)_8$ derivatives, and all of which result from substitution reactions at the cage silicon atoms. The first synthesis of $T_8(OMe)_8$ used the reaction of T_8Cl_8 with methyl nitrite [139]; however, due to the toxicity and potentially explosive nature of methyl nitrite, an alternative route was sought. Thus, T_8Cl_8 was found to react with trimethyl orthoformate to give $T_8(OMe)_8$ in 68% yield [140]. It is perhaps somewhat surprising that the preparation of further analogues of this compound have not been attempted, using other orthoesters. The remaining synthetic work in this area has all been based around the same reaction, that between T_8H_8 and an alcohol in the presence of catalytic amounts of NEt_2OH base. The first alkoxy-POS compounds prepared in this manner were $T_8[O(CH_2)_3CH{=}CH_2]_8$ and $T_8[O(CH_2)_5Me]_8$ [48], although no yields were reported. This has been followed by a report in the patent literature, using an optimized version of the same methodology, and preparing T_8R_8, where R is OEt, O-*i*-Pr, O-*t*-Bu, OOct and OCy in 54 to 87 % yields [141].

2.5 Synthesis of T_nR_n Compounds where R = Siloxy

2.5.1 *Corner Capping*

Two compounds, $T_8Cy_7OSiCl_3$, and the siloxy-bridged $O(T_8Cy_7)_2$ can be prepared in 62 or 69% yield, respectively, by the corner capping of $T_7Cy_7(OH)_3$ using either one or a half equivalent of $O(SiCl_3)_2$, and NEt_3 or pyridine as base [142].

2.5.2 *Substitution*

2.5.2.1 T_8 Compounds

Most new siloxy-POS compounds are prepared by modification of the substituents at a pre-existing siloxy centre rather than by introduction of a completely new siloxy group via a substitution reaction. Those that have been prepared by this method generally contain a simple siloxy substituent, or derive from a reactive group at silicon. Thus, $T_8(OSiMe_3)_8$ can be prepared either in quantitative yield by reaction of T_8H_8 with $Me_3SiOSbMe_4$ [143], or in 54% yield by reaction of T_8H_8 with Me_3SiOH in the presence of a base [141]. Four other compounds, $T_8[OSi(CD_3)_3]_8$ [144], $T_8(OSiMe_2CH{=}CH_2)_8$ and $T_8(OSiMe_2CH_2Cl)_8$ [145], and $T_8Cy_7OSiMe_3$. [36] have been prepared by a similar route, but using the trimethylamine-*N*-oxide adduct of the relevant chlorosilane. A limited number of other T_8R_7R'-siloxy POS, where R' is a bridging group, has been prepared by substitution at silicon. These reactions involve a monochloro POS, T_8R_7Cl, being treated with either a lithiated monohydroxy POS, or a lithiated dihydroxy-containing partial-cage POS [146]. These reactions result in the formation of the di-POS compounds $(T_8R_7)_2O$ (R = *c*-C_5H_9, Cy) and the tri-POS compound $(T_8Cy_7)_2$(*exo,exo*-$Si_8O_{11}Cy$) in 65% to 73% yield.

2.5.2.2 T_{10}, T_{12} and T_{14} Compounds

A series of larger siloxy-POS compounds has been prepared, as in the case of the T_8 analogue, via the trimethylamine-N-oxide adduct of the relevant chlorosilane. This has allowed the synthesis of $T_{10}(OSiMe_2CH{=}CH_2)_{10}$ and $T_{10}(OSiMe_2CH_2Cl)_{10}$ in 65% and 52% yields [145], and $T_{12}(OSiMe_3)_{12}$ [147], and $T_{14}(OSiMe_3)_{14}$ [147], both in 95% yield.

2.5.3 *Modification of R*

2.5.3.1 T_6 Compounds

The majority of the functional group modification reactions involving T_6 siloxy-POS compounds either involve formation from isolated $[T_6O_6]^{6-}$ salts [148], or the modification of these siloxy compounds. The silylation reactions include the formation of $T_6(OSiMe_2H)_6$, from the treatment of $[NEt_4]_6[T_6O_6]^{6-}$ with HMe_2SiCl [148], and $T_6(OSiMe_3)_6$, from treatment with Me_3SiCl [149]. The silane $T_6(OSiMe_2H)_6$ can be converted to the bromide $T_6(OSiMe_2Br)_6$, by catalytic bromination using $Rh(acac)_3$ and allyl bromide [148]. Both POS species have been used in the preparation of porous polymeric POS-based materials, with $T_6(OSiMe_2H)_6$ undergoing hydrosilylation with $T_8(CH{=}CH_2)_8$, and $T_6(OSiMe_2Br)_6$ being hydrolyzed [150].

2.5.3.2 T_8 Compounds

The most common functional group modification reaction of siloxy-POS compounds is hydrosilylation, but other methods are available, such as the reaction involving an alkene in T_8 siloxy-POS compounds, exemplified by the cross-metathesis of $T_8(OSiMe_2CH{=}CH_2)_8$ with either styrene or 1-pentene, using Schrock's catalyst [76].

The reactions of acid derivatives include examples of both hydrolysis and esterification, with $T_8[OSiMe_2(CH_2)_2C_6H_4\text{-4-OAc}]_8$ having been hydrolyzed to give the alcohol [151], and a number of different siloxy-POS being reacted with methacryloyl chloride to give the methacrylate derivatives, $T_8[OSiMe_2(CH_2)_3O(CH_2CH_2O)_nOC({=}O)CMe{=}CH_2]_8$ (n = 2, 3, 4, 6) [152], and $T_8[OSiMe_2(CH_2)_3OC(O)CMe{=}CH_2]_8$ [153].

A wide range of substitution reactions of siloxy-POS compounds has been observed and a selection of these is presented in Table 2.11. While some of these reactions are of very similar type to those observed for other POS, some of them take advantage of the additional distance between reacting centres and the POS cage, and the added steric shielding provided by the siloxy groups being between the rest of the functional group and the POS core, which allows the use of reagents, such as Grignard reagents, that have been known to attack the POS cage in less well protected systems. One particular compound of interest is $T_8(OSiMe_2C_6H_4\text{-4-}CO_2H)_8$ [154], containing a carboxylic acid group held in a comparatively constrained orientation by the $\text{-OSiMe}_2C_6H_4\text{-}$ linker. This compound has been prepared in two steps from $T_8(OSiMe_2C_6H_4\text{-4-Me})_8$, and opens up possibilities for a further range of carboxylate-functionalized POS compounds to be prepared.

Table 2.11 Siloxy-T_8R_8 compounds prepared by substitution reactions of functional groups

R	Starting materials	Yield (%)	Refs.
$\text{-OSiMe}_2CH(O)CH_2$	$T_8(OSiMe_2CH{=}CH_2)_8$ + *m*-CPBA	87	[155]
$\text{-OSiMe}_2CH(CO_2Et)_2$	$T_8(OSiMe_2H)_8$ + $BrCH(CO_2Et)_2$, Na	27	[156]
$\text{-OSiMe}_2(CH_2)_2SiMe(CH_2CH{=}CH_2)_2$	$T_8[OSiMe_2(CH_2)_2SiMeCl_2]_8$ + $CH_2{=}CHCH_2MgBr$	-	[157]
$\text{-OSiMe}_2(CH_2)_2SiMe(OCH_2CH{=}CH_2)_2$	$T_8[OSiMe_2(CH_2)_2SiMeCl_2]_8$ + $CH_2{=}CHCH_2OH$, pyridine	-	[157]
$\text{-OSiMe}_2(CH_2)_2Si(CH{=}CH_2)_3$	$T_8[OSiMe_2(CH_2)_2SiCl_3]_8$ + $CH_2{=}CHMgBr$	78	[158]
$\text{-OSiMe}_2(CH_2)_2Br$	$T_8(OSiMe_2CH{=}CH_2)_8$ + HBr, $(PhO)_2$	79	[156]
$\text{-OSiMe}_2C_6H_4\text{-4-CBr}_3$	$T_8(OSiMe_2C_6H_4\text{-4-Me})_8$ + NBS, AIBN	86[a]	[154]
$\text{-OSiMe}_2C_6H_4\text{-4-}CO_2H$	$T_8(OSiMe_2C_6H_4\text{-4-CBr}_3)_8$ + $AgNO_3$, HCOOH	90	[154]
-OSiMe_2Br	$T_8(OSiMe_2H)_8$ + $CH_2{=}CHCH_2Br$, $Rh(acac)_3$	99	[159]

[a] Contains a mixture of products with varying degrees of bromination.

Hydrosilylation has led to the preparation of a wide range of T_8 siloxy-POS compounds with various functional groups appended to them; a selection of these is presented in Table 2.12. The starting POS compounds are generally either

$T_8(OSiMe_2H)_8$ or $T_8(OSiMe_2CH{=}CH_2)_8$, with the former being the reagent of choice in recent years, probably due to both its ease of preparation, and commercial availability. Of those reactions starting from $T_8(OSiMe_2CH{=}CH_2)_8$, most of them are hydrosilylations with simple silane derivatives, introducing new silicon centres into the POS compounds. For the reactions starting from $T_8(OSiMe_2H)_8$, the hydrosilylation generally introduces functionalized organic groups to provide the desired chemical functionality. Alcohols and ethylene glycol oligomers have been commonly prepared by this method, but more exotic derivatives have included a carborane cage compound, $T_8[OSiMe_2(CH_2)_3C{\equiv}CMeB_{10}H_{10}]_8$ [160].

Table 2.12 Siloxy-T_8R_8 compounds prepared by hydrosilylation

R	Starting materials	Yield (%)	Refs.
$-OSiMe_2(CH_2)_2C_6H_4$-4-NH_2	$T_8(OSiMe_2H)_8 + CH_2{=}CHC_6H_4$-4-$NH_2$, Pt(dvs)	-	[161]
$-OSiMe_2(CH_2)_2SiMe_nCl_{3-n}$ (n=1,2)	$T_8(OSiMe_2CH{=}CH_2)_8 + HSiMe_nCl_{3-n}$, H_2PtCl_6	86,97	[160]
$-OSiMe_2(CH_2)_2SiMe(OSiMe_3)_2$	$T_8(OSiMe_2CH{=}CH_2)_8 + HSiMe(OSiMe_3)_2$, H_2PtCl_6	-	[162]
$-OSiMe_2(CH_2)_2Si(OEt)_3$	$T_8(OSiMe_2CH{=}CH_2)_8 + HSi(OEt)_3$, H_2PtCl_6	83	[160]
$-OSiMe_2(CH_2)_2SiCl_3$	$T_8(OSiMe_2CH{=}CH_2)_8 + HSiCl_3$, H_2PtCl_6	95	[158]
$-OSiMe_2(CH_2)_3C{\equiv}CMeB_{10}H_{10}$	$T_8(OSiMe_2H)]_8 + CH_2{=}CHCH_2C{\equiv}CMeB_{10}H_{10}$, Pt(dvs)	85	[160]
$-OSiMe_2(CH_2)_3CN$	$T_8(OSiMe_2H)_8 + CH_2{=}CHCH_2CN$, H_2PtCl_6	43	[163]
$-OSiMe_2(CH_2)_3OH$	$T_8(OSiMe_2H)_8 + CH_2{=}CHCH_2OH$, Pt(dvs)	86	[153]
$-OSiMe_2(CH_2)_3O(CH_2)_2OH$	$T_8(OSiMe_2H)_8 + CH_2{=}CHCH_2O(CH_2)_2OH$, Pt(dvs)	87	[153]
$-OSiMe_2(CH_2)_3OCH_2CH(O)CH_2$	$T_8(OSiMe_2H)_8 + CH_2{=}CHCH_2OCH_2CH(O)CH_2$, Pt(dcp)	91	[164]
$-OSiMe_2(CH_2)_3O(CH_2CH_2O)_nH$ (n = 2, 3, 4, 6)	$T_8(OSiMe_2H)_8 + CH_2{=}CHCH_2O(CH_2CH_2O)_nH$, Pt(dvs)	-	[165]
$-OSiMe_2(CH_2)_3O(CH_2CH_2O)_nMe$ (n = 2, 3, 4, 6, 8, 12.5)	$T_8(OSiMe_2H)_8 + CH_2{=}CHCH_2O(CH_2CH_2O)_nMe$, Pt(dvs)	-	[166]
$-OSiMe_2(CH_2)_3OSiMe_3$	$T_8(OSiMe_2H)_8 + CH_2{=}CHCH_2OSiMe_3$, Pt(dvs)	82	[153]
$-OSiMe_2(CH_2)_3Cl$	$T_8(OSiMe_2H)_8 + CH_2{=}CHCH_2Cl$, H_2PtCl_6	48	[167]
$-OSiMe_2(CH_2)_3Br$	$T_8[OSiMe_2H]_8 + CH_2{=}CHCH_2Br$, H_2PtCl_6	50	[168]

The other common reactions of T_8 siloxy-POS compounds are those where reaction occurs at the siloxy-silicon centres. These reactions are often silicon-specific substitution reactions, with silylation being most common. As well as reactions where additions to the siloxy groups occur, reactions resulting in the removal of part or all of a siloxy substituent are known, for example by desilylation or alcoholysis. A selection of compounds resulting from these reactions is presented in Table 2.13. There have been recent attempts to prepare highly silylated T_8 siloxy-POS compounds by treatment of simple starting materials, $[T_8O_8]^{8-}$ or T_8H_8, with either $(Me_3SiO)_3SiCl$ or $(Me_3SiO)_3SiOH$. In all cases the reaction proceeded, but not to completion, with two or three corners of the cube showing no substitution [169], presumably due to the steric bulk of the silylating agents. While the desired $T_8[OSi(OSiMe_3)_3]_8$ has not yet been produced, a closely related compound, $T_8[OSi(OSiMe_3)_3]_7OSi(OSiMe_3)_2OH$, has been obtained by silylation of $T_8[OSi(OEt)_3]_8$ with Me_3SiCl [169].

Table 2.13 Siloxy-T_8R_8 compounds prepared by reaction at cage-silicon atoms

R or T_8R_8	Starting materials	Yield (%)	Refs.
$[T_8O_8]^{8-}$	$T_8(OSiMe_2H)_8 + H_2O$	-	[170]
$-OSiMe_2H$	$[NMe_4]_8[T_8O_8] + HMe_2SiCl$	87	[168]
$-OSiMe_3$	$T_8(OSiMe_2H)_8 + Me_3SiCl$, AcOH	98	[171]
$-OSiMe_2CH_2Cl$	$T_8(OSiMe_2H)_8 + ClCH_2Me_2SiCl$, AcOH	94	[171]
$-OSiMe_2CH_2Br$	$T_8(OSiMe_2H)_8 + BrCH_2Me_2SiCl$, AcOH	90	[171]
$-OSiMe_2CH{=}CH_2$	$[NMe_4]_8[T_8O_8] + CH{=}CH_2Me_2SiCl$	87	[156]
$-OSiMe_2CH_2CH{=}CH_2$	$[NMe_4]_8[T_8O_8] + CH{=}CH_2CH_2Me_2SiCl$	52	[172]
$-OSiMe_2Ph$	$[NMe_4]_8[T_8O_8] + PhMe_2SiCl$	54	[173]
$-OSiMe_2(CH_2)_3OH$	$T_8[OSiMe_2(CH_2)_3OSiMe_2\text{-}t\text{-}Bu]_8 + NBu_4F$	81	[174]
$-OSiMe_2C_6H_4\text{-}4\text{-}Me$	$[NMe_4]_8[T_8O_8] + Me\text{-}4\text{-}C_6H_4Me_2SiCl$	40	[154]
$-OSiMe_2OH$	$T_8(OSiMe_2H)_8$ + Pd/C	-	[175]
$-OSiMe_n(OEt)_{3-n}$ (n = 2, 1, 0)	$[NMe_4]_8[T_8O_8] + (EtO)_{3-n}Me_nSiCl$	-	[176]
$-OSiMe_2OSiMe_2H$	$T_8(OSiMe_2OH)_8 + HMe_2SiCl$	-	[175]
$-OSiMe_2OSiMe_3$	$T_8(OSiMe_2H)_8 + (Me_3Si)_2$, HCl	97	[170]

2.5.3.3 *T_{10} Compounds*

Only two T_{10} siloxy POS compounds have been prepared by modification of their functional groups, both using the same methodology. These are $T_{10}(OSiMe_2H)_{10}$ [177], and $T_{10}(OSiMe_3)_{10}$ [178], prepared by the silylation of $[NBu_4]_{10}[T_{10}O_{10}]$ with HMe_2SiCl and Me_3SiCl, respectively.

2.6 Synthesis of T_nR_n Compounds where R = Metal Complex

2.6.1 *Hydrolysis*

There have been very few reports of fully condensed POS compounds containing metal ions directly prepared by hydrolysis and condensation. One of these,

$[NMe_4]_7Na[T_8O_8]$, can be prepared by the hydrolysis of sodium silicate in the presence of NMe_4OH and NaOH [179], while another is a more complicated system comprising $\{T_8[(CH_2)_3NH_3]_8\}^{8+}$ cations and a mixture of Cl^- and $[ZnCl_4]^{2-}$ anions, formed by the acidic hydrolysis of $H_2N(CH_2)_3Si(OEt)_3$ in the presence of $ZnCl_2$ [73].

2.6.2 Substitution

2.6.2.1 T_8 Compounds

A small range of metal-containing T_8 POS compounds has been prepared by substitution at their cage silicon atoms. Three alkyltin compounds, $T_8(OSnBu_3)_8$ and $T_8(OSnBu_3)_8{\cdot}4H_2O$ [180], and $T_8(OSnMe_3)_8$ [143], were all prepared by treating T_8H_8 with the appropriate hexaalkyldistannoxane. Reaction yield varied depending on the alkyl ligands, with butyltin products being obtained in up to 95% yield, and methyltin products in 54% yield. The cobalt compound, $T_8[Co(CO)_4]_8$ was obtained by reaction between T_8H_8 and dicobaltoctacarbonyl, in 87% yield [181] and a mono-ferrocenyl derivative was prepared by hydrosilylation of T_8H_8 with vinylferrocene to afford $T_8H_7(CH_2)_2Fc$ in 14% yield [182] (see also Section 1.8.2, Chapter 1).

2.6.2.2 T_{10} Compounds

Only one metal-containing T_{10} compound has been prepared by a substitution reaction at the cage silicon atoms. Thus, $T_{10}(OSnBu_3)_{10}{\cdot}4H_2O$ can be prepared, as for the T_8 analogue, in 93% yield by the direct reaction of $T_{10}H_{10}$ with hexamethyldistannoxane [180].

2.6.3 Modification of R

The most common method of preparing metal-containing POS compounds has been by modification of their functional groups. Of the metallated POS compounds, those containing eight metal atoms, as a T_8R_8 compound, are rarer than the T_8R_7R' compounds, as these latter are often employed as either model compounds for catalytic systems, or as catalysts in their own right (see Chapter 3). A selection of octametallated T_8 compounds is presented in Table 2.14 which includes examples of metals binding to suitable T_8 compounds such as a series of $T_8[(CH_2)_3NH_2]_8$ transition metal complexes [73] and a tetramethylstibine compound [143], as well as reaction with a vinylferrocene derivative [183].

Table 2.14 T_8R_8 metal complexes prepared by modification of functional groups

R or T_8R_8	Starting materials	Yield (%)	Refs.
$T_8[(CH_2)_3NH_2]_8M_2X_4$ [MX_2 = $Ni(NO_3)_2$, $Cu(NO_3)_2$, $CuBr_2$, $Zn(NO_3)_2$, ZnI_2, $Cd(BF_4)_2$, $Cd(NO_3)_2$]	$T_8[(CH_2)_3NH_2]_8 + MX_2{\cdot}nH_2O$	60-90	[73]
$T_8[(CH_2)_3NH_3]_8[MCl_4]_4$ (M = Co, Cu, Cd)	$T_8[(CH_2)_3NH_3]_8Cl_8 + MCl_2{\cdot}nH_2O$	27-82	[73]
$-OSiMe_2(CH_2)_2Fc$	$T_8(OSiMe_2H)_8 + CH_2{=}CHFc$, Pt(dvs)	57	[183]
$-OSiMe_2(CH_2)_2SiMe_2Fc$	$T_8(OSiMe_2CH{=}CH_2)_8 + HSiMe_2Fc$, H_2PtCl_6	70	[160]
$-OSiMe_2Co(CO)_4$	$T_8[OSiMe_2H]_8 + Co_2(CO)_8$	99	[159]
$-OSbMe_4$	$T_8(OSnMe_3)_8 + Me_3SiOSbMe_4$	60	[143]
$-OTiClCp_2$	$T_8(OSnMe_3)_8 + Cp_2TiCl_2$, CH_2Cl_2	84	[180]

A broad selection of complexes of T_8R_7R' POS derivatives has been prepared for use, for example, in the modelling of silica-anchored catalyst systems involving metals such as Ti, Zr, Mo, W, Mn, Os, Rh, Pd and Pt (see Chapter 3). Such a variety of introduced metals allows for an equally broad range of catalyst systems that have been modelled, or complexes that have been determined to be catalytically active in their own right. A selection of the T_8R_7R' metal complexes that have been prepared, both for catalytic studies and other uses, is presented in Table 2.15. Other complexes that have been prepared include osmium complexes, from the reaction of a dimethylcyclohexeneyl-POS derivative with $K_2OsO_2(OH)_4$, resulting in the formation of a chelated osmate ester complex [35].

Table 2.15 T_8R_7R' metal complexes prepared by modification of R'

T_8 derivative	Starting materials	Yield (%)	Refs.
$[(T_8Cy_7\text{-}\mu\text{-}O)Sc(acac)]_2$	$T_8Cy_7OH + Sc(acac)_2Cp^*$	61	[184]
$T_8(c\text{-}C_5H_9)_7OTiCl_2Cp'$	$T_8(c\text{-}C_5H_9)_7OH + Cp'TiCl_3$, *n*-BuLi	51	[185]
$T_8(c\text{-}C_5H_9)_7OTi(O\text{-}i\text{-}Pr)_3$	$T_8(c\text{-}C_5H_9)_7OH + Ti(O\text{-}i\text{-}Pr)_4$	80	[186]
$T_8(c\text{-}C_5H_9)_7OTi(CH_2Ph)_2C_5H_3\text{-}1,3\text{-}(SiMe_3)_2$	$T_8(c\text{-}C_5H_9)_7OH + Ti(CH_2Ph)_3C_5H_3\text{-}1,3\text{-}(SiMe_3)_2$	-	[187]
$[T_8(c\text{-}C_5H_9)_7O]_2[Mn(TMEDA)_2]$	$T_8(c\text{-}C_5H_9)_7OH + [Mn(CH_2\text{-}t\text{-}Bu)_2(TMEDA)]$	54	[188]
$T_8(c\text{-}C_5H_9)C_6H_4\text{-}4\text{-}CH{=}CHFc$	$[T_8(c\text{-}C_5H_9)_7C_6H_4\text{-}4\text{-}CH_2PPh_3]Cl + FcCHO$, NaOEt	89[a]	[189]
$T_8(c\text{-}C_5H_9)_7OMo({=}NH)(CH_2\text{-}t\text{-}Bu)_3$	$T_8(c\text{-}C_5H_9)_7OH + Mo({\equiv}N)(CH_2\text{-}t\text{-}Bu)_3$	76	[190]
$T_8(c\text{-}C_5H_9)_7OMo({=}NC_6H_3\text{-}2,6\text{-}i\text{-}Pr_2)(CH_2\text{-}t\text{-}Bu)({=}CH\text{-}t\text{-}Bu)$	$T_8(c\text{-}C_5H_9)_7OH + Mo({=}NC_6H_3\text{-}2,6\text{-}i\text{-}Pr_2)(CH_2\text{-}t\text{-}Bu)_2({=}CH\text{-}t\text{-}Bu)$	99	[191]
$[T_8(c\text{-}C_5H_9)_7O]_2Mo[NH(t\text{-}Bu)C_6H_3\text{-}3,5\text{-}(i\text{-}Pr)_2][N(t\text{-}Bu)C_6H_3\text{-}3,5\text{-}(i\text{-}Pr_2)]({\equiv}CEt)$	$T_8(c\text{-}C_5H_9)_7OH + Mo[N(t\text{-}Bu)C_6H_3\text{-}3,5\text{-}(i\text{-}Pr)_2]_3({\equiv}C\text{-}Et)$	60	[192]
$[T_8Cy_7OP(NMe_2)_2]_2Mo(CO)_4$	$T_8Cy_7OP(NMe_2)_2$ + norbornadienemolybdenumtetracarbonyl	95	[193]

T_8 derivative	Starting materials	Yield (%)	Refs.
$T_8Cy_7OSnMe_3$	$T_8Cy_7H + Me_3SnCl \cdot Me_3NO$	87	[36]
$T_8Cy_7OSbMe_4$	$T_8Cy_7H + Me_3SiCl \cdot Me_3NO$	83	[36]
$[T_8Cy_7\text{-}\mu\text{-}O]_2[SmCp^*_2][Li(THF)]$	$T_8Cy_7OH + Cp^*_2Sm(\mu\text{-}Cl)_2Li(THF)_2$, MeLi	68	[184]
$T_8(c\text{-}C_5H_9)_7OW[{=}NC_6H_3\text{-}2,6\text{-}(i\text{-}Pr_2)](CH_2\text{-}t\text{-}Bu)({=}CH\text{-}t\text{-}Bu)$	$T_8(c\text{-}C_5H_9)_7OH + W[{=}NC_6H_3\text{-}2,6\text{-}(i\text{-}Pr_2)](CH_2\text{-}t\text{-}Bu)_2({=}CH\text{-}t\text{-}Bu)$	-	[194]
$[T_8(c\text{-}C_5H_9)_7O]_2W[{=}NC_6H_3\text{-}2,6\text{-}(i\text{-}Pr_2)](CH_2\text{-}t\text{-}Bu)_2$	$T_8(c\text{-}C_5H_9)_7OH + W[{=}NC_6H_3\text{-}2,6\text{-}(i\text{-}Pr_2)](CH_2\text{-}t\text{-}Bu)_2({=}CH\text{-}t\text{-}Bu)$	-	[195]
$T_8Cy_7ORe(CO)_5$ and dimer	$T_8Cy_7OH + Re(CO)_5CF_3SO_3$, *n*-BuLi, CO	-	[196]
$T_8(c\text{-}C_5H_9)_7ORe(CH_2\text{-}t\text{-}Bu)({=}CH\text{-}t\text{-}Bu)({\equiv}C\text{-}t\text{-}Bu)$	$T_8(c\text{-}C_5H_9)_7OH + Re(CH_2\text{-}t\text{-}Bu)_2({=}CH\text{-}t\text{-}Bu)({\equiv}C\text{-}t\text{-}Bu)$	95	[197]
$[T_8Cy_7OP(NMe_2)_2]_2PtCl_2$	$T_8Cy_7OP(NMe_2)_2 + Pt(cod)Cl_2$	96	[193]

[a] Mixture of *E* and *Z* isomers.

Four mono- and one di-POS complex have been prepared with three different metals starting from the POS-fluorenyl compound $T_8(c\text{-}C_5H_9)_7CH_2Flu$ (Flu = 9-*H*-fluorene) [34]. These include trimethylsilyl and trimethylstannyl derivatives, sandwich compounds with $ZrCl_2Cp^*$ and $ZrCl_2Cp''$, and a di-POS sandwich compound with two fluorenyl rings sandwiching a $ZrCl_2$ fragment. Two titanium alkoxide complexes containing either one or three (1*R*,2*S*,5*R*)-2-isopropyl-5-methylcyclohexyloxy groups at the titanium centre have been prepared using $T_8(c\text{-}C_5H_9)_7OH$ as a precursor [198]. A second series of four titanium compounds was also prepared from $T_8(c\text{-}C_5H_9)_7OH$ [199]; reaction with Cp^*TiMe_3 gave the di-POS compound $[T_8(c\text{-}C_5H_9)_7O]_2TiCp^*Me$ which, when treated with either $B(C_6F_5)_3$ or $[C_6H_5NHMe_2][B(C_6F_5)_4]$, gave the adducts $[T_8(c\text{-}C_5H_9)_7O]_2TiCp^*(\mu\text{-}Me)B(C_6F_5)_3$ and $\{[T_8(c\text{-}C_5H_9)_7O]_2TiCp^*(\kappa\text{-}NMe_2Ph)\}[B(C_6F_5)_4]$, respectively. The $B(C_6F_5)_3$ adduct reacts with $[Ph_3C][B(C_6F_5)_4]$ to give $\{[T_0(c\text{-}C_5H_9)_7O]_2TiCp^*\}B(C_6F_5)_4$ along with several by-products [199]. In order to confirm the identity of the major product, this mixture was further treated with *N,N*-dimethylaniline, resulting in the quantitative formation of $\{[T_8(c\text{-}C_5H_9)_7O]_2TiCp^*(\kappa\text{-}NMe_2Ph)\}[B(C_6F_5)_4]$.

A series of three iridium-POS complexes has recently been prepared, starting from $T_8(c\text{-}C_5H_9)_7OSiMe_2H$, which undergoes hydrosilylation with three different iridium complexes containing ligands derived from 2-pyridylbenzene, and a ligand containing a pentene group [200] (see also Section 7.2.4, Chapter 7). The octa-silane $T_8(OSiMe_2H)_8$ was similarly hydrosilylated sequentially with one equivalent of one of the complexes, followed by an excess of *N*-allylcarbazole, to give mono-iridium, hepta(propylcarbazole) POS compounds.

An unusual series of metallo-POS compounds are phthalocyanines functionalised with eight POS thiol groups, $T_8(i\text{-}Bu)_7(CH_2)_3S$- [68, 201]. These were prepared by the reaction of the di-POS compound $1,2\text{-}(CN)_2C_6H_2\text{-}4,5\text{-}[S(CH_2)_3T_8(i\text{-}Bu)_7]_2$ with $CoCl_2\cdot 6H_2O$, CuCl, $Zn(OAc)_2\cdot 2H_2O$, $Lu(OAc)_3\cdot 6H_2O$ or $Gd(OAc)_3\cdot 6H_2O$, resulting in the formation of the phthalocyanine, and assembly of the single- [68] or double-layer complexes [201]. The single layer complexes, with Co, Cu and Zn, showed yields ranging from 33% to 71%, while the double layer complexes, with Lu and

Gd, showed 2% and 17% yields, respectively.

2.7 Synthesis of Miscellaneous T_nR_n Compounds

2.7.1 *Hydrolysis*

The contribution of the tetraalkylammonium cation to the stability of anionic POS oxides in aqueous solution has been shown to be more important than merely balancing the charge. The anionic species exist in solution as a large number of different silicate anions produced by the action of the base on the silicon source [202]. These silicate anions can be stabilized as the T_6, T_8, or larger POS anion by the presence of the appropriate tetraalkylammonium cation or non-aqueous co-solvents [203]. Molecular dynamics studies on $[T_8O_8]^{8-}$ / $[NMe_4]^+$ / H_2O solutions (1 $[T_8O_8]^{8-}$: 16 $[NMe_4]^+$: 450 H_2O) show that the $[NMe_4]^+$ ions are preferentially sited above the POS cage faces to form a coordination sphere that excludes water molecules and decreases the ease of cage hydrolysis [204]. This exclusion of water molecules by $[NMe_4]^+$ is much less significant for the $[T_6O_6]^{6-}$ ion, and $[T_8O_8]^{8-}$ is found to be ca. 290 kJ mol^{-1} more stable in an aqueous $[NMe_4]^+$ solution [205].

2.7.1.1 T_6 Compounds

The hexameric silicate anion, $[T_6O_6]^{6-}$ can be prepared and isolated as a salt by the action of an appropriate base, most commonly tetraethylammonium hydroxide, on a silicon source such as tetraethoxysilane [148], or fumed silica [206], both of which can give an isolable solid product in good yield.

2.7.1.2 T_8 Compounds

The octasilicate anion $[T_8O_8]^{8-}$ can be prepared by hydrolysis/condensation as a salt, usually with either $[NMe_4]^+$ or $[NMe_3(CH_2)_2OH]^+$ as counterion, and it is an important precursor to the vast majority of the octasiloxy POS compounds. It is commonly prepared and used in solution, where it exists as the major silicate product from the reaction of $Si(OEt)_4$ and NMe_4OH [207]. It can also be precipitated as a solid with a large amount of water of crystallization, either as the tetramethylammonium [159] or choline [208] (trimethylhydroxyethylammonium) salts. There are also a few T_8R_7Cl compounds that have been prepared by corner capping of the partially condensed cubic species, $R_7Si_7O_9(OH)_3$ with $SiCl_4$. These include the R = *i*-Bu

[209], c-C_5H_9 and Cy compounds [146], which are reported with yields around 90%.

2.7.1.3 T_{10} Compounds

The $[NBu_4]_{10}[T_{10}O_{10}]$ salt, which can be isolated as a solid, appears to be stabilized by the $[NBu_4]^+$ cation, much as $[NMe_4]^+$ stabilizes the T_8 and $[NEt_4]^+$ the T_6 cages [178].

2.7.2 Co-Hydrolysis

The co-hydrolysis and condensation of different chloro- or alkoxysilanes has proven only a marginally successful route for the preparation of POS compounds with different functional groups, as a variety of different products are generally formed, including POS of varying sizes, and substitutions, and different substitutional isomers (see also Section 1.8.1, Chapter 1). Even when relatively few products are formed during a particular reaction, separating individual products is often impractical. However, a few cases have been reported where individual products could be identified, and sometimes separated from the product mixture.

The co-hydrolysis of a 5:1 mixture of $HSiCl_3$ and $PhSiCl_3$ leads to a mixture comprising T_8H_8, T_8H_7Ph, $T_{10}H_9Ph$ and other unidentified silsesquioxane products [210]. Following the removal of the T_8H_8, size-exclusion HPLC led to a mixture of T_8H_7Ph and $T_{10}H_9Ph$, recrystallization of this led to pure T_8H_7Ph in 3 % yield, and $T_{10}H_9Ph$ in 0.5% yield. When the co-hydrolysis of $MeSiX_3$ and CH_2=$CHSiX_3$ (X = Cl or OAc) is carried out using varying ratios of starting materials, nine products $T_8Me_n(CH{=}CH_2)_{8-n}$ (n = 0–8) are formed, as identified by GLC and MS [211]. The co-hydrolysis of a combination of CH_2=$CHSiCl_3$ and $PhSiCl_3$ in a 5:3 ratio gave $T_8(CH{=}CH_2)_7Ph$ and $T_8(CH{=}CH_2)_6Ph_2$ as the main products with the expected $T_8(CH{=}CH_2)_5Ph_3$ being obtained in a lower yield [212]. The related co-hydrolysis of CH_2=$CHSiCl_3$ and $PhCH_2SiCl_3$ in a 6:2 ratio gave a surprising mixture of products, characterized by GCMS, including $T_8(CH{=}CH_2)_8$, $T_8(CH{=}CH_2)_7CH_2Ph$, and $T_8(CH{=}CH_2)_6(CH_2Ph)_2$, along with compounds containing hydroxy and alkoxy substituents [213]. Products from the co-hydrolysis of $RSiX_3$ and $R'SiX_3$ [R = Pr, R' = $(CH_2)_3Cl$, $(CH_2)_3I$, $(CH_2)_3SH$, $CH_2CH{=}CH_2$; R = Et, R' = $CH{=}CH_2$] in a constant 7:1 ratio, have been studied in detail [214]. The product mixtures comprised $T_8R_{8-n}R'_n$ (n = 0, 1 or 2) that can be separated by HPLC, and in addition, three isomeric disubstituted products can be readily distinguished both from each other and from other products by ^{29}Si NMR spectroscopy. A more complicated co-hydrolysis reaction, between $NH_2(CH_2)_3Si(OMe)_3$, i-$OctSi(OMe)_3$, and $CF_3(CF_2)_5(CH_2)_2Si(OEt)_3$ led to the formation of a product mixture with a nominal composition of $T_8[(CH_2)_3NH_2]_2$i-$Oct_4[(CH_2)_2(CF_2)_5CF_3]_2$ [215]. Further treatment of this product with $OCN(CH_2)_3Si(OEt)_3$ led to the recovery of a product with nominal composition of $T_8[(CH_2)_3NHC(O)NH(CH_2)_3Si(OEt)_3]_2$*i*-$Oct_4[(CH_2)_2(CF_2)_5CF_3]_2$.

2.7.3 *Substitution and Modification of Functional Groups*

The photochemical chlorination of T_8H_8 using chlorine gas was found to give rise to T_8Cl_8 in 95% yield [139], and there have also been attempts to convert simple T_8R_8 compounds, like T_8H_8 into specific $T_8R_{8-n}R'_n$ compounds by substitution at the cage silicon atoms, the main disadvantage to this method being the production of complicated product mixtures. There are, however, some simple synthetic routes to $T_8R_{8-n}R'_n$ compounds. For example, the disilanol $T_8[OC(=CH_2)Me]_6(OH)_2$ was serendipitously produced during an attempted synthesis of $T_8[OC(=CH_2)Me]_8$ by the $[Co_2(CO)_8]$ catalyzed reaction between T_8H_8 and acetone, the product being isolated as a single isomer in 35% yield [43]. A series of monoamine-containing T_8 POS compounds has also been prepared, generally via a substitution reaction at the corner of the cage [216]. Thus, T_8Cy_7Cl reacts with NH_3, $NH{=}C(NMe_2)_2$, and $NH{=}C(NHC_6H_4\text{-}2\text{-}Me)_2$ to give T_8Cy_7R [R = NH_2, $N{=}C(NMe_2)_2$, or $N{=}C(NHC_6H_4\text{-}2\text{-}Me)_2$] in greater than 80% yield. Furthermore, the $T_8Cy_7NH_2$ could be silylated using Me_3SiCl, to give $T_8Cy_7NHSiMe_3$ in 87% yield [216].

2.7.4 *Other Synthetic Methods*

2.7.4.1 T_4 Compounds

Evidence has been found for the existence of an unusual T_4 derivative detectable by mass spectrometry and prepared by laser ablation of porous silica targets using a XeCl excimer laser [217]. A variety of clusters was produced including an anionic cluster corresponding in molecular weight to $[T_4(OH)_3O]^-$, which appears as a more stable, magic-number cluster, as does the related T_8 cluster. The production of the T_4 cluster can be optimized such that it is the major cluster produced. While there is no direct evidence for this cluster to have the T_4 structure, calculations at the HF/6-31G** level show it to be the most stable form for this molecular formula [217].

2.7.4.2 T_8 Compounds

There are several unusual preparative methods for T_8 POS, one being the basic hydrolysis/condensation reaction of the silica extracted from rice hull ash, which was found to lead to the formation of the anionic $[T_8O_8]^{8-}$ POS core [218].

There have been a number of attempts to prepare T_8 cages in stepwise fashion, beginning with a cyclotetrasiloxane, but these are usually unsatisfactory. The reaction of $(MeClSiO)_4$ with CH_2=$CHSiCl_3$ was found to give $T_8Me_6(CH{=}CH_2)_2$ as the major product following purification; although there was evidence to suggest that the desired $T_8Me_4(CH{=}CH_2)_4$ might exist as a by-product [211]. Similarly, $Na_4[PhSiO(O)]_4$ was treated with $MeSiCl_3$, followed by a condensation reaction under acidic conditions to give a very insoluble product, which could be brominated on the phenyl rings, resulting in a more soluble derivative, readily identified as $T_8Me_4(C_6H_4Br)_4$ [219]. The $T_8Ph_4(CH{=}CH_2)_4$ analogue could be prepared in the same manner, but an attempt to produce $T_8Cy_4Ph_4$ produced predominantly polymeric species [219].

The attempted preparation of $T_8R_4R'_4$ by condensation of two cyclotetrasiloxanes is rare; early attempts used the reaction of $[PhSi(OH)O]_4$ and $[MeSi(OH)O]_4$ in low concentration in the presence of a base catalyst [220]. This reaction proceeded in 95% yield, giving the desired $T_8Me_4Ph_4$, but the susceptibility of the $[RSi(OH)O]_4$ species to other condensation reactions severely limits the wider applicability of this reaction. More recent studies have shown that both *n*-Bu_4NF and $PhCH_2Me_3NOH$ will catalyze the formation of symmetrical T_8 species from $[RSi(OH)O]_4$ (R = phenyl, *p*-tolyl, isobutyl and naphthyl) [137].

2.7.4.3 T_{10} Compounds

The reaction of polyphenyl and polyvinyl silsesquioxanes with *n*-Bu_4F leads to reorganization and rearrangement of the polymers to give mixed cage species such as T_{10} and T_{12} compounds with a mixture of phenyl and vinyl substituents, which may be identified using MALDI-TOF MS [96].

2.8 Synthesis of Endohedral T_8R_8 Compounds

A limited number of T_8 POS derivatives wherein the silsesquioxane cage encapsulates an atom or ion as an endohedral complex have been prepared. While there is little space within the cage, it is big enough to take up monoatomic species, both during synthesis and via post-synthetic modification. The most common endohedral species is the fluoride anion, which can be introduced using either method, generally through the use of *n*-Bu_4NF. Including *n*-Bu_4NF during some hydrolysis and condensation reactions may afford endohedral complexes, for example, formation of $[NBu_4][F@T_8Ph_8]$ [221], $[NBu_4][F@T_8(CH{=}CH_2)_8]$, and $[NBu_4][F@T_8(C_6H_4\text{-}4\text{-}Me)_8]$ [222]. Single crystal X-ray diffraction and multinuclear NMR studies have confirmed the presence of the fluoride anion within the cage. When T_8R_8 compounds with electron-withdrawing substituents are treated with n-Bu_4NF [R = $CH{=}CH_2$, $CH{=}CHPh$, $(CH_2)_2CF_3$, $(CH_2)_2(CF_2)_3CF_3$, $(CH_2)_2(CF_2)_5CF_3$, Ph] then endohedral fluoride complexes may be isolated [223], but with electron-donating substituents (R = Me, Et, Cy, *i*-Bu), no fluoride encapsulation was observed [223]. The presence of mixed electron donating and withdrawing substituents led to products resulting from cage opening and rearrangement.

The encapsulation of atomic hydrogen has been found to occur, first demonstrated using γ-irradiation of $T_8(OSiMe_3)_8$ in air to form $H@T_8(OSiMe_3)_8$ [224], and more recently by glow discharge using a Tesla coil for a series of different POS compounds to form $H@T_8R_8$ (R = Me, *i*-Bu, Cy, Ph, $OSiMe_3$) [225]. The latter method is both faster and safer than the use of γ-irradiation. These endohedral compounds have relatively long lifetimes, enough that the interactions of $H@T_8(OSiMe_3)_8$ with metal complexes have been studied by EPR Spectroscopy [226].

2.9 Introduction to the Physical Properties of POS Compounds

Most simple T_8R_8 and T_8R_7R' compounds are air-stable, white or off-white powders, colorless crystals, or oils, and low molecular weight compounds such as T_8H_8 and T_8Me_8 may be readily sublimed. The electronic effect of the T_8 cage in T_8R_8 and T_8R_7R' compounds has been shown by ^{13}C NMR chemical shift studies to be electron-withdrawing, similar to that of the CF_3 group [113] and this has an effect on reactions taking place on rings in T_8 compounds, see Section 2.3.2. The incompletely condensed trisilanols $Si_7O_9(OH)_3$ are closely related to the T_8R_8 and T_8R_7R' polyhedral cage compounds and their extensive use as models for silica-grafted catalysis has been reviewed recently [227].

2.10 NMR and EPR Spectroscopy of POS Compounds

2.10.1 *Solution ^{29}Si NMR Studies*

Solution NMR spectroscopy is a useful tool for investigating the structures of POS compounds, since the organic derivatives of the cages are often soluble in common NMR solvents and ^1H and ^{13}C spectra of substituents are not greatly affected by the POS cage and are usually as expected by comparison with simple, non-

POS analogues. The sensitivity of ^{29}Si chemical shifts to changes in substituents in POS cage silicon atoms make ^{29}Si NMR spectroscopy an important technique for compound identification, for monitoring reaction progress, for elucidating the composition of isomer mixtures, and for checking the purity of POS compounds. For example, the progress of the free radical polymerization of $T_8(CH{=}CH_2)_8$ and acetoxystyrene can readily be monitored by using the significant chemical shift difference of the $SiCH{=}CH_2$ and $SiCH_2CH_2$- at -79 and -66 ppm respectively [228]. The ^{29}Si NMR chemical shift difference can be used to show that the product from hydrosilylation of T_8H_8 with $HC{\equiv}CCMe_2OH$ is mainly the C-substituted $T_8(CH{=}CHCMe_2OH)_8$ and not $T_8(OCMe_2C{\equiv}CH)_8$ [49], and to determine α/β adduct ratio in the hydrosilylation reactions between T_8H_8 and alkynyl stilbenes [229], or with azobenzene derivatives [56, 230]. Monitoring the cage rearrangement reactions of various $T_8[(CH_2)_3X]_8$ (X = Cl, NCS etc.) compounds to give mixtures of $T_8[(CH_2)_3X]_8$, $T_{10}[(CH_2)_3X]_{10}$, and $T_{12}[(CH_2)_3X]_{12}$ compounds shows that upon going from compounds where the Si is just involved in 8-membered rings, $T_8[(CH_2)_3X]_8$, to two 8- and one 10-membered, $T_{10}[(CH_2)_3X]_{10}$, to one 8 and two 10-membered, $T_{12}[(CH_2)_3X]_{12}$, there is increased shielding of ca. 2-3 ppm for each change, see Table 2.16 [57].

Characteristic ^{29}Si chemical shifts for alkyl-T_8 compounds come in the range of ca. -65 to -70 ppm, as shown in Table 2.16, and similar shifts are found for the POS cages when such precursors are incorporated into polymeric materials. The chemical shifts for aryl-T_8 compounds are found in the range ca. -77 to -83 ppm, although for the substituted compounds such as $T_8(C_6H_4NH_2)_8$ many of these spectra are complicated by the presence of several isomers. The chemical shift range for T_8 compounds with siloxy substituents is usually observed at ca. -110 ppm, as is seen in Table 2.16.

POS cages containing a fluoride ion [223, 231], for example, $[NBu_4][F@T_8(CH{=}CH_2)_8]$ (δ ca. -83 ppm), show an upfield shift of a few ppm compared to their empty cage counterparts, Table 2.16, with only one chemical shift seen for the endohedral anion, consistent with the fluoride interacting equally with all eight of the Si atoms in the T_8 cage.

Table 2.16 Selected solution phase ^{29}Si NMR data for T_nR_n compounds[a]

R, or T_nR_n compound	^{29}Si NMR chemical shift (ppm from Me_4Si)	Refs.
	n = 6	
-*i*-Bu	-55.4	[10]
-*c*-C_6H_{11}	-56.6	[10]
-Ph	-66.9	[10]
$[n\text{-}Pr_4N]_6[T_6O_6]$[b]	-88.2	[232]
-$OSiMe_3$	-99.1, 14.5 ($SiMe_3$)	[233]
	n= 8	
-H	-84.12	[234]
-CH_2CH_3	-65.5	[82]
-CH_2CH_2Br	-70.45	[83]

R, or T_nR_n compound	^{29}Si NMR chemical shift (ppm from Me_4Si)	Refs.
-CH=CH_2	-80.21	[235]
-CH=CH_2[c,d]	-83.0	[223]
-CH=CHPh	-78.2	[223]
-CH=CHPh[c,d]	-81.0	[223]
-CH_2CH=CH_2	-70.77	[88]
-$CH_2CH_2SiMe_2OMe$[e]	-65.91, 18.62 ($SiMe_2OMe$)	[236]
-*i*-Pr	-66.26	[104]
-$(CH_2)_3Ph$	-66.8	[46]
-$(CH_2)_3NH_2$	-66.5	[237]
{$T_8[(CH_2)_3NH_3]_8$}Cl_8[f]	-66.4	[238]
-$(CH_2)_3N(CH_2CO_2H)_2$[b]	-67.2	[239]
-$(CH_2)_3N_3$	-67.0	[66]
-$(CH_2)_3PPh_2$[g]	-66.41	[23]
-$(CH_2)_3OH$[f]	-65.9	[17]
-$CHMeCH_2OCH_2CH(O)CH_2$	-67.6	[240]
-$(CH_2)_3SH$	-66.80	[241]
-$(CH_2)_3Cl$	-67.28	[242]
-$(CH_2)_3I$	-66.2	[243]
-$(CH_2)_2CF_3$[c]	-66.7	[223]
-$(CH_2)_2CF_3$[c,d]	-70.4	[223]
-$CH_2CH(O)CH_2$	-69.7	[88]
-$(CH_2)_2CO_2Me$	-67	[83]
-*i*-Bu	-67.9	[244]
-*c*-C_5H_9	-66.6	[11]
-Cy	-68.7	[11]
-Ph[c]	-78.1	[223]
-Ph[c,d]	-80.7	[223]
-C_6H_4-4-Me	-77.62	[245]
-$C_6H_4NO_2$[e]	-79.22	[115]
-C_6H_4-4-Br[h]	-78.3	[137]
-C_6H_4-4-$SiMe_3$	-78.4, -3.83 ($SiMe_3$)	[115]
-naphthyl[c]	-77.02	[137]
$[NMe_4]_8[T_8O_8]$[b,i]	-98.122	[202]
-OEt	-103	[141]
-$OSiMe_2H$	-112.48, -3.95 ($SiMe_2$)	[246]
-$OSiMe_2OSiMe_3$[c]	-108.986, -21.990 ($SiMe_2O$), 12.469 ($SiMe_3$)	[170]
-$OSiMe_2CH$=CH_2	-100.4, 8.58, 9.026 ($SiMe_2$)	[247]
-$OSiMe_2(CH_2)_3NMe_2$[e]	-107.3, 14.9 ($SiMe_2$)	[248]

R, or T_nR_n compound	^{29}Si NMR chemical shift (ppm from Me_4Si)	Refs.
$-OSiMe_2(CH_2)_3OH$	-100.4, 8.6 ($SiMe_2$)	[249]
$-OSiMe_2(CH_2)_4CH(O)CH_2$	-101, 21.3 ($SiMe_2$)	[250]
$-OSiMe_3$	-109, 13 ($SiMe_3$)	[141]
$-OSnMe_3$	-110.144	[180]
$-O[TiClCp_2]$	-101.9	[180]
$-SiMe_2$-*t*-Bu	-71.31, -12.67 ($SiMe_2$-*t*-Bu)	[251]
	n = 10	
-H	-86.25	[1]
$-CH{=}CH_2$	-81.48	[38]
$-(CH_2)_3N_3$	-68.94	[30]
$-(CH_2)_3Cl$	-68.97	[57]
-*c*-C_5H_9	-69.20	[38]
$-C_8H_{17}$	-68.68	[57]
$-C_6H_4$-4-Me	-79.5	[115]
$-OSiMe_2CH_2Cl$	-110.82, 6.96 ($SiMe_2$)	[180]
	n = 12	
-H	-85.78, -97.76	[1]
$-CH{=}CH_2$	-81.34, -83.35	[38]
$-(CH_2)_3N_3$	-68.69, -71.36 (1:2)	[30]
$-(CH_2)_3Cl$	-68.73, -71.37	[57]
-Cy	-71.29, -74.29	[112]
$-C_6H_4$-4-$SiMe_3$	-81.5, -79.4; -4.13, -4.06 ($SiMe_3$)	[115]
	n = 14	
-H	-87.89, -88.03, -89.71	[1]

[a] Unless otherwise stated, chemical shifts are for $CDCl_3$ solutions and are referenced with respect to Me_4Si. [b] H_2O/D_2O solution. [c] THF-d_8 solution. [d] Anionic endohedral fluoride, $[NBu_4]^+$ cation. [e] $(CD_3)_2CO$ solution. [f] $(CD_3)_2SO$ solution. [g] C_6D_6 solution. [h] CD_2Cl_2 solution. [i] Enriched in ^{29}Si.

The chemical shifts for T_8R_7R' compounds (Table 2.17) are similar to those found for the corresponding TR or TR' groups in T_8R_8 and $T_8R'_8$ compounds, as might be expected, but the three signals expected for the TR environments in T_8R_7R' compounds are often difficult to resolve and can appear as a single broad signal, along with the TR' signal. One of the few detailed NMR studies of an isomeric mixture of POS compounds showed that complicated mixtures of all possible isomers of $T_8(OPh)_n(R)_{8-n}$ [n = 0-8; R = *n*-$C_{11}H_{23}$ or $(CH_2)_2CMe_2CH_2CO_2Me$] were formed in reactions of T_8H_8 with PhOH and undec-1-ene or 3,3-dimethylpent-4-enoate. Computational models of the spectra suggest that the predominant products bear dissimilar substituents on adjacent Si atoms [54]. In a related study, products having ^{29}Si NMR signals at -108.84 (*Si*O_4), 12.67 [$Ph(CH_2)_3$*Si*Me_2O] and 12.78 ppm (C_6H_{11}*Si*Me_2O) corresponding to a mixture of isomers of $T_8[OSiMe_2(CH_2)_3Ph]_{8-n}(OSiMe_2C_6H_{11})_n$ (n = 0-6) are formed on hydrosilylation of $T_8(OSiMe_2H)_8$ with a 6:2 mixture of allylbenzene and 1,5-hexadiene [252]. The ^{29}Si NMR spectrum of the mixture formed by co-hydrolysis of $H_2N(CH_2)_3Si(OMe)_3$, *i*-OctSi(OMe)$_3$, and $F_3C(CF_2)_5CH_2CH_2Si(OEt)_3$ gives a broad signal at ca. -68 ppm with several shoulders indicating that several POS species are present, and provides an example of how ^{29}Si NMR spectroscopy is limited in distinguishing between closely related POS compounds [215]. The silicate anion, $[T_8(O)_4(OT)_4]^{16-}$, in which the T-unit substituents are at alternate corners of the T_8 cage is formed in silicate solutions containing $[NMe_3(CH_2CH_2OH)]^+$ and has ^{29}Si NMR chemical shifts attributable to the exocyclic Si, the cage T-O-Si, and the cage SiO^- silicon atoms at -82, -99 and -109 ppm respectively [253].

Table 2.17 ^{29}Si NMR data for T_8R_7R' compounds in solution[a]

R and R'		^{29}Si NMR chemical shift (ppm from Me_4Si)	Refs.
R	R'		
$-(CH_2)_2CF_3$	-Me[b]	-64.8, -67.5, -68.0	[32]
$-(CH_2)_2CF_3$	$-(CH_2)_2OH$	-67.65, -67.66, -67.84	[254]
$-(CH_2)_2CF_3$	$-(CH_2)_2Ph$[c]	-66.4, -67.4, -67.7	[32]
$-(CH_2)_2CF_3$	$-(CH_2)_2(CF_2)_5CF_3$[d]	-67.2, -67.4	[254]
$-(CH_2)_2CF_3$	$-C_6H_4$-4-OCF=CF_2[c]	-68, -71.5	[255]
-*i*-Bu	-CH=CH_2[e]	-66.9, -67.4, -80.9(SiCH=)	[78]
-*i*-Bu	$-(CH_2)_2OH$	-67.1, -67.3, -69.1 (3:4:1)	[256]
-*i*-Bu	$-(CH_2)_3Cl$	-67.57, -67.87, -68.10 (3:4:1)	[30]
-*i*-Bu	$-OSiMe_2H$	-2.98, -66.84, -67.76, -67.79, -108.97	[257]
-*c*-C_5H_9	-H	-66.47 (SiCH), -83.90 (Si-H)	[258]
-*c*-C_5H_9	-OH	-63.48, -64.18, -97.75 (4:3:1 ratio)	[185]
-*c*-C_5H_9	-Cl	-65.74, -66.32, -66.36 (3:3:1), -89.53 (SiCl)	[146]
-*c*-C_5H_9	$-(CH_2)_2OH$	-65.9, -68.1 (7:1)	[256]
-*c*-C_5H_9	$-(CH_2)_3Cl$	-66.52, -67.11	[259]
-*c*-C_5H_9	$-C_6H_4$-4-CH_2Cl[f]	-67.8, -68.2, -79.6	[260]
-*c*-C_5H_9	$-OSiMe_2C{\equiv}CH$	-15.79, -65.58, -66.24, -108.13 (1:3:4:1)	[261]
-*c*-C_5H_9	-OTiMeCp*$(OT_8(C_5H_9)_7$[e]	-110.8 (SiO_4), -66.0, -65.9, -65.8(3:1:3)	[199]
-Cy	$-(CH_2)_3SiMe_2C_6H_4$-4-$SiMe_2H$	-3.43 ($SiMe_2$), -17.03 ($SiMe_2H$), -68.38 (4 SiCy), -68.56 (3 SiCy), -66.47 ($SiCH_2$)	[262]
-Cy	$-C_6H_4$-4-CH_2Cl[f]	-67.8, -68.2, -79.6	[118]
-Cy	$-SiCl_3$	-62.0 ($SiSiCl_3$), -31.4 ($SiCl_3$), -67.7, -67.8, -67.9, -68.4	[142]
Cy	$-OSiMe_3$[e]	11.29, -67.54, -68.09, -68.12, -107.81 (1:3:3:1:1)	[36]
-Cy	$-OSiCl_3$	-66.0 ($SiOSiCl_3$), -56.6 ($OSiCl_3$), -67.8, -68.0, -68.6, -69.0, -69.6, -69.8, -70.2	[142]

R and R'		^{29}Si NMR chemical shift (ppm from Me_4Si)	Refs.
R	R'		
-Cy	-O[Re(CO)$_5$][g]	-68.06, -68.66, -99.40 (1:6:1)	[196]
-Ph	$-(CH_2)_2C_6H_4$-4-SO_2Cl	-66.69 ($SiCH_2$), -78.35, -78.41, -78.67	[263]
-Ph	$-(CH_2)_3NH_2$	-64.76, -78.30, -78.78 (1:4:3)	[37]
-Ph	$-(CH_2)_2OH$	-67.31 ($SiCH_2$), -78.42, -78.79	[263]
-Ph	$-(CH_2)_3SiMe_2C_6H_4$-4-$SiMe_2H$	-3.61 ($SiMe_2$), -17.22 ($SiMe_2H$), -78.23 (4 SiPh), -78.71 (3 SiPh), -65.25 ($SiCH_2$)	[262]
-OEt	$-(CH_2)_2CO_2C_{16}H_{33}$	-102.80, -102.75, -102.72, -65.26 ($SiCH_2$)	[264]
-OEt	$-C_{16}H_{33}$	-102.56, -102.63, -102.69, -64.12 ($SiCH_2$)	[53]

[a] Unless otherwise stated, all data are for $CDCl_3$ solutions or were not reported, and are referenced with respect to Me_4Si. [b] $(CD_3)_2CO/CDCl_3$ solution. [c] $(CD_3)_2CO$ solution. [d] $(CD_3)_2CO$/THF solution. [e] C_6D_6. [f] THF solution.

Besides ^{29}Si NMR, heteronuclear studies of POS compounds have sometimes been carried out when looking at applications of the POS species. For example, ^{11}B NMR spectra have been used to monitor the interaction of $T_8(c\text{-}C_5H_9)_7(OH)$ with $(C_6F_5)_3B$ via B···OH coordination [265], and ^{19}F NMR spectroscopy can be used to monitor enzymatic activity by using the high number of fluorine nuclei in POS species derived from $T_8[(CH_2)_3NH_2]_{8-n}[(CH_2)_3NHC(=O)CF_3]_n$ (where n = 3–5) when attached to the surface of silica nanoparticles [266]. Applications of T_8 POS compounds in MRI imaging studies have also recently become of interest, owing to their potential to bind a high concentration of Gd(III) as a relaxation agent [267] (see also Section 7.5.1, Chapter 7).

2.10.2 Solid State NMR Studies

The use of solid state ^{29}Si NMR spectroscopy for the analysis of both molecular and polymer materials containing POS cages has become widespread in the last 10 years. It is a technique used extensively to help determine whcther a polymerization or mechanical process has caused degradation of the POS cage, when the cage is a component in a polymeric or composite material. Chemical shifts for a range of molecular T_8R_8 and T_8R_7R' compounds are shown in Table 2.18, and show well-defined chemical shift ranges for alkyl (-65 to -70 ppm), aryl (-75 to -81 ppm), and siloxy (-95 to -120 ppm) substituted POS cages. The solid state chemical shift values for molecular species are generally close to those for analogous solution data (see Section 2.10.1) and can be used for comparison with spectra recorded on polymeric materials containing POS cages when confirmation is needed that a POS cage is still intact. For example, upon reaction of $T_8[CH_2CH(O)CH_2]_8$ with diamines [88], the product exhibits signals at -68.56, and -67.46 ppm and the radical-induced polymerization of methyl methacrylate with $T_8(OSiMe_2H)_8$ gives a product with signals at -108.6 (POS) and 13.3 ($OSiMe_2CH_2$-) ppm, both spectra indicating that little rearrangement has occurred [268]. Similarly, solid state ^{29}Si NMR spectra for nanocomposites prepared using $T_8[OSiMe_2(CH_2)_4CH(O)CH_2]_8$ or $T_8(C_6H_4NH_2)_8$ precursors indicate that the POS cores are largely unchanged during the process [269], and while broadening of the solid state ^{29}Si NMR signal from $\{T_8[(CH_2)_3NH_3]_8\}$ -Cl_8 occurs in the presence of Pd nanoparticles, the chemical shift is not much altered, suggesting that the POS core is not cleaved by the nanoparticles [270].

Table 2.18 Solid state ^{29}Si NMR chemical shifts for T_nR_n and T_8R_7R' compounds

R or T_nR_n		^{29}Si NMR chemical shift (ppm from Me_4Si)	Refs.
		n = 6	
-$OSiMe_3$		-98.9, -99.2, -100.0, -100.1, -100.6 (SiO_4); 15.0, 14.7, 13.7, 13.1, 12.4, 12.3 ($SiMe_3$)	[149]
		n = 8	
-H		-85.5	[271]
-Me[a]		-66, -66.7	[272]
-CH=CH_2		-79.5	[273]
-$CH_2CH(O)CH_2$		-69.76	[88]
-$(CH_2)_3NH_2$		-66.4	[274]
$\{T_8[(CH_2)_3NH_3]_8\}Cl_8$		-67.0	[275]
-$(CH_2)_3Cl$		-66.0, -67.1	[67]
-$(CH_2)_2(CF_2)_5CF_3$		-66.7, -67.4	[276]
-*c*-C_5H_9		-64, -68	[277]
-Ph[a]		-79.4	[278]
-C_6H_4-4-NO_2		-82.6	[279]
-$C_6H_4NH_2$		-66.7	[280]
-$OSiMe_2H$		-108.69, -2.89 ($SiMe_2$)	[281]
-$OSiMe_3$		-106.0, 14.2 ($SiMe_3$)	[282]
-$OSiMe_2(CH_2)_3CN$[a]		-118 ($SiMe_2$ not seen)	[163]
-$OSi(OSiMe_3)_3$		-111.9 (POS), -107.5 [$OSi(OSiMe_3)_3$], 9.0 ($SiMe_3$)	[169]
$[NMe_4]_8[T_8O_8]$		-96.5	[281]
-$SiMe_2$-*t*-Bu[a]		-70.76, -11.20, -12.90	[251]
R	R'		
-$(CH_2CH_2CF_3)_7$	-H	-68.6, -58.3 (SiH)	[283]
-*i*-Bu_7	-C_6H_4-4-CH=CH_2	-67.8 (SiBu), ca. -80 SiC_6H_4	[284]
-Ph_7	-$(CH_2)_3Cl$	-77.02, -64.06	[285]
		n = 10	
-$OSiMe_2H$		ca. -110, ca. -2 ($SiMe_2H$)	[286]

[a] ^{13}C Solid state spectrum also recorded.

Despite the widespread use of solid state ^{29}Si NMR spectra for POS characterization, solid state ^{13}C NMR spectra of POS compounds are rarely recorded, presumably due to their more complicated nature and the lack of information obtained about the POS core in such spectra. However, comparison of the solid state ^{13}C NMR spectrum of copolymers of vinylpyrrolidine with $T_8(i\text{-Bu})_7C_6H_4\text{-4-CH=CH}_2$ show that the POS cage is incorporated intact into the polymer [284], and spectra of star polymers comprising a T_8 core and poly(ε-caprolactone) substituents are consistent with the formation of inclusion complexes with cyclodextrins [174]. The spectrum of $T_8(CH=CH_2)_8$ has been used to set up the Hartmann-Hahn conditions for magic angle spinning $^{29}Si \rightarrow ^{13}C$ cross polarization transfer, a value of 136 Hz being found for $^1J_{Si\text{-}C}$ [287], and a detailed ^{13}C NMR study of $T_8(i\text{-Bu})_8$ has been used to investigate the motion of the substituents as a function of temperature [288]. Solid state ^1H and ^2H NMR spectra of T_8Et_8 and its partially deuterated derivatives are consistent with, and typical of, transitions occurring in a plastic crystal, a triclinic phase being present at low temperature, and a rhombohedral phase being present at high temperature [82].

2.10.3 *EPR Spectra*

EPR spectroscopy has rarely been used for the characterisation of POS compounds, but it has been used successfully for investigating the nature of certain endohedral T_8 species. Thus, a negative temperature coefficient for the ^{29}Si super hyperfine coupling constant in a mixture of the endohedral POS complexes $H@T_8(n\text{-Pr})_8$ and $D@T_8(n\text{-Pr})_8$ in toluene has been observed by EPR spectroscopy [289]. Solid and solution EPR spectra of $H@T_8[(CH_2)_nH]_8$ (where n = 0-3) species exhibit isotropic hyperfine interactions with ^{29}Si [290], and activation energies, E_a, for detrapping of H and D from T_8Me_8 are 126.7±1.2 and 127.4±1.2 kJ mol^{-1} respectively [289]. EPR spectra have also been used to monitor the rapid uptake of hydrogen atoms into the cage of T_8Me_8 upon its being subject to an electric discharge to form $H@T_8Me_8$ [225].

2.11 Vibrational Spectra of Polyhedral Oligomeric Silsesquioxane Compounds

Vibrational spectroscopy is often used in the identification of both simple molecular POS compounds and in polymeric materials as, in both cases, they give rise to a characteristic Si-O-Si band in the IR spectrum. There have been a range of values quoted for the ν_{as} band associated with the Si-O-Si linkages, even in simple POS compounds, since although it is strong (occurring at ca. 1100 cm^{-1}) it is also broad. Table 2.19 shows IR data for a range of POS derivatives, and more data can be found in references [119] and [120]. As the bands due to Si-O-Si linkages in ladder silsesquioxanes usually occur from 1030–1055 cm^{-1} [291, 292], IR spectroscopy can be used to distinguish POS cage derivatives from ladder siloxanes. The solid state IR spectrum of T_8H_8 shows bands for δ_s (O-Si-O) at 557.9 and δ_{as} (O-Si-O) at 389.5 cm^{-1} [271], while its Si-H stretch has been variously reported at 2144 cm^{-1} [293], and at 2275 cm^{-1} [240], in CCl_4 solution, and as solid KBr and CsI pellets, ν(Si-H) has been reported at 2277, 2294 and 2300 cm^{-1} respectively [294]. For $T_8(OSiMe_2H)_8$, the Si-H band has been reported at 2250 cm^{-1} [44], 2142 cm^{-1} [295], and at 2140 cm^{-1} [296], while its Raman spectrum shows a band due to Si-H at 2141 cm^{-1} [297]. The widely different Si-O-Si bond angles (136.35-172.13°) seen in the solid state structure of $T_8(OSnMe_3)_8{\cdot}4H_2O$ (see section 2.14.1) give rise to significant splitting of both the symmetric and antisymmetric Si-O-Si stretches [180].

Examples of the use of IR spectroscopy for monitoring whether the T_8 cage stays intact during a polymerization process can be seen for reactions of $T_8[(CH_2)_3OCH_2CH(O)CH_2]_8$ which has an Si-O-Si stretch at 1103-1110 cm^{-1}. This band can also be seen in the interpenetrating network formed in its reaction with $CMe_2(C_6H_4$-4-$OH)_2$ in the presence of poly(ethylene oxide) [298, 299], in crosslinked polymers it forms with poly(*N*-isopropylacrylamide) [300], and with poly(4-vinylpyridine) [301], and hybrids it makes with poly(ethylene imine) [283]. Similarly, the IR band at ca. 1109 cm^{-1} in $T_8(CH{=}CH_2)_8$ has been used to monitor the solution blending of the POS compound with polystyrene and with poly(4-acetoxystyrene) to show that the T_8 cage remains intact during the blending process [302].

Table 2.19 Infra-red stretching frequencies associated with the Si-O-Si bonds in T_8R_8 and T_8R_7R' compounds

R or T_8 derivative	Si-O frequencies (cm^{-1})	Refs.
-H	ν_{as}(Si-O-Si) 1117	[303]
-Me	ν_{as}(Si-O-Si) 1115, 1192, ν_s(Si-O-Si) 517, and δ(O-Si-O) 465, 379	[103]
$-CH{=}CH_2$	1110-1112	[235]
$-CH_2CH{=}CH_2$	1105	[88]
$-CH_2CH(O)CH_2$	1107	[88]
$-(CH_2)_3Cl$	ν_{as}(Si-O) 1230-940, δ(O-Si-O) 552	[304]
-*i*-Bu	1115	[244]
-Cy	1110	[305]
$-(CH_2)_3N_3$	1111	[66]
$-(CH_2)_3NH_2$	1030	[237]
$\{T_8[(CH_2)_3NH_3]_8\}Cl_8$	1105	[306]
$-(CH_2)_3OCH_2CH(O)CH_2$	1106	[283]
-Ph	1115	[279]
$-C_6H_4NO_2$[a]	ca. 1100	[307]

R or T_8 derivative		Si-O frequencies (cm^{-1})	Refs.
$-C_6H_4NH_2$[a]		1100	[307]
$[NMe_4]_8[T_8O_8]$		1037	[308]
$-OSiMe_2H$		1096	[309]
$-OSiMe_2(CH_2)_2Ph$		1100	[310]
R	R'		
-H	-*n*-C_6H_{13}	1139	[311]
$-(CH_2)_2CF_3$	$-(CH_2)_3Br$	1090-1000	[312]
-*i*-Bu	$-(CH_2)_2OH$	1082	[313]
-*i*-Bu	$-(CH_2)_3NH_2$	ca. 1100	[314]
-*i*-Bu	$-C_6H_4$-4-CH_2Cl	1110	[130]
-Cy	$-(CH_2)_3OC(=O)C(Me)=CH_2$	1110	[315]
-Ph	$-(CH_2)_3OH$	1134	[285, 316]

[a] Contains a mixture of isomers.

The IR bands for T_8R_7R' compounds are similar to those for the related T_8R_8 species, and the same problems associated with reporting the value for the broad Si-O-Si are found; typical values for Si-O-Si IR data are given in Table 2.19. Calculated, 2350 cm^{-1}, and experimental, 2274 cm^{-1}, values of the Si-H stretch for $T_8H_7(n\text{-}C_6H_{13})$ are significantly different [311], and the experimental value for the Si-H stretch in $T_8(i\text{-Bu})_7H$ is 2215 cm^{-1} [317]. The IR spectrum of $T_8(i\text{-Bu})_7OH$ in KBr has been reported to show a broad hydrogen-bonded SiOH band in the 3425 cm^{-1} region [318], a sharp band due to free SiOH groups at ca. 3680 cm^{-1} and a broad band at ca. 3500 cm^{-1} attributed to a hydrogen-bonded dimeric silanol species [319]. This difference in the degree of hydrogen bonding is probably due to variations in sample concentration. Only a single sharp band due to free SiOH at ca. 3680 cm^{-1} is seen in the spectrum of a 0.1 M solution of $T_8(i\text{-Bu})_7OH$ in CCl_4 [319], whereas silanols usually exhibit strong hydrogen bonding to each other [320, 321].

As for T_8R_8 compounds, the characteristic Si-O-Si stretches in T_8R_7R' compounds can be used to monitor whether processing of monomeric species into polymers causes breakdown of the cage structure. Thus, poly(ε-caprolactone) and poly(L,L-lactide) end-capped by $T_8(i\text{-Bu})_7(CH_2)_3NH$ show the expected Si-O-Si T_8 band at 1099 cm^{-1} [322], and a band at 1109 cm^{-1} characteristic of the Si-O-Si stretch in $T_8(i\text{-Bu})_7C_6H_4\text{-4-}CH{=}CH_2$ occurs in spectra of its copolymers with vinylpyrrolidone [284]. Conversely, the sharp absorption at 1130 cm^{-1} due to the T_8 cage in the IR spectrum of the copolymer poly(*N*-dodecylacrylamide-*co*-3-methacryloxypropylheptaphenyl POS) is lost on photooxidation and replaced by one at 1065 cm^{-1}, due to the formation of an SiO_2 film on breakdown of the cage [323]. Similar "SiO_2-like" layers have also been characterized by IR spectroscopy and shown to be formed by reaction of polymers of pendent T_8Et_7 cages with oxygen plasmas [324].

2.12 Mass Spectra of POS Compounds

Mass spectrometry has been used widely for the analysis of simple POS compounds with alkali metals or protons being readily bound by the π-electron density and electronegative atoms in the substituents on the POS cage to give positive ions. There may be difficulties in generating large ions from oligomers or polymers containing POS components without fragmentation, probably due to low ionization efficiency caused by the ability of the POS cage to delocalize electron density. It may, however, be possible to address this problem by preparing endohedral fluoride complexes having a higher ionization efficiency (see, for example, below) [223]. Mass spectrometry may be particularly useful in the analysis of complicated mixtures of POS-containing products derived from the co-hydrolysis of mixtures of simple monomers, owing to the widely differing masses of the components, the sizes of the POS cages and the number of each substituent present. Simple chemical ionization spectra can be recorded for small compounds such as $T_8[(CH_2)_3Cl]_8$ [325], but higher molecular weight T_8R_8 compounds require more sophisticated techniques such as MALDI-TOF MS and electrospray ionization (EI), possibly in conjunction with ion mobility mass spectrometry. Mass spectrometry is also important in the analysis of products arising from incomplete reaction of, for example, an octa-functional precursor, T_8R_8, not affording a clean $T_8R'_8$ product. The degree of reaction is usually easily assessed by taking into account the differing masses of the R and R' groups in the $T_8R_nR'_{8-n}$ (n = 0-8) products, but the distribution of the R and R' groups around a POS cage is better investigated by ^{29}Si NMR spectroscopy. FABMS (fast atom bombardment) has been used to characterize simple monomeric POS species such as $T_8(CH_2CH_2R)_8$ [R = Ph [84]; R = Cy, $(CH_2)_3Br$ or $(CH_2)_4Cl$ [326]].

MALDI-TOF spectra have been used to show that compounds containing from five to twenty "T" units are formed on hydrolysis of CH_2=$CHSiCl_3$ [40], to determine the ratios of T_8:T_{10}:T_{12} cages produced on the amine-catalyzed condensation of silanols such as $PhSi(OH)_3$ or $[Ph(HO)Si]_6$ derived from hydrolysis of $(MeO)_3SiR$ (R = CH=CH_2, Ph or C_6H_4-2-Me) [327], to determine the ratios of T_8:T_{10}:T_{12} derivatives formed on F^- catalyzed rearrangements of polysilsesquioxanes [106], and to investigate the structure and stabilities of the components of the mixture formed on hydrolysis of $CH_2(O)CHCH_2O(CH_2)_3Si(OMe)_3$ [328]. Similarly, UV-MALDI-TOF mass spectrometry has been used to investigate the complicated mixtures obtained by hydrolysis of $(EtO)_3Si(CH_2)_3NH_2$ in the presence of phenylglycidylether [329]. Incompletely condensed POS species as well as T_8-type compounds are shown to form as products from hydrolysis of MeC(=CH_2)C(=O)O$(CH_2)_3Si(OMe)_3$ by capillary electrophoresis-ion-trap mass spectometry [330], and LC/MS has been used to show that $T_8Me_{8-n}[(CH_2)_3OCH_2CH(O)CH_2]_n$ compounds (n = 4 or 6) are formed together with T_9 and T_{10} products from the co-hydrolysis of $MeSi(OEt)_3$ and $CH_2(O)CHCH_2O(CH_2)_3Si(MeO)_3$ [331]. The formation of silicate polyhedra such as $T_8(OH)_8$ from smaller silicate fragments has also been monitored in detail using electrospray mass spectrometry [332-334].

MALDI methods have been used to show that ruthenium-catalyzed metathesis reactions of $T_8(CH=CH_2)_8$ do afford octa-substituted products $T_8(CH=CHAr)_8$ (Ar = Ph, C_6H_4-4-Me, C_6H_4-4-OMe, C_6H_4-4-Cl, C_6H_4-4-Br, C_6H_4-4-NO_2) [335], to investigate the degree of substitution in the platinum-catalyzed hydrosilylation reaction between $T_8(CH=CH_2)_8$ and $(EtO)_3SiH$ [336], and in elucidating the degree of substitution in the hydrosilylation reaction between $T_8(OSiMe_2H)_8$ and combinations of allylbenzene and 1,5-hexadiene, giving an isomeric mixture of $T_8[OSiMe_2(CH_2)_3Ph]_{8-n}(OSiMe_2C_6H_{11})_n$ (n = 0-8) compounds [252]. The extent of bromination of T_8Ph_8 using different methods, and $AgNO_3$ or AgO_2CCF_3 as ion source has been investigated using MALDI-TOF, showing that up to 17 bromines per POS molecule may be introduced [125, 337]. The formation of the tetra-POS compound $[T_8(i\text{-Bu})_7(CH_2CH_2SiMe_2O)]_4Si$ (m/z = 3821.91) from the Pt-catalyzed reaction between $(HSiMe_2O)_4Si$ and $T_8(i\text{-Bu})_7(CH=CH_2)$ has also been confirmed by MALDI methods [338].

MALDI-TOF mass spectrometry can be used to characterize high molecular weight POS derivatives such as glycoclusters (up to ca. m/z = 5000) prepared using $T_8(CH=CH_2)_8$ [339], a bis-POS derivative of a phthalocyanine (m/z = 3656) [340], octakis-POS substituted metallophthalocyanines (m/z = 7676-7681) [68], poly-pyrene substituted POS species (m/z = 3436) containing up to 14 pyrene groups,derived from Heck coupling of $T_8(CH=CH_2)_8$ with 1-bromopyrene [341], and poly-fluorene POS substituted compounds (m/z up to 7818) derived from coupling of $T_8(CH=CH_2)_8$ with a bromofluorene [342]. MALDI-TOF has also been used to characterize dendrimer drug carriers such as $T_8[(CH_2)_3NH(\text{L-Lys})(\text{L-Lys})_2(\text{L-Lys})_4(\text{L-Lys})_8]_8$, ($m/z$ = 16314) [70], and 4,4'-dihydroxyoctafluorobenzene has been proposed as a new matrix for MALDI spectra of POS derivatives with masses of >13 000 being observed for ethyl isobutyrate POS oligomers [343]. Recently, the use of atmospheric pressure photoionization mass spectrometry has been found to be a useful method for the characterization of POS compounds such as $T_8(i\text{-Bu})_7CH=CH_2$ [78].

Mass spectrometry has also been used to determine the behaviour of POS cages towards electron impact; most POS cages are resistant to electron impact up to 3 keV in symmetrically substituted compounds, T_8R_8, but in $T_8(CH_2CH_2CF_3)_8$ the cage is cleaved much more easily, probably via transfer of F to a silicon atom [39]. The collision cross sections for T_8Ph_8 and $T_8(CH_2Ph)_8$ have been determined using electrospray ionization, MALDI and ion mobility methods, and, when compared with calculated data and those determined from single crystal X-ray data, show good agreement, suggesting that the POS cage is not significantly deformed in the gas phase, and that Na^+ does not cleave the cage in Na^+POS ions [344]. Similar studies have been carried out for the related $T_8(CH=CHPh)_8$ and $T_8(CH_2CH_2Ph)_8$ compounds [223, 345], and the epoxide derivatives $Na^+T_8(CH=CHPh)_x[CH(O)\text{-}CHPh]_{8-x}$ (where x = 5, 6 or 7) [346]. The collision cross sections for a range of negative POS ions $F^-@T_8R_8$ (R = $CH=CH_2$, Ph, CH=CHPh, $CH_2CH_2CF_3$ etc.) and for the related sodiated species $H^+[F^-@T_8R_8]Na^+$ have been calculated and determined experimentally by ESI or MALDI, the results from both methods being in good agreement [223].

2.13 Electronic Spectra of POS Compounds

The electronic spectra of POS compounds have only been studied intensively quite recently when their potential as tethers for photoluminescent, fluorescent and other organic fragments of optical interest was realized (see also Chapter 7). The electronic structures of T_8H_8, T_8H_7Ph, $T_8H_6Ph_2$, $T_8H_7(C_6H_4\text{-4-Ph})$ and T_8Ph_8 and crystalline T_8H_8 and $T_8Ph_8\cdot$acetone have been calculated and show that, in contrast to T_8H_8 which has a HOMO localized at the oxygen lone pairs, the aryl-substituted compounds have HOMO and LUMO orbitals localized on the aromatic rings [347, 348]. The absorption and emission spectra for POS derivatives $T_8(CH{=}CHC_6H_4\text{-}4\text{-}CH{=}CHC_6H_4\text{-}4\text{-}R)_8$ (R = H, Me, OMe or NH_2) have been compared with simple stilbene analogues and partial cage analogues and show that the emissions for the full cage derivatives show a red shift of 60 to 100 nm. This significant effect is thought to be due to interaction of the LUMO, centred in the cage with contributions from all the cage atoms, with the stilbene substituent π^* orbitals. As the T_8 cage is symmetrical, a three-dimensional conjugation might be implied, where the cage does not act as a traditional silica-like insulator [335, 349]. DFT (density functional theory) calculations on T_8H_8 and its derivatives substituted with either 4-carbazolylphenyl (electron-donating) or 4-cyanophenyl (electron-withdrawing) groups, with the cage being either empty or containing N_2, show that the HOMO and LUMO can be independently changed and tuned, and that the T_8 cage is partially conjugated, serving as an electron acceptor [350].

The alcohol, $T_8[CH{=}CHC_6H_4\text{-}4\text{-}C_6H_3\text{-}3,5(OH)_2]_8$, has a photoluminescent quantum yield of 10% at saturation, and a bathochromic emission shift when compared to $CH_2{=}CHC_6H_4\text{-}4\text{-}C_6H_3\text{-}3,5(OH)_2$ [81], while the UV spectrum of the unusual silyl-substituted POS, $T_8(SiMe_2\text{-}t\text{-}Bu)_8$, exhibits a significant bathochromic shift with a lowest transition energy absorption at 285 nm compared with simple R_3SiSiR_3 (R = alkyl) species which have corresponding absorptions at ca. 200 nm [251]. Poly-pyrene species derived from Heck-coupling of $T_8(CH{=}CH_2)_8$ with 1-bromopyrene show solution photoluminescence bathochromic shifts of ca. 50 nm from pyrene itself, probably due to the increased conjugation due to the Si-$CH{=}CH_2$ fragment [341]. Aryl-substituted T_8 derivatives such as **2** – **5** have been shown by photoluminescence excitation spectra, and other techniques, to be quantum dot materials in which the organic arms are isolated by the POS cage. The spectra indicate that the emissive centers are not the cage itself, but the substituents on the cage. The compounds are soluble and the photoluminescent yields are enhanced when the conjugated arms are attached to the POS core [124, 351]. The related aryl compounds $T_8Ph_n(C_6H_4Ar)_{8-n}$ (n = 2, 3, 4 or 5; Ar = Ph, biphenyl, naphthyl, 9,9-dimethylfluorenyl etc.) have UV-VIS and photoluminescence spectra similar to those for the simple untethered aromatic compound, with a small bathochromic shift of about 20-30 nm [352].

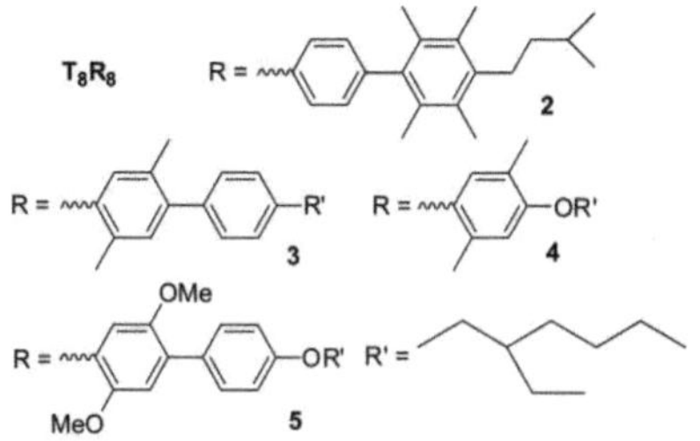

Electronic spectroscopy is also being increasingly used to probe the properties of polymeric materials containing POS groups, in particular to investigate the changes in materials induced by the presence of T_8 cage derivatives. Polyfluorenes having one or two $T_8(c\text{-}C_5H_9)_7$ substituents per fluorene unit emit blue light with high quantum efficiency in solution [353, 354], but green emission is seen for those with $T_8(c\text{-}C_5H_9)_7$ groups just as termini [355, 356], and photoluminescence studies show that polyfluorene derivatives containing $T_8(c\text{-}C_5H_9)_7OSiMe_2$- substituents may have reduced fluorescence quenching leading to increased fluorescence quantum yields [357] (see also Section 7.2, Chapter 7). As the POS content is increased in poly(phenylene vinylene)s with $T_8(c\text{-}C_5H_9)_7OSiMe_2(CH_2)_3O$- substituents their UV-vis absorption and photoluminescent emission maxima have increasing hypsochromic shifts [358], and introduction of $T_8[OSiMe_2(CH_2)_2C_6H_4\text{-}4\text{-}CH_2\text{–}]_8$ units as crosslinkers into poly(phenylene vinylene) materials has been shown to afford highly efficient green emitting materials [359].

2.14 Structural Studies of POS Compounds

2.14.1 Single Crystal X-Ray Diffraction Studies

Many T_6, T_8 and T_{10} POS derivatives have had their structures determined by single crystal X-ray diffraction and they have been reviewed in detail previously [119, 120]. Structural data for selected T_6R_6, T_8R_8 and T_8R_7R', and $T_{10}R_{10}$ compounds are given in Table 2.20. The Si-O bond lengths mostly fall in the range 1.60-1.63 Å, as expected from values in other siloxanes [360]. However, the Si-O-Si angles within a single T_8 or T_{10} POS cage may vary significantly, for example from 136.35-172.13° in $T_8(OSnMe_3)_8{\cdot}4H_2O$ [180, 361] and 140.76-160.68° in T_8Cy_8 [362]. Despite the wide variations in individual angles the average Si-O-Si angles for each compound in Table 2.20 fall in the narrow range of ca. 147.5–150.8°. The variations in angles arise from the well known flexibility of the siloxane linkage coupled with the ability of more flexible substituents to deform and adopt conformations that minimize the voids in the lattice that would form if they were to point ideally towards the vertices of a cube. The T_8 POS species with larger, flexible substituents form either a disc-like structure when substituents on two opposite faces of the core close up towards each other, or a rod-like structure when the substituents around a pair of opposite faces close up (see also Section 7.1, Chapter 7). The bond angles for the smaller, more rigid T_6R_6 compounds, are generally less varied but the structure of D_{2d} symmetric $T_{12}H_{12}$ shows Si-O-Si angles varying by up to ca. 11° [363]. Poor quality X-ray data may be obtained for T_8 POS compounds with flexible substituents, since they are prone to disorder within the lattice [83].

The potential of POS derivatives as nano-sized building blocks for materials construction has been widely described, and derives from the high symmetry of the cage, and the size of the cage, with Si···Si distances being ca. 3.11, 4.40 and 5.39 Å for distances along an edge, across a face, and across the body diagonal of a typical cage. These cage dimensions vary little over a wide range of POS derivatives, and their values have been tabulated and discussed previously [362]. An interesting sub-group of the structures in Table 2.20 are those containing an endohedral fluoride ion, in which the Si-O-Si angles are lower than those for the empty cages and close to 141.2°, see for example, Figure 1. Attraction between the fluoride and all eight Si atoms causes the cage to contract slightly and repulsions between the oxygen atoms and the fluoride cause the Si-O-Si angle to change [222]. Analyses of the unit cell data for several alkyl-substituted T_8 derivatives crystallising in rhombohedral space groups have been tabulated, and show that packing of the molecules within the hexagonal planes is closer than the separation between the planes. A c/a ratio of ca. 1.03 in T_8R_8 and T_8R_7R' compounds for a range of alkyl substituents of varying size and nature is found, implying a related molecular arrangement in these structures. However, the parent silane T_8H_8 has a c/a ratio of 1.68, and aryl-substituted T_8 compounds do not fit this pattern [364]. The solid state structure of $T_8H_7$$n$-$C_6H_{13}$ exhibits bilayers, owing to a preference for POS-POS and hydrocarbon-hydrocarbon interactions rather than hydrocarbon-POS interactions [311]. The preference for POS-POS interactions is often seen in more complicated materials (see WAXS and SAXS studies in Section 2.14.3) and is also shown to occur in molecular simulations of blends of $T_8(c\text{-}C_5H_9)_8$ and oligoethylene [365].

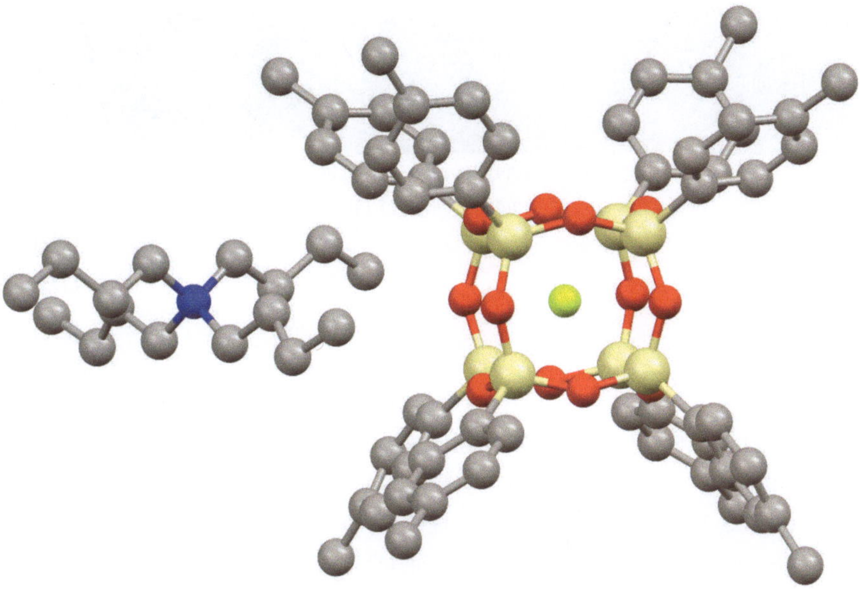

Fig. 1 The solid state structure of the endohedral POS salt $[Bu_4N][T_8(p\text{-tolyl})_8F]$, hydrogen atoms omitted for clarity [222]. (Figure drawn using data from the Cambridge Crystallographic Database, structure 229789)

Table 2.20 Selected structural data for T_nR_n, and related compounds[a]

Compound	Si-O range (Å)	Si-O-Si range (°) Si_4O_4 rings	Si-O-Si range (°) Si_3O_3/ Si_4O_4 rings (T_6) Si_5O_5 rings (T_{10})	Refs.
		n = 6		
T_6H_6[b]	1.634-1.643	139	131	[366]
T_6Cl_6[b]	-	136.6	126.9	[367]
T_6Me_6	1.633-1.642	141.278	131.321	[368]
$T_6(t\text{-Bu})_6$	1.623(2)-1.642(2)	137.1(1)-141.1(1)	129.7(1)-131.2(1)	[101]
T_6Cy_6	1.625(6)-1.649(6)	139.3(4)-144.7(5)	128.8(4)-130.8(5)	[9]
$T_6(2,4,6\text{-}i\text{-}Pr_3C_6H_2)_6$[c]	1.622(2)-1.646(4)	138.3(3)-141.7(3)	131.8(2)-134.0(2)	[98]
$T_6(OSiMe_3)_6$	1.56(1)-1.63(1)	136.8(8)-138.6(6)	128.9(8)-131.7(9)	[149, 369]
$T_6(OSiMe_3)_6$[d]	1.633-1.640	141.8(14)	129.2(7)	[233]
$Na_3YT_6(O^-)_6$	1.623(2)-1.651(1)	132.8(2)-135.8(2)	130.3(2)-134.1(1)	[370]
		n = 8		
T_8H_8	1.6168(11)-1.6195(7)	147.49(6)-147.60(7)		[371]
T_8H_8[d]	1.6143(3)	147.9(2)		[372]

Compound	Si-O range (Å)	Si-O-Si range (°) Si_4O_4 rings	Si-O-Si range (°) Si_3O_3/ Si_4O_4 rings (T_6) Si_5O_5 rings (T_{10})	Refs.
T_8Cl_8	1.595(4)-1.610(4)	148.0(3)-148.8(3)		[373]
T_8Me_8	1.610(2)-1.617(2)	148.9(1)-149.6(5)		[374]
T_8Me_8[d]	1.6174(5)	148.9(2)		[372]
$T_8(CH{=}CH_2)_8$	1.596(6)-1.617(7)	150.0(5) and 150.5(5)		[375]
$T_8(CH{=}CH)_8$[e]	1.6184(13)-1.6295(13)	138.67(9)-143.06(9)		[222]
$T_8[(CH_2)_2PMe_2]_8$	1.588(2)-1.598(2)	145.44(17)-152.37(17)		[89]
$T_8[(CH_2)_3C_6H_4\text{-}4\text{-}OMe]_8$	1.604(5)-1.627(4)	145.6(3)-156.6(3)		[362]
$T_8[(CH_2)_3Br]_8$	1.597(7)-1.633(7)	141.7(4)-153.8(5)		[25]
$T_8(CH_2CH_2CF_3)_8\cdot THF$	1.6164(10)-1.6232(10)	144.29(7)-154.95(8)		[32]
$T_8(CH_2CH{=}CH_2)_8$	1.607-1.618	145.0-152.7		[88]
$T_8(CH_2CHMe_2)_8$	1.611(5)-1.622(5)	143.7(3)-152.3(3)		[11]
$T_8(CH{=}CHPh)_8$	1.598(4)-1.627(3)	142.52(17)-153.15(19)		[77]
$\{Si_8O_{12}[(CH_2)_3NH_3]_8\}$-$[ZnCl_4]_{2.8}Cl_{2.4}$	1.605(6)-1.636(7)	144.0(5)-157.2(5)		[73]
$T_8[(CH_2)_2(CF_2)_7CF_3]_8$	1.614(2)-1.626(2)	144.30(15)-149.94(14)		[22]
$T_8(c\text{-}C_5H_9)_8$	1.615(3)-1.630(3)	140.5(2)-156.04(19)		[11]
T_8Cy_8	1.539(2)-1.6705(17)	140.76(8)-160.68(11)		[362]
$T_8Ph_8\cdot Me_2CO$	1.607-1.617	144.63-151.53		[376]
T_8Ph_8[e]	1.6199(15)-1.6294(16)	138.57(10)-143.87(10)		[221]
T_8Ph_8[d]	1.634(15)-1.645(19)	147.5(45)-149.8(24)		[245]
$T_8(C_6H_4\text{-}4\text{-}Me)_8$	1.610(2)-1.629(4)	144.0(2)-151.60(17)		[120]
$T_8(C_6H_4\text{-}4\text{-}Me)_8$[e]	1.6234(19)-1.628(2)	140.50(13)-142.47(13)		[222]
$T_8(C_6H_4\text{-}4\text{-}CH_2Cl)_8$	1.600(4)-1.634(4)	139.0(3)-164.3(3)		[120]
$T_8(C_6H_4\text{-}4\text{-}I)_8\cdot EtOAc$	1.603(4)-1.616(4)	144.7(4)-152.0(3)		[126]
$T_8(OMe)_8$	1.592-1.609	145.42-151.92		[139]
$T_8[OSi(CH{=}CH_2)_3]_8$	1.588(3)-1.617(3)	142.8(2)-151.8(2)		[377]
$T_8(OSnMe_3)_8$	1.580(3)-1.629(4)	148.8(2)-161.2(2)		[180, 361]
$T_8(OSnMe_3)_8\cdot 4H_2O$	1.5872(16)-1.6277(17)	136.35(10)-172.13(12)		[180, 361]
$T_8(SiMe_2\text{-}t\text{-}Bu)_8$	1.620(7)-1.630(6)	148.5(4)-152.2(4)		[251]
$T_8(Co(CO)_4)_8$	1.610(4)-1.637(4)	147.6(2)-154.1(3)		[181]
$[T_8(c\text{-}C_5H_9)_7O]_2Mn(TMEDA)_2$	1.608(2)-1.641(3)	143.95(17)-152.58(17)		[188]

Compound	Si-O range (Å)	Si-O-Si range (°) Si_4O_4 rings	Si-O-Si range (°) Si_3O_3/ Si_4O_4 rings (T_6) Si_5O_5 rings (T_{10})	Refs.
T_8Cy_7(fluorenyl)	1.596(3)-1.632(3)	142.8(2)-161.3(2)		[34]
$(T_8Cy_7)_2O$	1.6060(15)-1.6314(14)	137.88(9)-168.09(11)		[146]
		n = 10		
$T_{10}H_{10}$	1.591(3)-1.615(3)	147.8(2)-150.2(2)	152.8(2)-157.9(2)	[378]
$T_{10}H_{10}$ [e]	1.616(2) and 1.622(3)	153.9(7)	155.0(5)	[379]
$T_{10}Me_{10}$	1.582-1.626	147.3-152.7	148.0-160.5	[107]
$T_{10}(OSiMe_3)_{10}$	1.574-1.609(8)	149.25-154.99	149.31-157.89	[177]
$T_{10}H_9Ph$	1.571(4)-1.620(3)	145.2(2)-150.9(2)	150.5(2)-162.7(3)	[210]

[a] Unless otherwise stated, all data are from single crystal X-ray crystallography studies. [b] Calculated structure. [c] For the right-handed enantiomer. [d] Gas phase electron diffraction structure. [e] Anionic endohedral fluoride with $[NBu_4]^+$ cation.

2.14.2 Structures Derived from Computational and Gas-Phase Electron Diffraction Studies

The structure and properties of the parent hydride, T_8H_8, have been calculated and the results described in detail in several reviews [119, 120, 289, 372, 380-382]. In the gas phase, an electron diffraction study of T_8H_8 shows it to have O_h symmetry, an Si-O-Si angle of 147.9(2)°, and Si-O and Si-H distances of 1.6141(3) and 1.454(8) Å respectively, in agreement with MP2/6-311++G(3df,3pd) calculated values of 1.629, 1.453 Å and 147.8° [372]. MP2/6-311++G(3df,3pd) calculations on T_8Me_8 give Si-O and Si-C values of 1.632 and 1.837 Å, respectively, and an Si-O-Si angle of 148.6°, which agree well with the corresponding experimental values of 1.6174(5) and 1.829(3) Å and 148.9(2)° determined by gas-phase electron diffraction methods [372]. The calculated structure [245, 347] of T_8Ph_8 agrees well with that determined by gas-phase electron diffraction methods [245]. A comparison of force fields shows that the charge-transfer reactive, and universal force fields together with COMPASS and Hybrid-COMPASS force fields all give agreement between theory and experiment for crystal packing data for both T_8H_8 and T_8Me_8, while CTR and Hybrid-COMPASS methods give good agreement with experimental data for the melting point of T_8H_8 [381]. An ab initio study of T_8R_8 (R = H, Me, SiH_3, F or Li) shows that more electropositive substituents increase the Si-O-Si angle and the Si-O bond length, while electronegative substituents do the reverse when compared to T_8H_8. The window in any given side of the POS cage is important when considering the potential of POS cages for storing hydrogen, and the same study shows that the larger $T_{14}H_{14}$ (D_{3h} isomer) and $T_{16}H_{16}$ (D_{4d} isomer) cages can accommodate two or three H_2 molecules [383]. A detailed molecular mechanics and dynamics study of $T_8(i\text{-Bu})_8$ in the solid state has been used to describe its phase change at ca. 319 K in terms of a dynamic disordering of the cage substituents [288].

Hartree-Fock and MP2 calculations on T_8H_7R compounds (R = Et, *n*-Pr, *n*-Bu, Cy) show that small changes occur on substituting one H for an alkyl group on one of the corners of the cage, and that combining force fields for silsesquioxanes with those for hydrocarbons may obviate the need for force fields for individual POS compounds [380, 382]. Molecular simulations of $T_8H_7(n\text{-}C_9H_{19})$ show that it forms cylindrical structures in solvents that are good for the POS cage but poor for the alkyl group, but lamellar structures if the solvent for the cage is poor [384, 385]. Related calculations on POS cubes with four substituents around a single face indicate that hexagonally arranged cylinders are the preferred structure [385]. Molecular dynamics calculations on molecules in which a hydrocarbon chain connects two T_8Me_7 fragments show that if the chain is long they are able to move independently, but if short they move together through a solution [386].

Computational studies have been carried out on several T_8 cages containing endohedral fluoride ions, to complement the single crystal X-ray diffraction studies (see Table 2.20), with MP2 calculations of $[F@T_8R_8]^-$ ions (R = Me, CF_3, CH=CH_2, $CH_2CH_2CF_3$) showing binding energies for the fluoride from 293 to 1130 kJ mol^{-1} [223]. Calculations at the B3LYP/6-31G(d) and B3LYP/6-311++G(d,p) levels on the binding of cations (Li^+, Na^+, K^+), anions (F^-, Cl^-, Br^-), and noble gasses (He, Ne, Ar) to T_8H_8 show that endohedral binding for the anions is preferred, as seen experimentally, but that exohedral binding is preferred for the cationic and neutral species [387]. DFT methods show that for exohedral and endohedral transition metal complexes with T_8H_8 the insertion of a metal into the cage reduces the HOMO-LUMO gap from 8.1 eV for the empty cage to between 1.2 and 4.96 eV for the endohedral species, and that the endohedral complexes $M@T_8H_8$ for M = Cr, Fe, Co, Ni, Cu, Ru and Os are more stable than the separate components [388]. However, attempts to prepare molecular T_8 cages containing endohedral cations have, so far, been unsuccessful [73]. DFT calculations on endohedral complexes of $T_{10}H_{10}$ show that endohedral binding is preferred over the separated species for complexes with Li^+, Na^+, F^-, Cl^- and Br^- [389].

2.14.3 X-ray Diffraction Studies on Powders, Thin Films, etc.

X-ray diffraction data have been used widely to determine the degree of dispersion of POS species within a polymer or composite, and the degree of crystallinity of any POS domains present by noting common features attributable to the POS species in diffraction patterns recorded both before and after processing POS monomers. Much of this work has been reviewed [390-402]. Diffraction peaks typical of POS cubes occur at 2θ values of 8.3, 18.9 and 24.4° corresponding to *d*-spacings of ca. 11, 5, and 3 Å, and can be attributed to the overall dimensions of the POS molecule, the body diagonal of the POS cage, and the distance between opposite faces of the T_8 cage respectively [309]. Care needs to be exercised in comparing diffraction data, since the ideal cubic arrangement of the eight Si atoms in a T_8 cage is often distorted (see section 2.14.1) causing the body diagonal, face diagonal, and face-to-face distances across the POS cage to vary significantly. Several widely used T_8 POS compounds such as $T_8(CH{=}CH_2)_8$ [375] and $T_8(OSiMe_2H)_8$ [177] have rhombohedral structures, and the rhombohedral nature of many T_8 POS crystal structures has been commented on [364]. However, the majority of POS structures are in fact triclinic and monoclinic, and a more comprehensive discussion of the crystal structures has been published recently [120].

2.14.3.1 T_8R_8 Compounds

The powder X-ray diffraction pattern for T_8H_8 shows characteristic peaks at 2θ = 7.9, 8.4, 11.04, 18.9 and 24.13° [403]; while the pattern for T_8Me_8 is indicative of a rhombohedral structure [19, 103, 404], and has a characteristic peak at ca. 2θ = 10.8° [272, 405] which is also present in diffraction patterns from POS polyethylene nanocomposites containing >1 wt % POS, indicating that the T_8Me_8 crystallizes when dispersed in the polyethylene [404, 406, 407]. The WAXD pattern of $T_8(i\text{-Bu})_8$ shows several sharp peaks, e.g. at ca. 2θ =7.96, 8.84, and 10.86° corresponding to *d*-spacings of 11.1, 10.0 and 8.1 Å, which are also seen in blends with poly(methyl methacrylate), with phenolic resin, with styrene-maleic anhydride polymers, with poly(L-lactide), and with polyethylene/ethylene-vinylacetate copolymers [314, 406, 408-411] in blends with silicone rubber [412], and also in POS- epoxy-cyanate composites, in all cases indicative of POS phase segregation [413]. A variable temperature powder XRD study of $T_8(i\text{-Bu})_8$ showed a phase change to occur at ca. 319 K, from a low temperature $P\bar{1}$ to a high temperature $R\bar{3}m$ lattice [288]. The XRD pattern for $T_8(CH{=}CH_2)_8$ shows several sharp peaks at 2θ = 9.8, 20.1 and 29.9° [414-416], while the XRD pattern of $T_8[(CH_2)_3NH_2]_8$ shows several sharp peaks indicative of crystallinity and consistent with a rhombohedral unit cell, with a = 11.57 Å and α = 95.1° [417-420], and the WAXD pattern for T_8Ph_8 shows sharp diffraction peaks again indicative of crystallinity [278].

The XRD pattern for $T_8(OSiMe_2H)_8$ shows sharp peaks at ca. 2θ = 8.3, 18.9 and 24.4° corresponding to *d*-spacings of 11, 5, and 3 Å attributable to the distance between opposite Si_4O_4 faces of the POS core, the POS cage diagonal distance, and the size of the POS molecule respectively [151, 167, 309, 421]. WAXD studies of PEG derivatives of $T_8(OSiMe_2H)_8$ provide evidence for the presence of small disordered POS domains [421]. Conversely, although the XRD pattern for $T_8[OSiMe_2(CH_2)_3OH]_8$ is indicative of crystallinity, upon incorporation into a POS-polyimide film the crystallinity is lost [249, 422].

2.14.3.2 T_8R_7R' Compounds

As for T_8R_8 compounds, diffraction studies on T_8R_7R' compounds often concentrate on whether domains of POS cages are formed in a polymer or composite material by comparison of patterns with those from simple molecular species. The peaks in the powder X-ray diffraction pattern of $T_8(i\text{-Bu})_7(CH_2)_2C_6H_4CH_2Cl$ at 2θ = 8.3 and 11.0° are characteristic of a hexagonal POS structure and are retained when the POS compound is tethered to a dimethylimidazolium center and also when it is subsequently exchanged into a montmorillonite clay, clearly demonstrating the propensity of POS species to form crystalline domains [423]. POS domains from 1.3 to 2.7 nm across, corresponding to one to two POS units per domain are indicated from the WAXD patterns of a range of ethylene-propylene-POS polymers containing pendant $T_8(i\text{-Bu})_7$ groups [424]. Diffraction patterns also show that the mixing time has an effect on the crystallinity of the POS, for example, mixing $T_8(i\text{-Bu})_7(CH_2)_3NH$ with styrene-maleic anhydride copolymers gives a hybrid system showing the precursor diffraction pattern clearly after 2 min, but after mixing for 20 min no sharp peaks are seen [314]. The WAXD pattern of $T_8(i\text{-Bu})_7(CH_2)_3OC({=}O)$-$C({=}CH_2)Me$ shows sharp peaks at 2θ = 9.5, 12.9 and 22.4° corresponding to *d*-spacings of 10.8, 8.0 and 4.6 Å [425]; but these are largely lost on incorporation of the compound into dimethacrylate [426, 427], isobornyl methacrylate and diethylene glycol dimethacrylate networks [428], although at high loadings in copolymers with poly(octafluoropentyl acrylate) a peak at ca. 2θ = 8°, indicative of POS aggregation, is prominent [429]. Similarly, the sharp diffraction peaks associated with $T_8(i\text{-Bu})_7(CH_2)_3OH$ are lost on its incorporation into polylactide nanocomposites [430].

The WAXD pattern for $T_8(i\text{-Bu})_7OSiMe_2(CH_2)_3OCH_2CEt(CH_2OH)_2$ has significant reflections at ca. 2θ = 7.9, 10.52 and 18.6° corresponding to *d*-spacing of 11.2, 8.3 and 4.8 Å respectively and a rhombohedral or equivalent hexagonal unit cell. Some polycaprolactone networks containing this POS grouping also show these diffraction peaks indicating that the cage survives the polymerization reaction [431], and the 2θ = 7.9° peak is observed in $T_8(i\text{-Bu})_7OSiMe_2(CH_2)_3OCEt(CH_2OH)_2$– polyurethane copolymers and is attributable to POS crystals (of rhombohedral unit cell; *a* = 11.0 Å, α = 104°) in the polymer, rather than molecular level dispersion of the cages [432-435]. The XRD pattern for $[T_8(i\text{-Bu})_7(CH_2)_3NH_3]Cl$ shows reflections at 2θ = 7.9 and 8.8° [436], and for $T_8(i\text{-Bu})_7(CH_2)_3NH_2$, the highest intensity peak is seen at 2θ = 8.3° [437].

A range of compounds containing the $T_8(c\text{-}C_5H_9)_7$ group have been studied by diffraction methods, for example, the WAXD pattern for $T_8(c\text{-}C_5H_9)_7(CH_2)_2C_6H_4$-4-$CH_2Cl$ shows peaks at ca. 2θ = 8.3, 11.3, 12.0, 19.1 and 25.9° corresponding to *d*-spacings of 10.5, 7.2, 6.9, 4.6 and 3.3 Å respectively, the 10.5 Å spacing being due to the POS molecule [438, 439], and the pattern for $T_8(c\text{-}C_5H_9)_7(CH_2)_3CN$ shows sharp peaks including one at 2θ = 8.2° corresponding to a *d*-spacing of 10.2 Å which is also seen in cyanate ester composite materials derived from it [440]. The XRD pattern for $T_8(c\text{-}C_5H_9)_7(CH_2)_3OC({=}O)C({=}CH_2)Me$ shows peaks at 2θ = 8.2, 11.1, 12.3, 19.2 and 24.9° corresponding to *d*-spacings of 10.7, 7.9, 7.1, 4.6 and

3.6 Å respectively. The 10.7 Å peak is attributable to the size of the POS molecule, and the remaining peaks to a rhombohedral lattice [441-443]. WAXD has been used to show that the POS additives $T_8(c\text{-}C_5H_9)_7R$ [R = $(CH_2)_2(CF_2)_7CF_3$ or $CH_2CH(OH)CH_2OH$] in polystyrene segregate at the surface, forming crystallites of 10 and 19 nm sizes respectively [444].

Diffraction studies of materials containing T_8Cy_7 or T_8Ph_7 groups have also been carried out, for example, $T_8Cy_7OSiMe_2(CH_2)_3NCO$, and polymers derived from its reaction with poly(ethylene glycol), show strong reflections at 2θ = 7.8, 10.62 and 18.20° and corresponding rhombohedral unit cell parameters of a = 11.57 Å and α = 95.5° associated with a crystalline POS phases, in both the monomeric compound and the polymers [445]. The WAXS pattern of $T_8Ph_7(CH_2)_3OC(O)C(=CH_2)Me$ shows sharp peaks that are retained in the patterns derived from copolymers with poly(octafluoropentyl acrylate), again indicative of POS aggregation [429]. The WAXD patterns for a range of ethylene-propylene-POS polymers containing T_8Ph_7 groups show POS domains from 1.3 to 9.5 nm across, suggesting from three to seven T_8Ph_7 units per domain [424].

2.15 TGA, DSC and Related Studies of POS Compounds

2.15.1 T_8R_8 Compounds (R = H, Alkyl, Vinyl, Aryl or Silyl Derivatives)

The thermal stability and decomposition pathways of many simple POS compounds and materials containing POS components have been studied and have been reviewed [120], and this section concentrates only on studies of simple molecular POS species. TGA studies of low molecular weight compounds such as T_8R_8 (R = H, Me, *i*-Bu, *i*-Oct, Ph) may be complicated by their ready sublimation under nitrogen, and occasional partial sublimation in air. Thus, for T_8H_8 and T_8Me_8 incomplete evaporation has been reported to occur both in air and in a N_2 atmosphere, but for $T_8(i\text{-Bu})_8$ and $T_8(i\text{-Oct})_8$ near complete evaporation occurs in an inert atmosphere, at approximately 265 °C for $T_8(i\text{-Bu})_8$ [437], whereas oxidation occurs in air to give silica. However, TGA of T_8H_8 has also been reported to show rapid and near-complete mass loss at temperatures over ca. 200 °C due to sublimation [446], and TGA of T_8Me_8 and T_8Et_8 in N_2 has also been reported to show almost 100% evaporation at ca. 250-260 °C [405, 447]. POS compounds such as T_8R_8 (R = C_8H_{17} or $C_{18}H_{37}$) containing longer alkyl chains are thermally quite stable, showing 10% decomposition only at 343 and 366 °C respectively [448].

The DSC trace of $T_8(i\text{-Bu})_8$ shows two endotherms, a sharp one at 60 °C and a broader one at ca. 265 °C probably due to the melting point, while TGA shows that maximum evaporation occurs at 285 °C, and that complete mass loss occurs at ca. 375 °C due to evaporation of the sample in N_2, but in air a ceramic residue of 26% is formed as oxidation competes with evaporation [408, 449, 450]. The alkyl substituted cages T_8R_8 [R = $(CH_2)_5Br$, CH_2CH_2Cy, and $CH_2CH_2OCH_2CH_2Cl$] lose no mass on heating to 381, 442, and 338 °C respectively [51, 451], but for R = CH_2CH_2Cy and $(CH_2)_5Br$, significant mass loss is seen at 460 and 380 °C respectively by TGA [326]. The TGA analysis of the fluorinated derivatives, $T_8(CH_2CH_2R)_8$ [R = CF_3, $(CF_2)_3CF_3$, $(CF_2)_5CF_3$ or $(CF_2)_7CF_3$], in air shows that they all evaporate, the $(CF_2)_7CF_3$ derivative subliming at over 300 °C [452] and being stable up to about 350 °C in N_2 [453]. The TGA trace of $T_8(CH{=}CH_2)_8$ suggests stability to ca. 280 °C [454, 455], but the residual mass is significantly lower than what would be expected for complete conversion to silica, and so sublimation also appears to occur [456].

The TGA analysis of $T_8[(CH_2)_3NH_2]_8$ has led to a range of results, presumably due to its hygroscopic nature and sensitivity to back-biting reactions. Thus, degradation at between 425-500 °C of the POS cage has been reported [417], as has decomposition starting at ca. 320 °C [418], and rapid initial mass loss occurring at ca. 350 °C followed by steady mass loss up to 700 °C [419]. DSC studies of $T_8[(CH_2)_3NH_2]_8$ indicate a melting point of 196.6 °C and decomposition onset at 242 °C, i.e. lower than determined by TGA [18]. TGA of $\{T_8[(CH_2)_3NH_3]_8\}Cl_8$ shows a two-step decomposition corresponding to initial breakdown of the alkylammonium chloride group between 305 and 420 °C and a second step at 420-650 °C due to degradation of the propyl chains [275].

TGA of the aryl-POS compounds T_8R_8 (R = Ph, C_6H_4-2-Me, C_6H_4-3-Me, C_6H_4-4-Me) show melting temperatures of 472, 385, 424 and 407 °C, respectively, and onsets for decomposition at 486, 435, 431 and 413 °C, respectively [457]. The phenyl derivative is also reported to be stable to about 480 °C by TGA/DTA and to give SiO_2 at higher temperatures, but the dinitrated derivative $T_8[C_6H_3(NO_2)_2]_8$ detonated at 420 °C [122, 458]. The derivative thermogravimetry traces of T_8Ph_8 and $T_8(C_6H_4NO_2)_8$ have, however, both been reported to show rapid decomposition, at 420 and 380 °C respectively, but that $T_8(C_6H_4NH_2)_8$ undergoes a slow decomposition from 300 to 650 °C [459].

2.15.2 T_8R_8 Compounds (R = Siloxy Derivatives)

As for the case of simple alkyl-substituted POS compounds, TGA studies of simple siloxy derivatives are complicated by sublimation occurring as well as decomposition. TGA of $T_8(OSiMe_2H)_8$, $T_8[OSiMe_2(CH_2)_3CN]_8$ and $T_8[OSiMe_2(CH_2)_3Cl]_8$ shows a tendency for sublimation, and formation of SiO_2 at high temperatures [163, 460], but under nitrogen, $T_8(OSiMe_2H)_8$ sublimes between 188 and 280 °C [163, 460, 461]. The DSC traces of $T_8(OSiMe_2H)_8$ and $T_8[OSiMe_2(CH_2)_2C_6H_4\text{-4-OAc}]_8$ show the latter to have a single glass transition at -15 °C, but that $T_8(OSiMe_2H)_8$ shows no such transition [166, 462]. TGA traces for $T_8(OSiMe_2C_6H_4\text{-4-R})_8$ (R = Me, CBr_3, or

CO_2H) show 5 % mass loss at 370, 164 and 180 °C respectively and a residual mass associated, as with many POS compounds, for SiO_2 [154]. Methacrylate and epoxide POS monomers are common components used for incorporation into polymer materials and have good thermal stability. Thus, TGA analysis of $T_8(OSiMe_2R)_8$ [R = $(CH_2)_3(OCH_2CH_2CH_2)_2OC(=O)C(=CH_2)Me$ or $CH=CH_2CH_2OC(=O)C(=CH_2)$-Me] reveal 5 % mass loss at ca. 257 and 302 °C respectively [463], and TGA of $T_8[OSiMe_2(CH_2)_3OCH_2CH(O)CH_2]_8$ shows decomposition under nitrogen starting at ca. 350 °C [464]. TGA analysis of the macromonomers $T_8[OSiMe_2(CH_2)_3(OCH_2CH_2)_nOC(=O)CMe(=CH_2)]_8$ and $T_8[OSiMe_2(CH_2)_3(OCH_2CH_2)_nOH]_8$ (n = 2, 3, 4 or 6) shows that a two step decomposition takes place, that the expected ceramic yield of SiO_2 is formed, and that the T_g values for the methacrylate derivatives are ca. 10-20 °C higher than those for the analogous ethylene glycol derivatives [152].

2.15.3 T_8R_7R' Compounds

The DSC trace of $T_8(i\text{-Bu})_7(CH_2)_3OCH_2CH(O)CH_2$ shows melting peaks at ca. 112 and 133 °C with a total melting heat of 26.1 J g^{-1} and crystallization peaks at ca. 119 and 139 °C, with a total crystallization heat of 15.1 J g^{-1}, presumably indicative of two crystalline phases [465, 466]. In an inert atmosphere, TGA of $T_8(i\text{-Bu})_7(CH_2)_3OCH_2CH(O)CH_2$ gives a char yield of 24.4%, and in oxygen a char yield of 42.7%, the onset of decomposition being slightly less than 300 °C in both cases [465]. The TGA of $T_8(i\text{-Bu})_7(CH_2)_3OCH_2CH(O)CH_2$ has also been reported to show significant decomposition at 220 °C under argon [467]. TGA of $T_8(i\text{-Bu})_7(CH_2)_3OC(=O)C(=CH_2)Me$ shows 5% decomposition in air and in nitrogen at 265 and 282 °C respectively [425], while DSC shows that it melts at 110.4 °C without decomposing, and that at 140 °C it exhibits an exotherm attributable to self-polymerization [468]. TGA of $T_8(i\text{-Bu})_7(CH_2)_3NH_2$ shows an onset of decomposition in Ar and in O_2 at ca. 277 and 265 °C respectively [469], with rapid mass loss, possibly due to evaporation, at 293 °C [437]. Most studies on T_8R_7R'compounds are for R = *i*-Bu, but the T_g values for $T_8R_7(CH_2)_3OC(=O)C(=CH_2)Me$ compounds have been found to be 251 °C and 350 °C for R = Et and $c\text{-}C_5H_9$ respectively [470]. TGA of $T_8(c\text{-}C_5H_9)_7OSiMe_2(CH_2)_3NCO$ in air or in nitrogen shows decomposition starting at ca. 250 °C [471], and decomposition of $T_8Cy_7(CH_2)_3OC(=O)C(Me)=CH_2$ occurs in two stages, the first at 350 °C and the second starting at 533 °C [315].

2.16 Microscopy Studies of T_8 POS Compounds

2.16.1 T_8R_8 Compounds

STM (scanning tunnelling microscopy) has been used to provide images of individual molecules of T_8H_8 chemisorbed on a Si(100)-2x1 surface. These indicate that the preferred mode of chemisorption is via one of the Si vertices rather than through an arrangement in which the cage opens up along one edge to bind via both Si and O [472]. However, when deposited on a Si(111)-7x7 surface, the cage appears to break open along one edge to bind as a "cracked cluster", the difference in reactivity between the two surfaces possibly being due to the proximity of Si surface diradicals [473]. STM images of T_8H_8 on highly oriented pyrolytic graphite show that it can form two different types of monolayer, one in which the face of the cage is in contact with the graphite and a second in which an edge contacts the surface [348]. SEM and TEM have been used to monitor the progress of siloxane formation upon hydrolysis of $MeSi(OEt)_3$ and $EtSi(OEt)_3$ and show that self assembly of the final cubic crystalline material occurs via initial spherical particle formation, followed by chains of spherical particles and bundles of rod-like structures, resulting in regular cubic crystals smaller than ca. 5 μm [103, 405, 447]. SEM of T_8Me_8 prepared from swollen poly(2-hydroxyethylmethacrylate) shows that cubic crystals having edges up to 20-30 μm long may be formed [19]. Brewster angle microscopy of $T_8(i\text{-Bu})_8$ shows that it forms aggregates at the air/water interface at all concentrations [474], and AFM of $T_8(i\text{-Bu})_8$ blended in phenolic resins shows that evenly dispersed POS particles occur at 20% POS content, while at 60% POS content macrophase separation occurs [409]. SEM of poly(ε-caprolactam) blended with T_8Ph_8 shows that the POS molecules adhere poorly to the polymer and instead form POS domains of 1-20 μm across [475]. SEM of $\{T_8[(CH_2)_3NH_3]_8\}Cl_8$ shows it to form crystalline microrods and particles ranging in size from 1 – 80 μm [476], while SAXS and TEM of $\{T_8[(CH_2)_3NH_3]_8\}[n\text{-}C_{12}H_{25}SO_3]_8$ show that a lamellar structure is formed with an interlayer spacing of 3.1 nm [275]. SEM of $T_8(OSiMe_2H)_8$ shows that well defined cubic particles with a porous structure and of ca. 100 μm size may be formed from the reaction of $[T_8O_8]^{8-}$ with HMe_2SiCl [163], but the related derivatives $T_8[OSiMe_2(CH_2)_3CN]_8$ and $T_8[OSiMe_2(CH_2)_3Cl]_8$ show a much lower degree of porosity [163, 460].

2.16.2 T_8R_7R' Compounds

AFM of $T_8(i\text{-Bu})_7(CH_2)_3OCH_2CH(O)CH_2$ has been used to image crystals of size 9-10 μm across [465], and AFM studies of polyurethanes containing pendant $T_8(i\text{-Bu})_7$ groups show that the surface of the polymer is much rougher than when no POS species are present [477]. TEM and EDXS mapping have also been used to show that the $T_8(i\text{-Bu})_7$ segments in POS-poly(carbonate-urea)urethane nanocomposites can form crystalline areas [478].

Specular X-ray reflectivity studies on LB multilayer films of polymers containing a pair of $T_8Cy_7OSiMe_2(CH_2)_3NHC(=O)$- groups separated by poly(ethylene glycol) chains show them to have a double layer structure with a thickness of 17.6 Å [479], while similar studies of films of the type T_8Cy_7-PEG-T_8Cy_7 show that double layers are formed with a calculated POS diameter for each POS cage of ca. 12.4 Å [480]. It has also been shown that the $T_8(c\text{-}C_5H_9)_7$–containing blocks in block copolymers of poly(*n*-butyl acrylate) and poly(propyl methacrylate) can be selectively stained by RuO_4 for imaging in TEM studies [443].

2.17 X-Ray Photoelectron Spectra of POS Compounds

Materials containing POS cages have not been widely studied by XPS but the Si 2p photoemission spectrum of T_8H_8 chemisorbed on a gold surface is consistent with the POS cages bonding to the surface via a single vertex as it shows two peaks in a 1:7 ratio [481]. Most XPS studies of POS-containing materials have been used to determine the presence and distribution of T_8 cages within nanocomposite materials. Thus, POS enrichment at the surface of POS-capped poly(ethylene oxide) in an epoxy resin has been inferred from the Si 2s and Si 2p signals in its XPS spectrum [31], and a higher than expected POS concentration at the surface of acrylate copolymers with pendant T_8R_7 group (R = Et or $c\text{-}C_5H_9$) is shown in the XPS spectrum [482]. XPS has also been used to show the presence of Si in polymeric materials, for example, in thermoset polymers derived from $T_8Ph_7(CH_2)_3OH$ and caprolactone [483], and to determine its concentration, for example, in POS-crosslinked polyurethane networks [484].

2.18 Electrochemistry of POS Compounds

The presence of a T_8 core was found to have little effect on the electrochemical properties of carbazole centers [485, 486], and cyclic voltammetry studies indicate that, although the monomer **6** has initial electroactivity (E_{ox} = 1.45 V in $[NBu_4][BF_4]/CH_2Cl_2$), the current decreases on repeated cycling. Electrochemical polymerization of **6** with pyrrole affords a polypyrrole containing POS cages which shows optical contrast, switching time (0.4 versus 1.1 s) and color properties, better than those of polypyrrole itself [133]. Cyclic voltammetry studies on films cast from the lutetium(III) complex **7** show it to be electrochromic, having a green-blue transition and possible uses in electrochromic devices [201].

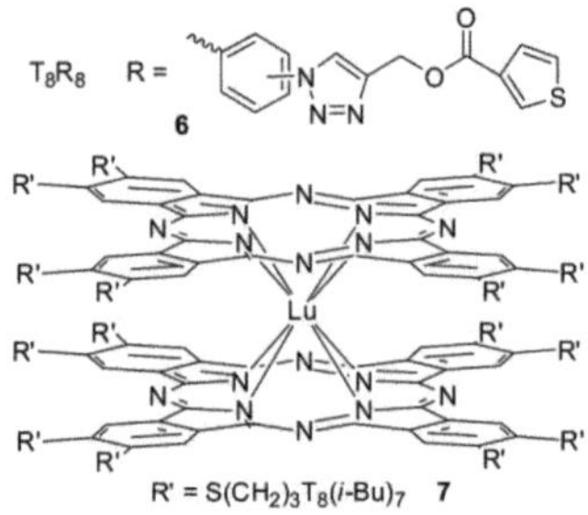

2.19 Chromatographic Methods Applied to POS Compounds

GPC has been used widely in the analysis of T_8 POS-containing materials, but studies can be complicated, since mixtures of compounds containing eight or fewer reacted substituents at a POS cage may have similar hydrodynamic radii, and, as POS materials tend to be globular compared to, for example, polystyrene, the choice of standards is important. Despite this, GPC analysis has been used to assess the purity of simple T_8 derivatives such as $T_8(C_6H_4NO_2)_8$ and $T_8(C_6H_4NH_2)_8$ [487], $T_8(c\text{-}C_5H_9)_7OSiMe_2(CH_2)_3NCO$ [471], $T_8(i\text{-}Bu)_7(CH_2)_3NH_2$ [29], and of $T_8[OSiMe_2(CH_2)_3OCH_2CH(O)CH_2]_8$ [488], and also to assess the nature of higher molecular weight materials containing POS cages. For example, GPC analysis indicates that the reaction between $T_8(C_6H_4NH_2)_8$ and $Me_2BrCC(=O)Br$ gives not only the expected $T_8[C_6H_4NHC(=O)CMe_2Br]_8$, but also higher molecular weight products [489], and that only about five of the Si-H groups per T_8 cage react in the hydrosilylation of $T_8(OSiMe_2H)_8$ with allyl-poly(ethylene oxide) [490]. It has also been used to show that the reaction between $T_8(C_6H_4Br)_8$, Mg, and 2-bromo-3-ethyl-hexylthiophene gives a fairly monodisperse product [131], to characterize the different isomers formed from the hydrosilylation of $T_8(OSiMe_2H)_8$ with a mixture of di(propylene glycol)allylethermethacrylate and 4-vinyl-cyclohexene-epoxide [463], and to determine that up to ca. 16 bromines may be introduced upon bromination of T_8Ph_8 [125].

SEC has been used to determine the purity of products derived from reactions of T_8 compounds, for example, a monomodal mass distribution was found for the polymer formed on hydrosilylation of $T_8(OSiMe_2H)_8$ with undecenyl polystyrene macromonomers [491], as was the case for a series of octa triphenylene-substituted T_8 compounds derived from $T_8[(CH_2)_3NH(CO)CH{=}CHCO_2H]_8$ [492]. SEC can also be used to confirm the nature of the siloxanes formed in the hydrolytic condensation reactions of trialkoxysilanes. For example, SEC methods show that T_8 species occur in the mixture formed by the hydrolysis/condensation reactions of $CH_2(O)CHCH_2O(CH_2)_3Si(OMe)_3$ [493], and the formation of $T_8[(CH_2)_3Cl]_8$ resulting from the hydrolysis of $Cl(CH_2)_3Si(OMe)_3$ [325]. SEC has also been used to show the purity of the isomeric tolyl derivatives, T_8R_8 ($R = C_6H_4$-2-Me, C_6H_4-3-Me or C_6H_4-4-Me) [457], and to characterize the different dendrimers formed in the preparation of globular MRI contrast agents based on $T_8[(CH_2)_3NH]_8$ cores surrounded by L-lysine [494].

2.20 Miscellaneous Physical Properties of POS Compounds

The mechanical properties (bulk modulus, averaged Young's modulus and isotropic averaged shear modulus) of $T_8(c\text{-}C_5H_9)_8$ have been calculated and suggest that, as crystalline aggregates, the POS cages can be considered as "rigid" entities when in a rubbery matrix [495]. A pristine film of $T_8(c\text{-}C_5H_9)_7CH_2CH_2C(=O)OMe$ has been shown to have a hardness and elastic modulus of 0.11 and 5 GPa respectively, but upon irradiation by Si ions it decomposes and forms an amorphous silicate-like material with hardness and elastic modulus of 6 and 65 GPa respectively [496].

Most POS derivatives are solids, but the kinematic viscosities of the viscous liquids T_8R_8 [R = $(CH_2)_7Br$, $(CH_2)_8Br$ $(CH_2)_6Cl$ and CH_2CH_2Ph] have been found to be 35, 38, ca. 42, and 48 cSt, respectively, and to have onsets for solidification of 0, +8, -5, and -20 °C [49]. The siloxy derivatives, $T_8(OSiMe_2R)_8$ where R = $(CH_2)_3OCH_2CH(O)CH_2$, $(CH_2)_2C(=O)C(=CH_2)Me$, and $OSiMe_2OSiMe_2H$, have dynamic viscosities of ca. 600, 220 and 200 mPa s respectively [497, 498].

The water contact angles for $T_8R_7(CH_2)_3OC(=O)C(=CH_2)Me$ (R = Et or $c\text{-}C_5H_9$) are 101 and 116° respectively [470], while for the fluorinated derivatives, T_8R_8 [where R = $(CH_2)_2(CF_2)_7CF_3$, $(CH_2)_2(CF_2)_5CF_3$, or $(CH_2)_2(CF_2)_3CF_3$], the surface energy decreases as the fluoroalkyl chain length increases such that the water contact angle for $T_8[(CH_2)_2(CF_2)_7CF_3]_8$ is 154°, i.e. higher than that of PTFE [22, 256] (see also Chapter 6). The related fluorinated derivatives, $T_8[OSiMe_2CH_2CH_2S(CH_2)_2(CF_2)_nCF_3]_8$ and $T_8[CH_2CH_2S(CH_2)_2(CF_2)_nCF_3]_8$ (where n = 5 or 7), when blended with poly(methyl methacrylate) give materials with water contact angles of 114-124°, which is much greater than the 71° for poly(methyl methacrylate) itself [85].

A plot of the dielectric loss of $T_8[(CH_2)_2Ph]_8$ as a function of temperature and frequency shows a dielectrically active relaxation process attributable to dynamic glass transition or α-relaxation [499], and blends of $T_8[(CH_2)_2Ph]_8$ with polystyrene and with poly(bisphenol A carbonate) have similar features [499, 500], with POS modification of polymers tending to lead to a reduction in dielectric constant [439, 441, 442, 501]. Dielectric relaxation spectroscopy and dynamic mechanical spectroscopy studies of $T_8[OSiMe_2(CH_2)_3OCH_2CH(O)CH_2]_8$ show that it undergoes two relaxation processes, both increasing in frequency as the temperature is increased [502].

2.21 Acknowledgments

We thank the UK Energy Research Centre for financial support to DBC.

2.22 References

1. Agaskar PA, Klemperer WG (1995) Inorg Chim Acta 229:355-364.
2. Bürgy H, Calzaferri G (1990) Journal of Chromatography A 507:481-486.
3. Wiberg E, Simmler W (1955) Z Anorg Allg Chem 282:330-344.
4. Sprung MM, Guenther FO (1955) J Am Chem Soc 77:3990-3996.
5. Sprung MM, Guenther FO (1955) J Am Chem Soc 77:3996-4002.
6. Feher FJ, Newman DA, Walzer JF (1989) J Am Chem Soc 111:1741-1748.
7. Zhinkin DY, Korneeva GK, Korneev NN, Sobolevskii MV (1966) Zh Obshch Khim 36:350-352.
8. Andrianov KA, Izmailov BA (1976) Zh Obshch Khim 46:329-335.
9. Behbehani H, Brisdon BJ, Mahon MF, Molloy KC (1994) J Organomet Chem 469:19-24.
10. Bassindale AR, MacKinnon IA, Maesano MG, Taylor PG (2003) Chem Commun:1382-1383.
11. Bassindale AR, Liu Z, MacKinnon IA, Taylor PG, Yang Y, Light ME, Horton PN, Hursthouse MB (2003) Dalton Trans:2945-2949.
12. Müller R, Köhne R, Sliwinski S (1959) J Prakt Chem 9:71-74.
13. Agaskar PA (1991) Inorg Chem 30:2707-2708.
14. Andrianov KA, Petrovnina NM, Vasil'eva TV, Shklover VE, D'Yachenko BI (1978) Zh Obshch Khim 48:2692-2695.
15. Dare EO, Liu L-K, Peng J (2006) Dalton Trans:3668-3671.

16. Gao J-g, Wang S-c, Zhang X-j, Run M-t (2005) Youjigui Cailiao 19:5-7.

17. Feher FJ, Wyndham KD, Soulivong D, Nguyen F (1999) J Chem Soc, Dalton Trans:1491-1497.

18. Gao J, Zhang X, Wang S, Run M (2005) Chemical Journal on Internet. http://www.chemistrymag.org/cji/2005/077048ne.htm. Accessed 13 May 2010.

19. Luo Q, Tang D, Li X, Wang Q, Wang Z, Zhen Z, Liu X (2006) Chem Lett 35:278-279.

20. Gu Y, Liang G, Zhang Z, Wang J, Lu T (2006), CN Pat. 1803808.

21. Kuehnle A, Jost C, Abbenhuis HCL, Gerritsen G (2006), DE Pat. 102004042815.

22. Mabry JM, Vij A, Iacono ST, Viers BD (2008) Angew Chem, Int Ed 47:4137-4140.

23. Dittmar U, Hendan BJ, Flörke U, Marsmann HC (1995) J Organomet Chem 489:185-194.

24. Dias Filho NL, Marangoni F, Costa RM (2007) J Colloid Interface Sci 313:34-40.

25. Lücke S, Stoppek-Langner K, Krebs B, Läge M (1997) Z Anorg Allg Chem 623:1243-1246.

26. Kudo T, Machida K, Gordon MS (2005) J Phys Chem A 109:5424-5429.

27. Brown JF, Jr., Vogt LH, Jr. (1965) J Am Chem Soc 87:4313-4317.

28. Okada T, Ikeda M (2004), JP Pat. 2004196958.

29. Hussain H, Tan BH, Mya KY, Liu Y, He CB, Davis TP (2010) J Polym Sci, Part A: Polym Chem 48:152-163.

30. Ervithayasuporn V, Wang X, Kawakami Y (2009) Chem Commun:5130-5132.

31. Zeng K, Zheng S (2007) J Phys Chem B 111:13919-13928.

32. Iacono ST, Vij A, Grabow W, Smith DW, Jr., Mabry JM (2007) Chem Commun:4992-4994.

33. Severn JR, Duchateau R, van Santen RA, Ellis DD, Spek AL (2002) Organometallics 21:4-6.

34. Severn JR, Duchateau R, Van Santen RA, Ellis DD, Spek AL, Yap GPA (2003) Dalton Trans:2293-2302.

35. Pescarmona PP, Masters AF, van der Waal JC, Maschmeyer T (2004) J Mol Catal A: Chem 220:37-42.

36. Feher FJ, Weller KJ (1990) Organometallics 9:2638-2640.

37. Wheeler PA, Fu BX, Lichtenhan JD, Weitao J, Mathias LJ (2006) J Appl Polym Sci 102:2856-2862.

38. Liu Z-h, Bassindale AR, Taylor PG (2004) Chem Res Chin Univ 20:433-436.

39. Lavrent'ev VI (2004) Russ J Gen Chem (Engl Trans) 74:1188-1193.

40. Zhang X, Hu L, Huang Y, Sun D, Sun Y (2004) Sci China, Ser B: Chem 47:388-395.

41. Bärtsch M, Bornhauser P, Calzaferri G, Imhof R (1994) J Phys Chem 98:2817-2831.

42. Bolln C, Tsuchida A, Frey H, Mülhaupt R (1997) Chem Mater 9:1475-1479.

43. Said MA, Roesky HW, Rennekamp C, Andruh M, Schmidt H-G, Noltemeyer M (1999) Angew Chem, Int Ed 38:661-664.

44. Chen K-B, Chen H-Y, Yang S-H, Hsu C-S (2006) J Polym Res 13:237-245.

45. Kim K-M, Ouchi Y, Chujo Y (2003) Polym Bull (Berlin) 49:341-348.

46. Kim K-M (2006) Polymer (Korea) 30:380-384.

47. Adachi K, Tamaki R, Chujo Y (2004) Bull Chem Soc Jpn 77:2115-2119.

48. Bassindale AR, Gentle T (1996) J Organomet Chem 521:391-393.

49. Dare EO (2006) Turk J Chem 30:585-593.

50. Crivello JV, Malik R (1997) J Polym Sci, Part A: Polym Chem 35:407-425.

51. Liu L-K, Dare EO (2004) J Chin Chem Soc (Taipei, Taiwan) 51:175-182.

52. Tsuchida A, Bolln C, Sernetz FG, Frey H, Mülhaupt R (1997) Macromolecules 30:2818-2824.

53. Shimojima A, Goto R, Atsumi N, Kuroda K (2008) Chem–Eur J 14:8500-8506.

54. Bassindale AR, Parker DJ, Taylor PG, Watt AC (2003) Can J Chem 81:1341-1349.

55. Xiaodong H, Yan Z, Xuening S, Farong H, Du L (2008) Yuhang Caliao Gongyi 38:54-57.

56. Su X, Guang S, Xu H, Liu X, Li S, Wang X, Deng Y, Wang P (2009) Macromolecules 42:8969-8976.

57. Rikowski E, Marsmann HC (1997) Polyhedron 16:3357-3361.

58. Krüger R-P, Much H, Schulz G, Rikowski E (2000) Synthesis of polyhedral hexa-, octa-, deca- and dodeca-silsesquioxanes separation by NP-HPLC, SEC and LAC characterization by MALDI-TOF-MS. In Auner N, Weis J (eds) Organosilicon Chemistry IV: From Molecules to Materials, Wiley-VCH, Weinheim, Germany, pp. 545-550.

59. Ihara N, Kurisawa M, Chung JE, Uyama H, Kobayashi S (2005) Appl Microbiol Biotechnol 66:430-433.

60. Cheng G, Vautravers NR, Morris RE, Cole-Hamilton DJ (2008) Org Biomol. Chem 6:4662-4667.

61. Stengel BF, Ridland J, Hearshaw M, Cole-Hamilton DJ, Tooze R, Morris RE (2004), WO Pat. 2004022231.

62. Ropartz L, Foster DF, Morris RE, Slawin AMZ, Cole-Hamilton DJ (2002) J Chem Soc, Dalton Trans:1997-2008.

63. Ropartz L, Morris RE, Schwarz GP, Foster DF, Cole-Hamilton DJ (2000) Inorg Chem Commun 3:714-717.

64. Coupar PI, Jaffres P-A, Morris RE (1999) J Chem Soc, Dalton Trans:2183-2188.

65. Vautravers NR, Cole-Hamilton DJ (2009) Chem Commun:92-94.

66. Trastoy B, Pérez-Ojeda ME, Sastre R, Chiara JL (2010) Chem–Eur J 16:3833-3841.

67. Liu Y, Yang X, Zhang W, Zheng S (2006) Polymer 47:6814-6825.

68. Ceyhan T, Yüksek M, Yağlıoğlu HG, Salih B, Erbil MK, Elmali A, Bekaroğlu Ö (2008) Dalton Trans:2407-2413.

69. Miyake J, Chujo Y (2008) Macromol Rapid Commun 29:86-92.

70. Kaneshiro TL, Wang X, Lu Z-R (2007) Mol Pharmaceutics 4:759-768.

71. Naka K, Fujita M, Tanaka K, Chujo Y (2007) Langmuir 23:9057-9063.

72. Tanaka K, Inafuku K, Naka K, Chujo Y (2008) Org Biomol Chem 6:3899-3901.

73. Goodgame DML, Kealey S, Lickiss PD, White AJP (2008) J Mol Struct 890:232-239.

74. Gravel M-C, Zhang C, Dinderman M, Laine RM (1999) Appl Organomet Chem 13:329-336.

75. Feher FJ, Wyndham KD, Scialdone MA (1998) Chem Commun:1469-1470.

76. Feher FJ, Soulivong D, Eklund AG, Wyndham KD (1997) Chem Commun:1185-1186.

77. Itami Y, Marciniec B, Kubicki M (2004) Chem–Eur J 10:1239-1248.

78. Żak P, Pietraszuk C, Marciniec B, Spólnik G, Danikiewicz W (2009) Adv Synth Catal 351:2675-2682.

79. Sulaiman S, Brick C, Roll M, Laine RM (2007) Polym Prepr 48:911.

80. André P, Cheng G, Ruseckas A, van Mourik T, Früchtl H, Crayston JA, Morris RE, Cole-Hamilton DJ, Samuel IDW (2008) J Phys Chem B 112:16382-16392.

81. Vautravers NR, André P, Cole-Hamilton DJ (2009) J Mater Chem 19:4545-4550.

82. Poliskie GM, Haddad TS, Blanski RL, Gleason KK (2005) Thermochim Acta 438:116-125.

83. Drylie EA, Andrews CD, Hearshaw MA, Jimenez-Rodriguez C, Slawin A, Cole-Hamilton DJ, Morris RE (2006) Polyhedron 25:853-858.

84. Dare EO, Olatunji GA, Ogunniyi DS, Lasisi AA (2005) Pol J Chem 79:109-114.

85. Xu J, Li X, Cho CM, Toh CL, Shen L, Mya KY, Lu X, He C (2009) J Mater Chem 19:4740-4745.

86. Jaffrès P-A, Morris RE (1998) J Chem Soc, Dalton Trans:2767-2770.

87. Feher FJ, Wyndham KD (1998) Chem Commun:323-324.

88. Lee L-H, Chen W-C (2005) Polymer 46:2163-2174.

89. Lücke S, Stoppek-Langner K, Kuchinke J, Krebs B (1999) J Organomet Chem 584:11-15.

90. Chan JW, Hoyle CE, Lowe AB (2009) J Am Chem Soc 131

91. Xu H, Yang B, Gao X, Li C, Guang S (2006) J Appl Polym Sci 101:3730-3735.

92. Feher FJ, Wyndham KD, Baldwin RK, Soulivong D, Lichtenhan JD, Ziller JW (1999) Chem Commun:1289-1290.

93. Petraru L, Binder WH (2005) Polym Prepr 46:841-842.

94. Binder WH, Petraru L, Sachenshofer R, Zirbs R (2006) Monatsh Chem 137:835-841.

95. Chiacchio MA, Borrello L, Di Pasquale G, Pollicino A, Bottino FA, Rescifina A (2005) Tetrahedron 61:7986-7993.

96. Asuncion MZ, Laine RM (2010) J Am Chem Soc 132:3723-3736.

97. Clark TS, Hoyle CE, Nazarenko S (2008) J Coat Technol Res 5:345-351.

98. Unno M, Imai Y, Matsumoto H (2003) Silicon Chem 2:175-178.

99. Unno M, Alias SB, Saito H, Matsumoto H (1996) Organometallics 15:2413-2414.

100. Unno M, Suto A, Takada K, Matsumoto H (2000) Bull Chem Soc Jpn 73:215-220.

101. Spirk S, Nieger M, Belaj F, Pietschnig R (2009) Dalton Trans:163-167.

102. Voronkov MG, Lavrentyev VI, Kovrigin VM (1981) J Organomet Chem 220:285-293.

103. Handke B, Jastrzębski W, Mozgawa W, Kowalewska A (2008) J Mol Struct 887:159-164.

104. Unno M, Takada K, Matsumoto H (1998) Chem Lett:489-490.

105. Unno M, Alias SB, Arai M, Takada K, Tanaka R, Matsumoto H (1999) Appl Organomet Chem 13:303-310.

106. Ronchi M, Sulaiman S, Boston NR, Laine RM (2010) Appl Organomet Chem 24:551-557.

107. Baidina IA, Podberezskaya NV, Borisov SV, Alekseev VI, Martynova TN, Kanev AN (1980) Zh Strukt Khim 21:125-129.

108. Ni Y, Zheng S, Nie K (2004) Polymer 45:5557-5568.

109. Olsson K (1958) Ark Kemi 13:367-378.

110. Brown JF, Jr., Vogt LH, Jr., Prescott PI (1964) J Am Chem Soc 86:1120-1125.

111. Olsson K, Gronwall C (1961) Ark Kemi 17:529-540.

112. Feher FJ, Budzichowski TA (1989) J Organomet Chem 373:153-163.

113. Feher FJ, Budzichowski TA (1989) J Organomet Chem 379:33-40.

114. Tacke R, Lopez-Mras A, Sheldrick WS, Sebald A (1993) Z Anorg Allg Chem 619:347-358.

115. Miyazato A, Pakjamsai C, Kawakami Y (2010) Dalton Trans 39:3239-3244.

116. Olsson K, Axen C (1964) Ark Kemi 22:237-244.

117. Song XY, Geng HP, Li QF (2006) Chin Chem Lett 17:427-430.

118. Leu C-M, Reddy GM, Wei K-H, Shu C-F (2003) Chem Mater 15:2261-2265.

119. Lickiss PD, Rataboul F (2008) Fully Condensed Polyhedral Oligosilsesquioxanes (POSS): From Synthesis to Application. In Hill AF, Fink MJ (eds) Advances in Organometallic Chemistry, Academic Press, Oxford, U.K., pp. 1-116.

120. Cordes DB, Lickiss PD, Rataboul F (2010) Chemical Reviews 110:2081-2173.

121. Tamaki R, Tanaka Y, Asuncion MZ, Choi J, Laine RM (2001) J Am Chem Soc 123:12416-12417.

122. Takahashi K, Sulaiman S, Katzenstein JM, Snoblen S, Laine RM (2006) Aust J Chem 59:564-570.

123. Chen H-J, Meng F (2007) Macromolecules 40:2079-2085.

124. He C, Xiao Y, Huang J, Lin T, Mya KY, Zhang X (2004) J Am Chem Soc 126:7792-7793.

125. Brick CM, Tamaki R, Kim SG, Asuncion MZ, Roll M, Nemoto T, Ouchi Y, Chujo Y, Laine RM (2005) Macromolecules 38:4655-4660.

126. Roll MF, Asuncion MZ, Kampf J, Laine RM (2008) ACS Nano 2:320-326.

127. Hartmann-Thompson C, Merrington A, Carver PI, Keeley DL, Rousseau JL, Hucul D, Bruza KJ, Thomas LS, Keinath SE, Nowak RM, Katona DM, Santurri PR (2008) J Appl Polym Sci 110:958-974.

128. Subianto S, Mistry MK, Choudhury NR, Dutta NK, Knott R (2009) ACS Appl Mater Interfaces 1:1173-1182.

129. Liu L, Song L, Zhang S, Guo H, Hu Y, Fan W (2006) Mater Lett 60:1823-1827.

130. Clarke DJ, Matisons JG, Simon GP, Samoc M, Samoc A (2010) Appl Organomet Chem 24:184-188.

131. Xiao Y, Lu X, Tan L-W, Ong KS, He C (2009) J Polym Sci, Part A: Polym Chem 47:5661-5670.

132. Feher FJ, Schwab JJ, Phillips SH, Eklund A, Martinez E (1995) Organometallics 14:4452-4453.

133. Ak M, Gacal B, Kiskan B, Yagci Y, Toppare L (2008) Polymer 49:2202-2210.

134. Asuncion MZ, Roll MF, Laine RM (2008) Macromolecules 41:8047-8052

135. Zou Q-C, Yan Q-J, Song G-W, Zhang S-L, Wu L-M (2007) Biosens Bioelectron 22:1461-1465.

136. Li X, Du Y, Dai J, Wang X, Yang P (2007) Catal Lett 118:151-158.

137. Tateyama S, Kakihana Y, Kawakami Y (2010) J Organomet Chem 695:898-902.

138. Rattay M, Jutzi P, Fenske D (2000) Highly functionalized octasilsesquioxanes synthesis, structure and reactivity. In Auner N, Weis J (eds) Organosilicon Chemistry IV: From Molecules to Materials, Wiley-VCH, Weinheim, Germany, pp. 526-530.

139. Day VW, Klemperer WG, Mainz VV, Millar DM (1985) J Am Chem Soc 107:8262-8264.

140. Brevett CS, Cagle PC, Klemperer WG, Millar DM, Ruben GC (1991) J Inorg Organomet Polym 1:335-342.

141. Abe Y, Gunji T, Arimitsu K, Ueda N (2008), JP Pat. 2008063228.

142. Gießmann S, Fischer A, Edelmann FT (2004) Z Anorg Allg Chem 630:1982-1986.

143. Feher FJ, Weller KJ (1991) Inorg Chem 30:880-882.

144. Päch M, Stösser R (1997) J Phys Chem A 101:8360-8365.

145. Agaskar PA (1990) Synth React Inorg Met-Org Chem 20:483-493.

146. Anderson SE, Mitchell C, Haddad TS, Vij A, Schwab JJ, Bowers MT (2006) Chem Mater 18:1490-1497.

147. Agaskar PA, Day VW, Klemperer WG (1987) J Am Chem Soc 109:5554-5556.

148. Harrison PG, Kannengiesser R, Hall CJ (1997) Main Group Met Chem 20:137-141.

149. Hoebbel D, Engelhardt G, Samoson A, Újszászy K, Smolin YI (1987) Z Anorg Allg Chem 552:236-240.

150. Harrison PG, Kannengiesser R (1996) Chem Commun:415-416.

151. Sheen Y-C, Lu C-H, Huang C-F, Kuo S-W, Chang F-C (2008) Polymer 49:4017-4024.

152. Markovic E, Clarke S, Matisons J, Simon GP (2008) Macromolecules 41:1685-1692.

153. Zhang C, Laine RM (2000) J Am Chem Soc 122:6979-6988.

154. Liu H, Kondo S-i, Takeda N, Unno M (2008) J Am Chem Soc 130:10074-10075.

155. Zhang C, Laine RM (1996) J Organomet Chem 521:199-201.

156. Yuchs SE, Carrado KA (1996) Inorg Chem 35:261-262.

157. Mutluay A, Jutzi P (2000) Synthesis and characterization of chloro-, allyl- and ferrocenyl-substituted silsesquioxanes. In Auner N, Weis J (eds) Organosilicon Chemistry IV: From Molecules to Materials, Wiley-VCH, Weinheim, Germany, pp. 531-535.

158. Muller E, Edelmann FT (1999) Main Group Met Chem 22:485-488.

159. Harrison PG, Hall C (1997) Main Group Met Chem 20:515-529.

160. Jutzi P, Batz C, Mutluay A (1994) Z Naturforsch, B: Chem Sci 49:1689-1692.

161. Létant SE, Maiti A, Jones TV, Herberg JL, Maxwell RS, Saab AP (2009) J Phys Chem C 113:19424-19431.

162. Hoebbel D, Pitsch I, Heidemann D, Jancke H, Hiller W (1990) Z Anorg Allg Chem 583:133-144.

163. do Carmo DR, Paim LL, Dias Filho NL, Stradiotto NR (2007) Appl Surf Sci 253:3683-3689.

164. Dias Filho NL, Adolfo de Aquino H, Pereira DS, Rosa AH (2007) J Appl Polym Sci 106:205-213.

165. Markovic E, Ginic-Markovic M, Clarke S, Matisons J, Hussain M, Simon GP (2007) Macromolecules 40:2694-2701.

166. Maitra P, Wunder SL (2004) Electrochem Solid-State Lett 7:A88-A92.

167. do Carmo DR, Guinesi LS, Dias Filho NL, Stradiotto NR (2004) Appl Surf Sci 235:449-459.

168. Provatas A, Luft M, Mu JC, White AH, Matisons JG, Skelton BW (1998) J Organomet Chem 565:159-164.

169. Kawahara K, Hagiwara Y, Shimojima A, Kuroda K (2008) J Mater Chem 18:3193-3195.

170. Hasegawa I, Imamura W, Takayama T (2004) Inorg Chem Commun 7:513-515.

171. Hasegawa I, Niwa T, Takayama T (2005) Inorg Chem Commun 8:159-161.

172. Hoebbel D, Pitsch I, Reiher T, Hiller W, Jancke H, Mueller D (1989) Z Anorg Allg Chem 576:160-168.

173. Hasegawa I, Ishida M, Motojima S (1994) Synth React Inorg Met-Org Chem 24:1099-1110.

174. Chan S-C, Kuo S-W, Chang F C (2005) Macromolecules 38:3099-3107.

175. Hao J, Palmieri F, Stewart MD, Nishimura Y, Chao H-L, Collins A, Willson CG (2006) Polym Prepr 47:1158-1159.

176. Hagiwara Y, Shimojima A, Kuroda K (2008) Chem Mater 20:1147-1153.

177. Auner N, Ziemer B, Herrschaft B, Ziche W, John P, Weis J (1999) Eur J Inorg Chem:1087-1094.

178. Hoebbel D, Wieker W, Franke P, Otto A (1975) Z Anorg Allg Chem 418:35-44.

179. Wiebcke M, Koller H (1992) Acta Crystallogr, Sect B: Struct Sci 48:449-458.

180. Clark JC, Saengkerdsub S, Eldridge GT, Campana C, Barnes CE (2006) J Organomet Chem 691:3213-3222.

181. Rattay M, Fenske D, Jutzi P (1998) Organometallics 17:2930-2932.

182. Calzaferri G, Imhof R (1992) J Chem Soc, Dalton Trans:3391-3392.

183. Moran M, Casado CM, Cuadrado I, Losada J (1993) Organometallics 12:4327-4333.

184. Lorenz V, Fischer A, Edelmann FT (2002) J Organomet Chem 647:245-249.

185. Duchateau R, Abbenhuis HCL, Van Santen RA, Thiele SKH, Van Tol MFH (1998) Organometallics 17:5222-5224.

186. Fraile JM, García JI, Mayoral JA, Vispe E (2005) J Catal 233:90-99.

187. Duchateau R, Dijkstra TW, van Santen RA, Yap GPA (2004) Chem–Eur J 10:3979-3990.

188. Riollet V, Quadrelli EA, Copéret C, Basset J-M, Andersen RA, Köhler K, Böttcher R-M, Herdtweck E (2005) Chem–Eur J 11:7358-7365.

189. Song XY, Geng HP, Li QF (2006) Polymer 47:3049-3056.

190. Blanc F, Chabanas M, Copéret C, Fenet B, Herdweck E (2005) J Organomet Chem 690:5014-5026.

191. Blanc F, Copéret C, Thivolle-Cazat J, Basset J-M, Lesage A, Emsley L, Sinha A, Schrock RR (2006) Angew Chem, Int Ed 45:1216-1220.

192. Cho HM, Weissman H, Wilson SR, Moore JS (2006) J Am Chem Soc 128:14742-14743.

193. Fei Z, Schmutzler R, Edelmann FT (2003) Z Anorg Allg Chem 629:353-356.

194. Rhers B, Salameh A, Baudouin A, Quadrelli EA, Taoufik M, Copéret C, Lefebvre F, Basset J-M, Solans-Monfort X, Eisenstein O, Lukens WW, Lopez LPH, Sinha A, Schrock RR (2006) Organometallics 25:3554-3557.

195. Rhers B, Quadrelli EA, Baudouin A, Taoufik M, Copéret C, Lefebvre F, Basset J-M, Fenet B, Sinha A, Schrock RR (2006) J Organomet Chem 691:5448-5455.

196. Lucenti E, D'Alfonso G, Macchi P, Maranesi M, Roberto D, Sironi A, Ugo R (2006) J Am Chem Soc 128:12054-12055.

197. Chabanas M, Baudouin A, Copéret C, Basset J-M, Lukens W, Lesage A, Hediger S, Emsley L (2003) J Am Chem Soc 125:492-504.

198. Pérez Y, Quintanilla DP, Fajardo M, Sierra I, del Hierro I (2007) J Mol Catal A: Chem 271:227-237.

199. Varga V, Pinkas Ji, Císařová I, Horáček M, Mach K (2009) Organometallics 28:6944-6956.

200. Yang X, Froehlich JD, Chae HS, Li S, Mochizuki A, Jabbour GE (2009) Adv Funct Mater 19:2623-2629.

201. Ceyhan T, Özdağ MA, Salih B, Erbil MK, Elmah A, Özkaya AR, Bekaroğlu Ö (2008) Eur J Inorg Chem:4943-4950.

202. Knight CTG, Balec RJ, Kinrade SD (2007) Angew Chem, Int Ed 46:8148-8152.

203. Kinrade SD, Knight CTG, Pole DL, Syvitski RT (1998) Inorg Chem 37:4272-4277.

204. Caratzoulas S, Vlachos DG, Tsapatsis M (2005) J Phys Chem B 109:10429-10434.

205. Caratzoulas S, Vlachos DG, Tsapatsis M (2006) J Am Chem Soc 128:596-606.

206. Wiebcke M, Felsche J (2001) Microporous Mesoporous Mater 43:289-297.

207. Hasegawa I, Motojima S (1992) J Organomet Chem 441:373-380.

208. Hasegawa I, Sakka S (1988) Chem Lett:1319-1322.

209. Waehner J, Marciniec B, Pawluć P (2007) Eur J Inorg Chem 2007:2975-2980.

210. Calzaferri G, Marcolli C, Imhof R, Toernroos KW (1996) J Chem Soc, Dalton Trans:3313-3322.

211. Martynova TN, Chupakhina TI (1988) J Organomet Chem 345:11-18.

212. Lavrent'ev VI, Sheludyakova LA (1983) Izv Akad Nauk SSSR, Ser Khim:1883-1885.

213. Lavrent'ev VI (1997) Russ J Gen Chem (Engl Trans) 67:239-244.

214. Hendan BJ, Marsmann HC (1994) J Organomet Chem 483:33-38.

215. Jerman I, Orel B, Šurca Vuk A, Koželj M, Kovač J (2010) Thin Solid Films 518:2710-2721.

216. Fei Z, Ibrom K, Edelmann FT (2002) Z Anorg Allg Chem 628:2109-2112.

217. Xu C, Wang W, Zhang W, Zhuang J, Liu L, Kong Q, Zhao L, Long Y, Fan K, Qian S, Li Y (2000) J Phys Chem A 104:9518-9524.

218. Asuncion MZ, Hasegawa I, Kampf JW, Laine RM (2005) J Mater Chem 15:2114-2121.

219. Asuncion MZ, Ronchi M, Abu-Seir H, Laine RM (2010) C R Chim 13:270-281.

220. Andrianov KA, Tikhonov VS, Makhneva GP, Chernov GS (1973) Izv Akad Nauk SSSR, Ser Khim:956-957.

221. Bassindale AR, Pourny M, Taylor PG, Hursthouse MB, Light ME (2003) Angew Chem, Int Ed 42:3488-3490.

222. Bassindale AR, Parker DJ, Pourny M, Taylor PG, Horton PN, Hursthouse MB (2004) Organometallics 23:4400-4405.

223. Anderson SE, Bodzin DJ, Haddad TS, Boatz JA, Mabry JM, Mitchell C, Bowers MT (2008) Chem Mater 20:4299-4309.

224. Sasamori R, Okaue Y, Isobe T, Matsuda Y (1994) Science (Washington, DC, US) 265:1691-1693.

225. Harima Y, Komaguchi K, Oka K, Maruoka T, Imae I, Ooyama Y (2010) Chem Commun 46:2076-2078.

226. Okaue Y, Isobe T (2003) Kidorui 42:186-187.

227. Quadrelli EA, Basset J-M (2010) Coord Chem Rev 254:707-728.

228. Xu H, Yang B, Wang J, Guang S, Li C (2005) Macromolecules 38:10455-10460.

229. Su X, Guang S, Li C, Xu H, Liu X, Wang X, Song Y (2010) Macromolecules 43:2840-2845.

230. Su X, Guang S, Xu H, Yang J, Song Y (2010) Dyes Pigm 87:69-75.

231. Mabry JM, Haddad TS (2008) Polym Prepr 49:391.

232. Hoebbel D, Garzo G, Engelhardt G, Ebert R, Lippmaa E, Alla M (1980) Z Anorg Allg Chem 465:15-33.

233. Wann DA, Reilly AM, Rataboul F, Lickiss PD, Rankin DWH (2009) Z Naturforsch, B: Chem Sci 64:1269-1275.

234. Lin W-J, Chen W-C, Wu W-C, Niu Y-H, Jen AKY (2004) Macromolecules 37:2335-2341.

235. Chen D, Nie J, Yi S, Wu W, Zhong Y, Liao J, Huang C (2010) Polym Degrad Stab 95:618-626.

236. Yim J-H, Lyu Y-Y, Jeong H-D, Mah SK, Hyeon-Lee J, Hahn J-H, Kim GS, Chang S, Park J-G (2003) J Appl Polym Sci 90:626-634.

237. Ni C, Ni G, Zhang S, Liu X, Chen M, Liu L (2010) Colloid Polym Sci 288:469-477.

238. Tanaka K, Inafuku K, Adachi S, Chujo Y (2009) Macromolecules 42:3489-3892.

239. Tanaka K, Kitamura N, Naka K, Morita M, Inubushi T, Chujo M, Nagao M, Chujo Y (2009) Polym J (Tokyo, Jpn) 41:287-292.

240. Liu Y, Zheng S, Nie K (2005) Polymer 46:12016-12025.

241. Fu J, Shi L, Chen Y, Yuan S, Wu J, Liang X, Zhong Q (2008) J Appl Polym Sci 109:340-349.

242. Marciniec B, Dutkiewicz M, Maciejewski H, Kubicki M (2008) Organometallics 27:793-794.

243. Shen J, Zheng S (2006) J Polym Sci, Part B: Polym Phys 44:942-952.

244. Jerman I, Koželj M, Orel B (2010) Sol Energy Mater Sol Cells 94:232-245.

245. Zakharov AV, Masters SL, Wann DA, Shlykov SA, Girichev GV, Arrowsmith S, Cordes DB, Lickiss PD, White AJP (2010) Dalton Trans 39:6900-6966.

246. Madhavan K, Gnanasekaran D, Reddy BSR (2009) J Appl Polym Sci 114:3659-3667.

247. Zhang X, Tay SW, Hong L, Liu Z (2008) J Membr Sci 320:310-318.

248. Majumdar P, Lee E, Gubbins N, Stafslien SJ, Daniels J, Thorson CJ, Chisholm BJ (2009) Polymer 50:1124-1133.

249. Abdul Wahab M, Mya KY, He C (2008) J Polym Sci, Part A: Polym Chem 46:5887-5896.

250. Huang J, Xiao Y, Mya KY, Liu X, He C, Dai J, Siow YP (2004) J Mater Chem 14:2858-2863.

251. Unno M, Matsumoto T, Mochizuki K, Higuchi K, Goto M, Matsumoto H (2003) J Organomet Chem 685:156-161.

252. Seino M, Kawakami Y (2004) Polym J (Tokyo, Jpn) 36:422-429.

253. Samadi-Maybodi A, Goudarzi N (2006) Spectrochim Acta, Part A 65:753-758.

254. Koh K, Sugiyama S, Morinaga T, Ohno K, Tsujii Y, Fukuda T, Yamahiro M, Iijima T, Oikawa H, Watanabe K, Miyashita T (2005) Macromolecules 38:1264-1270.

255. Ligon SC, Jr., Iacono ST, Mabry JM, Vij A, Smith DW, Jr. (2006) Polym Prepr 47:450-451.

256. Iacono ST, Budy SM, Mabry JM, Smith DW (2007) Polymer 48:4637-4645.

257. Mammeri F, Douja N, Bonhomme C, Ribot F, Babonneau F, Dirè S (2005) Mater Res Soc Symp Proc 847:363-368.

258. Randriamahefa S, Lorthioir C, Guégan P, Penelle J (2009) Polymer 50:3887-3894.

259. Kim K-M, Keum D-K, Chujo Y (2003) Macromolecules 36:867-875.

260. Chou C-H, Hsu S-L, Yeh S-W, Wang H-S, Wei K-H (2005) Macromolecules 38:9117-9123.

261. Wada K, Yano K, Kondo T, Mitsudo T-a (2006) Catal Lett 112:63-67.

262. Drazkowski DB, Lee A, Haddad TS, Cookson DJ (2006) Macromolecules 39:1854-1863.

263. Ohno K, Sugiyama S, Koh K, Tsujii Y, Fukuda T, Yamahiro M, Oikawa H, Yamamoto Y, Ootake N, Watanabe K (2004) Macromolecules 37:8517-8522.

264. Goto R, Shimojima A, Kuge H, Kuroda K (2008) Chem Commun:6152-6154.

265. Millot N, Santini CC, Lefebvre F, Basset J-M (2004) C R Chim 7:725-736.

266. Tanaka K, Kitamura N, Naka K, Chujo Y (2008) Chem Commun:6176-6178.

267. Tan MQ, Wu XM, Jeong EK, Chen QJ, Lu ZR (2010) Biomacromolecules 11:754-761.

268. Toepfer O, Neumann D, Choudhury NR, Whittaker A, Matisons J (2005) Chem Mater 17:1027-1035.

269. Huang J, He C, Liu X, Xu J, Tay CSS, Chow SY (2005) Polymer 46:7018-7027.

270. Létant SE, Herberg J, Dinh LN, Maxwell RS, Simpson RL, Saab AP (2007) Catal Commun 8:2137-2142.

271. Desmartin Chomel A, Jayasooriya UA, Babonneau F (2004) Spectrochim Acta, Part A 60:1609-1616.

272. Vannier A, Duquesne S, Bourbigot S, Alongi J, Camino G, Delobel R (2009) Thermochim Acta 495:155-166.

273. Yang B, Li J, Wang J, Xu H, Guang S, Li C (2009) J Appl Polym Sci 111:2963-2969.

274. Zhang Z, Liang G, Ren P, Wang J (2008) Polym Compos 29:77-83.

275. Liu L, Hu Y, Song L, Chen H, Nazare S, Hull TR (2007) Mater Lett 61:1077-1081.

276. Vilčnik A, Jerman I, Vuk AŠ, Koželj M, Orel B, Tomšič B, Simončič B, Kovač J (2009) Langmuir 25:5869-5880.

277. Helmy R, Wenslow RW, Fadeev AY (2004) J Am Chem Soc 126:7595-7600.

278. Nagendiran S, Alagar M, Hamerton I (2010) Acta Mater 58:3345-3356.

279. Huang F, Rong Z, Shen X, Huang F, Du L, Li Z (2008) Polym Eng Sci 48:1022-1028.

280. Choi J, Kim SG, Laine RM (2004) Macromolecules 37:99-109.

281. Dias Filho NL, Adolfo de Aquino H (2006) e-Polymers. http://www.e-polymers.org/journal/papers/nldfilho_280406.pdf. Acessed 13 May, 2010.

282. Yan F, Lu B-Z, Hu G-F, Guo C-X, Xu J-H (2004) Bopuxue Zazhi 21:57-60.

283. Zeng K, Liu Y, Zheng S (2008) Eur Polym J 44:3946-3956.

284. Xu H, Kuo S-w, Huang C-f, Chang F-c (2004) J Appl Polym Sci 91:2208-2215.

285. Ni Y, Zheng S (2007) J Polym Sci, Part A: Polym Chem 45:1247-1259.

286. Mya KY, He C, Huang J, Xiao Y, Dai J, Slow Y-P (2004) J Polym Sci, Part A: Polym Chem 42:3490-3503.

287. Bonhomme C, Coelho C, Azaïs T, Bonhomme-Coury L, Babonneau F, Maquet J, Thouvenot R (2006) C R Chim 9:466-471.

288. Croce G, Carniato F, Milanesio M, Boccaleri E, Paul G, van Beek W, Marchese L (2009) Phys Chem Chem Phys 11:10087-10094.

289. Päch M, Macrae RM, Carmichael I (2006) J Am Chem Soc 128:6111-6125.

290. Schoenfeld RS, Harneit W, Paech M (2006) Phys Status Solidi B 243:3008-3012.

291. Desmartin Chomel A, Dempsey P, Latournerie J, Hourlier-Bahloul D, Jayasooriya UA (2005) Chem Mater 17:4468-4473.

292. Baney RH, Itoh M, Sakakibara A, Suzuki T (1995) Chem Rev 95:1409-1430.

293. Dias Filho NL, Adolfo de Aquino H, Pires G, Caetano L (2006) J Braz Chem Soc 17:533-541.

294. Wang J, Kuimova MK, Poliakoff M, Briggs GAD, Khlobystov AN (2006) Angew Chem, Int Ed 45:5188-5191.

295. Soh MS, Yap AUJ, Sellinger A (2008) J Biomed Mater Res, Part B 85:78-86.

296. Kolel-Veetil MK, Dominguez DD, Keller TM (2008) J Polym Sci, Part A: Polym Chem 46:2581-2587.

297. Huang J-M, Huang H-J, Wang Y-X, Chen W-Y, Chang F-C (2009) J Polym Sci, Part B: Polym Phys 47:1927-1934.

298. Liu Y, Zheng S (2006) J Polym Sci, Part A: Polym Chem 44:1168-1181.

299. Mu J, Liu Y, Zheng S (2007) Polymer 48:1176-1184.

300. Mu J, Zheng S (2007) J Colloid Interface Sci 307:377-385.

301. Liu Y, Meng F, Zheng S (2005) Macromol Rapid Commun 26:920-925.

302. Feng Y, Jia Y, Guang S, Xu H (2010) J Appl Polym Sci 115:2212-2220.

303. Liu Z, Joung S-K, Okazaki T, Suenaga K, Hagiwara Y, Ohsuna T, Kuroda K, Iijima S (2009) ACS Nano 3:1160-1166.

304. Tamaki R, Choi J, Laine RM (2003) Chem Mater 15:793-797.

305. Lu T-L, Liang G-Z, Cheng Q, Guo Z-A (2006) Cailiao Kexue Yu Gongyi 14:527-531, 534.

306. Liu H, Zhang W, Zheng S (2005) Polymer 46:157-165.

307. Choi J, Tamaki R, Kim SG, Laine RM (2003) Chem Mater 15:3365-3375.

308. Liu L, Song L, Zhang S, Hu Y (2006) Zhongguo Kexue Jishu Daxue Xuebao 36:29-33.

309. Takamura N, Viculis L, Zhang C, Laine RM (2007) Polym Int 56:1378-1391.

310. Lee R-H, Lai H-H (2007) Eur Polym J 43:715-724.

311. Zhou J, Kieffer J (2008) J Phys Chem C 112:3473-3481.

312. Zeng K, Wang L, Zheng S (2009) J Phys Chem B 113:11831-11840.

313. Chan S-C, Kuo S-W, She H-S, Lin H-M, Lee H-F, Chang F-C (2007) J Polym Sci, Part A: Polym Chem 45:125-135.

314. Monticelli O, Fina A, Ullah A, Waghmare P (2009) Macromolecules 42:6614-6623.

315. Amir N, Levina A, Silverstein MS (2007) J Polym Sci, Part A: Polym Chem 45:4264-4275.

316. Ni Y, Zheng S (2007) J Polym Sci, Part B: Polym Phys 45:2201-2214.

317. Lee Y-J, Huang J-M, Kuo S-W, Chen J-K, Chang F-C (2005) Polymer 46:2320-2330.

318. Liu YR, Huang YD, Liu L (2007) J Mater Sci 42:5544-5550.

319. Liu H, Kondo S-i, Tanaka R, Oku H, Unno M (2008) J Organomet Chem 693:1301-1308.

320. Lickiss PD (1995) The Synthesis and Structure of Organosilanols. In Sykes AG (ed) Advances in Inorganic Chemistry, Academic Press, Oxford, UK, pp. 147-262.

321. Lickiss PD (2001) Polysilanols. In Rappoport Z, Apeloig Y (eds) The Chemistry of Organic Silicon Compounds, Wiley, Weinheim, Germany, pp. 695-744.

322. Goffin A-L, Duquesne E, Moins S, Alexandre M, Dubois P (2007) Eur Polym J 43:4103-4113.

323. Kim Y, Zhao F, Mitsuishi M, Watanabe A, Miyashita T (2008) J Am Chem Soc 130:11848-11849.

324. Eon D, Raballand V, Cartry G, Cardinaud C, Vourdas N, Argitis P, Gogolides E (2006) J Vac Sci Technol, B: Microelectron Nanometer Struct–Process, Meas, Phenom 24:2678-2688.

325. Chojnowski J, Fortuniak W, Rościszewski P, Werel W, Lukasiak J, Kamysz W, Halasa R (2006) J Inorg Organomet Polym Mater 16:219-230.

326. Dare EO, Olatunji GA, Ogunniyi DS (2005) Pol J Chem 79:101-107.

327. Kawakami Y, Yamaguchi K, Yokozawa T, Serizawa T, Hasegawa M, Kabe Y (2007) Chem Lett 36:792-793.

328. Zhang X, Hu L, Sun D, Zhao W (2008) J Mol Struct 872:197-204.

329. dell' Erba IE, Fasce DP, Williams RJJ, Erra-Balsells R, Fukuyama Y, Nonami H (2003) J Organomet Chem 686:42-51.

330. Morin CJ, Geulin L, Desbène A, Desbène PL (2004) J Chromatogr, A 1032:327-334.

331. Gao J, Kong D, Li S (2008) Int J Polym Mater 57:940-956.

332. Eggers K, Eichner T, Woenckhaus J (2005) Int J Mass Spectrom 244:72-75.

333. Pelster SA, Schüth F, Schrader W (2007) Anal Chem (Washington, DC) 79:6005-6012 .

334. Pelster SA, Weimann B, Schaack BB, Schrader W, Schüth F (2007) Angew Chem, Int Ed 46:6674-6677.

335. Sulaiman S, Bhaskar A, Zhang J, Guda R, Goodson T, III, Laine RM (2008) Chem Mater 20:5563-5573.

336. Zhang L, Abbenhuis HCL, Yang Q, Wang Y-M, Magusin PCMM, Mezari B, van Santen RA, Li C (2007) Angew Chem, Int Ed 46:5003-5006.

337. Erben C, Grade H, Goddard GD (2006) Silicon Chem 3:43-49.

338. Pan G, Mark JE, Schaefer DW (2003) J Polym Sci, Part B: Polym Phys 41:3314-3323.

339. Gao Y, Eguchi A, Kakehi K, Lee YC (2004) Org Lett 6:3457-3460.

340. Cheng G, Peng X, Hao G, Kennedy VO, Ivanov IN, Knappenberger K, Hill TJ, Rodgers MAJ, Kenney ME (2003) J Phys Chem A 107:3503-3514.

341. Lo MY, Zhen C, Lauters M, Jabbour GE, Sellinger A (2007) J Am Chem Soc 129:5808-5809.

342. Sellinger A, Tamaki R, Laine RM, Ueno K, Tanabe H, Williams E, Jabbour GE (2005) Chem Commun:3700-3702.

343. Anderson SE, Somogyi A, Haddad TS, Coughlin EB, Gadodia G, Marten DF, Ray J, Bowers MT (2010) Int J Mass Spectrom 292:38-47.

344. Gidden J, Kemper PR, Shammel E, Fee DP, Anderson S, Bowers MT (2003) Int J Mass Spectrom 222:63-73.

345. Baker ES, Gidden J, Fee DP, Kemper PR, Anderson SE, Bowers MT (2003) Int J Mass Spectrom 227:205-216.

346. Baker ES, Gidden J, Anderson SE, Haddad TS, Bowers MT (2004) Nano Lett 4:779-785.

347. Lin T, He C, Xiao Y (2003) J Phys Chem B 107:13788-13792.

348. Shieh D-L, Chen F-C, Lin J-L (2006) Appl Surf Sci 252:2171-2177.

349. Laine RM, Sulaiman S, Brick C, Roll M, Tamaki R, Asuncion MZ, Neurock M, Filhol JS, Lee CY, Zhang J, Goodson T, Ronchi M, Pizzotti M, Rand SC, Li Y (2010) J Am Chem Soc 132:3708-3722.

350. Zhen C-G, Becker U, Kieffer J (2009) J Phys Chem A 113:9707-9714.

351. Xiao Y, Liu L, He C, Chin WS, Lin T, Mya KY, Huang J, Lu X (2006) J Mater Chem 16:829-836.

352. Brick CM, Ouchi Y, Chujo Y, Laine RM (2005) Macromolecules 38:4661-4665.

353. Lee J, Cho H-J, Jung B-J, Cho NS, Shim H-K (2004) Macromolecules 37:8523-8529.

354. Takagi K, Kunii S, Yuki Y (2005) J Polym Sci, Part A: Polym Chem 43:2119-2127.

355. Zhao W, Cao T, White JM (2004) Adv Funct Mater 14:783-790.

356. Xiao S, Nguyen M, Gong X, Cao Y, Wu H, Moses D, Heeger AJ (2003) Adv Funct Mater 13:25-29.

357. Lee J, Cho H-J, Cho NS, Hwang D-H, Kang J-M, Lim E, Lee J-I, Shim H-K (2006) J Polym Sci, Part A: Polym Chem 44:2943-2954.

358. Kang J-M, Cho H-J, Lee J, Lee J-I, Lee S-K, Cho N-S, Hwang D-H, Shim H-K (2006) Macromolecules 39:4999-5008.

359. Lee CW, Josse Y, Hsu CH, Nguyen TP (2008) Eur Phys J Appl Phys 42:213-218.

360. Kaftory M, Kapon M, Botoshansky M (1998) The Structural Chemistry of Organosilicon Compounds. In Rappoport Z, Apeloig Y (eds) The Chemistry of Organic Silicon Compounds, Wiley, Weinheim, Germany, pp. 181-265.

361. Jiao J, Lee M-Y, Barnes CE, Hagaman EW (2008) Magn Reson Chem 46:690-692.

362. Bassindale AR, Chen H, Liu Z, MacKinnon IA, Parker DJ, Taylor PG, Yang Y, Light ME, Horton PN, Hursthouse MB (2004) J Organomet Chem 689:3287-3300.

363. Törnroos KW, Bürgi H-B, Calzaferri G, Bürgy H (1995) Acta Crystallogr, Sect B: Struct Sci B51:155-161.

364. Waddon AJ, Coughlin EB (2003) Chem Mater 15:4555-4561.

365. Capaldi FM, Kalra A, Manevitch O, Rutledge GC (2006) PMSE Prepr 94:804-805.

366. Earley CW (1994) J Phys Chem 98:8693-8698.

367. Jug K, Wichmann D (1997) J Mol Struct THEOCHEM 398-399:365-379.

368. Shen J, Cheng WD, Wu DS, Li XD, Lan YZ, Zhang H, Gong YJ, Li FF, Huang SP (2005) J Chem Phys 122:204709/204701-204709/204708.

369. Smolin YI, Shepelev YF, Hobbel D (1994) Kristallografiya 39:558-560.

370. Haile SM, Maier J, Wuensch BJ, Laudise RA (1995) Acta Crystallogr, Sect B: Struct Sci B51:673-680.

371. Auf der Heyde TPE, Bürgi H-B, Bürgy H, Törnroos KW (1991) Chimia 45:38-40.

372. Wann DA, Less RJ, Rataboul F, McCaffrey PD, Reilly AM, Robertson HE, Lickiss PD, Rankin DWH (2008) Organometallics 27:4183-4187.

373. Törnroos KW, Calzaferri G, Imhof R (1995) Acta Crystallogr, Sect C: Cryst Struct Commun 51:1732-1735.

374. Koellner G, Müller U (1989) Acta Crystallogr, Sect C: Cryst Struct Commun 45:1106-1107.

375. Bonhomme C, Tolédano P, Maquet J, Livage J, Bonhomme-Coury L (1997) J Chem Soc, Dalton Trans:1617-1626.

376. Hossain MA, Hursthouse MB, Malik KMA (1979) Acta Crystallogr, Sect B: Struct Sci 35:2258-2260.

377. Slebodnick C, Angel RJ, Hanson BE, Agaskar PA, Soler T, Falvello LR (2008) Acta Crystallogr, Sect B: Struct Sci 64:330-337.

378. Bürgi HB, Törnroos KW, Calzaferri G, Bürgy H (1993) Inorg Chem 32:4914-4919.

379. Wann DA, Rataboul F, Reilly AM, Robertson HE, Lickiss PD, Rankin DWH (2009) Dalton Trans:6843-6848.

380. Li H-C, Lee C-Y, McCabe C, Striolo A, Neurock M (2007) J Phys Chem A 111:3577-3584.

381. Ionescu TC, Qi F, McCabe C, Striolo A, Kieffer J, Cummings PT (2006) J Phys Chem B 110:2502-2510.

382. McCabe C, Glotzer SC, Kieffer J, Neurock M, Cummings PT (2004) J Comput Theor Nanosci 1:265-279.

383. Kudo T (2009) J Phys Chem A 113:12311-12321.

384. Zhang X, Chan ER, Glotzer SC (2005) J Chem Phys 123:184718/184711-184718/184716.

385. Zhang X, Chan ER, Ho LC, Glotzer SC (2005) Mater Res Soc Symp Proc 847:369-373.

386. Striolo A, McCabe C, Cummings PT (2006) J Chem Phys 125:104904/104901-104904/104911.

387. Park SS, Xiao C, Hagelberg F, Hossain D, Pittman CU, Jr., Saebo S (2004) J Phys Chem A 108:11260-11272.

388. Hossain D, Pittman CU, Jr., Hagelberg F, Saebo S (2008) J Phys Chem C 112:16070-16077.

389. Hossain D, Pittman CU, Saebo S, Hagelberg F (2007) J Phys Chem C 111:6199-6206.

390. Pielichowski K, Njuguna J, Janowski B, Pielichowski J (2006) Adv Polym Sci 201:225-296.

391. Li G, Pittman CU, Jr. (2005) Polyhedral oligomeric silsesquioxane (POSS) polymers, copolymers, and resin nanocomposites. In Abd-El-Aziz AS, Carraher CE, Jr., Pittman CU, Jr., Zeldin M (eds) Macromolecules Containing Metal and Metal-Like Elements, Wiley, Weinheim, Germany, pp. 79-131.

392. Laine RM (2005) J Mater Chem 15:3725-3744.

393. Phillips SH, Haddad TS, Tomczak SJ (2004) Curr Opin Solid State Mater Sci 8:21-29.

394. Wahab MA, Kim I, Ha C-S (2005) Silica- and silsesquioxane-containing polymer nanohybrids. In Abd-El-Aziz AS, Carraher CE, Jr., Pittman CU, Jr., Zeldin M (eds) Macromolecules Containing Metal and Metal-Like Elements, Wiley, Weinheim, Germany, pp. 133-160.

395. Kickelbick G (2003) Prog Polym Sci 28:83-114.

396. Bourbigot S, Duquesne S, Jama C (2006) Macromol Symp 233:180-190.

397. Joshi M, Butola BS (2004) J Macromol Sci Part C Polym Rev 44:389-410.

398. Chattopadhyay DK, Raju KVSN (2007) Prog Polym Sci 32:352-418.

399. Kannan RY, Salacinski HJ, Butler PE, Seifalian AM (2005) Acc Chem Res 38:879-884.

400. Morris RE (2005) J Mater Chem 15:931-938.

401. DeArmitt C, Wheeler P (2008) Plastics, Additives & Compounding 10:36-39.

402. Kawakami Y, Lee DW, Pakjamsai C, Seino M, Takano A, Miyasato A, Imae I (2007) ACS Symp Ser 964:301-312.

403. Dias Filho NL, Costa RM, Marangoni F, Pereira DS (2007) J Colloid Interface Sci 316:250-259.

404. Hato MJ, Ray SS, Luyt AS (2008) Macromol Mater Eng 293:752-762.

405. Kowalewska A, Rózga-Wijas K, Handke M (2008) e-Polymers. http://www.e-polymers.org/journal/papers/akowalewska_101108.pdf. Acessed May 13, 2010.

406. Fina A, Tabuani D, Frache A, Camino G (2005) Polymer 46:7855-7866.

407. Joshi M, Butola BS, Simon G, Kukaleva N (2006) Macromolecules 39:1839-1849.

408. Kopesky ET, Haddad TS, Cohen RE, McKinley GH (2004) Macromolecules 37:8992-9004.

409. Lee Y-J, Kuo S-W, Huang W-J, Lee H-Y, Chang F-C (2004) J Polym Sci, Part B: Polym Phys 42:1127-1136.

410. Qiu Z, Pan H (2010) Compos Sci Technol 70:1089-1094.

411. Scapini P, Figueroa CA, Amorim CL, Machado G, Mauler RS, Crespo JS, Oliveira RV (2010) Polym Int 59:175-180.

412. Zhang W, Liu D, Wang Y, Geng H, Zhang L (2005) Hecheng Xiangjiao Gongye 28:369-373.

413. Ma J, Li Q (2006) Advanced Materials Research 11-12:323-326.

414. Feng Y, Jia Y, Xu H (2009) J Appl Polym Sci 111:2684-2690.

415. Zhou Z, Cui L, Zhang Y, Zhang Y, Yin N (2008) Eur Polym J 44:3057-3066.

416. Wang R, Wang S, Zhang Y (2009) J Appl Polym Sci 113:3095-3102.

417. Zhang Z, Liang G, Lu T (2007) J Appl Polym Sci 103:2608-2614.

418. Seçkin T, Köytepe S, Adıgüzel Hİ (2008) Mater Chem Phys 112:1040-1046.

419. Seçkın T, Gültek A, Köytepe S (2005) Turk J Chem 29:49-59.

420. Gültek A, Seçkın T, Adigüzel Hİ (2005) Turk J Chem 29:391-399.

421. Markovic E, Matisons J, Hussain M, Simon GP (2007) Macromolecules 40:4530-4534.

422. Kim K-M, Chujo Y (2003) J Polym Sci, Part A: Polym Chem 41:1306-1315.

423. Fox DM, Maupin PH, Harris RH, Jr., Gilman JW, Eldred DV, Katsoulis D, Trulove PC, De Long HC (2007) Langmuir 23:7707-7714.

424. Seurer B, Coughlin EB (2008) Macromol Chem Phys 209:1198-1209.

425. Wang J, Ye Z, Joly H (2007) Macromolecules 40:6150-6163.

426. Bizet S, Galy J, Gérard J-F (2006) Macromolecules 39:2574-2583.

427. Zhang H-X, Lee H-Y, Shin Y-J, Lee D-H, Noh SK (2008) Chin J Polym Sci 26:533-537.

428. Zucchi IA, Galante MJ, Williams RJJ (2009) Eur Polym J 45:325-331.

429. Seurer B, Coughlin EB (2008) Macromol Chem Phys 209:2040-2048.

430. Lee JH, Jeong YG (2010) J Appl Polym Sci 115:1039-1046.

431. Lee KM, Knight PT, Chung T, Mather PT (2008) Macromolecules 41:4730-4738.

432. Knight PT, Lee KM, Qin H, Mather PT (2008) Biomacromolecules 9:2458-2467.

433. Jung YC, So HH, Cho JW (2006) J Macromol Sci, Part B: Phys 45:453-461.

434. Qin H, Mather PT (2006) PMSE Prepr 94:127-128.

435. Wu J, Ge Q, Burke KA, Mather PT (2005) Mater Res Soc Symp Proc 847:93-98.

436. Yei D-R, Kuo S-W, Su Y-C, Chang F-C (2004) Polymer 45:2633-2640.

437. Fina A, Tabuani D, Peijs T, Camino G (2009) Polymer 50:218-226.

438. Pu K-Y, Zhang B, Ma Z, Wang P, Qi X-Y, Chen R-F, Wang L-H, Fan Q-L, Huang W (2006) Polymer 47:1970-1978.

439. Leu C-M, Chang Y-T, Wei K-H (2003) Chem Mater 15:3721-3727.

440. Liang K, Toghiani H, Li G, Pittman CU, Jr. (2005) J Polym Sci, Part A: Polym Chem 43:3887-3898.

441. Chen Y, Kang E-T (2004) Mater Lett 58:3716-3719.

442. Chen Y, Chen L, Nie H, Kang ET (2006) J Appl Polym Sci 99:2226-2232.

443. Pyun J, Matyjaszewski K, Wu J, Kim G-M, Chun SB, Mather PT (2003) Polymer 44:2739-2750.

444. Hosaka N, Otsuka H, Hino M, Takahara A (2008) Langmuir 24:5766-5772.

445. Kim B-S, Mather PT (2006) Macromolecules 39:9253-9260.

446. Nie WY, Li G, Li Y, Xu HY (2009) Chin Chem Lett 20:738-742.

447. Jung CY, Kim HS, Hah HJ, Koo SM (2009) Chem Commun:1219-1221.

448. Tanaka K, Adachi S, Chujo Y (2009) J Polym Sci, Part A: Polym Chem 47:5690-5697.

449. Hamilton K, Misra R, Morgan SE (2006) Polym Prepr 47:603-604.

450. Fina A, Abbenhuis HCL, Tabuani D, Frache A, Camino G (2006) Polym Degrad Stab 91:1064-1070.

451. Dare EO, Olatunji GA, Ogunniyi DS (2004) J Appl Polym Sci 93:907-910.

452. Mabry JM, Vij A, Iacono ST (2007) Polym Prepr 48:970-971.

453. Mabry JM, Vij A, Viers BD, Blanski RL, Gonzalez RI, Schlaefer CE (2004) Polym Prepr 45:648-649.

454. Xue Y-H, Gu X-P, Feng L-F, Wang J-J, Hu G-H (2007) Zhejiang Daxue Xuebao, Gongxueban 41:679-682.

455. Zhou Z, Zhang Y, Zeng Z, Zhang Y (2008) J Appl Polym Sci 110:3745-3751.

456. Yu H, Ren W, Zhang Y (2009) J Appl Polym Sci 113:17-23.

457. Pakjamsai C, Kawakami Y (2005) Des Monomers Polym 8:423-435.

458. Chen H-J (2004) Chem Res Chin Univ 20:42-45.

459. Iyer P, Coleman MR (2008) J Appl Polym Sci 108:2691-2699.

460. do Carmo DR, Dias Filho NL, Stradiotto NR (2004) Mater Res (São Carlos, Brazil) 7:499-504.

461. Isayeva IS, Kennedy JP (2004) J Polym Sci, Part A: Polym Chem 42:4337-4352.

462. Kuo S-W, Lin H-C, Huang W-J, Huang C-F, Chang F-C (2006) J Polym Sci, Part B: Polym Phys 44:673-686.

463. Soh MS, Yap AUJ, Sellinger A (2007) Eur Polym J 43:315-327.

464. Lee Y-J, Huang J-M, Kuo S-W, Lu J-S, Chang F-C (2005) Polymer 46:173-181.

465. Ramírez C, Abad MJ, Barral L, Cano J, Díez FJ, López J, Montes R, Polo J (2003) J Therm Anal Calorim 72:421-429.

466. Ramírez C, Vilariño JML, Abad MJ, Barral L, Bouza R, Cano J, Díez FJ, García-Garabal S, López J (2004) J Appl Polym Sci 92:1576-1583

467. Abad MJ, Barral L, Fasce DP, Williams RJJ (2003) Macromolecules 36:3128-3135.

468. Lo Schiavo S, Mineo P, Cardiano P, Piraino P (2007) Eur Polym J 43:4898-4904.

469. Ramírez C, Rico M, Barral L, Díez J, García-Garabal S, Montero B (2007) J Therm Anal Calorim 87:69-72.

470. Tegou E, Bellas V, Gogolides E, Argitis P, Eon D, Cartry G, Cardinaud C (2004) Chem Mater 16:2567-2577.

471. Cardoen G, Coughlin EB (2004) Macromolecules 37:5123-5126.

472. Chen Y, Schneider KS, Banaszak Holl MM, Orr BG (2004) Phys Rev B: Condens Matter Mater Phys 70:085402/085401-085402/085407.

473. Schneider KS, Nicholson KT, Owens TM, Orr BG, Banaszak Holl MM (2003) Ultramicroscopy 97:35-45.

474. Hottle JR, Deng J, Kim H-J, Farmer-Creely CE, Viers BD, Esker AR (2005) Langmuir 21:2250-2259.

475. Ricco L, Russo S, Monticelli O, Bordo A, Bellucci F (2005) Polymer 46:6810-6819.

476. Perrin FX, Chaoui N, Margaillan A (2009) Thermochim Acta 491:97-102.

477. Yaseen M, Salacinski HJ, Seifalian AM, Lu JR (2008) Biomed Mater (Bristol, UK) 3:034123/034121-034123/034112.

478. Kannan RY, Salacinski HJ, De Groot J, Clatworthy I, Bozec L, Horton M, Butler PE, Seifalian AM (2006) Biomacromolecules 7:215-223.

479. Lee W, Ni S, Kim B-S, Satija SK, Mather PT, Esker AR (2006) Polym Prepr 47:1214-1215.

480. Lee W, Ni S, Deng J, Kim B-S, Satija SK, Mather PT, Esker AR (2007) Macromolecules 40:682-688.

481. Owens TM, Nicholson KT, Fosnacht DR, Orr BG, Banaszak Holl MM (2006) Langmuir 22:9619-9622.

482. Tegou E, Bellas V, Gogolides E, Argitis P, Dean KR, Eon D, Cartry G, Cardinaud C (2003) Proc SPIE-Int Soc Opt Eng 5039:453-461.

483. Ni Y, Zheng S (2007) Macromolecules 40:7009-7018.

484. Liu Y, Ni Y, Zheng S (2006) Macromol Chem Phys 207:1842-1851.

485. Imae I, Kawakami Y (2005) J Mater Chem 15:4581-4583.

486. Imae I, Kawakami Y (2005) Proc SPIE-Int Soc Opt Eng 5937:1N/1-1N/8.

487. Kim SG, Choi J, Tamaki R, Laine RM (2005) Polymer 46:4514-4524.

488. Choi J, Yee AF, Laine RM (2003) Macromolecules 36:5666-5682.

489. Hussain H, Mya KY, Xiao Y, He C (2008) J Polym Sci, Part A: Polym Chem 46:766-776.

490. Mya KY, Li X, Chen L, Ni X, Li J, He C (2005) J Phys Chem B 109:9455-9462.

491. Catari E, Harris H, Lutz PJ (2006) Polym Prepr 47:714-715.

492. Miao J, Zhu L (2010) J Phys Chem B 114:1879-1887.

493. Matějka L, Dukh O, Meissner B, Hlavatá D, Brus J, Strachota A (2003) Macromolecules 36:7977-7985.

494. Kaneshiro TL, Jeong E-K, Morrell G, Parker DL, Lu Z-R (2008) Biomacromolecules 9:2742-2748.

495. Capaldi FM, Boyce MC, Rutledge GC (2006) J Chem Phys 124:214709/214701-214709/214704.

496. Foerster CE, Serbena FC, Garcia ITS, Lepienski CM, Roman LS, Galvão JR, Zawislak FC (2004) Nucl Instrum Methods Phys Res, Sect B 218:375-380.

497. Schmid GM, Stewart MD, Wetzel J, Palmieri F, Hao J, Nishimura Y, Jen K, Kim EK, Resnick DJ, Liddle JA, Willson CG (2006) J Vac Sci Technol, B: Microelectron Nanometer Struct–Process, Meas, Phenom 24:1283-1291.

498. Mya KY, Nakayama N, Takaki T, Xiao Y, Lin TT, He C (2008) J Appl Polym Sci 108:181-188.

499. Hao N, Böhning M, Goering H, Schönhals A (2007) Macromolecules 40:2955-2964.

500. Hao N, Böhning M, Schönhals A (2007) Macromolecules 40:9672-9679.

501. Leu C-M, Chang Y-T, Wei K-H (2003) Macromolecules 36:9122-9127.

502. Bian Y, Pejanović S, Kenny J, Mijović J (2007) Macromolecules 40:6239-6248.

Chapter 3
Metallasilsesquioxanes: Molecular Analogues of Heterogeneous Catalysts

Antony J. Ward, Anthony F. Masters and Thomas Maschmeyer

3.1 Introduction

As soluble molecular analogues of silica surfaces, polyhedral oligosilsesquioxanes (POS or POSS®) have been used extensively, along with alkylsilanols, to mimic the structure and chemistry of silica surfaces (Figure 3.1). The silsesquioxane family represents the best molecular equivalent of the silica surface, since they mimic the environments of the different types of silanol groups (isolated, vicinal, or geminal), the dispositions of near-surface silicon and oxygen atoms, and have pK_a values for⇦SiOH units that are close to those of silica [1]. In addition, pure silica surfaces, particularly those containing grafted metal complexes, have also been successfully investigated. Important characteristics that can be replicated in silsesquioxane chemistry include electron-withdrawing bonding sites [2] and interactions with adjacent oxygen donors [3], which both contribute to the stability of grafted metal complexes [4]. Thus, silsesquioxane-based homogeneous models for heterogeneous catalysts offer a unique opportunity to understand heterogeneous catalysis on a molecular level. Furthermore, some of these model complexes exhibit catalytic activities that are comparable to, or even better than, those of commercially used heterogeneous silica-supported catalysts [5].

A.J. Ward, A.F. Masters and T. Maschmeyer
Laboratory of Advanced Catalysis for Sustainability
School of Chemistry F11, University of Sydney, New South Wales, Australia

C. Hartmann-Thompson (ed.), *Applications of Polyhedral Oligomeric Silsesquioxanes*, Advances in Silicon Science 3, DOI 10.1007/978-90-481-3787-9_3,

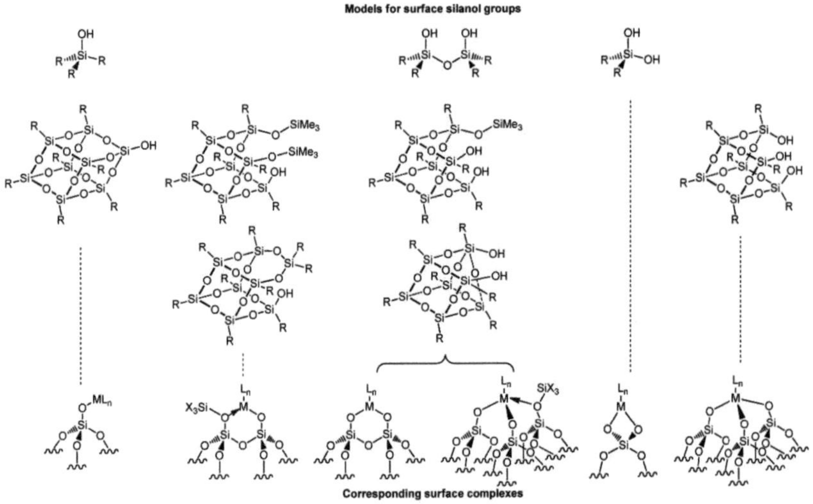

Fig. 3.1 Molecular models that mimic the various types of silanol groups of a silica surface. Modified from Copéret et al. [6]

In this chapter we will review metallasilsesquioxanes, or metal-complexes of incompletely condensed POS, which have demonstrated catalytic activity. Surprisingly, many of the POS complexes advanced as catalyst models in the literature have never been catalytically evaluated, and hence fall outside the scope of this work. Equally, the structural assignment of metallasilsesquioxanes can be difficult. To avoid ambiguity, we have, for the most part, restricted ourselves to structures obtained from single crystal X-ray diffraction data.

3.2 Metallasilsesquioxanes

3.2.1 Group 4 – Ti, Zr, Hf

The reaction of the cubic trisilanols $(c\text{-}C_5H_9)_7Si_7O_9(OH)_3$ and $(c\text{-}C_6H_{11})_7Si_7O_9(OH)_3$ with $Ti(O^iPr)_4$ in tetrahydrofuran (THF) afforded the titanosilsesquioxanes $\{[(c\text{-}C_5H_9)_7Si_7O_9]Ti(O^iPr)\}_n$ (**1a**) and $\{[(c\text{-}C_6H_{11})_7Si_8O_9]Ti(O^iPr)\}_n$ (**1b**), respectively [7]. Nuclear magnetic resonance (NMR) studies showed that these compounds were in equilibrium between a four-coordinate monomer (**1**$_M$) and a five-coordinate dimer (**1**$_D$) with a relative monomer-dimer ratio of *ca.* 2:1 (Fig. 3.2). Addition of methanol to a solution of (**1**) led to the rapid and exclusive formation of the 6-coordinate dimer (R = $c\text{-}C_5H_9$ (**2a**), $c\text{-}C_6H_{11}$ (**2b**)). A single crystal suitable for X-ray diffraction studies was obtained for (**2b**). Compound (**1a**) displayed catalytic activity for the epoxidation of cyclohexene (40 °C, $CHCl_3$, catalyst:^{t}BuOOH:substrate 0.05:1:12), achieving 13% conversion after 1 h, 98% selectivity to the epoxide, and a turnover frequency (TOF) of 0.7 mmol(product) (g Ti)$^{-1}$ min^{-1}. Under identical conditions, (**2b**) achieved a conversion of 93% in 1 h with 98% selectivity to the epoxide with a TOF of 4.8 mmol(product) (g Ti)$^{-1}$ min^{-1}. These complexes were prepared as model compounds for the grafted Ti heterogeneous catalyst system Ti↑MCM-41 [8]. The grafted Ti catalyst achieves a TOF of 3 mmol(product) (g Ti)$^{-1}$ min^{-1} for the epoxidation of cyclohexene under identical conditions to those employed for (**1a**) and (**2b**). Comparison of the TOFs shows that the dimer, (**2b**), displays the greatest activity of all three catalysts. This result contradicted the hypothesis of the authors, in that they had expected the monomer (**1a**) to have the greatest activity, owing to the greater number of available coordination sites relative to the dimer. Testing the nature of the group in the fourth coordination site of the dimer revealed that the reactivity of the catalyst was dependent upon the alkoxy group in the following order of decreasing reactivity: OMe > O^{t}Bu >> O^{i}Pr.

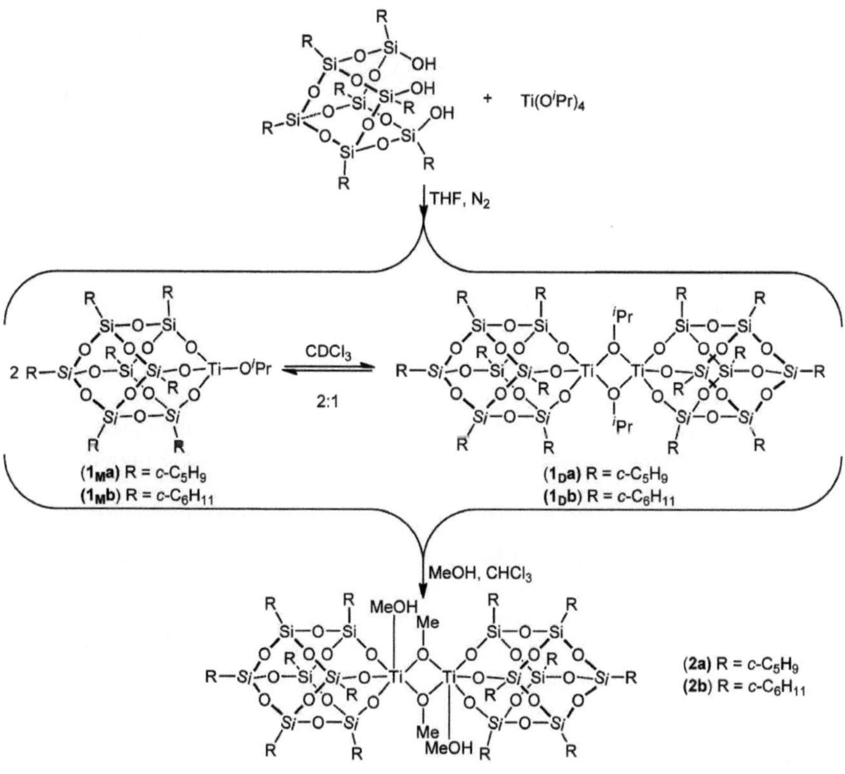

(a)

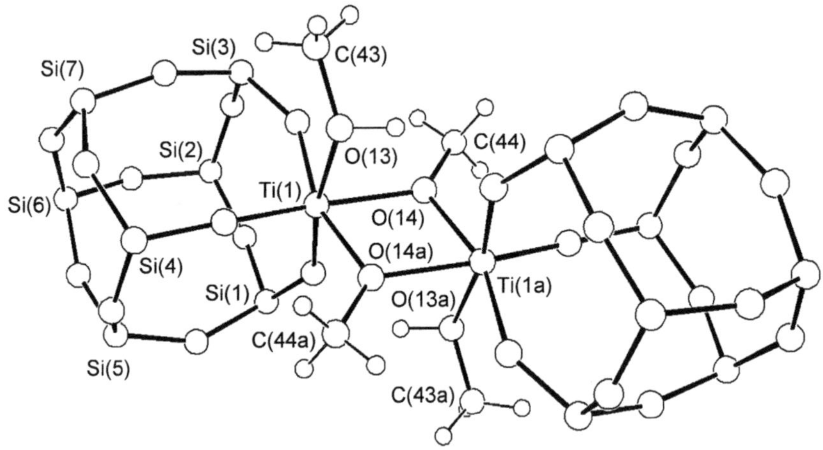

(b)

Fig. 3.2 (a) Synthesis of the titanasilsesquioxanes (**1**) and (**2**) (redrawn from Maschmeyer *et al.* [7]) (b) Crystal structure of (**2b**). Reproduced from [7] with permission from the Royal Society of Chemistry.

Reaction of homoleptic titanium(IV) complexes [TiL_4] with the incompletely condensed silsesquioxane $(c\text{-}C_6H_{11})_7Si_7O_9(OH)_3$ afforded the titanium silsesquioxane complexes $\{[(c\text{-}C_6H_{11})_7Si_7O_{12}]TiL\}$ (L = CH_2Ph (**3a**), NMe_2 (**3b**), $OSiMe_3$ (**3c**), O^iPr (**3d**)) via a protonolysis reaction [9,10]. The tripodal TiL(silsesquioxane) species formed were monomeric when freshly prepared, but when L = O^iPr, dimerization occurred upon standing (affording a monomer:dimer ratio of ca. 5:2)[7]. In the case of L = $OSiMe_3$ (**3c**), the monomeric nature of the titanasilisequioxane species was confirmed by single crystal X-ray diffraction. The titanium centre in (**3c**) has pseudo-tetrahedral geometry, and is coordinated to one tridentate silsesquioxane ligand and one siloxy ligand such that the molecule has exact C_3 symmetry (Figure 3.3). Reaction of the disilanol silsesquioxane $(c\text{-}C_5H_9)_7Si_7O_9(OH)_2(OSiMe_3)$ with half an equivalent of $TiCl_4$ afforded the bis(silsesquioxane) complex $\{[c\text{-}C_5H_9)_7Si_7O_{11}(OSiMe_3)]_2Ti\}$ (**3d**). The epoxidation of oct-1-ene with ^{t}BuOOH was chosen as a test of catalytic activity for the complexes (**3a** – **d**) (80 °C, no solvent, substrate:Ti 3342:1). The second order rate constants for the epoxidation were found to be 1.23 (**3a**), 0.63 (**3b**), 0.97 (**3c**) and 1.49 (**3d**) $dm^3\ mol^{-1}\ s^{-1}$ for the complexes. The selectivity of the reaction for all catalysts was >93%. The rate constant for the dimer (**3d**) (0.047 $dm^3\ mol^{-1}\ s^{-1}$) was an order of magnitude less than the other complexes tested with a selectivity of 83%. Comparison with Ti-MCM-41 under the same reaction conditions revealed that the silsesquioxane-based catalysts were an order of magnitude more active based on the rate constants. Hence, the authors concluded that the most active sites in the heterogeneous materials (i.e. Ti-MCM-41) were the tripodal, open-lattice titanium sites. These results suggest that the lower activity of the Ti-MCM-41, prepared by sol-gel means, can be ascribed to the fact that only a small fraction of the surface titanium sites possesses the correct coordination environment for catalysis.

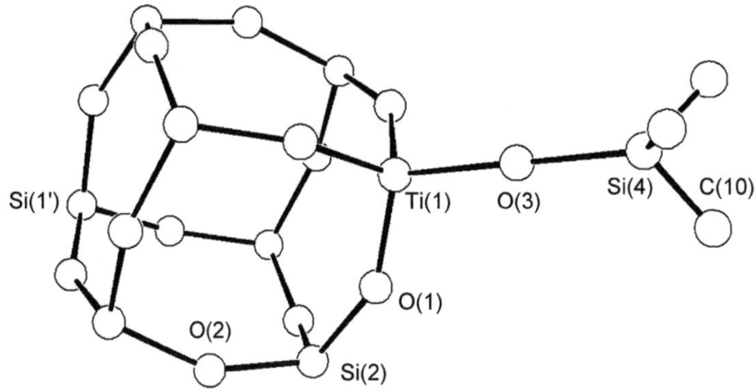

Fig. 3.3 Molecular structure of $\{[(c\text{-}C_6H_{11})_7Si_7O_{12}]TiO^iPr\}$ (**3d**) [10] (reproduced from [10] with permission from the Royal Society of Chemistry)

The titanium(IV) silsesquioxane $\{[(c\text{-}C_6H_{11})_7Si_7O_{12}]Ti(\eta^5\text{-}C_5H_5)\}$ (**4**) can be synthesized by the reaction of $(\eta^5\text{-}C_5H_5)TiCl_3$ with $(c\text{-}C_6H_{11})_7Si_7O_9(OSbMe_4)_3$ [11] or $[(c\text{-}C_6H_{11})_7Si_7O_{12}]^{3-}$ (Fig. 3.4) [12]. The latter method proved to be efficient for the synthesis of substituted-cyclopentadienyl titanium silsesquioxane complexes, such as $\{[(c\text{-}C_6H_{11})_7Si_7O_{12}]Ti(\eta^5\text{-}C_5Ph_5)\}$ which was synthesized by Field and co-workers [13]. This titanium-silsesquioxane complex is an efficient and robust epoxidation catalyst (50 °C, hexane, substrate:Ti ~100:1) for oct-1-ene, cyclohexene and norbornylene [14]. In the case of cyclohexene and norbornylene, conversions of 90% and selectivities of >95% were obtained in 3 h.

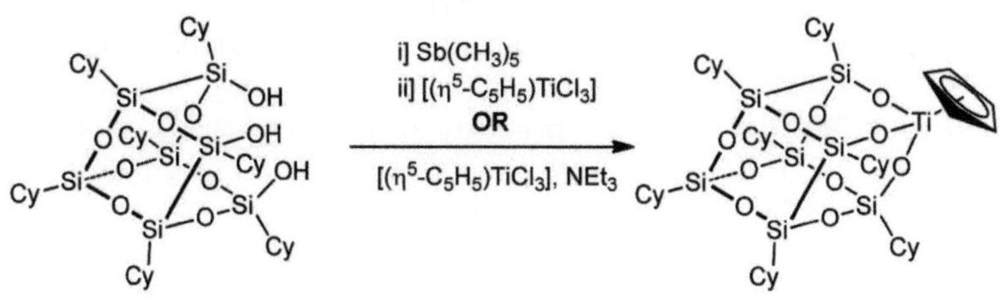

(4)

Figure 3.4 Synthetic route to the titanium(IV) silsesquioxane complex $\{[(c\text{-}C_6H_{11})_7Si_7O_{12}]Ti(\eta^5\text{-}C_5H_5)\}$ (4) [11,12]

The stability of (**4**) under the catalytic conditions employed led Krijnen and co-workers to immobilize the complex in the mesoporous silica MCM-41 (MCM-41 with Si/Al ratios of 42, 125, and ∞) [15]. The adsorption was performed simply and quantitatively by stirring a hexane solution of the complex with the MCM-

41. The adsorption kinetics for the complex were typical for 'single-file diffusion', which indicated that the adsorption was occurring within the channels of the silica. The maximum loading of titanium silsesquioxane in the MCM-41 host with a Si/Al ratio of 42 was found to be 41 mg per gram of silica; however, for the catalysis, loadings of 1.2 mg Ti per gram of MCM-41 were used. The titanium silsesquioxane–MCM-41 materials effectively catalyzed the epoxidation of cyclooctene with *t*-butyl hydroperoxide (50 °C, hexane, substrate:Ti 1200:1). In the case of the aluminium-free silica, high catalytic activity and heterogeneity were observed (84% conversion, >95% selectivity after 22 h). However, the presence of aluminium in the silica resulted in a reduction in activity of the catalysts, and, under the conditions employed, the catalyst was quantitatively leached from the silica. For the aluminium-containing catalysts, the leaching was completely stopped by treating the silica with the silylating agent Ph_2SiCl_2 prior to catalysis [16]. However, when these silylated MCM-41 catalysts were used for the epoxidation reaction, they exhibited lower activity than the aluminium-free catalyst, and the activity decreased over the course of the reaction owing to leaching. When the smaller silylating agent Me_2SiCl_2 was used, the catalyst obtained was active and resistant to leaching. This indicated that it was the apolarity of the surface that prevented loss of the catalyst, rather than physical entrapment by the silylating agent.

When a conventional silica gel was used to adsorb the titanium complex, a significant degree of leaching was observed during catalysis. This indicated that channel-type aluminium-free pores were required for irreversible adsorption.

The reaction of $M(CH_2Ph)_4$ (M = Zr, Hf) with the trisilanol silsesquioxane $(c\text{-}C_5H_9)_7Si_7O_9(OH)_3$ affords the dimeric species $\{[(c\text{-}C_5H_9)_7Si_7O_{12}]MCH_2Ph\}_2$ (M = Zr (**5a**), Hf (**5b**)) (Fig. 3.5a) [17]. A crystal of (**5a**) suitable for single crystal X-ray diffraction was grown from hexane. The structure of (**5a**) was found to be an asymmetric dimer formed by two corner-capped zirconasilsesquioxane moieties connected by two bridging siloxy functionalities (Fig. 3.5b). In order to increase the coordination number of the electron-deficient metal centres, the siloxy groups are μ^2-bridged. Preliminary ethylene polymerization experiments revealed that the dimers were active catalysts (25 °C, pentamethylheptane, 30 μmol catalyst, 0.3 MPa ethylene). The activity was low, affording yields of poly(ethylene) of 0.6 and 0.5 g for (**5a**) and (**5b**) after 15 min, respectively. However, when (**5a**) and (**5b**) were activated with $B(C_6F_5)_3$, the ethylene polymerization activity of both complexes increased dramatically. The reaction with $B(C_6F_5)_3$ resulted in the removal of one of the coordinated benzyl groups, and the formation of the corresponding dimeric zirconasilsesquioxane cation and $[B(C_6F_5)_3(CH_2Ph)]^-$. When (**5a**)-$B(C_6F_5)$ was used for the ethylene polymerization (80 °C, 7 min, pentamethylheptane, 20 μmol catalyst, 0.5 MPa ethylene), a yield of 5.5 g of poly(ethylene) was obtained (M_w = 6,600) with an activity of 2400 g(product) $mmol^{-1}$ h^{-1}. Under the same conditions, (**5b**)-$B(C_6F_5)$ afforded a yield of 11.1 g poly(ethylene) (M_w = 82,000) with an activity of 4800 g(product) $mmol^{-1}$ h^{-1}. Thus, it can be seen that the Hf complex formed a more active polymerization catalyst than its Zr analogue. The authors ascribed this greater activity to the higher thermal stability of the Hf complex, as NMR studies showed that the Zr complex was prone to decomposition at the higher reaction temperatures. For comparison, the polymerization activity of (**5a**) immobilized on silica, pretreated with poly(methylalumoxane) (MAO) was tested; after 7 min at 29 °C, a yield of 9.7 g of poly(ethylene) was obtained (M_w = 65,000) with an activity of 8300 g(product) $mmol^{-1}$ h^{-1}.

Duchateau and co-workers employed the hydroxysilsesquioxane (c-$C_5H_9)_7Si_8O_{12}(OH)$ as a model support for silica-grafted olefin polymerization catalysts [1,18]. Treatment of Cp''MCl_3 (M = Ti, Zr; Cp'' = $C_5H_3(SiMe_3)_2$) with the silsesquioxane afforded the dichloride complex {[(c-$C_5H_9)_7Si_8O_{13}$]MCp''Cl_2} (**6a**) *via* either chloride metathesis or protonolysis. Similarly, treatment of [Cp''MR_3] (R = CH_3, CH_2Ph) with the silanol (c-$C_5H_9)_7Si_8O_{12}(OH)$ afforded the complex {[(c-$C_5H_9)_7Si_8O_{13}$]MCp''R_2} (**6b**). These complexes were then treated with $B(C_6F_5)_3$ to form active polymerization catalysts (80 °C, toluene, 0.5 MPa ethylene). The complexes {[(c-$C_5H_9)_7Si_8O_{13}$]TiCp''$(CH_2Ph)_2$]} and {[(c-$C_5H_9)_7Si_8O_{13}$]TiCp''$(CH_3)_2$} achieved activities of 880 and 840 kg(PE) $mmol^{-1}$ h^{-1} atm^{-1} respectively. It is interesting to note that MAO was unsuitable for the activation of these catalysts, since the aluminium had a high affinity for the silsesquioxane oxo functionalities, and this led to the cleavage of M-O bonds [19].

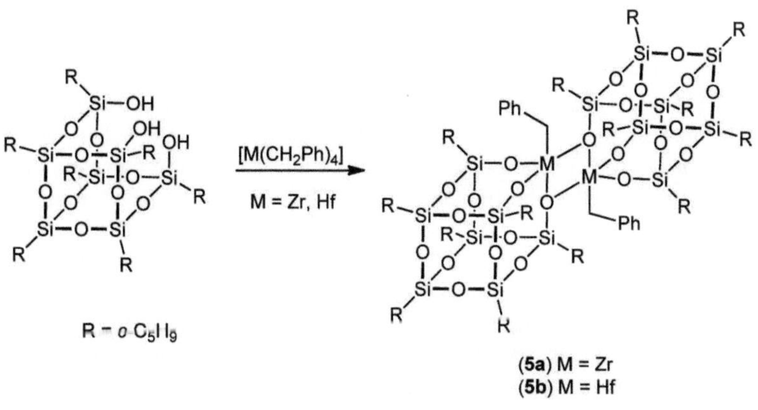

(a)

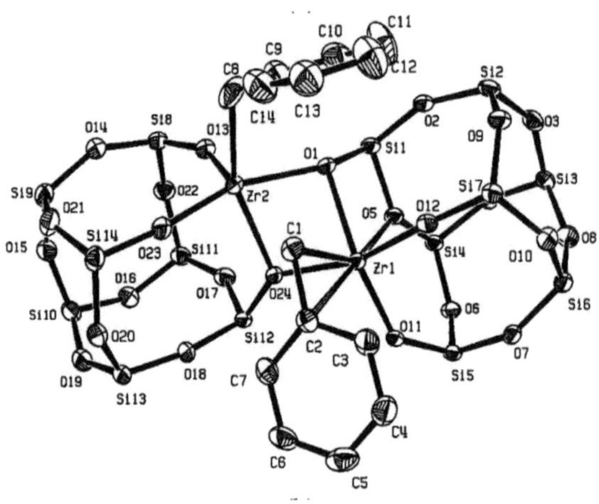

(b)

Fig. 3.5 (a) Preparation of the dimeric metallasilsesquioxanes $\{[(c\text{-}C_5H_9)_7Si_7O_{12}]MCH_2Ph\}_2$ (**5**), and (b) crystal structure of (**5a**) [17]. Reproduced from [17] with permission from the American Chemical Society.

The pentamethylcyclopentadienyl (Cp*) analogues of (**4**) were prepared by Kim and co-workers [20] by the reaction of Cp^*MCl_3 (M = Ti, Zr) with $(c\text{-}C_5H_9)_7Si_7O_9(OH)_3$ in the presence of NEt_3 as previously described by Buys *et al.* [12]. The resulting complexes $\{[(c\text{-}C_5H_9)_7Si_7O_{12}]MCp^*\}$ (M = Ti (**7a**), Zr (**7b**)) were then employed for the polymerization of ethylene in the presence of the co-catalyst modified methylaluminoxane (MMAO) (0, 30, 50 or 70 °C, toluene, Al:M = 1000:1, 0.1 MPa ethylene). The optimum activity for (**7a**)/MMAO and (**7b**)/MMAO was observed at 30 °C and 70 °C, respectively. The Ti catalyst (**7a**) afforded polymers with an M_w in the range of 23,200 – 563,200, whilst the Zr catalyst gave polymers with an M_w in the range of 3,600 – 224,300. The polydispersity index (PDI) obtained by the (**7a**)/MMAO catalyst system ranged from 3.06 (at 70 °C) to 5.25 (at 30 °C). In the case of the (**7b**)/MMAO system, the PDIs obtained ranged from 5.55 (at 70 °C) to 10 (at 50 °C). NMR analysis of the polymers produced by both catalysts systems revealed highly linear polyethylenes.

The synthesis of titanasilsesquioxanes that possess higher metal contents than the mono-metallated polyhedral silsesquioxanes discussed thus far was achieved by Roesky and co-workers by the controlled hydrolysis of monomeric silanetriols and alkoxytitanium complexes [21]. The preparation of the cubic titanosiloxane $[RSiO_3Ti(O^iPr)]_4$ (R = 2,6-$^iPr_2C_6H_3N(SiMe_3)$ (**8**)) was achieved by the addition of $Ti(O^iPr)_4$ or $TiCl(O^iPr)_3$ to a THF solution of *N*-(trimethylsilyl)-*N*-(trihydroxysilyl)-2,6-*iso*-propylaniline (Fig. 3.6a). A crystal structure of (**8**) has subsequently been published (Fig. 3.6b) and shows the cubic core to consist of a Ti-O-Si framework containing alternating titanium and silicon centres [22]. Complex (**8**) was shown to be an active epoxidation catalyst for the epoxidation of cyclohexene (65 °C, toluene, catalyst:cumenyl hydroperoxide:substrate 1:204 :852). After 23 h, a conversion of

15% was achieved with 100% selectivity to the epoxide [22].

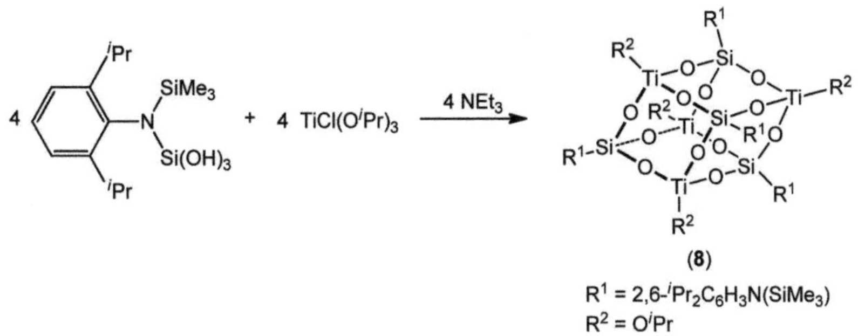

(a)

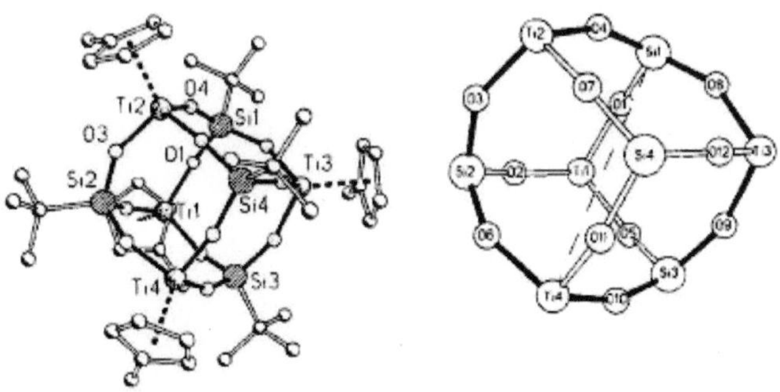

(b)

Fig. 3.6 (a) Preparation of complex $[RSiO_3Ti(O^iPr)]_4$ ($R = 2,6\text{-}^iPr_2C_6H_3N(SiMe_3)$ (**8**), and (b) crystal structure of (**8**) [21] (reproduced from [21] with permission from Wiley)

A series of metallocene-containing silsesquioxanes with alkenylsilyl and trimethylsilyl groups of the general formula $\{[(c\text{-}C_5H_9)_7Si_7O_{11}]\text{-}M(Cp)_2(OSiMe_2R)\}$ ($Cp = \eta^5\text{-}C_5H_5$; M = Ti (**9**), Zr (**10**); R = methyl (**a**), vinyl (**b**), allyl (**c**)) was prepared by Wada and co-workers [23]. The metallasilsesquioxanes were prepared in moderate yield by the reaction of the disilanol silsesquioxane, $(c\text{-}C_5H_9)_7Si_7O_9(OH)_2(OSiMe_2R)$, with the corresponding metallocene dichloride in the presence of excess NEt_3 (Fig. 3.7a). A crystal structure of (**10c**) was obtained (Fig. 3.7b). Several of the metallasilsesquioxanes were used as catalysts for the epoxidation of cyclohexene (50 °C, toluene, catalyst:^{t}BuOOH:cyclohexene 1:100:100). With the catalysts (**9a-c**) yields of 22 to 28% were achieved after 4 h with the activity increasing in the order Me < vinyl < allyl. With complex (**10a**) conversion could be observed only at trace level. The catalytic activities of the complexes (**9**) and (**10**) were compared to the activities of the previously reported complexes (**3d**) and (**4**); complex (**4**), one of the most active silsesquioxane catalysts known, achieved a conversion to the epoxide of 66% under the same conditions, while the dimer (**3d**) achieved a yield of only 15%.

Thus, as was reported by Crocker, Herold and Orpen [10], the catalytic activities of the titanasilsesquioxanes decreased in the following order: tripodal (**4**) > bipodal (**10a**) > tetrapodal (**3d**).

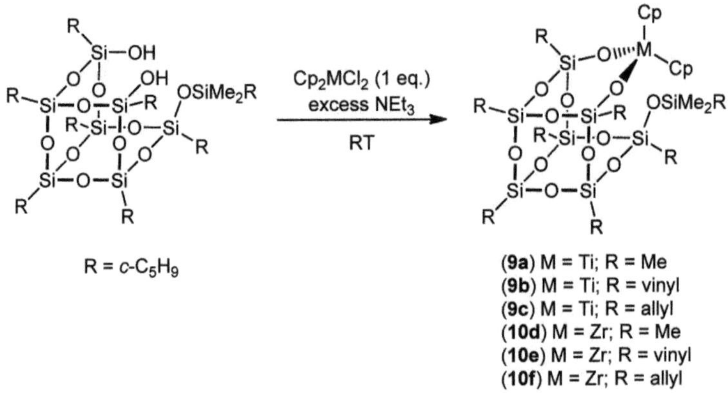

(a)

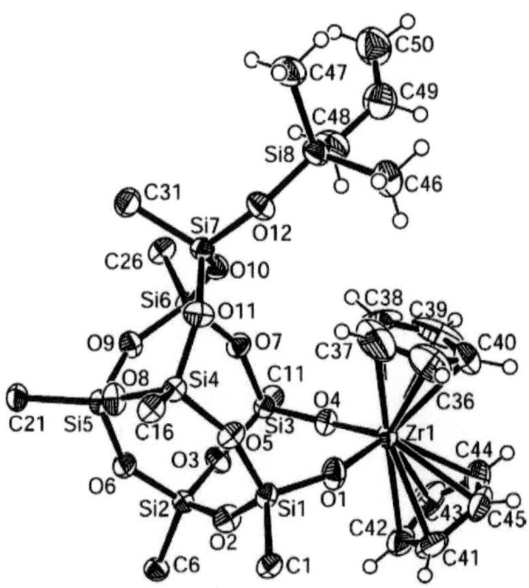

(b)

Fig. 3.7 (a) The synthesis of the metallocene-containing silsesquioxanes (**9**) and (**10**) prepared by Wada [23], and (b) the crystal structure of (**10c**) with ellipsoids at 50% thermal probability (cyclopentyl rings, except for *ipso* carbons, are omitted for clarity) (reproduced from [23] with permission from the American Chemical Society)

The reaction of the silsesquioxane $(c\text{-}C_5H_9)_7Si_8O_{12}(OH)$ with the titanium(IV) alkoxo complex $Ti(O^iPr)_2(OMent)_2$ (OMent = (1*R*,2*S*,5*R*)-(–)-menthoxo) in THF afforded the titanosilsesquioxane $\{[(c\text{-}C_5H_9)_7Si_8O_{13}]Ti(O^iPr)_2(OMent)\}$ (**11**) [24]. When this complex was used to catalyze the epoxidation of cinnamyl alcohol (-20 °C, CH_2Cl_2, catalyst:TBHP:substrate 1:20:2 or 1:20:10) yields of 58% (1:2 ratio) or 62% (1:10 ratio) were obtained. In both cases, an enantiomeric excess (ee) of approximately 15% was achieved. A comparison of the results achieved by (**11**) with those achieved by $Ti(O^iPr)_2(OMent)_2$ (13% yield, 58% ee) indicated that hydrolysis of the silsesquioxane had probably not occurred in the case of (**11**).

The reaction of $Ti(O^iPr)_4$ with the silsesquioxane $(CH_2CH(CH_3)_2)_7Si_7O_9(OH)_3$ in CH_2Cl_2 afforded the dimeric titanium-silsesquioxane complex $\{[CH_2CH(CH_3)_2]_7Si_7O_{12}]_2Ti_2(\mu\text{-}O^iPr)_2\}$ (**12**) [25], which was analogous to the intermediate complex ((**1**$_D$) in Fig. 3.2) described by Maschmeyer et al. [7]. Recrystallization from hexane afforded crystals suitable for single crystal X-ray diffraction, which revealed the dimeric nature of the complex (Fig. 3.8). The complex was tested for activity for ring opening polymerization in a melt of *rac*-lactide (130 °C, no solvent, initiator:*rac*-lactide 1:300). After 0.5 h, complex (**12**) had achieved a yield of 83% with an M_w = 506,200 and a PDI of 2.5. Heterogeneous catalysts were prepared by reacting silica (with pore sizes of 40 or 60 Å) with $Ti(O^iPr)_4$ to afford materials with loadings of 3.40 wt % Ti. Testing of the heterogeneous catalysts for the ring opening polymerization of *rac*-lactide revealed that the catalyst with a 60 Å pore diameter achieved higher yields (78% after 2 h) than the catalyst with a 40 Å pore diameter (52% after 2 h). The 60 Å pore diameter silica produced a polymer with M_W = 48,950 with a PDI of 1.02.

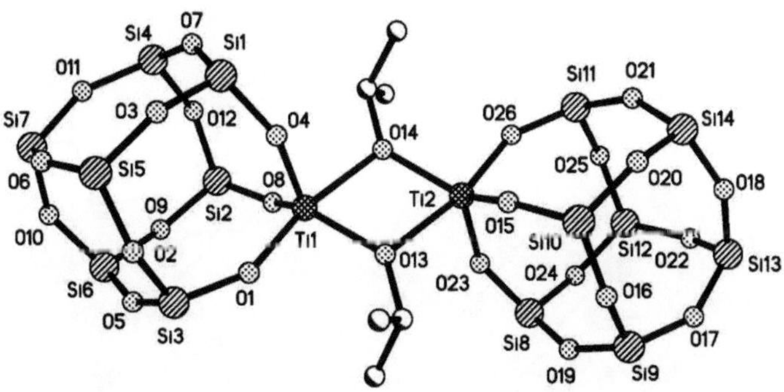

Fig. 3.8 Crystal structure of the dimeric silsesquioxane complex $\{[CH_2CH(CH_3)_2\}_7Si_7O_{12}]_2Ti_2(\mu\text{-}O^iPr)_2\}$ (**12**) [25] (reproduced from [25] with permission from the Royal Society of Chemistry). The isobutyl groups have been omitted for clarity.

3.2.2 Group 5 – V

The vanadate-silsesquioxane complex $\{[(c\text{-}C_5H_9)_7Si_7O_{12}]VO\}$ (**13**) was employed by Wada and co-workers for the photo-assisted catalytic oxidation of cyclohexane [26,27]. The oxidations of cyclohexane were performed in a Pyrex reactor using a mercury arc lamp (32 °C, 0.1 MPa O_2, catalyst:cyclohexane 1:14). After a 6 h irradiation, the silsesquioxane (**13**) achieved yields of 165 μmol cyclohexanol and 91 μmol cyclohexanone, which represented a turnover number (TON) of 20, and no formation of higher oxidation products was observed. A mixture of $VO(acac)_2$ and $(c\text{-}C_5H_9)_7Si_7O_9(OH)_3$ gave yields of 172 μmol and 171 μmol for cyclohexanol and cyclohexanone, respectively, to achieve a TON of 26. Analysis of the products of the reaction of $VO(acac)_2$ with $(c\text{-}C_5H_9)_7Si_7O_9(OH)_3$ showed that a different product to the vandate-silsesquioxane complex (**13**) was initially formed, but after irradiation, the predominant species in solution was (**13**). By comparison, the V_2O_5/silica catalyst achieved a TON of 0.3, and $VO(acac)_2$ achieved a TON of 2.0 under the same reaction conditions.

Two new oxovanadasilsesquioxanes were prepared by Ohde and co-workers [28]. The first, the dimeric $\{[(c\text{-}C_5H_9)_7Si_8O_{11}(OSiMePh_2)]\}_2\{VO_2(O)O^iPr)\}_2$ (**14**), was prepared by the reaction of the silsesquioxane $(c\text{-}C_5H_9)_7Si_7O_{11}(OH)_2(OSiMe_2Ph)$, which possesses a geminal $Si(OH)_2$ unit, with 1 equivalent of $(^iPrO)_3V{=}O$. The second, $[PPh_4]\{[(c\text{-}C_5H_9)_7Si_7O_{12}]VO_2\}$ (**15**), was prepared by reaction of the trisilanol $(c\text{-}C_5H_9)_7Si_7O_9(OH)_3$ with $PPh_4[VO_2Cl_2]$. In both cases, crystal structures were obtained (Fig. 3.9). The new complexes were then tested for their potential to act as catalysts for the oxidation of benzene under photocatalytic conditions using a mercury arc lamp (25 °C, no additional solvent, O_2 atmosphere, catalyst:benzene 1:451). After 6 h of irradiation, 2.0 and 3.4 turnovers were achieved by (**14**) and (**15**), respectively. By comparison, $OV(acac)_2$ and $\{[(c\text{-}C_5H_9)_7Si_7O_{12}]VO\}$ (**13**) achieved 1.6 and 2.6 turnovers in 6 h under the same reaction conditions.

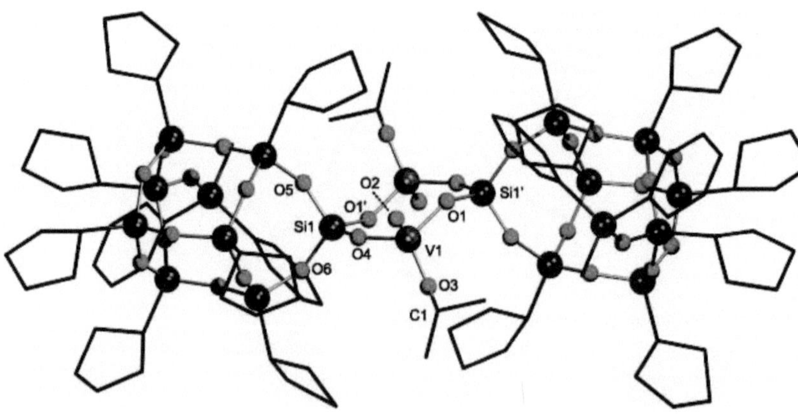

(a)

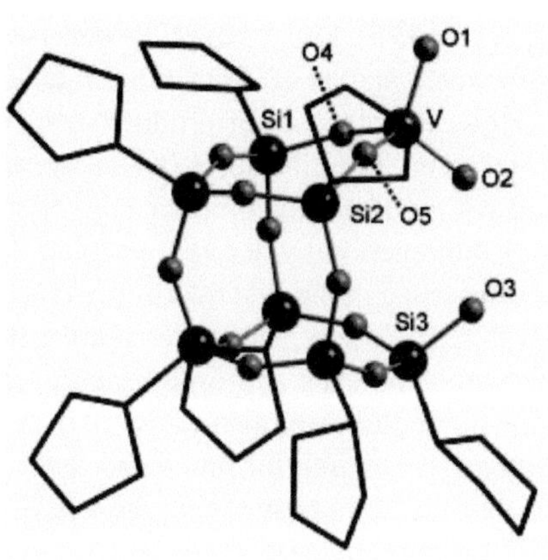

(b)

Fig. 3.9 Crystal structures of (a) the dimeric silsesquioxane $\{[(c\text{-}C_5H_9)_7Si_8O_{11}(OSiMePh_2)]\}_2\{VO_2(O)O^iPr)\}_2$ (**14**) and (b) $[PPh_4]\{[(c\text{-}C_5H_9)_7Si_7O_{12}]VO_2\}$ (**15**) (with the PPh_4^+ counterion omitted for clarity) [28] (reproduced from [28] with permission from the Royal Society of Chemistry)

The vanadate-silsesquioxane complex $\{[(CH_3CH(CH_3)CH_2)_7Si_7O_{12}]VO\}$ (**16**) has been shown to be stable toward alkyl hydroperoxides and was able to activate cumyl hydroperoxide (CHP) [29]. This result is surprising, since cyclopentyl- and cyclohexyl-substituted analogues (*e.g.* (**13**)) are unstable in the presence of hydroperoxides, alcohols and water. The ability of (**16**) to activate CHP was applied to the oxidation of methyl-*p*-tolyl sulfide to methyl-*p*-tolyl sulfoxide (28 °C, $CHCl_3$, catalyst:substrate:CHP 1:100:100). It was found that (**16**) achieved quantitative conversion of the sulfide to the sulfoxide with the formation of a small amount of methyl-*p*-tolyl sulfone (2%). The half-life ($t_{1/2}$) for this reaction was found to be 6000 min. It was noticed that the presence of an electron rich co-ligand, such as trioctylphosphine oxide or dimethylhexyamine *N*-oxide, resulted in dramatic enhancement of activity: the presence of 1 equivalent trioctylphosohine oxide resulted in $t_{1/2}$ of 628 min for the reaction, whilst for dimethylhexyamine *N*-oxide the $t_{1/2}$ is reduced to 250 min when 1 equivalent was added. When trioctylphosphine oxide was used as a co-catalyst, a greater selectivity to the sulfoxide was attained than that observed when using dimethylhexyamine *N*-oxide.

3.2.3 *Group 6 – Mo*

Reaction of the metathesis catalyst precursor $[Mo(CHCMe_2Ph)(NAr)(OSO_2CF_3)_2(DME)]$ (Ar = 2,6-diisopropylphenyl; DME = 1,2-dimethoxyethane) with the incompletely condensed thallate-substituted silsesquioxane $(c\text{-}C_6H_{11})Si_7O_9(OSiMe_3)(OTl)_2$ (Tl = thallium) afforded a quantitative yield of the molybdenum-containing silsesquioxane $\{[(c\text{-}C_6H_{11})Si_7O_9(OSiMe_3)(O)_2]$-

$Mo(CHCMe_2Ph)(NAr)\}$ ((**17a**, **17b**), Fig. 3.10) [2]. When 1-octene was added to the complex (25 °C, neat, substrate:catalyst 4800:1), more than 150 turnovers occurred within 20 s of mixing. However, the rate of productive metathesis slowed as the ethylene concentration increased. Equilibrium between 1-octene and 7-tetradecenes (in a ~1:1 ratio) occurred within 2 h, with complete conversion accomplished more than 4 h after the start of the reaction by actively purging the system of ethylene. Only 100 turnovers were observed in the first 60 s of the reaction when *cis*-2-octene was added to the catalyst (25 °C, neat, substrate:catalyst 9300:1,), but complete equilibrium of 2-octene with 2-butenes and 6-dodecanes was achieved within 10 min. Complete conversion to products was also achieved once purging of the butenes was undertaken. Unfortunately, meaningful rate data were not able to be obtained for this catalytic system, but the authors state that the activity is comparable to the activity observed for the complexes $[M\{(CF_3)_2(CH_3)CO\}_2(CHR)(NAr)]$ (M = Mo, W; Ar = 2,6-diisopropyl-phenyl) [30-33].

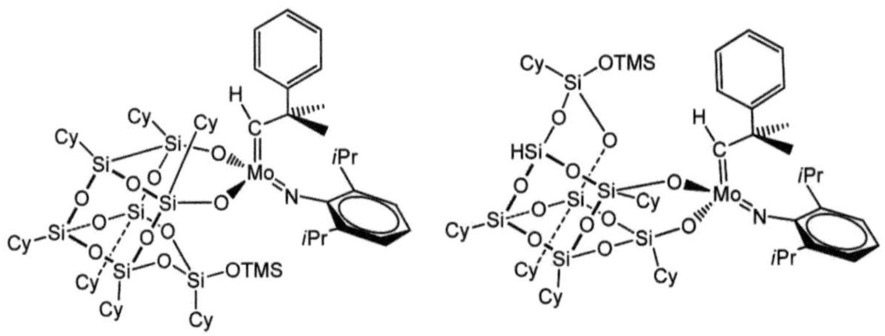

(17a) **(17b)**

Fig. 3.10 The molybdenum alkylidene silsesquioxane complexes (**17a**) and (**17b**) employed as catalysts for olefin metathesis [2]

3.2.4 Group 8 – Fe

A series of anionic iron(III) silsesquioxanes of the general formula $[Bu_4N][R_7Si_7O_{12}FeCl]$ (R = ^{i}Bu (**18a**), R = Et (**18b**), R = Ph (**18c**), R = c-C_5H_9 (**18d**)) was prepared by the reaction of $[Bu_4N][FeCl_4]$ with the incompletely condensed silsesquioxanes $R_7Si_7O_9(OH)_3$ in THF [34,35]. All the metallasilsesquioxanes were characterized crystallographically, and the structure of the ethyl-substituted species is shown in Fig. 3.11 [35]. The complexes were investigated as potential catalysts for the aerobic oxidation of THF (25 °C, 5 days, catalyst:THF 1:200) [35]. After 5 days, the major product was γ-butyrolactone, with two minor products, 2-hydroxytetrahydrofuran and 4-hydroxybutanal, also observed. The most active complex was found to be the ethyl-substituted species **(18b)** which achieved a total oxidation TON of 170 (± 10), of these 114 (± 9) being specific for γ-butyrolactone. The least active complex, the isobutyl-substituted metallasilsesquioxane (**18a**), achieved a total oxidation TON of 100 (± 8).

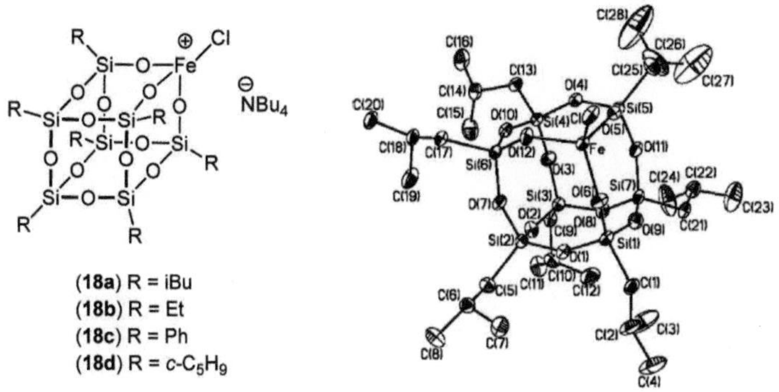

(a) (b)

Fig. 3.11 (a) Structure of the complexes (**18**), and (b) molecular structure of (**18a**) [34] (reproduced from [34] with permission from Elsevier)

3.2.5 *Group 12 – Zn*

The reaction of the silsesquioxane disilanol $(c\text{-}C_5H_9)_7Si_7O_9(OSiMePh_2)(OH)_2$ with 2 equivalents of Me_2Zn afforded the dimeric silsesquioxane complex $\{[(c\text{-}C_5H_9)_7Si_7O_{11}(OSiMePh_2)]_2ZnMe_4\}$ (**19**) (Fig. 3.12a) [36]. The crystal structure of (**19**) (Fig. 3.12b) showed that the complex consisted of two silsesquioxane cages bridged by four Zn atoms. The complex was tested as a catalyst for the copolymerization of cyclohexene oxide and CO_2 (50 – 120 °C, toluene, 1 – 8 MPa CO_2, catalyst:substrate 1:1480). At 120 °C, in the absence of CO_2, complex (**19**) was moderately active for the homo polymerization of cyclohexene oxide to poly(cyclohexene oxide), achieving a conversion of 13% after 24 h and a TOF of 4.0 h^{-1}. The introduction of 0.1 MPa CO_2 at 50 °C considerably inhibited the formation of the homopolymer, instead resulting in the formation of the co-polymer poly(cyclohexene carbonate) (carbonate content ~80%) with 9% conversion of the monomer after 24 h (a TOF of 2.8 h^{-1}). Increasing the pressure to 0.8 MPa CO_2 at 80 °C resulted in 34% conversion of the monomer in 24 h (TOF = 10.1 h^{-1}) and the formation of co-polymer with M_w = 10,600 and 92% carbonate content. When the temperature was increased to 120 °C, the conversion of the monomer increased to 44% with a carbonate incorporation of 98%. From this series of experiments, it was found that the dependence of the activity of the catalyst on CO_2 pressure closely resembled the behaviour of the zinc-phenoxide catalyst $[(F_2C_6H_3O)_2Zn.THF]_2$. The zinc-phenoxide complex was tested (80 °C and 0.8 MPa CO_2) and achieved 19% conversion of the monomer. The polymer produced had M_w = 25,800 and a carbonate incorporation of 98% at a TOF of 15.4 h^{-1}. By comparison, Et_2Zn immobilized on silica was tested for catalytic activity; after 24 h at 120 °C, 6% of monomer had been consumed at TOF 0.7 h^{-1} to afford a polymer with M_w 9,000 and a carbonate content of 93%.

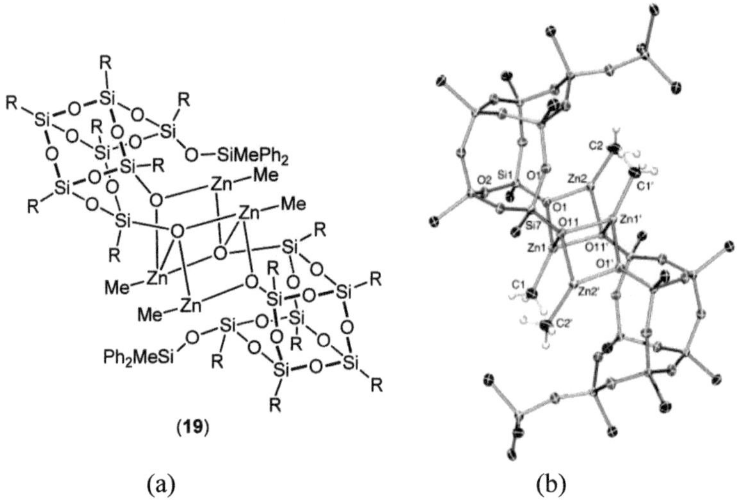

Fig. 3.12 (a) The dimeric zinc-silsesquioxane complex $\{[(c\text{-}C_5H_9)_7Si_7O_{11}(OSiMePh_2)]_2ZnMe_4\}$ (**19**), and (b) crystal structure [36] (reproduced from [36] with permission from the American Chemical Society)

3.2.6 Group 13 – Al

An aluminosilsesquioxane gel was prepared by Abbenhuis and co-workers [37] by the reaction of $AlMe_3$ with a bifunctional silsesquioxane in which the two hydroxyl groups occupied '*exo*' positions above the interconnected Si_4O_4 rings of the Si_8O_{11} skeleton ((**20**), Fig. 3.13). Upon addition of the $AlMe_3$ to a toluene solution of the silsesquioxane, methane was evolved, and the formation of the colourless aluminosilsesquioxane gel (**21**) resulted in the occlusion of the entire volume of solvent. Drying of the gel in vacuo resulted in the loss of all solvent to form a powder. Analysis of the gel with ^{29}Si MAS NMR spectroscopy revealed that >80% of all the silanol units of the starting silsesquioxane had been functionalized in the gel forming reaction. Since the gel can be considered a polymeric Lewis acid, the authors tested (**21**) as a catalyst for the Diels-Alder reactions of enones with butadienes to form substituted cyclohexenes (20 °C, $CDCl_3$, catalyst:diene:dienophile 0.01: 200: 200). It was found that (**21**) increased the rate of the reactions by up to three orders of magnitude compared to the non-catalyzed thermal reactions. Additionally, when the reactions were catalyzed by (**21**), the amount of by-product produced by the homo-Diels-Alder reaction (to form 3,4-dihydropyrans) was dramatically reduced to less than 12%, compared to the 30% obtained under thermal conditions.

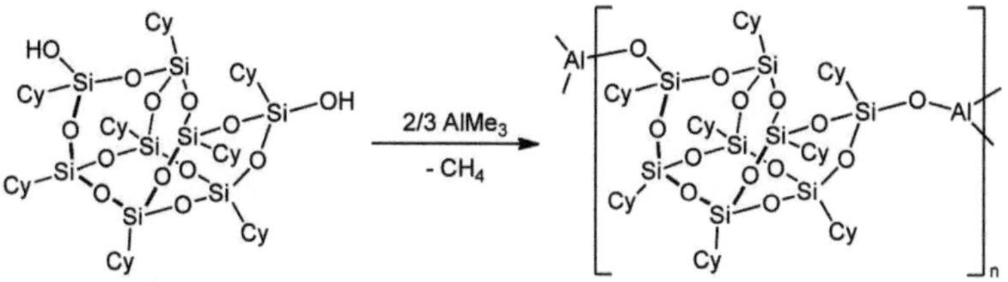

Fig 3.13 Formation of the aluminosilsesquioxane gel (**21**) [37]

In addition to the formation of the titanium dimer (**12**), Jones and co-workers produced the aluminium dimer $\{[CH_2CH(CH_3)_2\}_7Si_7O_{12}]_2Al_2\}$ (**22**) [25] by the reaction of $AlMe_3$ with the silsesquioxane $(CH_2CH(CH_3)_2)_7Si_7O_9(OH)_3$. Recrystallization from hexane afforded crystals suitable for X-ray diffraction (Fig. 3.14). The complex was tested for activity for the ring opening polymerization of *rac*-lactide in the melt (130 °C, no solvent, initiator:*rac*-lactide 1:300). After 24 h, only 5% conversion had been achieved, and a polymer was produced with M_w = 81,400 and PDI = 3.3. By comparison, the heterogeneous system Al-SiO_2 (with 60 Å pore size) achieved a 10% conversion after 48 h and produced a polymer with an M_w = 4,050 and a PDI of 1.13.

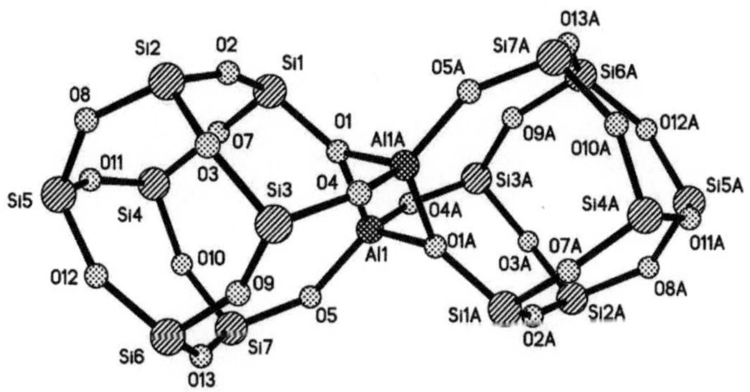

Fig 3.14 Crystal structure of the dimeric aluminosilsesquioxane complex $\{[CH_2CH(CH_3)_2\}_7Si_7O_{12}]_2Al_2\}$ (**22**) [25] (reproduced from [25] with permission from the Royal Society of Chemistry). The isobutyl groups have been omitted for clarity

3.2.7 Group 14 – Si

Recent work within our laboratory has revealed that the completely condensed silsesquioxane $(c\text{-}C_6H_{11})_6Si_6O_9$, when heated in the presence of methyl acrylate (70 °C, hexane, catalyst:substrate 1:1777), affords poly(methyl acrylate) in 53% yield after 24 h [38]. The average molecular weight of the polymer obtained was 14,000 with a polydipsersity index (PDI) of 3.9. When silica calcined at 600 °C,

$SiO_{2(600)}$, was used under identical conditions, the poly(methyl acrylate) obtained had a molecular weight of 27,000 with a PDI of 1.4. This result represents the first example of a purely siliceous silsesquioxane acting as a catalyst or initiator for a chemical process.

Initiation of radicals is suggested to occur by the homolytic cleavage of the siloxane rings (shown in Fig. 3.15). Cleavage of the Si-O bond of the siloxane ring results in the formation of silicon-based and oxygen-based radicals which can then initiate the polymerization process by reaction with the available monomer. The siloxane rings on the silica surface are generated on the surface of silica, by the controlled dehydration of the surface silanol groups, on heating to elevated temperatures. The optimum activity for the polymerization reactions was achieved when the silica was calcined at 600 °C prior to use. At temperatures lower than 600 °C, it is believed that the siloxane rings formed are too large, and hence, more difficult to cleave to generate radicals, whilst at temperatures above 600 °C sintering of the silica surface results in a relaxation of the surface, leading to a reduced number of active siloxane rings. In order to confirm this, the fully condensed silsesquioxane *a6b0* and the incompletely condensed silsesquioxane *a7b3* were used as models of the silica surface. The silsesquioxane *a6b0* consists of two 6– and three 8–-membered siloxane rings, and represents the highly strained surface that might possibly be achieved when the silica is calcined at 600 °C. On the other hand, the silsesquioxane *a7b3* consists of three 8–membered siloxane rings with a corner missing from the cubic structure, and represents a silica surface that has been calcined at lower temperatures and has not undergone full dehydration. Contrary to the more strained *a6b0*, silsesquioxane *a7b3* is not active in the polymerization of methyl acrylate under these conditions, consistent with the hypothesis that neither surface silanols, nor surface four-rings are the active centres, but that the strained three rings are.

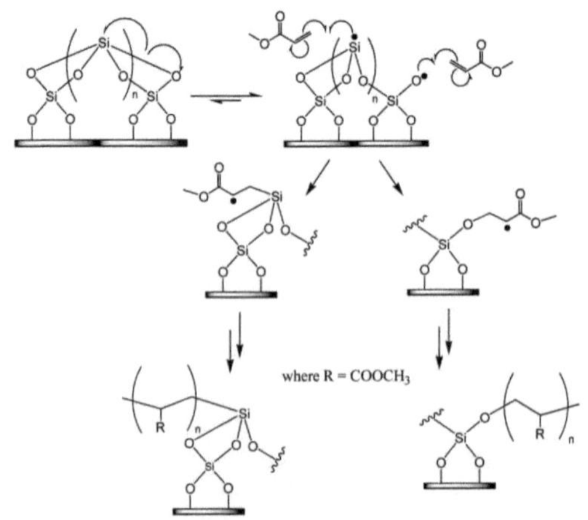

Fig. 3.15 Proposed mechanism for the polymerization of methyl acrylate using radicals generated on the surface of $SiO_{2(600)}$ Reproduced from Lesic et al. [38]

3.2.8 Lanthanides – Nd

The reaction of anhydrous $NdCl_3$ with NaO^{i}Pr in benzene/isopropanol followed by addition of the incompletely condensed silsesquioxane (*iso*-$C_4H_9)_7Si_7O_9(OH)_3$ afforded the blue tetrameric cage complex {[(*iso*-$C_4H_9)_7Si_7O_{12}Nd]_4NaCl$} (**23**) (Fig. 3.16a) [39]. The solid state structure of (**23**) was determined by X-ray crystallography (Fig. 3.16b). The complex consists of four silsesquioxane-Nd units in which the silsesquioxane siloxy anions chelate one Nd atom and, additionally, two of the siloxy oxygen atoms bridge two adjacent Nd atoms to form the tetrameric structure. Finally, a chloride ion sits in the centre of the cage with close contacts to the four Nd atoms. A sodium ion (with half site occupancy) balances the charge. When the Nd-silsesquioxane complex (**23**) was tested for catalytic activity for isoprene polymerization in the presence of 25 equivalents of $AlEt_3$, Al^{i}Bu$_3$ or MAO no polymerization was observed. However, the addition of Me_3SiCl, as a chloride source, resulted in moderate polymerization activity. The catalyst system achieved polymer yields >84% after 18.5 h with molecular weight distributions between 1.33 – 1.42 and *cis*-1,4 selectivity of between 92% and 93% (60 °C, toluene, Al:Nd:Cl:isoprene 25:1:10:1481) [39,40].

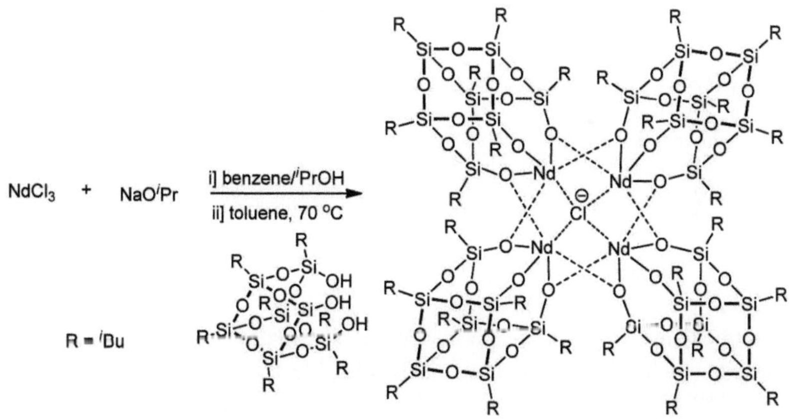

(**23**)

(a)

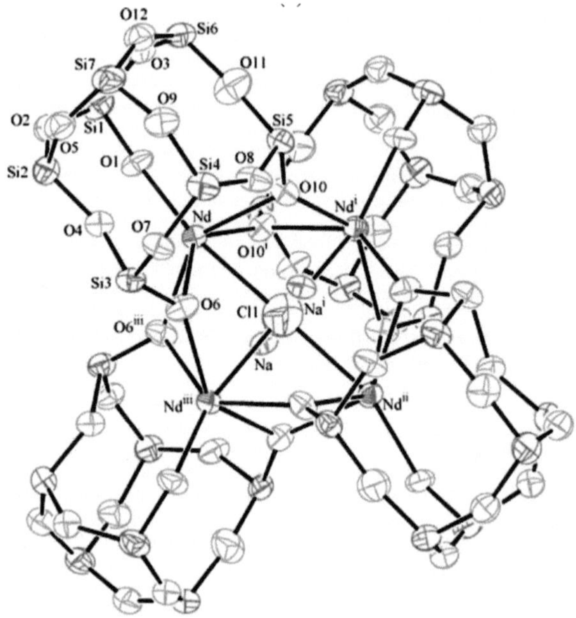

(b)

Fig. 3.16 (a) Synthesis of the tetrameric Nd silsesquioxane complex $\{[(iso\text{-}C_4H_9)_7Si_7O_{12}Nd]_4NaCl\}$ (**23**) [39], and (b) molecular structure with 25% displacement ellipsoids with C and H atoms omitted for clarity (reproduced from [39] with permission of Elsevier)

3.2.9 Hetero-bimetallic Systems

Feher and Blanski synthesized a range of silsesquioxane-supported vanadium complexes, which, after treatment with $Al(CH_2SiMe_3)_3$ afford active polymerization catalysts [19,41,42]. The vanadate-silsesquioxane complex (**24**), prepared by the reaction of $V(O)(O^iPr)_3$ with $(c\text{-}C_6H_{11})_7Si_7O_9(OH)_3$ [43], reacts rapidly with $Al(CH_2SiMe_3)_3$ at -50 °C to form the Lewis acid adduct (**25**). Upon warming (**25**) to -20 °C, a rapid rearrangement occurs to form the mono-alkyl vanadium complex (**26**). Upon warming the solution of (**26**) above -10 °C a second alkyl transfer occurs to form a new oxo-vanadium(V) alkyl complex (**27**), which is stable indefinitely at room temperature. Fig. 3.17 shows the formation of (**27**). Exposure of (**27**) to ethylene (34 μmol V, 25 °C, benzene or hexane, 0.1 MPa) resulted in the formation of polyethylene (typically 1000 – 1500 turnovers) which continued until gelation of the reaction mixture occurred, preventing further uptake of the monomer [42]. When polymerization was attempted using the complexes $[V(O)(CH_2TMS)_n(OSiPh_3)_{3\text{-}n}]$ (n = 0 - 3) in the presence of 3 equivalents of $Al(CH_2TMS)_3$ (25 °C, benzene, 91 μmol V, 2 MPa ethylene) only low yields of polyethylene were obtained (with all the catalysts achieving less than 380 turnovers after 3 h) before deactivation of the catalyst [44].

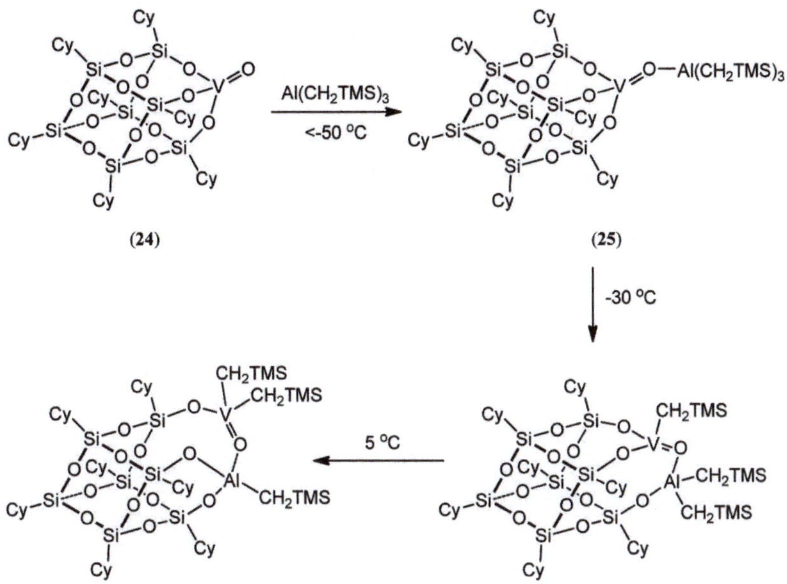

Fig. 3.17 The chemistry of the vanadate-silsesquioxane complex (**24**) with $Al(CH_2SiMe_3)_3$ to form the active ethylene polymerization catalyst (**27**) (redrawn from Feher et al. [41])

The hetero-bimetallic silsesquioxane $\{[c\text{-}(C_6H_{11})_7Si_7O_{12}]MgTiCl_3\}_n$ (n = 1,2 (**28**)) was prepared by a two-step synthesis which is outlined in Fig. 3.18 [45,46]. The first step involved the reaction of butylethylmagnesium with the $(c\text{-}C_6H_{11})_7SiO_9(OH)_3$ to afford the magnesium-silsesquioxane complex $\{[c\text{-}(C_6H_{11})_7Si_7O_{11}(OH)]Mg\}_n$ (n = 1,2 (**29**)) with the evolution of butane and ethane. Spectroscopic studies indicate that (**29**) is formed by the reaction between two hydroxyl groups of the starting silsesquioxane and the alkyl groups of the Grignard reagent and that one OH remains unreacted. Complex (**29**) is proposed to exist as a mixture of the monomer ($\mathbf{29_M}$) and dimer ($\mathbf{29_D}$). Reaction of (**29**) with $TiCl_4$ resulted in the formation of the desired hetero-bimetallic species (**28**) as an equilibrium mixture of the monomer ($\mathbf{28_M}$) and dimer ($\mathbf{28_D}$) in a 3:1 ratio. The complex (**28**) demonstrated high catalytic activity for the polymerization of ethylene (90 °C, 2-methylpropane, 17 μmol catalyst, 435 μmol Et_3Al, 180 mmol H_2, 3.45 MPa ethylene). The activity of the catalyst was 110.8 kg(PE) g(Ti)$^{-1}$ h^{-1} with an M_w = 140,000 and a PDI of 5.4. By comparison, a commercial Ti/Mg/SiO_2 catalyst with the same bimetallic content had an activity of 60.0 kg(PE) g(Ti)$^{-1}$ h^{-1}.

3.3 Phosphasilsesquioxanes as Ligands

Several phosphite ligands based on the silsesquioxane backbone were prepared by Vogt and co-workers [47]. The mono-phosphite ligand (**30**) was prepared *via* a four-step synthetic scheme starting from the incompletely condensed silsesquioxane $(c\text{-}C_5H_9)_7Si_7O_9(OH)_3$, shown in Fig. 3.19. The diphosphite ligand (**31**) was obtained in two steps from the same silsesquioxane (Fig. 3.18). These were then tested for their suitability as ligands for the rhodium-catalyzed hydroformylation of 1-octene (80 °C, toluene, ligand:Rh 5:1, 1-octene:Rh 4000:1, 2 MPa H_2/CO (1:1)). When the monophosphite was used, a conversion of 41.7% was achieved after 20 min, with an initial turnover frequency of 5100 mol(product) mol(Rh)$^{-1}$ h^{-1}. When the ligand:metal ratio was increased to 10:1, the TOF increased to 6800 mol(product) mol(Rh)$^{-1}$ h^{-1} to afford a yield of 55.2% after 20 min. Application of the diphosphite ligand (**31**) resulted in a less active catalyst than the monophosphite analogue: in this case, the yield was 11.3% after 60 min, which corresponded to an initial TOF of 500 mol(product) mol(Rh)$^{-1}$ h^{-1}. While the activity was lower for the diphosphite ligand, significantly higher selectivity (80.1%) was achieved than that obtained for the monophosphite ligand (68.7%).

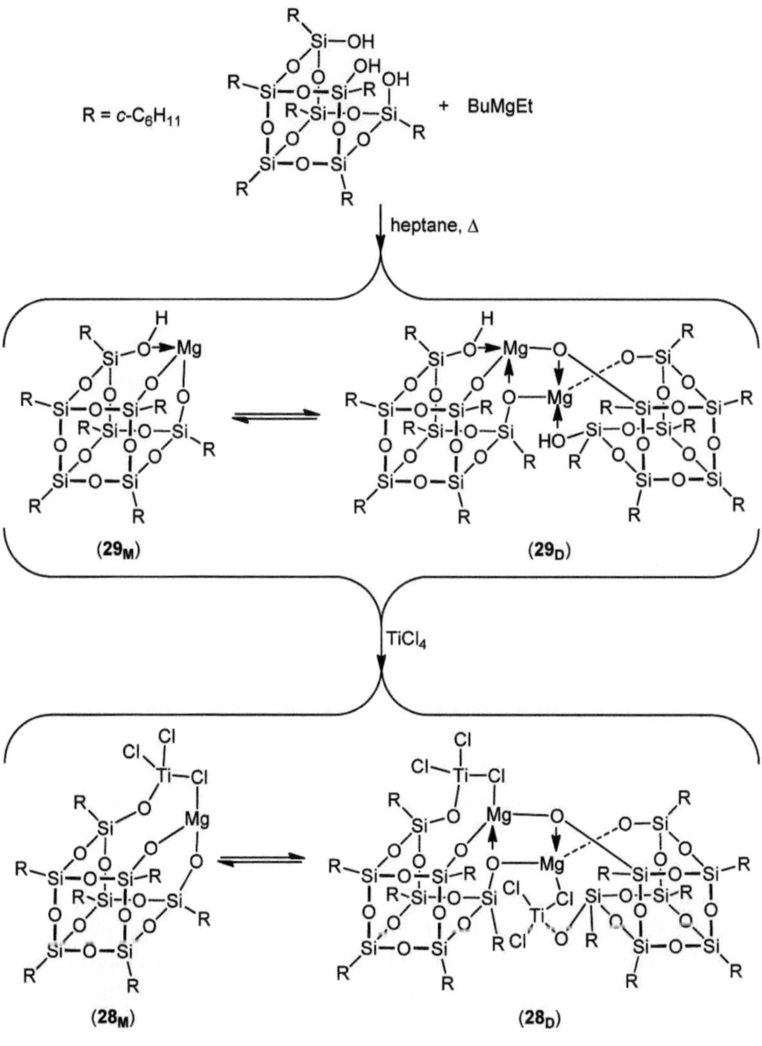

Figure 3.18 Synthesis of the hetero-bimetallic silsesquioxane $\{[c\text{-}(C_6H_{11})_7Si_7O_{12}]MgTiCl_3\}_n$ (n = 1,2 (**28**)) [45,46]

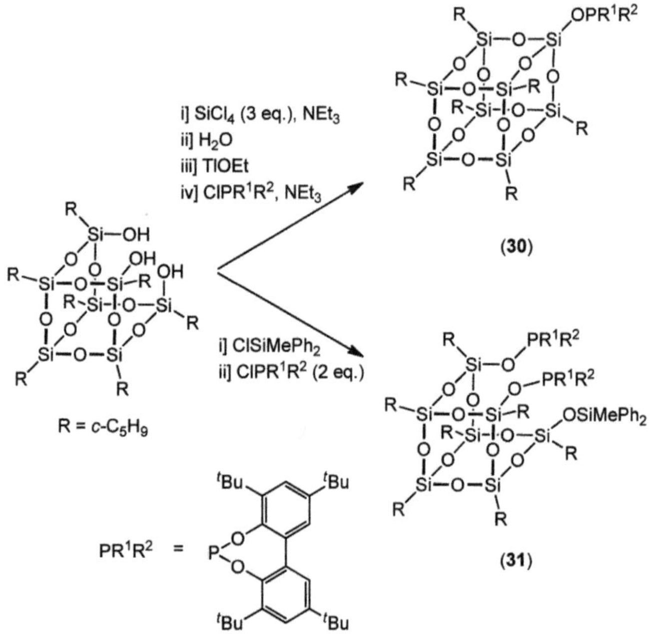

Figure 3.19 Synthesis of the mono- and di-phosphite silsesquioxane ligands (**30**) and (**31**)

Using a similar synthetic methodology to that shown in Fig. 3.18, the same research group prepared a series of phosphite silsesquioxane ligands possessing the chiral binaphthyl group (**32**), (**33**) and (**34**) (Fig. 3.20) [48]. The suitability of these as ligands in the rhodium-catalyzed hydroformylation of vinyl acetate was then tested (15 h, 60 °C, benzene, ligand:Rh 4:1, substrate:Rh 1500:1, 2 MPa CO/H_2 (1:1)). For the potentially tridentate ligands (**32**), moderate enantioselectivities were observed (30 – 38%), and in the case of (**32c**) a yield of 99% was achieved. In the case of these ligands, the branched to linear ratio of the products obtained ranged from 6 – 11. The bidentate ligands (**33**) and (**34**) gave lower enantioselectivities (13 – 16%) but higher branched to linear ratios (21 – 22) while achieving yields of 82 – 88%.

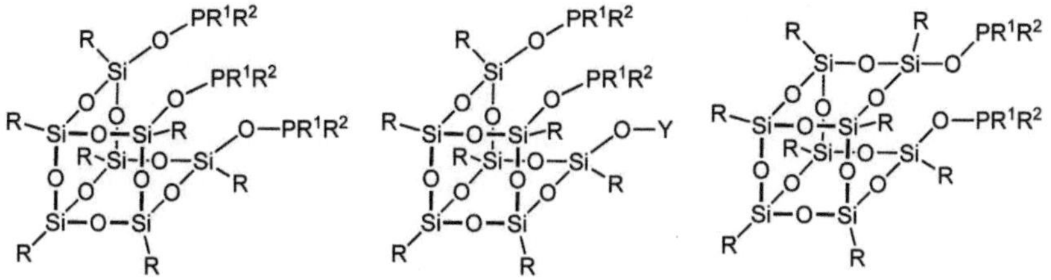

(32a) R = ^{i}Bu
(32b) R = c-C_6H_{11}
(32c) R = c-C_5H_9

(33a) R = ^{i}Bu; Y = $SiMe_3$
(33b) R = c-C_5H_9; Y = $SiMePh_2$

(34) R = c-C_5H_9

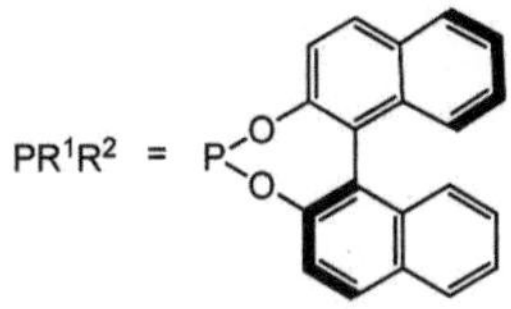

Fig. 3.20 The chiral phosphite silsesquioxane ligands (**32**) – (**34**) [48]

3.4 Catalytic Materials Derived From Metalla-Silsesquioxanes

This section, while not strictly dealing with the use of metallasilsesquioxanes as molecular analogues of heterogeneous catalysts, represents an interesting and relatively unexplored new use of discrete metal-containing silsesquioxanes as precursors for the preparation of catalytic materials. This method of preparing heterogeneous metal-containing materials has a distinct advantage over traditional methods of preparing metal impregnated siliceous materials in that the metal content can be carefully controlled and pre-determined. In addition, the metal atoms remain highly dispersed throughout the material resulting in catalysts with potentially higher activity than traditionally prepared catalysts, owing to the smaller nanoparticle size.

Wada and co-workers [49] prepared a range of silica materials by the incipient wetness adsorption of the vanadate silsesquioxane complex {[(c-$C_5H_9)_7Si_7O_{12}$] VO} into mesoporous silica followed by heating in air at 523 and 723 K for 2 h to afford the materials (**35**) and (**36**), respectively. Each material had a vanadium loading of 1.9 mol%. Analysis of the materials by X-ray photoelectron spectroscopy (XPS) revealed that heating of (**35**) at 523 K had not resulted in the oxidation of the organic substituents of the silsesquioxane precursor, whilst (**36**) was shown to have only traces of carbonaceous material present after treatment at 723 K. The resultant materials were then tested for the catalytic photo-assisted oxidation of methane to methanal (220 °C, 50 mg catalyst, CH_4:O_2:He 12:1:11, total gas flow 20 cm^3 min^{-1}, UV irradiation). When (**36**) was used as the catalyst, methanal was produced with a yield of 139 μmol(product) h^{-1} (0.56% of fed methane) with a selectivity >80%, which corresponds to a TOF of ca. 9 h^{-1}. In the case of (**35**), the yield and selectivity were low. For comparison, the V_2O_5/SiO_2 was tested, and this catalyst achieved a yield of methanal of 61 μmol(product) h^{-1} under the same conditions.

The controlled calcination of the metallasilsesquioxane [(*c*-$C_5H_9)_7Si_7O_9(OSiMe_3)O_2CrO_2$] (**37**) at 450 and 550 °C for 4 h and flushed by an 20% O_2/Ar afforded microporous silicates with Cr loadings of 10.2 wt % [50]. Nitrogen sorption analysis showed microporous materials with high surface area ((**37**$_{450}$) 614 m^2/g; (**37**$_{550}$) 519 $m^2\ g^{-1}$) and narrow pore size distribution, with an average pore diameter of 6.8 – 6.9 Å. Analysis of the samples by XPS revealed the presence of predominantly Cr^{6+} and a small amount of Cr^{3+}. Diffuse reflectance UV-Vis spectroscopy revealed the presence of a small amount of Cr_2O_3, which was postulated to be the result of the hydrolysis of the monochromate groups by the water formed by the combustion of the organic groups of the starting silsesquioxane. The resulting chromate silicates were tested as catalysts for the oxidation of ammonia in a continuous plug flow reactor (200 °C, flow 50 mL min^{-1}, 95 mg catalyst, 1000 ppm NH_3, 10% O_2). When (**37**$_{550}$) was used as the catalyst, 56% conversion of the ammonia was achieved with 68% selectivity to N_2. For comparison, a chromium oxide silica reference catalyst with the same metal loading achieved 45% conversion with 76% selectivity under the same conditions.

Microporous metallosilicate materials possessing Mg and Al were prepared by Maxim and co-workers by the thermolysis of metallasilsesequioxanes in the presence of metal-free silsesquioxane at 500 °C under a flow of Ar/O_2 for 4 h [51]. The metallasilsesquioxanes (**38**) – (**40**) used are shown in Fig. 3.21. The magnesium-containing silsesquioxane [(*c*-$C_5H_9)_7Si_7O_{12}]_2Mg_4Cl_2(THF)_2$ (**38**) was prepared by the reaction of (*c*-$C_5H_9)_7Si_7O_9(OH)_3$ with CH_3MgCl in THF [52]. The aluminium silsesquioxane complexes {[(*c*-$C_5H_9)_7Si_7O_{12}]Al\}_n$ (**39**) and [(*c*-$C_5H_9)_7Si_7O_{11}(OSiMe_3)]Al[$(*c*-$C_5H_9)_7Si_7O_{10}(OSiMe_3)(OH)$] (**40**) were prepared by reaction of $AlEt_3$ with the trisilanol silsesquioxane in THF and the disilanol silsesquioxane (*c*-$C_5H_9)_7Si_7O_9(OSiMe_3)(OH)_2$ in toluene, respectively [53]. The materials obtained had surface areas of 347 (**41**), 583 (**42**) and 641 (**43**) $m^2\ g^{-1}$ for the silicates derived by (**38**), (**39**) and (**40**), respectively, with average pore diameters of 6.0 – 6.4 Å. Extensive examination of the materials revealed that the metal dispersion was nearly homogeneous in all cases. The metallosilicates were tested for the dehydrogenation - dehydration of 1-butanol at 200 and 300 °C in a continuous plug-flow reactor (using 0.7 g of the catalyst). When (**41**) was used, 1-butanol conversion was 45% at 200 °C and reached 100% at 300 °C. In the first 30 min at 200 °C butanal was the main product, but after this time butanal and 1-butene were formed in equal amounts. At 300 °C, the dehydration product was the only one observed. Both of the aluminium silicates (**42**) and (**43**) achieved 100% conversion at 200 °C with the dehydration product 1-butene being predominant with cracking products observed. At 300 °C, the cracking reaction became the major pathway.

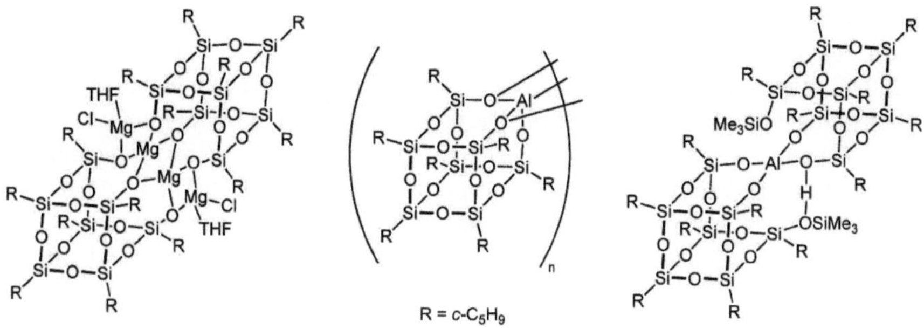

(38) (39) (40)

Fig. 3.21 The metallasilsesquioxanes (**38**) – (**40**) used for the preparation of Mg- and Al-containing microporous silicates (redrawn from Maxim et al. [51])

Using a similar methodology the same research group produced a range of iron containing microporous silicates [54]. The iron silicates were prepared with metal contents of up to 11% by varying the concentration of the metallasilsesquioxane [(c-$C_5H_9)_7Si_7O_{12}$Fe(tmeda)] (tmeda = *N*,*N*,*N*',*N*'-tetramethylethylenediamine, (**44**)) present in a mixture with the trisilanol (c-$C_5H_9)_7Si_7O_9(OH)_3$ during calcination in 20% O_2/Ar at 500 °C for 4 h. The materials (**45**) were shown by ICP-OES to have metal contents between 0.9 wt % to 10.6 wt %, which were slightly lower than the expected values. Nitrogen physisorption gave Type Ib isotherms characteristic of microporous materials, and revealed surface areas ranging from 445 – 628 m^2 g^{-1}. The materials had a narrow pore size distribution, with average pore sizes ranging from 5.9 – 7.2 Å. Spectroscopic analysis and TEM revealed that the iron was present in the iron silicates as isolated and clustered Fe^{3+} species, as well as particles of iron oxide. The authors postulated that the formation of the iron oxide particles resulted from the hydrolysis of the Fe-O-Si bonds by the water liberated from the combustion of the organic component of the silsesquioxane and subsequent migration of the metal species on the nascent silica surface. The iron silicates were tested for the catalytic oxidation of NH_3 in a continuous plug-flow reactor (400 °C, 0.1 g catalyst, flow rate 50 N mL min^{-1}, NH_3 1 vol %, O_2 10 vol %). The 11% iron silicate (**45**$_{11\%}$) catalyzed the selective oxidation of NH_3 to N_2 and H_2O with 76% conversion and 95% selectivity. When this catalyst was pre-treated with H_2 prior to the reaction, 84% conversion was achieved with 97% selectivity. The reduced 5% iron silicate (**45**$_{5\%}$) gave 58% conversion and 96% selectivity. Using the same gas mixture, Fe/Al_2O_3 achieved a 100% conversion of ammonia with a *ca.* 90% selectivity at 500 °C [55].

Murugavel and co-workers subjected the cubic titanosiloxane (**8**) (Fig. 3.6) to solid state thermolysis at temperatures of 450, 600, 800, 900, 1000 and 1200 °C in air to afford a range of titanosilicate materials (**46**) [22]. Characterization of these materials revealed that the Ti–O–Si linkages remained when (**8**) was subjected to thermolysis between 450 – 800 °C, whilst X-ray diffraction studies showed the formation of anatase and rutile forms of TiO_2 at temperatures above 900 °C. TEM electron diffraction patterns of those titanosilicates obtained at lower temperatures revealed a polycrystalline structure, while the material obtained at 1000 °C was the most crystalline. The TEM electron diffraction pattern of the material obtained at 1200 °C revealed the presence of amorphous, polycrystalline and single-crystalline titanosilicates in the same material. The materials obtained by thermolysis at 600, 800, 900 and 1000 °C were tested as catalysts for the epoxidation of cyclohexene (65 °C, toluene, catalyst:cumenyl hydroperoxide:substrate 1:207:852). After 23 h, a conversion of 11% was achieved with 100% selectivity to the epoxide for the materials (**46**$_{600}$) and (**46**$_{800}$). In the case of those materials (**46**$_{900}$) and (**46**$_{1000}$), conversions of <1% were obtained [22]. The authors attribute the decrease in activity to the observed phase separation of the TiO_2 in the materials at the higher temperatures.

3.5 Conclusions and Future Prospects

Despite the great number of metallasilsesquioxanes described in the literature, there is a distinct paucity of structurally unequivocally characterized examples with proven catalytic activity. In fact, the number of such metallasilsesquioxane catalysts is so few that all of them are presented in this chapter. By definition the catalysts modeled are heterogeneous in nature, therefore the modeling of the catalysts is usually targeted towards a better understanding of the active site and not a better catalyst as such. Hence, the structural information of a metallasilsesquioxane is of primary interest; a catalytic test is simply needed to establish that the structure derived from these model compound studies actually displays activity, and therefore has some justification in being claimed to resemble an active surface species.

However, these considerations must be treated with caution – even if one can model, say 95%, of all surface species, this does not mean that one has modeled the active surface species! Indeed, the chemistry of a silica surface when a metal ion or complex is deposited is very complex. An insight into this complexity is shown in Figure 3.1, revealing many different binding modes. As a consequence, when seeking a homogenous silsesquioxane analogue as a model for a heterogeneous catalyst, most commonly only the most readily available silsesquioxane geometries are used. It would seem that this method often affords a metal complex that has no catalytic activity, and hence the nature of the binding of the metal centre to the silica core of the silsesquioxane must, by inference, not resemble that found in the heterogeneous catalyst. In addition, due to the less rigid environment around the complexed metal, the catalyst models can usually be expected to perform less well than those anchored on the fixed, solid surface.

Therefore, we suggest that for the field to move forward and to realize its full potential, a fruitful if complicated step would be to link one, two, or even three silsesquioxanes, to obtain a slightly larger section of a silica surface, resulting in a model species of greater complexity and rigidity.

Aiming to develop model compounds that are more active than the heterogeneous catalysts themselves is most likely a less rewarding avenue for the reasons outlined above. However, using well-defined metallasilsesquioxanes as heterogeneous catalyst precursors in a variety of synthetic approaches (e.g., thermolytic decomposition and sol-gel synthesis) shows much more promise.

Overall the insights into catalytic structure-functionality relationships gained via chemical modeling using metallasilsesquioxanes are unique, are unobtainable using any other technique, and have been tremendously valuable. As heterogeneous catalysts are increasingly used in fine chemistry applications, where even greater insight into their mode of action is needed, we predict a rich future for research into metallasilsesquioxanes as catalyst models and catalysts precursors.

3.6 References

1. Duchateau R, Cremer U, Harmsen R, Mohamud S, Abbenhuis H, van Santen R, Meetsma A, Thiele S-H, van Tol M, Kranenburg M (1999) Organometallics 18:5447-5459.
2. Feher F, Tajima T (1994) J Am Chem Soc 116:2145-2146.
3. Herrmann W, Anwander R, Dufaud V, Scherer W (1994) Angew Chem Int Ed 33:1285-1286.
4. Liu J-C, Wilson S, Shapely J, Feher F (1990) Inorg Chem 29:5138-5139.
5. Abbenhuis H (2000) Chem Eur J 6:25-32.
6. Coperet C, Chabanas M, Petroff Saint-Arroman P, Basset J-M (2003) Angew Chem Int Ed 42:156-181.
7. Maschemeyer T, Klunduk M, Martin C, Shephard D, Thomas J, Johnson B (1997) Chem Commun:1847-1848.
8. Maschemeyer T, Rey F, Sankar G, Thomas J (1995) Nature 378:159-162
9. Crocker M, Herold R (1997) WO 97/24333.
10. Crocker M, Herold R, Orpen A (1997) Chem Commun: 2411-2412.
11. Feher F, Budzichowski T, Rahimian K, Ziller J (1992) J Am Chem Soc 114:3859-3866.
12. Buys I, Hambley T, Houlton D, Maschemeyer T, Masters A, Smith A (1994) J Mol Catal 86:309-318.

13. Field L, Lindall C, Maschemeyer T, Masters A (1994) Aust J Chem 47:1127-1132.

14. Abbenhuis H, Krijnen S, van Santen R (1997) Chem Commun:331-332.

15. Krijnen S, Mojet B, Abbenhuis H, van Hooff J, van Santen R (1999) Phys Chem Chem Phys 1:361-365.

16. Krijnen S, Abbenhuis H, Hanssen R, van Hooff J, van Santen R (1998) Angew Chem Int Ed 37:356-358.

17. Duchateau R, Abbenhuis H, van Santen R, Meetsma A, Thiele SK-H, van Tol M (1998) Organometallics 17:5663-5673.

18. Duchateau R, Abbenhuis H, van Santen R, Thiele S-H, van Tol M (1998) Organometallics 17:5222-5224.

19. Feher F, Blanski R (1992) J Am Chem Soc 114:5886-5887.

20. Kim Y, Han Y, Lee M, Yoon S, Choi K, Song B, Do Y (2001) Macromol Rapid Commun 22:573-578.

21. Winkhofer N, Voigt A, Dorn H, Roesky H, Steiner A, Stalke D, Reller A (1994) Angew Chem Int Ed 33:1352-1354.

22. Murugavel R, Davis P, Shete V (2003) Inorg Chem 42:4696-4706.

23. Wada K, Itayama N, Watanabe N, Bundo M, Kondo T, Mitsudo T (2004) Organometallics 23:5824-5832.

24. Pérez Y, Pérez Quintanilla D, Fajardo M, Sierra I, del Hierro I (2007) J Mol Catal A Chem 271:227-237.

25. Jones M, Davidson M, Keir C, Wooles A, Mahon M, Apperley D (2008) Dalton Trans:3655-3657.

26. Wada K, Nakashita M, Yamamoto A, Wada H, Mitsudo T (1997) Chem Lett:1209-1210.

27. Wada K, Nakashita M, Yamamoto A, Wada H, Mitsudo T (1998) Res Chem Intermed 24:515-527.

28. Ohde C, Brandt M, C, Döbler J, Zeimer B, Sauer J (2008) Dalton Trans:326-331.

29. Lovat S, Mba M, Abbenhuis H, Vogt D, Zonta C, Licini G (2009) Inorg Chem 48:4724-4728.

30. Oskam J, Schrock R (1993) J Am Chem Soc 115:11831-11845.

31. Schaverien C, Dewan J, Schrock R (1986) J Am Chem Soc 108:2771-2773.

32. Schrock R, DePue R, Feldman J, Schaverien C, Dewan J, Liu A (1988) J Am Chem Soc 110:1423-1435.

33. Schrock R, Murdzek J, Bazan G, Robbins J, DiMare M, O'Regan M (1990) J Am Chem Soc 112:3875-3886.

34. Hay M, Hainaut B, Geib S (2003) Inorg Chem Commun 6:431-434.

35. Hay M, Geib S, Pettner D (2009) Polyhedron 28:2183-2186.

36. Duchateau R, van Meerendonk W, Huijser S, Staal B, van Schilt M, Gerritsen G, Meetsma A, Koning C, Kemmere M, Keurentjes J (2007) Organometallics 26:4204-4211.

37. Abbenhuis H, van Herwijnen H, van Santen R (1996) Chem Commun 1941-1942.

38. Lesic R, Ward A, Masters A, Maschemeyer T (2010) submitted to Angew Chem Int Ed.

39. Wu G, Chen Y, Xu D-J, Liu J-C, Sun W, Shen Z (2009) J Organomet Chem 694:1571-1574.

40. Yu J, Wu G, Huang J, Sun W, Shen Z (2009) Chin J Polym Sci 27:597-600.

41. Feher F, Blanski R (1993) Makromol Chem Macromol. Symp 66:95-108.

42. Feher F, Walzer J, Blanski R (1991) J Am Chem Soc 113:3618-3619.

43. Feher F, Walzer J (1991) Inorg Chem 30:1689-1694.

44. Feher FJ, Blanski RL (1993) Organometallics 12:958-963.

45. Liu J-C (1996) Chem.Commun: 1109-1110.

46. Liu J-C (1999) Appl Organometal Chem 13:295-302.

47. van der Vlugt J, Ackerstaff J, Dijkstra T, Mills A, Kooijman H, Spek A, Meetsma A, Abbenhuis H, Vogt D (2004) Adv Synth Catal 346:399-412.

48. Ionescu G, van der Vlugt J, Abbenhuis H, Vogt D (2005) Tetrahedron-Asymmetr 16:3970-3975.

49. Wada K, Nakashita M, Yamamoto A, Mitsudo T (1998) Chem Commun: 133-134.

50. Maxim N, Abbenhuis H, Stobbelaar P, Mojet B, van Santen R (1999) Phys

Chem Chem Phys 1:4473-4477.

51. Maxim N, Magusin P, Kooyman P, van Wolput J, van Santen R, Abbenhuis H (2001) Chem Mater 13:2958-2964.

52. Hanssen R, Meetsma A, van Santen R, Abbenhuis H (2001) Inorg Chem 40:4049-4052.

53. Duchateau R, Harmsen R, Abbenhuis H, van Santen R, Meetsma A, Thiele S-H, Kranenburg M (1999) Chem Eur J 5:3130-3135.

54. Maxim N, Overweg A, Kooyman P, Van Wolput J, Hanssen R, van Santen R, Abbenhuis H (2002) J Phys Chem B 106:2203-2209.

55. Amblard M, Burch R, Southward B (1999) Appl Catal B 22:L159-L166.

Chapter 4
Polymers and Copolymers Containing Covalently Bonded Polyhedral Oligomeric Silsesquioxanes Moieties

Katherine Grace Williams, Samuel Paul Gido and Edward Bryan Coughlin

4.1 Introduction

The use of polyhedral oligomeric silsesquioxanes (POS or POSS®) as a moiety in polymers and copolymers is an area of research that has gained tremendous popularity in recent years. A number of excellent review articles have been published [1-8], as well as articles that elaborate on the larger field of silsesquioxanes [9-10]. In this chapter, we discuss T_8 POS compounds, which are represented by the general formula $Si_8O_{12}R_8$, where R is an organic group, or group suitable for polymerization or grafting. For ease of illustration, we have chosen to schematically represent the POS moiety as a cubic structure in many of the subsequent schemes that will appear in this chapter; this representation is shown in Fig. 4.1.

Fig. 4.1 Chemical structure and schematic representation of T_8 POS

K.G. Williams, S.P Gido, E.B. Coughlin
Department of Polymer Science and Engineering, University of Massachusetts Amherst
120 Governors Drive Amherst MA 01002, USA
E-mail: coughlin@mail.pse.umass.edu

C. Hartmann-Thompson (ed.), *Applications of Polyhedral Oligomeric Silsesquioxanes*, Advances in Silicon Science 3, DOI 10.1007/978-90-481-3787-9_4,

4.2 Synthetic Strategies

4.2.1 Free Radical Polymerization

Random POS copolymers with both linear [11-15], and star [16] architectures have been synthesized by free radical polymerization. A variety of monomers have been used, including styrene [11-13], vinyl pyrrolidone [14], methyl methacrylate [15], and acetoxystyrene[14-15]. This class of polymers can be synthesized in bulk [12-13, 15-16] or solution [11, 14]. An example of a linear random POS copolymer synthesized in bulk is shown in Fig. 4.2.

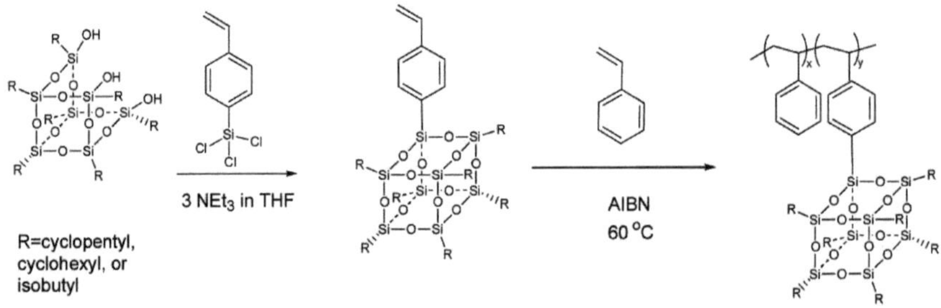

Fig. 4.2 Synthesis of POS macromers by corner capping silanation followed by polymerization in bulk styrene

Three monostyryl POS macromers, isobutyl-POS, cyclopentyl-POS, and cyclohexyl-POS, were synthesized by the reaction of styryltrichlorosilane with a POS-trisilanol in tetrahydrofuran (THF) solvent with triethylamine (TEA) as the base [12]. Phenyl-POS was also synthesized, but was found to be too insoluble in neat styrene. The POS macromers were bulk copolymerized in styrene with azobisisobutyronitrile (AIBN) as the radical initiator. For polymerizations with 30 wt % of POS monomer, the resulting polymers were composed of approximately 4 mol % POS. Other authors chose to synthesize POS copolymers in solution, again with AIBN as the radical initiator[17]. Xu et al. synthesized poly(acetoxystryrene)-POS materials with a star architecture, as shown in Fig. 4.3 [16]. Other star-like materials are discussed in Section 4.4.3.

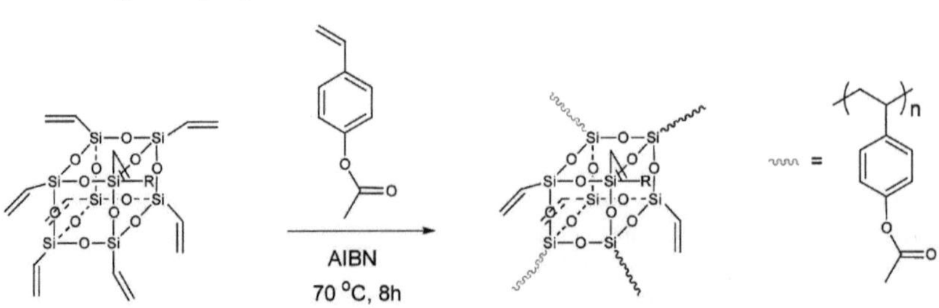

Fig. 4.3 Synthesis of Poly(acetoxystryrene)-POS star structure

4.2.2 *Living Radical Polymerization (ATRP, RAFT and NMP)*

Examples of techniques utilized to synthesize well-defined polymers and block copolymers are atom transfer radical polymerization (ATRP)[18], and radical addition-fragmentation chain transfer (RAFT) [19-20], and nitroxide-mediated polymerization (NMP) [21].

Mather, Matyjaszewski, and colleagues synthesized linear triblock copolymers of methyl methacrylate POS (MA-POS) and butyl acrylate (BA) by ATRP [22-23]. The authors prepared a difunctional macroinitiator by performing the ATRP of butyl acrylate on the difunctional initiator dimethyl 2,6 dibromoheptanedioate. Cu(I)Br and pentamethyldiethyltriamine (PMDETA) were chosen as the transition metal compound and the ligand respectively. This macroinitiator was chain extended by ATRP of MA-POS with Cu(I)Cl or Cu(I)Br and PMDETA. One such synthesis is shown in Fig. 4.4.

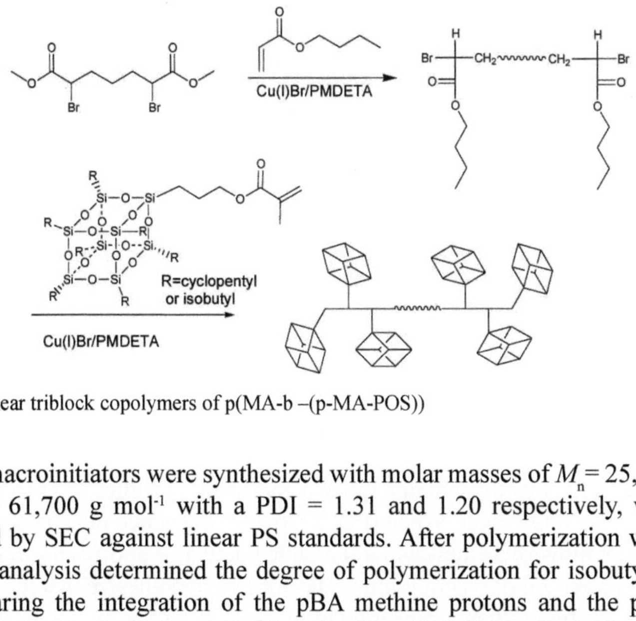

Fig. 4.4 Linear triblock copolymers of p(MA-b –(p-MA-POS))

Two macroinitiators were synthesized with molar masses of M_n = 25,800 g mol^{-1} and M_n = 61,700 g mol^{-1} with a PDI = 1.31 and 1.20 respectively, which were calculated by SEC against linear PS standards. After polymerization with POS, a ^{1}H NMR analysis determined the degree of polymerization for isobutyl MA-POS by comparing the integration of the pBA methine protons and the p(MA-POS) methylene proton. A degree of polymerization of 10 for each block of POS was calculated. Cyclopentyl MA-POS shows a degree of polymerization of 9 for each POS block, which was calculated by comparing the integration of the methoxy protons from the pBA to the methylene protons adjacent to the POS cage.

Star block polymers were also created using this technique. Matyjaszewski and colleagues created three- and four armed p(MA-*b* –(p-MA-POS)) copolymers by starting with a trifunctional initiator (1,1,1-tris(4-(2-isobutyryloxy)phenyl)-ethane) or a tetrafunctional initiator (pentaerythritol tetrakis(2-bromoisobutyrate)) [24-25]. From these cores, p(MA) blocks were synthesized by ATRP, with Cu(I)Br/PMDETA as the catalyst and ligand, to form a macroinitiator for the ATRP of cyclopentyl MA-

POS. By ^{1}H NMR, the trifunctional- initiated copolymers were determined to be 25 mol % p(MA-POS)and 75 mol % p(MA) with a M_n = 30,270 g mol^{-1} and a PDI = 1.30, based on SEC with linear PMMA standards The block copolymers initiated by the tetrafunctional initiator had a M_n = 70,000 g mol^{-1} with a PDI = 1.08, determined by SEC using linear PS standards.

More recently, surface-initiated ATRP was used to covalently bond isobutyl MA-POS to silicon wafers, as shown Fig. 4.5 [26]. Triclorosilane-based ATRP initiating groups were grafted onto the surface of the silicon wafer by reaction with hydroxy groups. After the grafting of these groups, the authors tried two different approaches for the ATRP: adding a free initiator, and adding a deactivator (persistent radical). Both methods will cause an increase in the concentration of the deactivator species. Ethyl 2-bromoisobutyrate (EBIB) was selected as the free initiator, while $CuBr_2$ was added as the deactivating species. Both approaches used CuCl and hexamethyltriethylenetetramine (HMTETA) as the catalyst/ligand system.

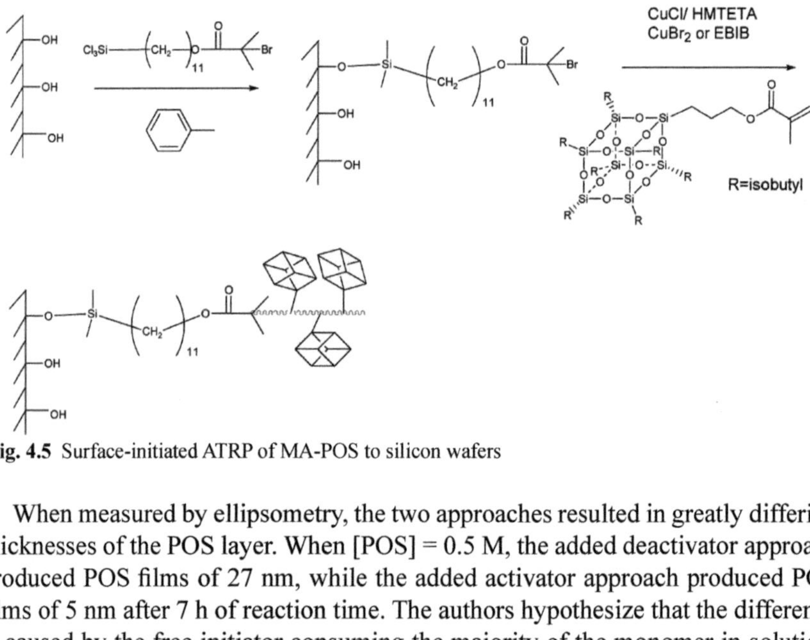

Fig. 4.5 Surface-initiated ATRP of MA-POS to silicon wafers

When measured by ellipsometry, the two approaches resulted in greatly differing thicknesses of the POS layer. When [POS] = 0.5 M, the added deactivator approach produced POS films of 27 nm, while the added activator approach produced POS films of 5 nm after 7 h of reaction time. The authors hypothesize that the difference is caused by the free initiator consuming the majority of the monomer in solution, thus reducing the concentration available to the chain propagating from the silicon surface.

Radical addition-fragmentation chain transfer (RAFT) was used to synthesize POS block copolymers with ‘tadpole’-shaped architecture. Zhang and colleagues synthesized a series of these polymers [27-29], and a representative reaction is shown in Fig. 4.6 [28].

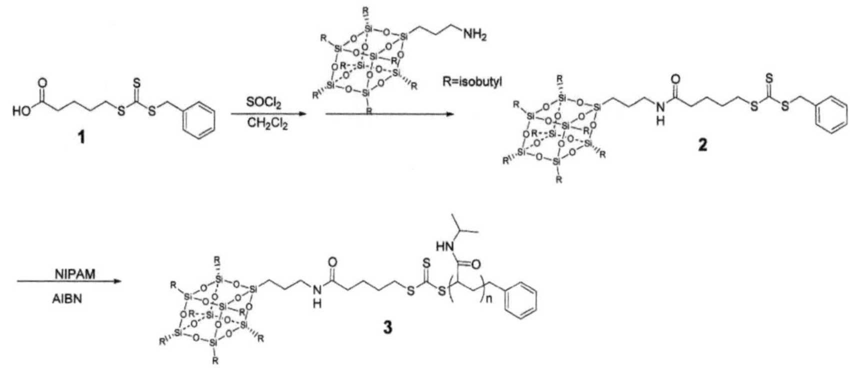

Fig. 4.6 RAFT synthesis of polymers with 'tadpole'-shaped architecture [28]

The authors synthesized the RAFT agent **1**. The carboxylic acid was converted to an acid chloride by reaction with thionyl chloride. The aminoisobutyl POS was then attached to the RAFT agent **1** via a nucleophillic substitution to form compound **2**. The monomer N-isopropyl acrylamide (NIPAM) was polymerized in the presence of AIBN (a traditional free radical initiator) and the RAFT agent **2** to form compound **3**. The resulting series of copolymers had M_n (GPC) ranging from 9,600-17,800 g mol^{-1}, with PDI =1.07-1.16. Properties of this class of polymers will be discussed in Section 4.4.3.

Nitroxide-mediated polymerization (NMP) was utilized to synthesize hemitelechelic POS-PS, as shown Fig. 4.7 [30]. Isocyanate functionalized POS was reacted with a TEMPO-based molecule in the presence of the catalyst dibutyl tin dilaurate (DBTDL) to form the POS-based NMP initiator. Styrene was subsequently polymerized, and the resulting PS oligomers had molecular weights of 2,100 or 2,500 g mol^{-1} with low PDIs of 1.11 or 1.05, measured against linear PS GPC standards.

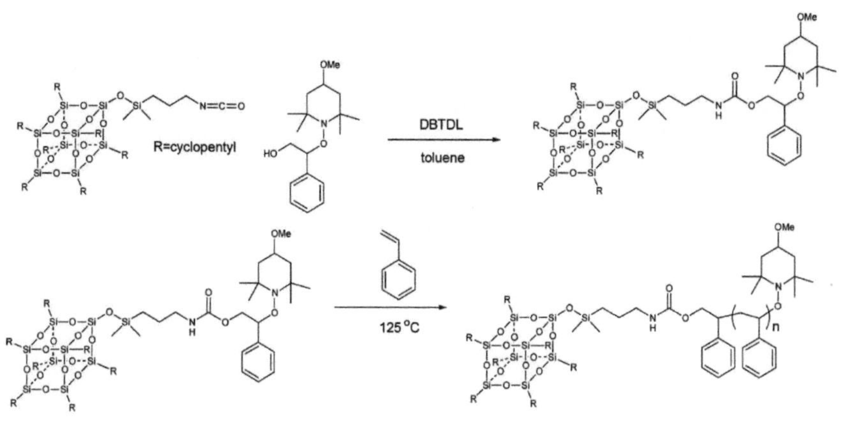

Fig. 4.7 Synthesis of hemitelechelic PS-POS by nitroxide-mediated polymerization (NMP)

4.2.3 *Anionic Polymerization*

Anionic polymerization is one synthetic technique that enables the synthesis of polymers with well-defined molecular weights and narrow molecular weight distributions. Recently, Gopalan and co-workers utilized this technique to create block copolymers with MA-POS, as seen in Fig. 4.8 [31-32].

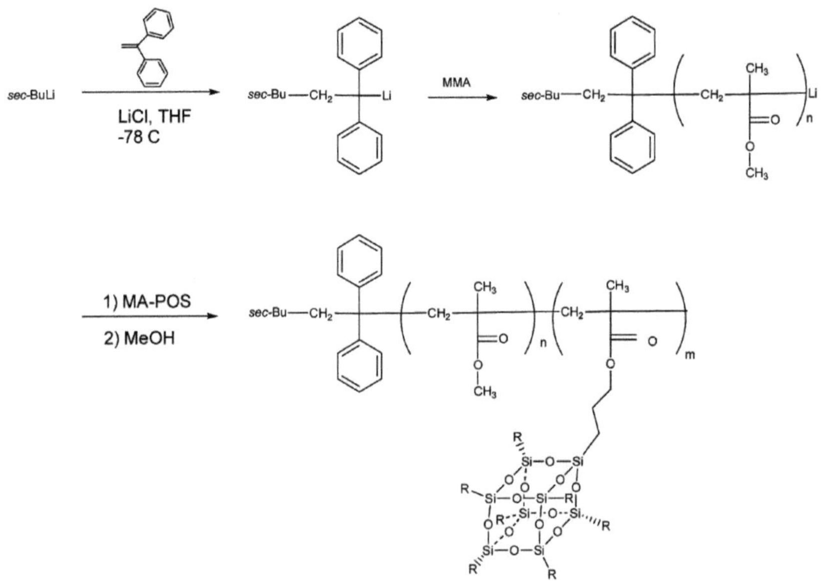

Fig. 4.8 Anionic synthesis of diblock PMMA-POS copolymer

A common anionic initiator, *sec*-butyl lithium, is endcapped by 1,1-diphenylethylene (DPE) in the presence of LiCl salt in THF at reduced temperature. The DPE limits the reactivity of the initiator, impeding its ability to attack the methyl methacrylate carbonyl group. The MMA monomer is then added. After the polymerization of the PMMA, the anionic chain is still active, which allows the second monomer, POS MA, to be added. After the POS polymerization, the active chain-end is quenched by the addition of methanol. The resulting polymers had narrow molecular weight distributions of less than 1.09. These polymers formed microphase-separated structures, which are dependent on POS content and will be discussed in Section 4.1.

4.2.4 *Ring-Opening Metathesis Polymerization (ROMP)*

Ring-opening metathesis polymerization (ROMP) is a synthetic technique that initiates polymerization of cycloalkenes by a metal-alkylidene complex [33].

To polymerize POS macromers by this technique, one or more of the eight peripheral groups should be replaced by a functionality susceptible to ROMP polymerization; often the peripheral group will have norbornenyl functionality [34-38]. An example is seen in Fig. 4.9 [34].

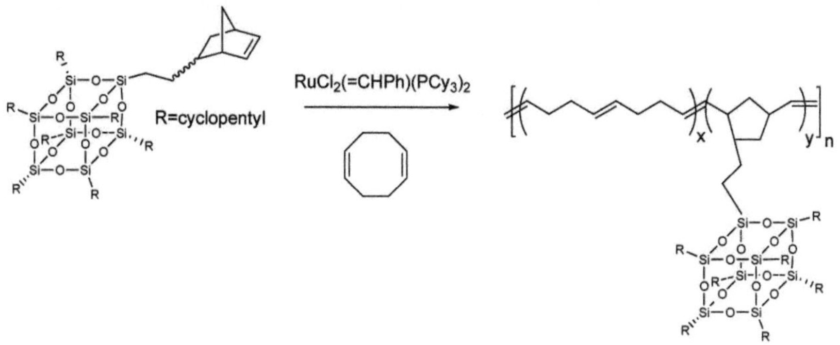

Fig. 4.9 Synthesis of PBD-POS copolymers

Zheng et al. used cyclopentyl POS, a co-monomer of cyclooctadiene (COD), and a Grubbs first generation catalyst to create a series of copolymers with POS content ranging from 12-53 wt% [34]. Analysis by ^{1}H NMR showed the absence of characteristic peaks from a POS homopolymer, suggesting that the POS-PDB copolymers were randomly sequenced. The authors speculate that despite the POS having higher ROMP activity than COD, long reaction times allowed for cross-metathesis reactions to occur on an interchain basis [39].

Poly(dicyclopentadiene) (PDCPD) is a commercial thermoset polymer that forms a cross-linked material after undergoing ROMP, and often serves as a matrix material in composites[40]. Monofunctional [35, 41], trifunctional [35], and octa-functional [41] norbornenyl-POS compounds have been investigated for use as a potential reinforcing component for PDCPD.

The synthesis of a PDCPD network with monofuctional norbornenyl-POS reinforcements is depicted in Fig. 4.10 [35].

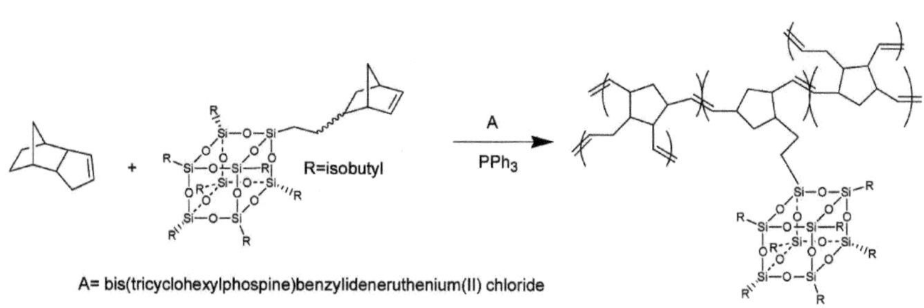

Fig. 4.10 ROMP polymerization of PDCPD network with norbornenyl-POS [35]

Work by Pittman et al. found that, compared to the pure PDCPD polymer, incorporation of 0.1% octa-functional POS increased the T_g of the material, while monofunctional POS decreased the T_g of the material [41]. The authors attributed this property change to changes in cross-link density. Work by Constable et al. also showed a decrease in the T_g for monofunctional POS [35]. However, the authors found no change in T_g when trifunctional norbornenyl-POS was added. More details on the properties of this and other POS networks are discussed in Section 4.6.

4.2.5 *Metallocene-Catalyzed Polymerization*

Metallocene-catalyzed polymerizations produce stereoregular products when the appropriate ligands and reaction conditions are chosen [42]. Zheng et al. synthesized isotactic polypropylene-POS copolymers using the catalyst shown in Fig. 4.11 Reaction A [43]. The resulting copolymers had relatively high isotacticity, with the pentad sequences (mmmm) at 88-89%. The PDIs of the polymers were between 1.42 and 1.54.

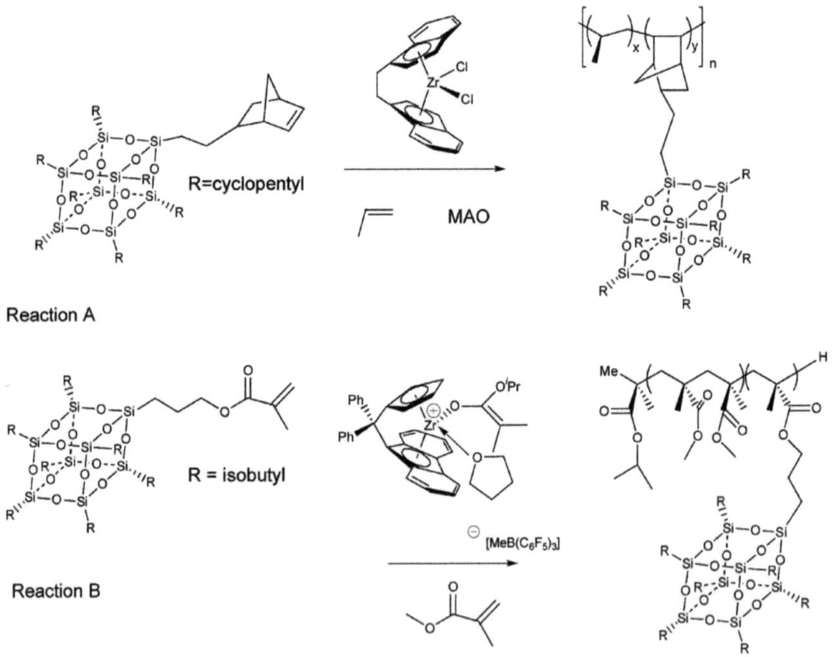

Fig. 4.11 Metallocene-catalyzed polymerization of stereoselective POS copolymers

Escudé and Chen synthesized syndiotactic methyl methacrylate-POS copolymers [44], as shown in Fig. 4.11 Reaction B. The resulting copolymers had 94 % syndiotacticity, based on triad measurement (rr) with PDI's of 1.21 and 1.33.

In work by Seurer and Coughlin, metallocene-catalyzed polymerizations were also used to synthesize unique thermoplastic ethylene-propylene-POS elastomers, as shown in Fig. 4.12 [45].

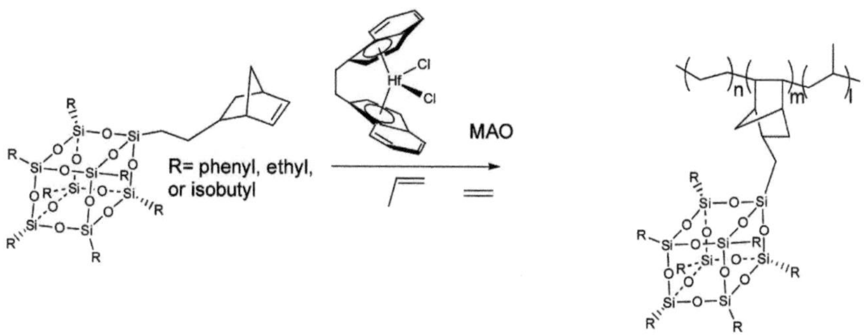

Fig. 4.12 Synthesis of ethylene-propylene-POS thermoplastic elastomers

Tensile tests indicated that the addition of POS to ethylene-propylene elastomers increased the storage modulus, as well as the maximum elongation at break.

Polymers with a high degree of syndiotacticity can also be synthesized using half-metallocene catalysts. In work by Zheng et al., random copolymers of PS and styryl-POS were synthesized using a half-metallocene titanium catalyst and MAO, as shown in Fig. 4.13 [37].

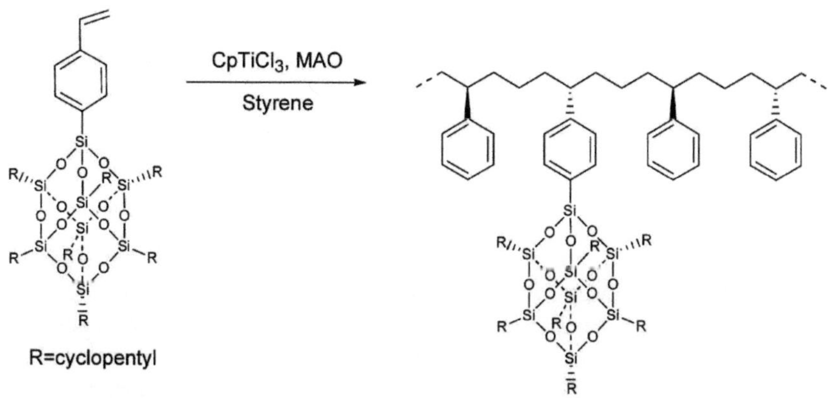

Fig. 4.13 Synthesis of syndiotactic random copolymers utilizing a half-metallocene catalyst

The resulting copolymers had PDIs around 1.5 and POS content between 3 and 24 wt %.

4.2.6 *Step-Growth Polymerization*

Three major classes of POS copolymers are prepared in step-growth polymerizations: POS-polyurethanes [46-48], POS-epoxies [49-51], and POS-polyimides [52-55].

POS copolymers with polyurethanes have been investigated for their mechanical [46], thermal [56], and biocompatible properties [47-48]. An example of a POS-polyurethane synthesis can be found in the work of Wang et al. as shown in Fig. 4.14 [48].

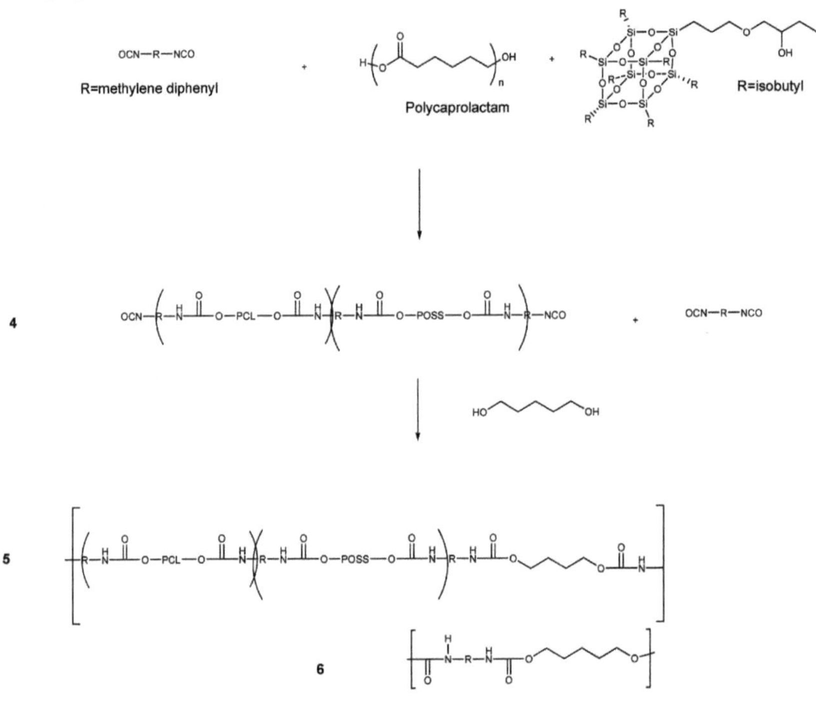

Fig. 4.14 Synthesis of poly(ester-urethane)-POS copolymer

The authors began the synthesis by reacting a diisocyanate, methylene diphenyldiisocyanate (MDI), with two polyols: polycaprolactam and POS diol. The copolymer **4** was chain extended with 1,4 butanediol in the presence of additional MDI to form poly(ester-urethane)-POS copolymer **5**, with a by-product of polymer **6.** The properties of POS-polyurethane copolymers is discussed in Section 4.5.2.

POS was studied as a nanoreinforcement for epoxy resins, which are common engineering thermosets [49-51]. A variety of mono- and poly-epoxy POS monomers have been investigated, as shown in Fig. 4.15 [50]. POS can be incorporated into the epoxy network as either a junction point or as a pendant group.

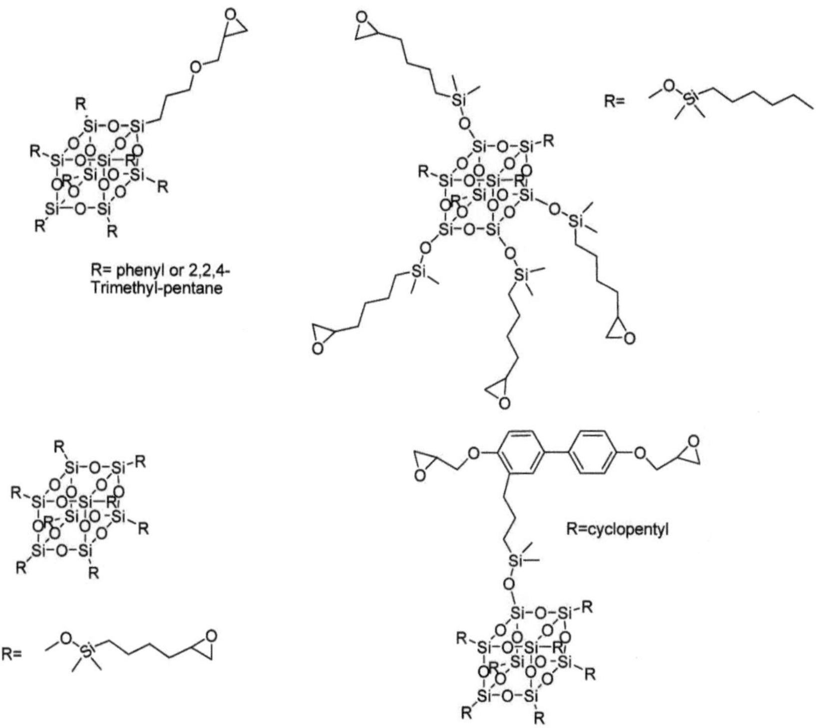

Fig. 4.15 Examples of mono- and multi-functional epoxy-POS monomers[50]

Fig. 4.16 is an example of a POS-reinforced epoxide network found in work by Lee and Lichtenhan [49]. An epoxide functionalized POS and 1,4-butanediol diglycidyl ether (BDGE) were cured with a diamine-terminated poly(propylene oxide) (JEFFFAMINE D-230) for 3 h at 60 °C. The resulting epoxy-POS prepolymer was reacted with the diglycidol ether of Bisphenol A for 24 h at 100° C to form the POS-reinforced resin. The properties of this polymer and other POS-epoxy polymers will be discussed in Section 4.6.1.

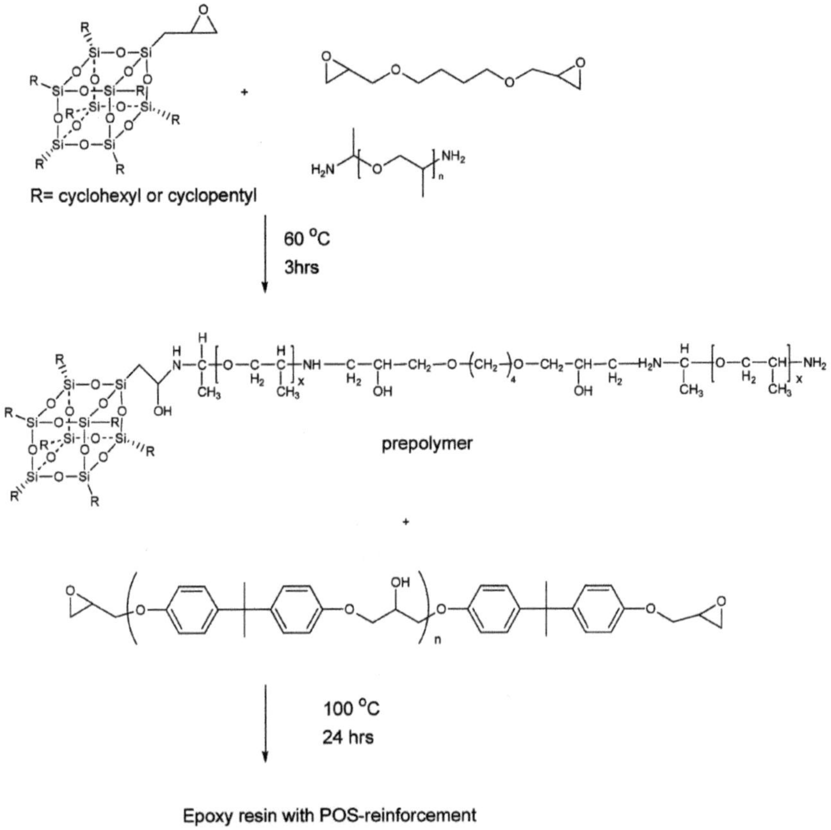

Fig. 4.16 Synthesis of a POS-epoxy resin

Polyimides are an important class of polymer; their properties include a low dielectric constant, low coefficient of thermal expansion, high temperature stability and mechanical strength. Aromatic polyimides are formed by a reaction between diamines and dianhydrides. POS moieties can be incorporated into an aromatic polyimide in two ways: directly into the main chain, or as a pendant group. An example of a polyimide with POS moieties as pendant groups is seen in Fig. 4.17.

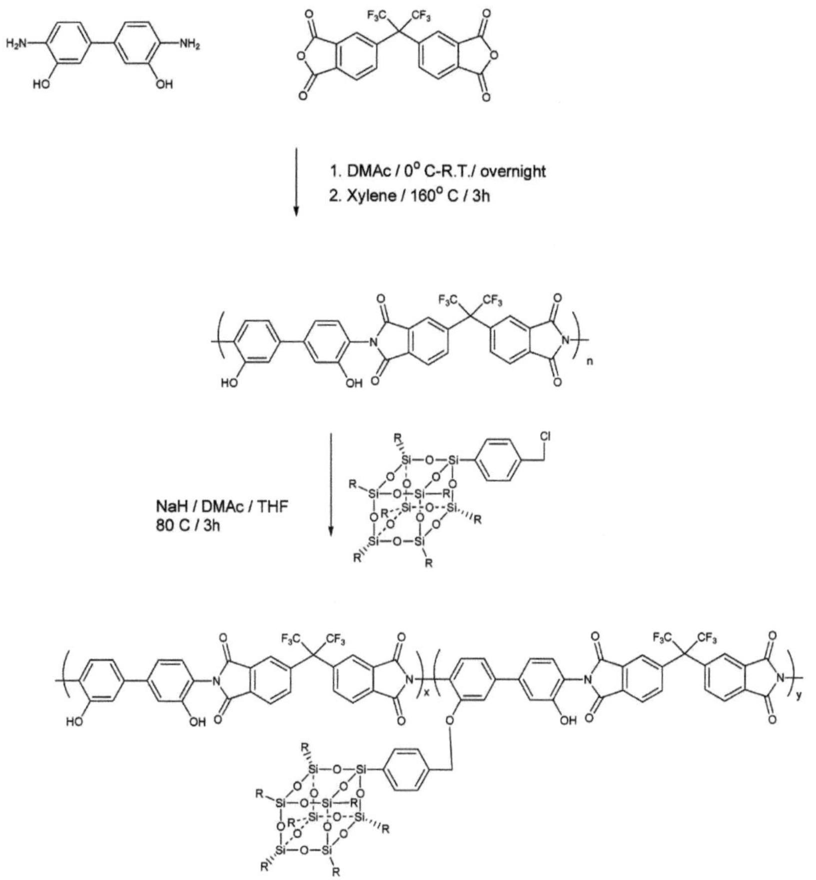

Fig. 4.17 Synthesis of aromatic polyimide with pendant POS moieties [57]

In work by Leu et al. the preceding POS-polyimide was synthesized by reacting a dihydroxy diamine with a fluorinated anhydride [57]. The resulting aromatic polyimide was deprotonated with the base NaH, and chlorobenzyl-POS was then grafted onto the deprotonated site.

POS can be incorporated into the main chain by using an open cage POS moiety. An example from the work of Brunsvold et al. is seen in Fig. 4.18 [52]. The properties of these polymers will be discussed in Section 4.5.1.

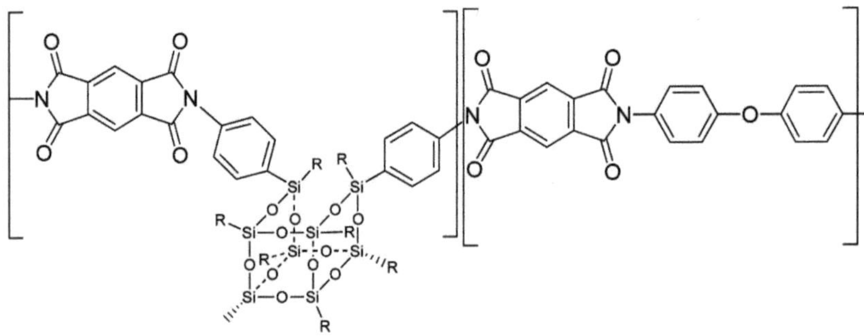

Fig. 4.18 Open cage POS moiety incorporated into polyimide chain

4.2.7 Grafting

Numerous efforts have focused upon grafting as a synthetic technique to attach POS moieties [58-66]. There are two major synthetic routes: using POS as a grafting site to initiate chain growth, or grafting POS to previously synthesized polymer chains.

In work by Kim et al., the first approach was taken [63]. POS was used as a grafting site to initiate polymer chain growth. The authors began with a POS core octa-functionalized with an alkyl iodide that initiates the ring-opening of oxazoline. The synthesis is seen in Fig. 4.19.

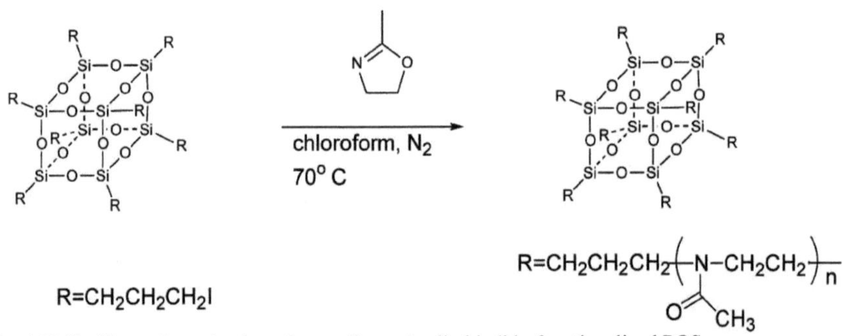

Fig. 4.19 Grafting polymerization of oxazoline onto alkyl iodide-functionalized POS core

By integrating the ^1H NMR peaks associated with the methylene protons attached to the POS core, the methyl protons of the polymer repeat unit, and the protons of the oxzolinium end group, the authors determined that not all of the eight possible grafting sites had initiated, and that the grafted chains were of differing lengths. The authors attributed the incomplete initiation to steric hindrance.

Drazkowski et al. employed the second approach, i.e., POS moieties grafted to previously synthesized polymers [64]. This chemistry is shown in Fig. 4.20.

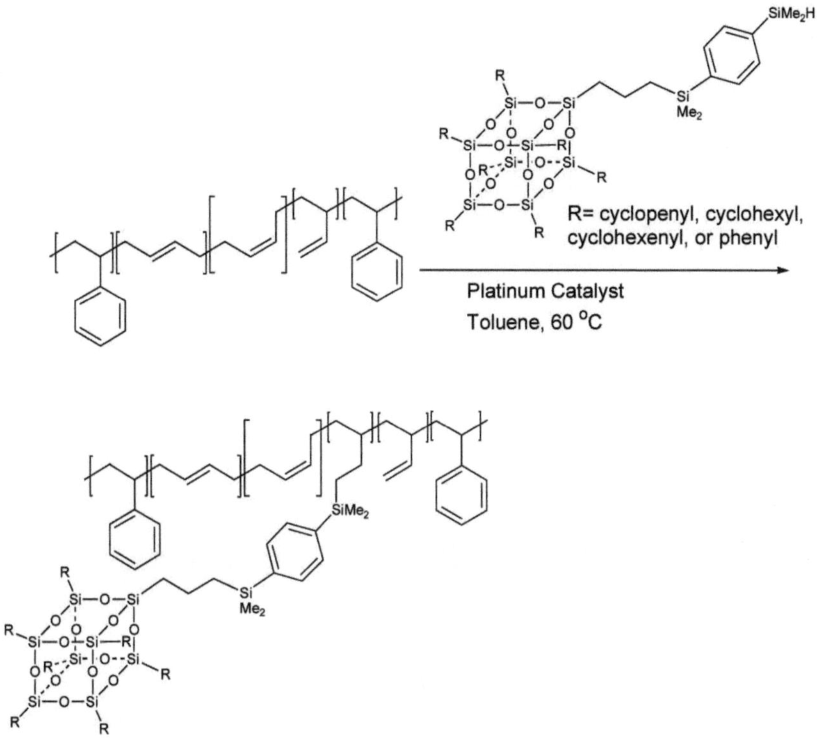

Fig. 4.20 Grafting of a POS moiety to a previously synthesized polymer chain

The authors synthesized a styrene-butadiene-styrene triblock polymer. Butadiene polymerizes by an anionic mechanism that results in both 1,2 and 1,4 repeat units. The former repeat unit results in a pendent vinylic group that serves as a site for grafting chemistry. POS hydrides were grafted to the vinyl group via hydrosilylation chemistry. The morphological effect of POS on this, and other triblock copolymers, will be discussed in Section 4.4.2.

Another possibility for the grafting of POS, which to the best of our knowledge has not been presented in the literature, is grafting a $T_7Si(OH)_3$ open-cage POS moiety to a previous synthesized polymer with pendent trichlorosilane groups. A proposed synthesis is shown in Fig. 4.21.

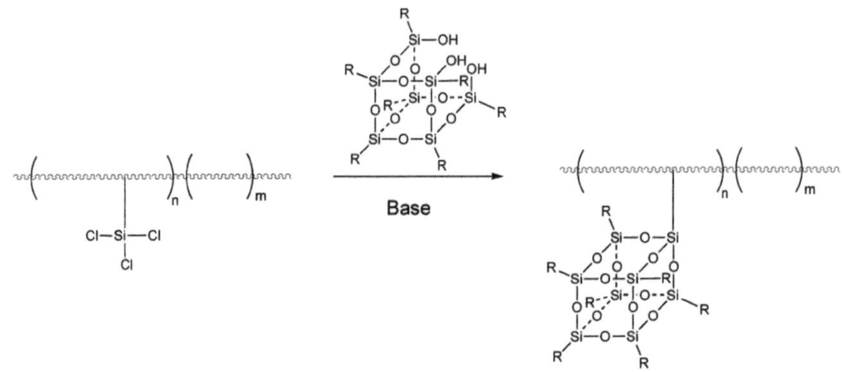

Fig. 4.21 Proposed synthesis for grafting $T_7Si(OH)_3$ to a polymer pendant group

4.3 POS Pendant-Random Copolymers

4.3.1 Glass Transition Temperature

The glass transition temperature T_g of random copolymers is affected by pendant POS moieties, and is governed by several competing factors: steric restriction, intermolecular interactions and free volume. In many polymers, POS has been shown to increase the T_g of polymers [14, 67-68]. In these polymers, POS and POS aggregates act to restrict the motion of the polymer chains, increasing the T_g of the copolymer relative to the homopolymer [14, 67]. The introduction of pendant POS moieties also can reduce the T_g of the polymer if the presence of the POS moieties reduces the number of intermolecular interactions, or increases the free volume relative to the homopolymer. Xu et al. reported this effect while studying POS-polyacetoxystyrene (POS-PAS) copolymers [14]. PAS homopolymers exhibit strong dipole-dipole interactions which restrict polymer chain motion. With the addition of POS moieties, this dipole-dipole interaction is diluted, leading to a decrease in T_g relative to the homopolymer. At higher concentrations of POS, the T_g was found to increase, owing to increasing numbers of POS moieties and aggregates restricting chain motion. Wu et al. also reported on random copolymers with a depressed T_g, attributed to the POS moieties introducing additional free volume [13, 67].

The T_g of a POS random copolymer also depends on the peripheral groups of the POS moieties [67, 69]. POS-PS copolymers were synthesized, having cyclopentyl (Cp), cyclohexyl (Cy) or isobutyl (^{i}Bu) peripheral groups [67]. Favorable interactions between PS segments and Cp/Cy groups led to an increase in T_g, while the poor interactions between PS and ^{i}Bu groups led to a decrease in T_g with increasing concentration of POS. Computer simulation studies of norbornene-POS copolymers also predict that material properties depend on POS peripheral groups [70].

4.3.2 Mechanical Properties

Viscoelastic behavior of POS random copolymers was studied in the work of Romo-Uribe et al. A series of copolymers of 4-methylstyrene and styryl-POS was synthesized [71]. This work reported that the properties were POS concentration dependent. Copolymers with a low POS concentration behaved similarly to 4-methylstyrene homopolymers. At higher POS concentrations, they found that the copolymers exhibited rubber-like behavior at increased temperature, despite having molecular weights below the entanglement molecular weight. This behavior was attributed to the bulkiness of the POS moiety, and to the cooperative rearrangement that would be required to relax the structure.

Wu et al. synthesized styrene-based copolymers with higher molecular weight than the previously mentioned copolymers synthesized by Romo-Uribe [67]. The authors report that the copolymers exhibit rubbery plateaus. The plateau modulus (G^0_N) was found to be inversely dependent upon the concentration of POS in the copolymer; the G^0_N also depended upon the peripheral group.

4.3.3 Crystallinity in POS Pendant-Random Copolymers

The crystal structure of untethered POS was analyzed by X-ray diffraction in the work of Larsson [72-73], and in later work by Waddon and Coughlin [74]. POS moieties with alkyl peripheral groups packed hexagonally with an ABCA stacking sequence, as shown in Fig. 4.22.

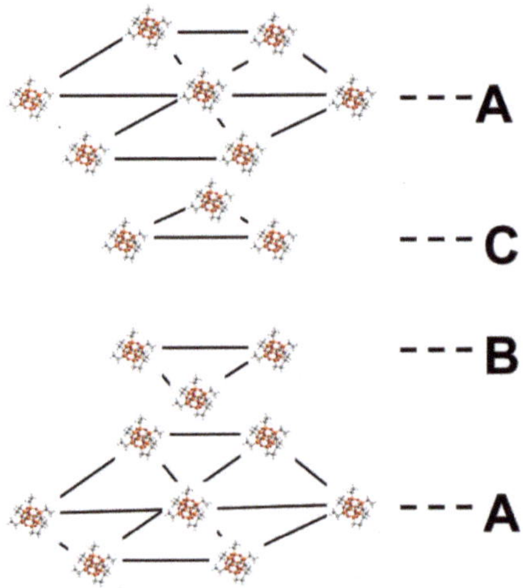

Fig. 4.22 Schematic of hexagonal-packed POS moieties with ABCA stacking. Reproduced from [74] with permission from the American Chemical Society.

The peripheral groups are located in the interstices and prevent close-packing of the moieties. The choice of peripheral groups was found to affect the hexagonal lattice vectors, a and c. Larsson reported significantly smaller values (a=9.13Å, c=15.36 Å) for POS with hydrogen peripheral groups than Waddon and Coughlin (a=16.06 Å, c=17.14 Å) reported for POS with one norbornyl group and seven cyclopentyl groups. When tethered, as POS moieties are in copolymers, their crystallization behavior differs, which is the topic of this subsection.

Zheng et al. studied a series of copolymers of norbornyl-POS and PE by WAXD spectroscopy, with POS incorporation ranging from19-56 wt % [75]. Both components were found to crystallize.

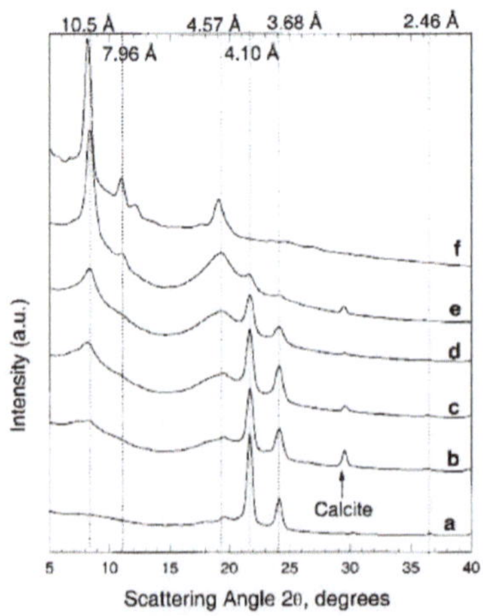

Fig. 4.23 WAXS data for PE-POS copolymers Profiles b-e represent copolymers of increasing POS content Profiles a and f are PE and POS moieties respectively. Reproduced from [75] with permission from the American Chemical Society.

At lower POS content, PE crystallinity dominates, although broad peaks corresponding to POS are present. With increasing POS content, peaks corresponding to PE crystallinity broaden, while POS peaks sharpen and intensify; PE crystallinity is decreasing as POS forms a crystal lattice (Fig. 4.23.) The authors also report d-spacings similar to untethered POS moieties, suggesting that the packing is similar. However, in copolymers, the POS moiety is tethered to a polymer chain, which is unlikely to be accommodated into the lattice. With this information, the authors proposed that the POS moieties self-assemble into a 2-D lattice structure with the host polymer (Fig. 4.24).

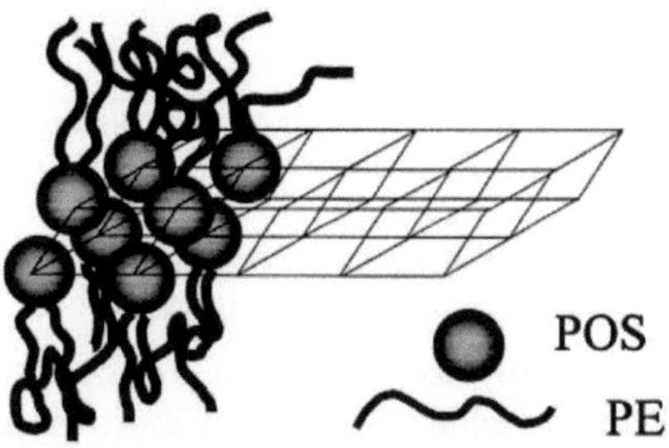

Fig. 4.24 Model of a 2-dimensionl POS lattice self-assembled from PE-POS copolymers. Reproduced from [75] with permission from the American Chemical Society.

The morphology of random copolymers was further studied by analyzing norbornyl-POS and 1,5 cyclooctadiene (Fig. 4.29) copolymers [76]. These studies revealed that with low POS content the 2-D lattice structures were randomly orientated with dimensions of about 50 nm. With increasing POS content, the copolymers self-assembled into a continuous lamellar morphology, with lamellae microns in length. The authors also noted that the POS crystals disrupted the crystallinity of the 1, 4 butadiene, lowering its melting point by up to 25 °Copolymer structure can be altered by the crystallization procedure. Waddon et al. crystallized previously synthesized norbornyl-POS and PE by both solution crystallization and melt crystallization techniques [77]. With solution crystallization the PE components precipitate more readily than the POS segment, forming a PE 'scaffold'. Since POS is covalently linked to the PE chains, it is limited in its ability to self-assemble. After melt processing, the crystallization of PE was found to be constrained by POS moieties; the resulting material had POS crystals with less ordered PE crystalline domains.

4.4 POS Pendant-Block Copolymers

4.4.1 Diblocks

Block copolymers are of interest owing to their ability to self-assemble into different morphologies. Several excellent reviews are available [78-80]. Block copolymers with POS as a pendant moiety have been synthesized. Pyun and Matyjaszewski synthesized p(MA)-*b*-p(MA-POS) via ATRP, and Xu et al. synthesized POS-norbornene-based block copolymers by ROMP; both varieties of block copolymer had low polydispersities.

Recently, several authors studied the self-assembly, morphology, and structure-property relationships of block copolymers containing pendant POS. Hussain et al. synthesized poly(ethylene glycol)-*b*-poly(pentafluorostyrene) (PEG-*b*-PPFS) [59]. Subsequently, aminopropylisobutyl POS moieties were grafted to a fraction of the reactive *para*-F functionality of the PFS unit by an aromatic nucleophillic substitution reaction. The copolymers, with or without grafted POS moieties, were found to self-assemble into micellar structures in aqueous solutions. Increasing the number of POS moieties per chain was found to increase both the hyrdrodynamic radius (R_h) and the radius of gyration (R_g), determined by dynamic light scattering (DLS) and static light scattering (SLS), respectively. The authors state that the increase of the radii was caused by the hydrophobicity of the POS moieties; polymer chains with an increased number of moieties led to higher aggregation numbers. Analysis by DSC showed that the POS units acted as a plasticizer, lowering the T_g of the PPFS by more than 30° C. This plasticizing effect has also been observed in perfluorocyclobutyl aryl ether –POS random copolymers [81]. The presence of POS also disrupted PEO crystallization.

Gopalan and co-workers [31-32] anionically synthesized two diblock copolymers: PS-*b*-PMA-POS and PMMA-*b*-PMA-POS. The synthesis was shown previously in Fig. 4.8. Bulk samples were prepared by slow evaporation from solution. Analysis by TEM and Transmission Small Angle X-ray Spectroscopy (TSAXS) showed that the morphology was dependent upon the volume fraction of POS (Fig. 4.25). ....

When the volume fraction of PMA-POS was low (13%), the sample exhibited spherical morphology that lacked long-range order. When the volume fraction of PMA-POS was increased to 44% or 77%, the material had a lamellar or hexagonally-packed cylindrical morphology; both morphologies exhibited long-range order. When these phase separated structures were subject to thermal annealing, hierarchical structures were formed. POS crystalline domains were formed within the phase separated structures. Additionally, the authors have proposed a helix-like structure for the POS crystalline domains.

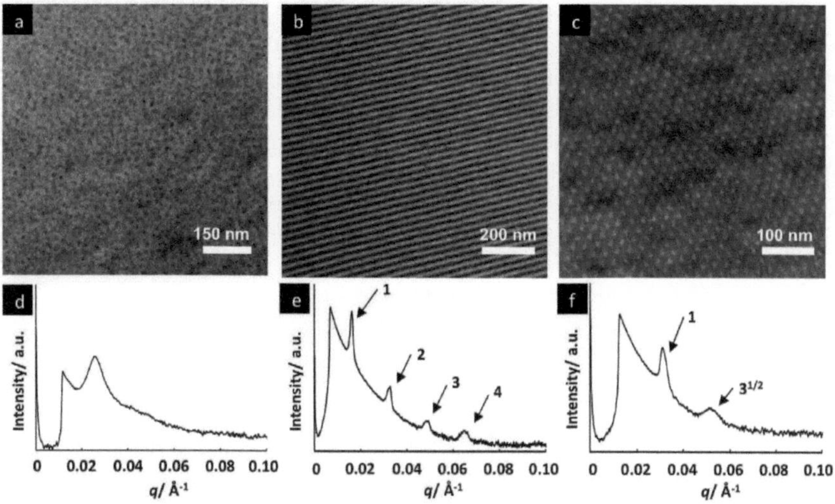

Fig. 4.25 TEM Micrographs and TSAXS profiles of (a, d) PMMA450-*b*-PMA-POS7, (b, e) PMMA262-*b*-PMA-POS23, (c, f) PMMA52-*b*-PMA-POS18. Reproduced from [31] with permission from the American Chemical Society.

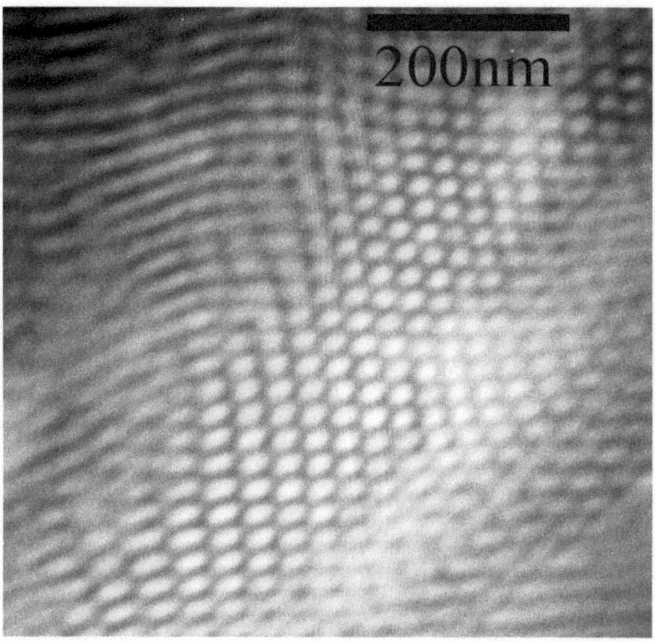

Fig. 4.26 Inverse cylindrical morphology in p((ethylene-butylene)-*b*-MA-POS) diblock copolymers

In work by Gadodia, p((ethylene-butylene)-*b*-MA-POS) diblock copolymers were synthesized by anionic polymerizations followed by ATRP [82]. Depending on the volume fraction of POS, spherical, cylindrical, and lamellar morphology was observed; both blocks also had the ability to crystallize. The authors observed inverse cylindrical morphology at 60 vol % EB shown in Fig. 4.26. The majority block (EB) formed the cylindrical phase, while the minority phase (MA-POS) formed the matrix around the cylindrical phases. The authors attribute this unusual morphology to the concept of conformational asymmetry [83-84].

4.4.2 Triblocks

In work by Fu et al. and Drazkowski et al., POS-hydride moieties were grafted onto unsaturated vinylic pendant groups of a styrene-butadiene-styrene (SBS) triblock copolymer [64-65, 85]. An example synthesis was shown in Fig. 4.20. Fu et al. grafted cyclopentyl-POS-hydrides and reported that minimal POS crystallization was visible by WAXD, suggesting that POS was well-dispersed throughout the matrix material (PBD) [85]. The authors analyzed both the non-grafted SBS and grafted SBS by SAXS. While both block copolymers exhibited cylindrical morphology, the grafting of POS to the PBD phase led to a decrease in the long-range order. Mechanical testing at elevated temperatures showed that the grafting of 20 wt% POS to SBS permitted the material to be deformed to a strain of 350%; unmodified SBS deformed to a strain of 150%.

Drazkowski and colleagues investigated how the morphologies of SBS triblock copolymers were affected by different peripheral groups (cyclopentyl, cyclohexyl, cyclohexenyl and phenyl) attached to POS using SAXS and rheological techniques [64]. Phenyl-POS interacted favorably with the PS blocks and become partially solvated, plasticizing the SBS. This interaction decreased the order-to-disorder transition temperature T_{ODT} and lowered the lamellar spacing (d) with increasing numbers of POS moieties grafted to SBS. POS moieties carrying the other cyclic peripheral groups (cyclopentyl, cyclohexyl, cyclohexenyl) showed less interaction with the PS phase, reducing the dependence of d and T_{ODT} on the number of grafted POS moieties. The authors also investigated SBS triblocks grafted with isobutyl-POS [65]. The grafting of ^{i}Bu-POS did not affect the local or long range morphological order, indicating that the ^{i}Bu-POS did not interact with the PS domains.

Pyun et al. synthesized MA-POS-*b*-BA-*b*-MA-POS triblock copolymers and observed phase separation dependent upon volume fraction of POS [23]. The authors report that after annealing samples near the T_g of the POS block, physical aging was apparent. WAXS studies suggest that POS domains became more ordered. The authors also report that T_g associated with the POS domain in the phase-separated triblock copolymer is more than 20° C higher than in POS homopolymers of identical molecular weight; the authors attribute this increase to a confinement-based enhancement of T_g.

4.4.3 Hemitelechelic ('Tadpole'-Shaped) Polymers

Hemitelechelic organic/inorganic molecules, also referred to as 'tadpole'-shaped molecules, have one of their two peripheral groups replaced by a polymer chain. A schematic of the molecular architecture and an example polymer are seen in Fig. 4.27.

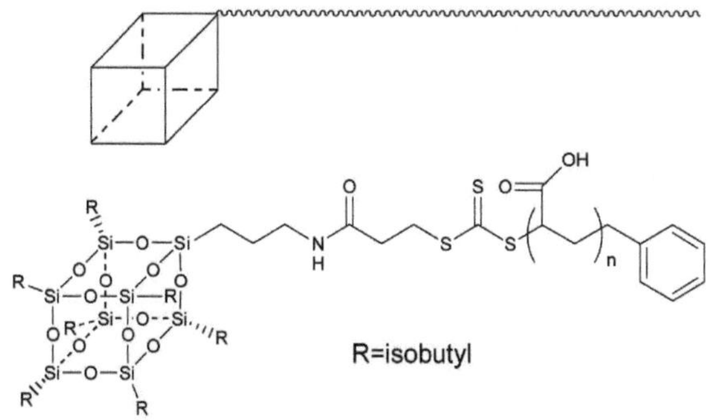

Fig. 4.27 Schematic 'tadpole'-shaped architecture of hemitelechelic polymer and example polymer

Hemitelechelic polymers are of interest owing to their often amphiphilic nature. As a result, a number of interesting morphologies have been observed, including micelles [28, 86], vesicles [86], and aggregates [29, 87]. Cardoen and Coughlin reported aggregate POS domains while studying hemitelechelic POS-PS hybrid polymers [87]. By monitoring a major diffraction peak for POS (2θ=8.6°) with Wide Angle X-ray Spectroscopy (WAXS), the authors found that the peak increased with increasing weight percent POS. Work by Müller and colleagues also observed aggregation with hemitelechelic hybrid polymers, using POS-poly(acrylic acid) polymers, as shown in Fig. 4.28 [29]. The self-assembly of POS-PAA was observed by TEM and the aggregates formed were polydisperse, with radii ranging from 15-42 nm. Using cryogenic TEM, the authors showed that the aggregates do not have a uniform density or core-shell structure.

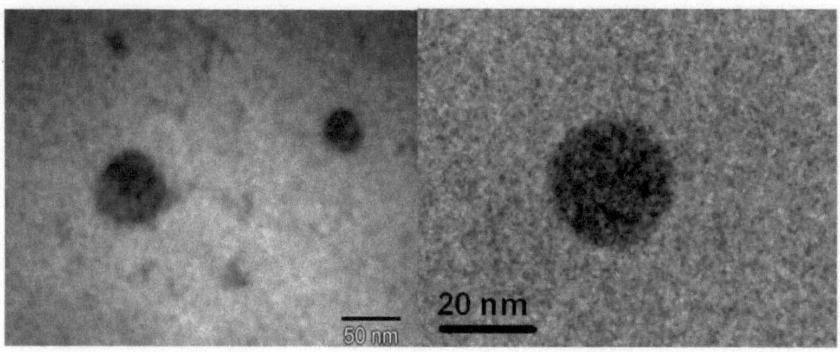

A B

Fig. 4.28 The self-assembly of POS-PAA aggregates by (a) TEM and (b) cryogenic TEM, reproduced from [29] with permission from the American Chemical Society

Müller et al. prepared POS-*p*-(*N*-isopropylacrylamide) (NIPAM) hemitelechelic polymers, of increasing molecular weight of NIPAM [29]. At temperatures below its LCST of 32° C, *p*-(NIPAM) is a hydrophilic molecule, owing to hydrogen bonding. Using AFM, the authors found that the polymers self-assembled into micellar structures in water, and that the average radius of the core is inversely proportional to the M_w of the *p*-(NIPAM). Self-assembled structures were not observed for solutions of POS-*p*-(NIPAM) prepared in chloroform.

Knischka et al. reported the self-assembly of structures with vesicular or micellar morphology [86]. POS-PEO hemitelechelic polymers were synthesized and allowed to self-assemble in water. At elevated pH, POS cores were condensed to form extended silicate domains. The authors studied the resulting structures by TEM and AFM, and observed vesicular-type structures (Fig. 4.29).

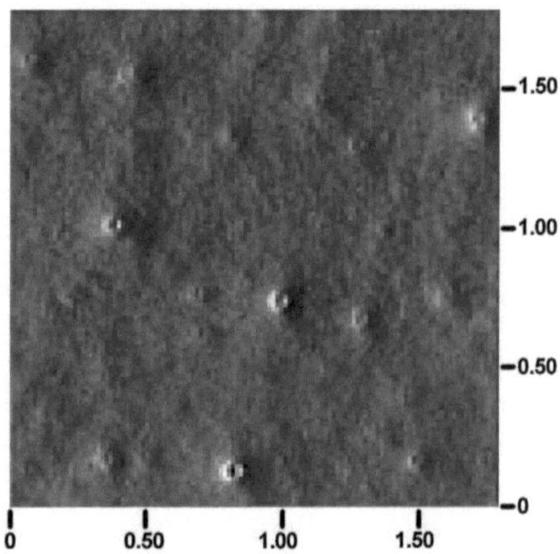

Fig. 4.29 AFM image showing vesicular-type structures. Reproduced from [86] with permission from the American Chemical Society.

Other authors have examined the effect of POS in hemi-telechelics upon properties such as thermal degradation temperature [27] and glass transition temperature [27, 87]. In a study by Zhang et al., a series of PS-POS polymers were synthesized by RAFT and it was found that the incorporation of a single POS unit increased the T_g by 4 °C relative to pure PS of similar molecular weight [27]. When the molecular weight of the PS-POS chain was increased, the increase in T_g was still observed. Analysis by TGA showed an increase in the thermal degradation temperature compared to pure PS. Cardoen and Coughlin synthesized hydroxy-terminated PS, and tethered a POS unit to the polymer using an isocyanate functionalized POS [87]. The authors found that the presence of POS did not affect the T_g relative to the pure PS analog. The only exception was the lowest molecular weight PS chain (873 g mol^{-1}), where the addition of the POS unit raised the T_g by 17° C.

Molecular simulations of hemi-telechelic molecules for the prediction of self-assembly behavior were created by Glotzer and colleagues [88]. They investigated various parameters that affect self-assembly, including choice of solvent, concentration, chain length, and temperature. The model molecule was a POS cage tethered to alkyl chains of varying length. The model predicts differing structures depending on the parameters. When the molecule is in a neutrally poor solvent, the system will undergo an order-disorder transition (ODT) as the volume fraction (φ) decreases, or as the length of the tether chain decreases. If the molecule is solvated in selectively good solvent for the POS moiety, the molecule will undergo an order-order transition, from cylindrical ordering at low temperatures to lamellar ordering at high temperatures.

4.4.4 Telechelic (Dumbbell-Shaped) Polymers

Telechelic organic/inorganic molecules, also referred to as dumbbell-shaped molecules, consist of a polymer chain linking two POS moieties. A schematic of the molecular architecture is shown in Fig. 4.30.

Fig. 4.30 Schematic of a telechelic organic/inorganic polymer

Like hemi-telechelic molecules, telechelic molecules are also studied for their amphiphilic properties. Mather and colleagues created a series of POS-PEG-POS telechelic polymers by reacting the diol end groups of poly(ethylene glycol) oligomers (M_n = 1, 2,3.4,8,10 kg mol^{-1}) with two monoisocyanate-functionalized cyclohexyl-POS moieties, and analyzed their behavior in water [89]. The authors found little difference between the behavior of the amphiphiles with the higher molecular weight PEG component (8 or 10 kg mol^{-1}) and that of their pure PEG analogs when dissolved into solution. However there was a significant difference between the behavior of the amphiphiles with the lower molecular weight components and that of their PEG analogues, the latter forming monolayers at the air-water interface. The hybrid polymers with 1 kg mol^{-1} PEG content were used to create Y-type Langmuir-Blodgett films on silicon substrates at ambient temperatures. X-ray reflectivity studies showed that a double-layer structure was formed with a double layer spacing of 3.52 nm.

Mather and his colleagues have also studied the thermal properties of these polymers. By studying thermal decomposition behavior by TGA, the authors found that the POS-PEG-POS amphiphiles underwent two-stage decomposition, instead of the one-stage decomposition of their PEG analogues [90]. The amphiphiles also exhibited different crystallization behavior than their PEG analogues when analyzed by DSC. In the lower PEG molecular weight amphiphiles (1 and 2 kg mol^{-1}), no melting endotherms were present. The authors speculate that the bulky POS group disrupts the orderly folding of PEG lamellae. In the higher molecular weight amphiphiles, melting endotherms are present, but are broader, and the calculated percent crystallinity is lower than their PEG analogues. Further studies with polarized optical microscopy showed irregular birefringence patterns for telechelic polymers, possibly due to the POS forming a separate phase that elicits PEG nucleation. Whereas the PEG analogues showed a Maltese cross birefringence pattern, diffraction peaks characteristic to POS are present. This was supported by WAXD patterns.

In order to better predict the properties of POS telechelics, scientists have turned to simulations. In work by Striolo et al., the authors model the self-organization of POS moieties connected by a hydrocarbon backbone in solution. Their model was based upon telechelic molecules dissolved in *n*-hexane at infinitely dilute concentrations in the temperature range of 300-600K [91]. In the low temperature range, the model predicts the POS moieties are more likely to be located near each other than in the higher temperature range. For hydrocarbon backbones of shorter lengths (8 carbons), the POS moieties tend to have larger separations than POS moieties connected by longer backbones (20 carbons), owing to the rigidity of the short chain backbone.

4.5 POS-Polyimide and POS-Urethanes

4.5.1 POS-Polyimide

Polyimides are a class of polymer with a low dielectric constant, and are utilized as an interlayer in microelectronics. Several approaches, including introducing porosity or fluorinated groups, have been shown to reduce the dielectric constant of polyimide materials [92-94]. The inclusion of POS moieties has also been shown to reduce the dielectric constant of polyimides; the dielectric constant of POS macromolecules is 2.1-2.7 [95]. In addition to dielectric properties, mechanical and thermal properties are also affected by the inclusion of POS moieties.

Seckin et al. synthesized anhydride terminated polyimides, which were subsequently attached to an octa(aminopropyl)-POS core [96]. The incorporation of POS lowered the dielectric constant compared to the neat polyimides. The authors attribute this effect to an increase in free volume. Improved thermal properties were also reported.

Leu and colleagues synthesized two POS-polyimide materials: polyimides with 10-35 mol % POS pendant groups (Fig. 4.17) and polyimide chains end-capped by 2.5 mol % POS moieties. Polyimides with POS pendant groups formed ordered structures with POS crystallites [53]. The authors attribute the POS aggregation to the difference in van der Waals forces between the cyclopentyl peripheral groups of the POS moieties and the imide segments. Addition of POS pendant groups decreased the dielectric constant of the polyimide materials by as much as 30% with the addition of 35 mol % of POS. Addition of POS decreased the T_g and thermal degradation temperatures T_d slightly. Polyimide materials with high POS content (20-35 mol %) exhibited a reduction in the Young's modulus. The polyimides end-capped by POS moieties were found by TEM to self-assemble into POS-rich and polyimide-rich domains that were formed perpendicular to the direction that the film was cast [57]. A reduction in the dielectric constant (3.09 vs. 3.40) also was observed relative to the non-modified polyimide analogues. The mechanical properties were not affected by the addition of POS moieties, but POS had a slightly detrimental effect upon the thermal properties. Incorporation of POS into polyimide has shown promise in the field of space-survivable materials (see also Chapter 8) and lithographic resists (see also Section 7.4, Chapter 7). Kapton and other polyimide films are used frequently in spacecraft applications but often face erosion by atomic oxygen. Brunsvold et al. [52] and Wright et al. [97] synthesized polyimide polymers with covalently attached POS moieties and found that upon exposure to the atomic oxygen, a passivating oxide layer formed, significantly reducing further erosion. Argitis and colleagues created copolymers containing POS moieties, which, upon exposure to oxygen plasma, also formed a passivating silicon oxide surface layer [98-99]. This oxide surface allowed for etch selectivity when paired with an organic underlayer in a bilayer lithography system.

4.5.2 POS-Urethane

Polyurethanes (PUs) are an important class of plastic owing to their mechanical and thermal properties. Modifying polyurethanes with POS moieties has resulted in improved and unique material properties. POS-polyurethanes can possess biodegradable or biocompatible character suitable for biomedical application with an appropriate choice of monomers. In work by Knight et al., polyurethanes were synthesized with a biodegradable poly(lactide) as the soft block, and POS moieties were incorporated into the hard block segments [100]. The inclusion of POS gave the resulting material good elastic properties, owing to the physical cross-linking provided by the POS crystallites, as well as increased material rigidity. In vitro degradation studies showed that significant mass loss (20%) did not occur until two months had elapsed.

Raghunath et al. also studied biodegradable POS-PUs based on polycaprolactone (PCL, another biodegradable polymer) and polycarbonate, with potential application as tissue engineering scaffolds [47] (see also Chapter 9). Non-PU based POS-PCL materials have also been synthesized for biodegradable applications [101]. Degradation studies were performed in vitro and showed that while the POS-PU material underwent degradation, it largely retained its mechanical properties when exposed to hydrolytic enzymes. The authors note a potential biocompatibility issue with the chosen diisocyanate, methylene diphenyl isocyanate, which can release a carcinogen. Wang et al. also synthesized a POS-PU system based on PCL, previously shown in Fig. 4.14. The incorporation of POS provided improved mechanical properties and increased resistance to degradation. The material was also shown to be a suitable supportive medium for cell growth, without toxicity [48].

To reduce the amount of volatile organic compounds (VOCs) released during the application of polyurethane-based products, aqueous-based PU dispersions were developed [102-103]. Nanda et al. synthesized aqueous PU dispersions covalently attached to 4-10 wt% pendant POS moieties [104]. The authors report that the pendant POS moieties do not affect the properties of the dispersion when compared to 0 wt% POS PU dispersion. Solution viscosity and dispersive particle size were not significantly altered, but mechanical properties were affected. Up to a tenfold increase in tensile strength at break, as well as a decrease in maximum elongation at break, was found by tensile testing of films formed from PU-POS dispersions. Analysis by DMA shows an increase in the T_g associated with the hard segments, indicating that the POS moieties are being incorporated into the hard segments of the PU, rather than the soft segments.

Turri and Levi synthesized aqueous PU dispersions with 6-20 wt % pendant POS moieties in a prepolymer emulsification process [105]. The dispersions were found to be stable when the POS content was less than 10%. In contrast to the work of Nanda et al., Turri and Levi observed crystallinity in films formed from PU dispersions with the POS reinforcement occurring in the soft segments.

Elastomers of POS-PU were synthesized by Janowski and colleagues; the incorporation of 4-6 wt % of POS increased the thermal stability of the material [106]. Later studies detailed the resulting morphology [107]. The authors report that at low concentrations of POS (4 wt %), large spherulitic structures were formed, while at higher concentrations (10%), smaller and more regular structures were formed. This difference in morphology affected molecular mobility and the associated relaxations.

Unlike the previous examples, where POS was included as a pendant moiety, POS has also been included as an octa-functional site in the creation of polyurethane networks. Mya et al. synthesized PU hybrid materials using two different POS moieties: one having eight isocyanate functional groups, the other possessing eight hydroxyl groups. The resulting films were amorphous, with low surface roughness and good optical transparency [108]. Work by Liu et al. utilized octa-functional cyanate POS as a cross-linker, and the resulting materials showed improved thermal stability compared to analogous materials cross-linked solely by a diamine [109].

4.6 Multifunctional POS in Network or Core Structures

4.6.1 Epoxy Networks

Epoxies are thermosetting polymers with many applications, including adhesives, coatings and composite materials [110]. Incorporating POS moieties into epoxy resins affects the thermal, mechanical and morphological properties of the material (see also Section 5.5.4, Chapter 5).

Inclusion of POS often increases the T_g of epoxy resins owing to its restriction of molecular motion. Lee and Lichtenhan synthesized an epoxy network with inclusion of monofunctional epoxy-POS moieties (Fig. 4.16) [111]. They report an increase of T_g with increasing POS content. Work by Abad et al. also shows an increase in T_g when a monofunctional epoxy-POS is incorporated into an epoxy network [112].

Incorporation of multifunctional POS also results in an increase in the T_g of the resin. Zhang et al. reported on an epoxy network synthesized from octa(aminopropyl)-POS and the diglycidyl ether of bisphenol A (DGEBA) [113]. When the wt % of POS was less than 10%, they observed a strong dependence of the T_g on the wt % of POS. For additional POS content above 10 wt %, the T_g showed a slight increase. In work by Teo et al., epoxide networks of tetraglycidyl-4,4'-diaminodiphyenyl methane (TGDDM) cured with hexahydrophthalic anhydride (HHPA) were compared to TGDDM/HHPA-based networks that incorporated POS moieties functionalized with eight epoxide groups, and increases of up to 20 °C in T_g were observed for the POS systems [114].

Studies of mechanical properties were also performed. Zhang et al. investigated the effect of octa(aminopropyl)-POS on the impact and flexural strength of epoxy networks, and found that strength was maximized by the addition of 10 wt % POS [113]. Strachota et al. showed that octa-functionalized epoxide POS served as a cross-linker and increased the rubber modulus [115]. Laine et al. studied epoxy systems based upon octa-functionalized epoxide POS reacted with a diamine, and found that the mechanical and thermal properties, as well as the network structure itself, were dependent upon the ratio of amine to epoxide groups [116]. The authors reported that optimal mechanical properties occurred at a ratio of 0.5.

The morphology of POS-epoxy networks depends on whether POS is incorporated as a pendant moiety, or whether it is incorporated as a network junction point. Matejka and colleagues synthesized epoxy networks based on DGEBA and cross-linked by poly(oxypropylene) diamine (Jeffamine) [50]. Pendant moieties were incorporated using monofunctional epoxy-POS, or an epoxy-POS molecule based on DGEBA (Fig. 4.15). Multifunctional epoxy-POS moieties (Fig. 4.15) were used to create junction points. Networks that incorporated multi-functional POS were found to be non-crystalline. Aggregation of POS was found to be inversely related to the functionality of the POS moiety.

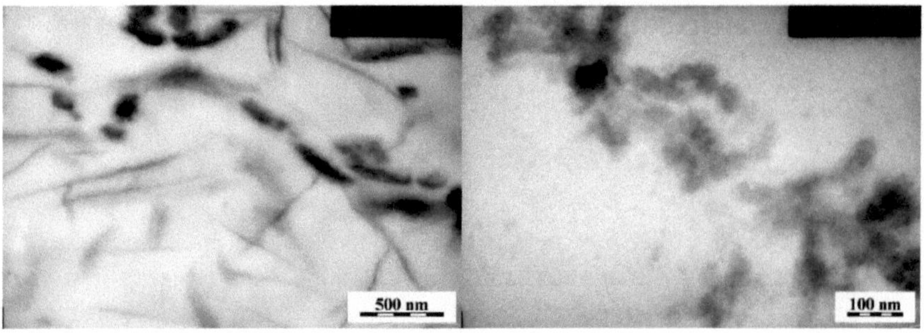

Fig. 4.31 TEM micrographs of epoxy networks with pendant POS moieties synthesized from (a) monofunctional epoxy-POS with phenyl groups (b) POS DGEBA. Reproduced from [50] with permission from the American Chemical Society.

Incorporation of pendant moieties led to crystallinity. Networks with epoxy-POS DGEBA formed anisotropic crystals (Fig. 4.31a). The authors proposed a model for this morphology where the anisotropic POS crystalline domains are linked by the Jeffamine chains.

In contrast, the morphology resulting from incorporating monofunctional epoxy-POS was dependent upon the peripheral groups. Incorporation of POS with phenyl peripheral groups resulted in roughly spherical isotropic crystalline domains (Fig. 4.31b). When monofunctional epoxy-POS with aliphatic peripheral groups were incorporated into the network, WAXD showed no crystallinity, only aggregation due to poor compatibility with the epoxy matrix. This incompatibility of a POS with aliphatic peripheral groups in an epoxy was also seen in work by Abad and colleagues [112]. In this work, the authors incorporated monofunctional epoxy-POS with isobutyl groups into an epoxy network and found that the system macrophase-separated into epoxy-rich and POS-rich phases.

The inclusion of POS has also affected other properties of epoxy networks. Incorporation of monofunctional epoxy-POS into epoxy networks has also been shown to be an effective fire retardant, but the extent of the effect was dependent upon the nature of the organic peripheral groups [117]. In addition, incorporation of octa-functionalized-epoxy POS into an epoxy network was found to substantially lower the dielectric constant compared to a neat epoxy network [118], and incorporation of POS has also been shown to hamper the physical aging process [111].

4.6.2 Other POS Networks

POS in non-epoxy networks has also been studied. In work by Isayeva and Kennedy, POS served as a cross-linking and reinforcing agent in membranes with potential biomedical applications [119]. Divinylic PDMS and diallylic PEG underwent a co-hydrosilylation reaction with octasilane POS, followed by hydrolysis/condensation

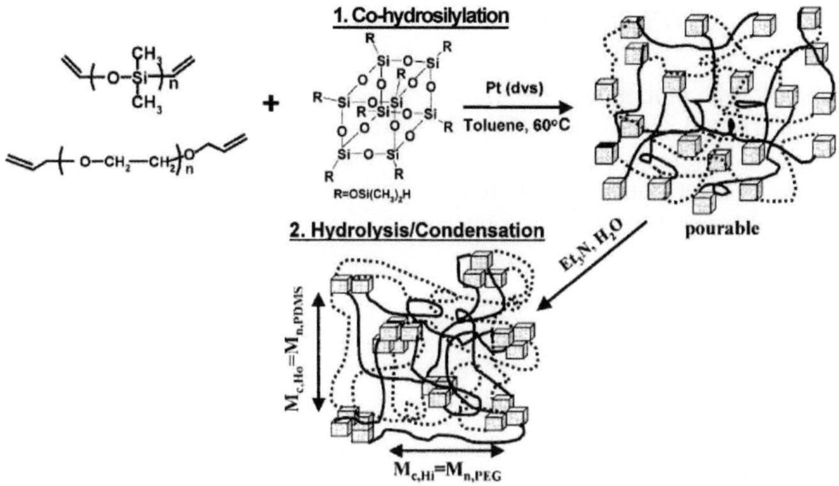

Fig. 4.32 Synthesis of PEG/PDMS/POS membranes [119]. Reproduced from [119] with permission from the American Chemical Society.

reactions to link the POS moieties, as seen in Fig. 4.32.

The authors tested the oxygen permeability of these water swollen membranes, and found that their oxygen permeability was intermediate between that of poly(2-hydroxylethylmethacrylate) (PHEMA) and that of pure PDMS. The thermal stability of the membranes was found to be sufficient to withstand the standard sterilization conditions that hydrogel membranes undergo in autoclaves. Finally, mechanical testing of the membranes showed that the cross-linked POS membranes exhibited significantly higher tensile strength and modulus compared to pure PDMS membranes, and that with increasing POS content in the membranes additional increases of tensile strength and modulus were observed.

Mesoporous POS networks were synthesized by a hydrosilylation reaction between silyl-functionalized POS and vinyl-functionalized POS [120-121]. In subsequent work, titanium-functionalized mesoporous POS networks with potential application as catalytic sites were prepared [122]. Other examples of POS used in catalysis applications can be found in a review article by Hanssen [123] and in Chapter 3.

Mather and colleagues are developing POS-PCL networks for use as biodegradable shape memory polymers [124]. Two PCL chains grown from a single POS moiety were subsequently attached to a tetrafunctional molecule. Analysis by DSC showed two thermal transitions, which corresponded to the melting of POS aggregates and a glass transition of the PCL; dynamic mechanical analysis (DMA) showed two rubbery plateaus. This evidence led the authors to conclude that the material had a "double network structure", processing both physical and chemical cross-links due to the POS crystalline phases and the tetrafunctional molecule respectively. The material underwent testing for one-way and two-way shape memory behavior, and was shown to recover deformation after repeated cycles.

4.6.3 POS Star or Core Structures

POS has been utilized as a multifunctional core [64, 125-131] or as a chain-end group[132-133] to synthesize hybrid molecules with star-shaped architectures, often with beneficial properties (see also Section 7.2.3, Chapter 7).

Star linear and unique quatrefoil star-cyclic structures have been synthesized from POS cores. This molecular architecture is seen in Fig. 4.33.

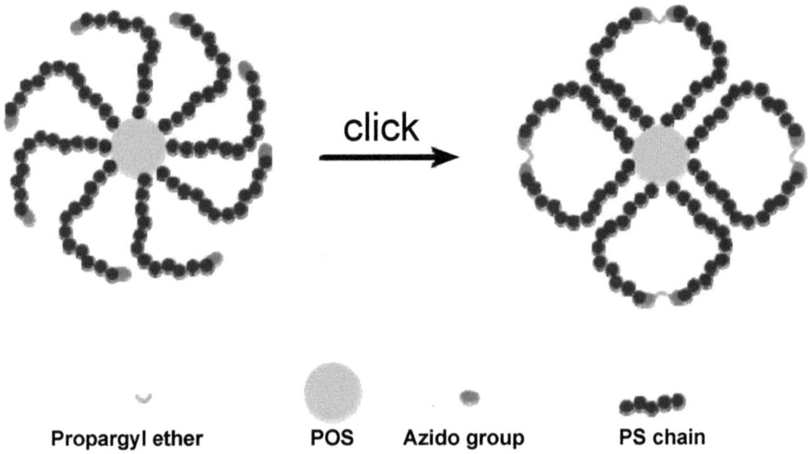

Fig. 4.33 Architecture of star-cyclic POS-PS

Wang et al. began with a POS core functionalized with eight ATRP-initiating groups [128]. The eight arms of polystyrene were synthesized under ATRP conditions. The bromine-terminated PS chains underwent a substitution with sodium azide, being terminated by azide groups, and forming star linear POS-PS as seen in Fig. 4.33 (left). The star cyclic structure is then formed by the intermolecular reaction of an azide-terminated PS chain with another chain in a 'Click' reaction with propargyl ether molecule to form a triazole linkage. The authors reported some inter- and intra-chain coupling, but the structures were relatively monodisperse, having a PDI of 1.14.

POS-core materials have been investigated for use as light-emitting materials [126-127] (see also Section 7.2.3, Chapter 7). In two separate studies, hydrosilylation reactions were used to react a octakis(dimethylsiloxy)-POS and an allyl-functionalized chromophore. The synthesized compounds exhibited good solution processing properties. In work by Chen et al., the synthesized materials were electro-oxidatively stable and possessed good thermal properties [126]. Light emitting diodes constructed from these materials exhibited bright green to blue-green emission. In work by Cho et al., the chromophores chosen were fluorene-based [127]. In a single-layer device constructed using the synthesized material, external quantum efficiencies were greater than the efficiencies exhibited by devices constructed with the more conventional linear poly(dihexylfluorene).

PEO is used as a solid polymer electrolyte in lithium ion batteries. However, because of its semicrystalline nature, conductivity at ambient temperatures is low. In work by Maitra et al., the authors sought to improve ion mobility by grafting oligomeric PEO chains to a POS core to suppress the crystallinity normally associated with PEO [134] (see also Section 7.7, Chapter 7). The resulting crystallization properties of the POS-PEO star were chain length-dependent. For attachment of PEO (n=4) to POS, the crystallization normally present in PEO is completely suppressed. Higher molecular weight PEO chains attached to POS did not suppress crystallization.

Hybrid star architecture also has an effect on glass transition temperature. In molecules with polyoxazoline arms attached to a POS core (seen in Fig. 4.19), the T_g was increased in comparison to polyoxazoline polymers synthesized by a monofunctional initiator, methyltosylate [63]. The increase was attributed to a decrease in the mobility of the polyoxazoline segments because of their attachment to the POS core.

The effects of POS star architecture on crystal structure and morphology have been investigated. In work by Liu et al., star-shaped poly(ε-caprolactone) polymers with POS cores were analyzed by WAXD. The POS cores did not affect the crystalline structure of the PCL; all diffraction peaks were present when compared against linear PCL [129]. Miao and Zhu synthesized a series of liquid crystalline polymers with POS cores linked to eight triphenylene units by alkyl chains of varying lengths [135]. The chain length strongly affected the resultant morphology of the crystalline mesophase, which was analyzed by TEM and X-ray diffraction.

Recently, star polymers with an organic core and POS end-groups were synthesized. In work by Gungor et al., three monoazide-functionalized POS moieties were attached by the 'Click' reaction to a trisalkyne core to form a three-armed POS star [132]. DSC analysis showed two thermal transitions at 201.8 and 217.9 °C. Zhang and Müller synthesized five-arm stars with POS end-groups [133]. A penta-bromofunctionalized core based on glucose was used to initiate ATRP of PS. The penta-brominated PS was reacted with NaN_3 to form azide end-groups, which were subsequently reacted with alkyne-functionalized POS. Some inter-star coupling occurred, and higher weight molecular shoulders were observed by GPC unless conversions remained low. The effect of the star architecture on morphology and properties is under investigation.

4.7 Conclusion

Copolymers containing covalently bonded POS moieties show promise in a wide variety of practical applications. Several of these applications were discussed in this chapter: photoresists for lithography [136] (see also Chapter 7), coatings for spacecraft [52] (see also Chapter 8), light-emitting diodes [126] (see also Chapter 7), membranes [119], and tissue scaffolds [101] (see also Chapter 9). Many other possible applications are currently under investigation, and include dental composites [137] (see Chapter 9) and superoleophobic surfaces [138] and superhydrophobic surfaces (see Chapter 6). However, to fully exploit all the potential benefits of POS copolymers, much greater understanding of structure-property relationships is required.

4.8 References

1. Joshi M, Butola BS (2004) J Macromol Sci Part C Polym Rev 44(4):389-410.

2. Li G, Wang L, Ni H, Pittman CU (2001) J Inorg Organomet Polym 11(3):123-154.

3. Kannan RY, Salacinski HJ, Butler PE, Seifalian AM (2005) Acc Chem Res 38(11):879-884.

4. Phillips S, Haddad TS, Tomczak S (2004) Current Opinion Solid State Mater Sci 8:21-29.

5. Wu J, Mather PT (2009) J Macromol Sci Part C Polym Rev 49:25-63.

6. Cordes DB, Lickiss PD, Rataboul F (2010) Chem Rev 110:2081-2173.

7. Pielichowski K, Njuguna J, Janowski B, Pielichowski J (2006) Adv Polym Sci 201: 225–296.

8. Schwab JJ, Lichtenhan J (1998) Appl Organomet Chem 12:707-713.

9. Baney RH, Cao X (2000) Polysilsesquioxanes. In Jones G (ed) Silicone-Containing Polymers, Kluwer Academic Publishers, pp. 157-184.

10. Loy DA, Rahimian K (2003) Building Hybrid Organic-Inorganic Materials Using Silsesquioxanes. In Nalwa HS (ed) Hybrid Materials, American Scientific Publishers, Vol. 1, pp. 125-163.

11. Haddad TS, Lichtenhan JD (1996) Macromolecules 29(22):7302-7304.

12. Haddad TS, Viers BD, Phillips SH (2002) J Inorg Organomet Polym 11(3):155-164.

13. Wu J, Haddad TS, Kim G, Mather PT (2007) Macromolecules 40:544-554.

14. Xu H, Kuo S, Lee J, Chang F (202) Macromolecules 35(23):8788-8793

15. Zhang W, Fu BX, Seo Y, Schrag E, Hsiao B, Mather PT, Yang N, Xu D, Ade H, Rafailovich M, Sokolov J (2002) Macromolecules 35:8029-8038.

16. Xu H, Yang B, Wang J, Guang S, Li C (2005) Macromolecules 38:10455-10460.

17. Lichtenhan JD, Otonari YA, Carr MJ (1995) Macromolecules 28:8435-8437.

18. Matyjaszewski K, XIa J (2001) Chem Rev 101:2921-2990.

19. Moad G, Rizzardo E, Thang SH (2005) Aust J Chem 58:379-410.

20. Moad G, Rizzardo E, Thang SH (2008) Polymer 49:1079-1131.

21. Hawker CJ, Bosman AW, Harth E (2001) Chem Rev 101(12):3661-3688.

22. Mather PT, Chun SB, Pyun J, Matyjaszewski K, Jeon HG (2000) Polym Prepr 41(1):582.

23. Pyun J, Matyjaszewski K, Wu J, Kim GM, Chun SB, Mather PT (2003) Polymer 44(9):2739-2750.

24. Pyun J, Matyjaszewski K (2000) Macromolecules 33:217-220.

25. Matyjaszewski K, Miller PJ, Pyun J, Kickelbick G, Diamanti S (1999) Macromolecules 32:6526-6535.

26. Chen R, Feng W, Zhu S, Botton G, Ong B, Wu Y (2006) Polymer 47:1119-1123.

27. Zhang W, Zhuang X, Li X, Lin Y, Bai J, Chen Y (2008) React Funct Mater 69:124-129.

28. Zhang W, Liu L, Zhuang X, Li X, Bai J, Chen Y (2008) J Polym Sci Part A Polym Chem 46:7049-7061.

29. Zhang W, Fang B, Walther A, Müller A (2009) Macromolecules

42:2563-2569.

30. Miyamoto K, Hosaka N, Otsuka H, Takahara A (2006) Chem Lett 3 (10):1098-1099.

31. Hirai T, Leolukman M, Jin S, Goseki R, Ishida Y, Kakimoto M, Hayakawa T, Ree M, Gopalan P (2009) Macromolecules 42:8835–8843.

32. Hira T, Leolukman M, Hayakawa T, Kakimoto M, Gopalan P (2008) Macromolecules 41:4558-4560.

33. Piotti ME (1999) Current Opinion Solid State Mater Sci 4:539–547.

34. Zheng L, Hong S, Cardoen G, Burgaz E, Gido S, Coughlin EB (2004) Macromolecules 37:8606-8611.

35. Constable GS, Lesser AJ, Coughlin EB (2004) Macromolecules 37:1276-1282.

36. Mather PT, Jeon HG, Romo-Uribe A (1999) Macromolecules 32:1194-1203.

37. Zheng L, Kasi RM, Farris RJ, Coughlin EB (2002) J Polym Sci Part A Polym Chem 40:885-891.

38. Xu W, Chung C, Kwon Y (2007) Polymer 48(21):6286-6293.

39. Zheng L, Farris RJ, Coughlin EB (2201) J Polym Sci Part A Polym Chem 39(17):2920-2928.

40. Kelsey RD, Handlin DL, Narayana M, Scardinao BM (1997) J Polym Sci Part A Polym Chem 35:3027-3047.

41. Pittman CU, Li GZ, Ni HL (2003) Macromol Symp 196:301-325.

42. Mohring PC, Coville NJ (1994) J Organomet Chem 479(1-2):1-29.

43. Zheng L, Farris RJ, Coughlin EB (2001) Macromolecules 34(23):8034-8039.

44. Escude NC, Chen E (2009) Chem Mater 21:5743-5753.

45. Seurer B, Coughlin EB (2008) Macromol Chem Phys 209:1198-1209.

46. Fu BX, Hsiao BS, Pagola S, Stephens P, White H, Rafailovich M, Sokolov J, Mather PT, Jeon HG, Lichtenhan J, Schwab J (2001) Polymer 42:599-611.

47. Raghunath J, Georgiou G, Armitage D, Nazhat SN, Sales KM, Butler PE, Seifalian AM (2008) J Biomed Mater Res Part A 91A(3):834-844.

48. Wang W, Guo Y, Otaigbe JU (2009) Polymer 50:5749-5757.

49. Lee A, Lichtenhan JD (1998) Macromolecules 31(15):4970-4974.

50. Matejka L, Strachota A, Plestil J, Whelan P, Steinhart M, Slouf M (2004) Macromolecules 37:9449-9456.

51. Strachota A, Kroutilova I, Kovarova J, Matejka L (2004) Macromolecules 37(25):9457-9464.

52. Brunsvold A, Minton T, Gouzman I, Grossman E, Gonzalez R (2004) High Perform Polym 16:303.

53. Leu C, Chang Y, Wei K (2003) Chem Mater 15:3721-3727.

54. Wright ME, Petteys BJ, Guenthner AJ, Fallis S, Yandek GR, Tomczak SJ, Minton TK, Brunsvold A (2006) Macromolecules 39:4710-4718.

55. Wright ME, Schorzman DA, Feher FJ, Jin R (2003) Chem Mater 15:264-268.

56. Cattopadhyay DK, Webster DC (2009) Prog Polym Sci 34:1068-1133.

57. Leu CM, Reddy GM, Wei K-H, Shu C-F (2003) Chem Mater 15(11):2261-2265.

58. Fang Y, Chen L, Chen S (2009) J Polym Sci Part A Polym Chem 47:1136.

59. Hussain H, Tan BH, Mya KY, Liu Y, He CB, Davis TP (2010) J Polym Sci Part A Polym Chem 48:152-163.

60. Kang J-M, Cho H-J, Lee J, Lee J-I, Lee S-K, Cho N-S, Hwang D-H, Shim H-K (2006) Macromolecules 39:4999-5008.

61. Xu Y, Yuan J, Muller AHE (2009) Polymer 50:5933-5939.

62. Sheikh FA, Barakat NAM, Kim B, Aryal S, Khil M-S, Kim H-Y (2009) Mater Sci Eng C 29:869-876.

63. Kim KM, Ouchi Y, Chujo Y (2003) Polym Bull 49:341-348.

64. Drazkowski DB, Lee A, Haddad TS, Cookson DJ (2006) Macromolecules 39:1854-1863.

65. Drazkowski DB, Lee A, Haddad TS (2007) Macromolecules 40(8):2798-2805.

66. Carroll JB, Waddon AJ, Nakade H, Rotello VM (2003) Macromolecules 36:6289-6291.

67. Wu J, Haddad TS, Mather PT (2009) Macromolecules 42(4):1142-1152.

68. Haddad TS, Lichtenhan JD (1996) Macromolecules 29(22):7302-7304.

69. Haddad TS, Viers BD, Phillips SH (2002) J Inorg Organometal Polym 11(3):155-164.

70. Bharadwaj RK, Berry RJ, Farmer BL (1998) Polymer 41:7209-7221.

71. Romo-Uribe A, Mather PT, Haddad TS, Lichtenhan JD (1998) J Polym Sci Part B Polym Phys 36(11):1857-1872.

72. Larsson K (1961) Ark Kemi 16(3):215-219.

73. Larsson K (1961) Ark Kemi 16(3):203-208.

74. Waddon AJ, Coughlin EB (2003) Chem Mater 15:4555-4561.

75. Zheng L, Waddon AJ, Farris RJ, Coughlin EB (2002) Macromolecules 35(6):2375-2379.

76. Zheng L, Hong S, Cardoen G, Burgaz E, Gido S, Coughlin EB (2004) Macromolecules 37:8606-8611.

77. Waddon AJ, Zheng L, Farris RJ, Coughlin EB (2002) Nano Lett 2(10):1149-1155.

78. Matsen MW, Bates FS (1996) Macromolecules 29:7641-7644.

79. Muthukumar M, Ober CK, Thomas EL (1997) Science 277:1225-1232.

80. Whitesides GM, Mathias JP, Seto CT (1991) Science 254:1312-1319.

81. Iacono ST, Budy SM, Mabry JM, Smith DW (2007) Macromolecules 40:9517-9522.

82. Gadodia GA, Synthesis and Study of Hybrid Organic-Inorganic Polyhedral Oligomeric Silsesquioxanes (POSS), University of Amherst

Amherst, MA, 2009.

83. Bates FS, Schulz MF, Khandpur AK, Forster S, Rosedale JH, Almdal K, Martensen K (1994) Faraday Discuss 98:7-18.

84. Milner ST (1994) Macromolecules 27:2333-2335.

85. Fu BX, Lee A, Haddad TS (2004) Macromolecules 37:5211-5218.

86. Knischka R, Dietsche F, Hanselmann R, Frey H, Mulhaupt R (1999) Langmuir 15:4752-4756.

87. Cardoen G, Coughlin EB (2004) Macromolecules 37:5123-5126.

88. Zhang X, Chan E, Glotzer S (2005) J Chem Phys 123:184718.

89. Lee W, Ni S, Deng J, Kim B-S, Satija SK, Mather PT, Esker AR (2007) Macromolecules 40:682-688.

90. Kim B-S, Mather PT (2002) Macromolecules 35:8378-8384.

91. Striolo A, McCabe C, Cummings PT (2006) J Chem Phys 125:104904.

92. Hedrick, JL, Carter KR, Richter R, Miller RD, Russell TP, Flores V, Meccereyes D, Dubois P, Jerome R (1998) Chem Mater 10:39-49.

93. Chung CL, Hsiao S-H (2008) Polymer 49(10):2476-2485

94. Carter KR, DiPietro RA, Sanchez MI, Russell TP, Lakshmanan P, McGrath JE (1997) Chem Mater 9(1):105-118.

95. Su RQ, Muller TE, Prochazka J, Lercher JA (2002) Adv Mater 14:1369-1373

96. Seckin T, Koytepe S, Adiguzel HI (2008) Mater Chem Phys 112:1040-1046.

97. Wright ME, Petteys BJ, Guenthner AJ, Fallis S, Yandek GR, Tomczak SJ, Minton TK, Brunsvold A (2006) Macromolecules 39:4710-4718.

98. Douvas AM, Van Roey F, Goethals M, Papadokostaki KG, Yannakopoulou K, Niakoula D, Gogolides E, Argitis P (2006) Chem Mater 18:4040-4048.

99. Eon D, Raballand V, Cartry G, Cardinaud C, Vourdas N, Argitis P, Gogolides E (2006) J Vac Sci Technol 24(6):2678-2688.

100. Knight PT, Lee KM, Qin H, Mather PT (2008) Biomacromolecules 9:2458–2467.

101. Pan H, Qiu Z (2010) Macromolecules 43:1499-1506.

102. Dieterich D (1981) Prog Org Coat 9:281-340.

103. Madbouly SA, Otaigbe JU (2009) Prog Polym Sci 34:1283-1332.

104. Nanda AK, Wicks DA, Madbouly SA, Otaigbe JU (2006) Macromolecules 39(20):7037-7043.

105. Turri S, Levi M (2005) Macromolecules 38(13):5569-5574.

106. Janowski B, Pielichowski K (2008) Thermochim Acta 478:51-53.

107. Raftopoulos KN, Pandis C, Apekis L, Pissis P, Janowski B, Jaczewska J (2010) Polymer 51:709-718.

108. Mya KY, Wang Y, Shen L, Xu J, Wu Y, Lu X, He C (2009) J Polym Sci Part A Polym Chem 47:4602-4616.

109. Liu H, Zheng S (2005) Macromol Rapid Commun 26(3):196-200.

110. Dewprashad B, Eisenbraun EJ (1994) J Chem Educ 71(4):290-294

111. Lee A, Lichtenhan JD (1998) Macromolecules 31(15):4970-4974.

112. Abad MJ, Barral L, Fasce DP, Williams RJJ (2003) Macromolecules 36:3128-3135.

113. Zhang Z, Liang G, Wang J, Ren P (2007) Polym Composites:175-179.

114. Teo JKH, Teo KC, Pan B, Xiao Y, Lu X (2007) Polymer 48:5671-5680.

115. Strachota A, Kroutilova I, Kovarova J, Matejka L (2004) Macromolecules 37(25):9457-9464.

116. Choi J, Harcup J, Yee AF, Zhu Q, Laine RM (2001) J Am Chem Soc 123:11420-11430.

117. Franchini E, Galy J, Gerard J-F, Tabuani D, Medici A (2009) Polym Degrad Stab 94:1728-1736.

118. Chen W-Y, Wang Y-Z, Kuo S-W, Huang C-F, Tung P-H, Chang F-C (2004) Polymer 45:6897-6908.

119. Isayeva IS, Kennedy JP (2004) J Polym Sci Part A:Polym Chem 42:4337-4352.

120. Harrison PG, Kannengisser R (1996) Chem Commun:415.

121. Zhang C, Babonneau F, Bonhomme C, Laine RM, Soles CL, Hristov HA, Yee AF (1998) J Am Chem Soc 120:8380.

122. Morrison JJ, Love CJ, Manson BW, Shannon IJ, Morris RE (2002) J Mater Chem 12:3209-3212.

123. Hanssen RWJM, van Santen RA, Abbenhuis HCL (2004) Eur J Inorg Chem:675-683.

124. Lee KM, Knight PT, Chung T, Mather PT (2008) Macromolecules 41:4730-4738.

125. Pyun J, Matyjaszewski K (2000) Macromolecules 33:217-220.

126. Chen K-B, Chang Y-P, Yang S-H, Hsu C-S (2006) Thin Sold Films 514:103-109.

127. Cho H-J, Hwang D-H, Lee J-I, Jung Y-K, Park J-H, Lee J, Lee S-K, Shim H-K (2006) Chem Mater 18(16):3780–3787.

128. Ge Z, Wang D, Zhou Y, Liu H, Liu S (2009) Macromolecules 42(8):2903-2910.

129. Liu Y, Yang X, Zhang W, Zheng S (2006) Polymer 47:6814-6825.

130. Maitra P, Wunder SL (2002) Chem Mater 14(11):4494-4497.

131. Miao J, Cui L, Lau HP, Mather PT, Zhu L (2007) Macromolecules 40:5460-5470.

132. Gungor EBC, Durmaz H, Hizal G, Tunca U (2009) J Polym Sci.Part A Polym Chem 47:5947-5953.

133. Zhang W, Muller AHE (2010) Macromolecules 43:3148-3152.

134. Maitra P, Wunder SL (2002) Chem Mater 14(11):4494-4497.

135. Miao J, Zhu L (2010) J Phys Chem B 114:1879-1887.

136. Douvas AM, Van Roey F, Goethals M, Papadokostaki KG, Yannakopoulou K, Niakoula D, Gogolides E, Argitis P (2006) Chem Mater 18:4040-4048.

137. Dodiuk-Kenig H, Maoz Y, Lizenboim K, Eppelbaum I, Zalsman B, Kenig S (2006) J Adhes Sci Technol 20(12):1401–1412.

138. Tuteja A (2007) Science 318:1618.

Chapter 5
Polyhedral Oligomeric Silsesquioxanes in Plastics

Chris DeArmitt

5.1 Introduction

Polyhedral oligomeric silsesquioxanes (POS or POSS®) were first synthesized as long ago as the 1960s. Since that time, global interest has grown considerably as indicated by the many articles and patents in the field [1,2,3,4,5]. The reason being that POS, with a unique three dimensional cage structure composed of silicon and oxygen (Fig. 5.1), is an unusual type of molecule with correspondingly unusual properties. The discovery of the cage form of carbon, called buckminsterfullerene (Fig. 5.1), was awarded the 1996 Nobel Prize for Chemistry, and other cage structure hydrocarbons such as adamantane and other diamondoids have attracted considerable interest. However, time has shown that POS is far more commercially important for several reasons. Firstly, POS is extremely versatile. It is possible to attach any chemistry to the silica-like core. Such customization of POS is simple and can be easily controlled, whereas in the case of C_{60}, there are many equivalent carbons so controlled derivatization is more challenging. Secondly, POS is rather straightforward to make using standard chemical reactors, which facilitates scale-up. Thirdly, because the raw materials are relatively inexpensive and yields can be as high as 90% or more, the price of POS is low enough to place it under consideration for multitudinous application areas from packaging [6] through to sensors [7,8,9] and fuel cell membranes [10,11,12].

Chris DeArmitt
Phantom Plastics, Hattiesburg, MS 39402, USA

C. Hartmann-Thompson (ed.), *Applications of Polyhedral Oligomeric Silsesquioxanes*, Advances in Silicon Science 3, DOI 10.1007/978-90-481-3787-9_5,

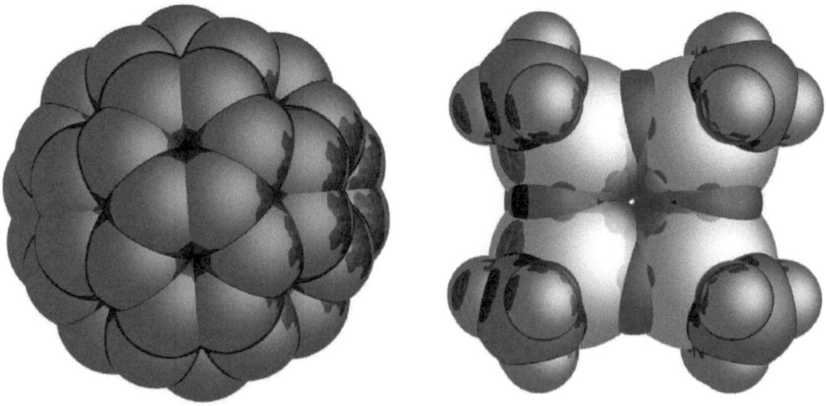

Fig. 5.1 C_{60} (left) and octamethyl POS (right) showing both are ~1 nm in diameter

This chapter describes the varieties of POS and how they can be incorporated into polymers to enhance the polymer properties, along with a description of applications already extant and expected. Hundreds of different POS molecule types have been reported, and naturally each has its own unique set of properties. An overview of the range of typical properties is therefore helpful (Table 5.1).

Table 5.1 Typical properties of POS

Property	Value or typical behavior
Density range	0.9-1.3 gcm^{-3} typical (up to 1.82 $g\ cm^{-3}$)
Refractive index range	1.40-1.65
Molecular size	1-5 nm
Form	Colorless, odorless crystalline solids, some waxes and liquids
Polarity	Very low (fluoroalkyl), low (alkyl), phenyl (medium) to polyionic (high)
Chemical and pH stability	Molecular silicas (closed cage) very stable, trisilanols good stability
Thermal stability	250-350 °C typical (> 400 °C for some types)
Safety	All testing performed has shown POS to be safe
Purity	Standard purity >97% (higher purity and electronics grades are available)

5.2 POS are Molecules

Similarly to norbornane, bornane, adamantine and the fullerenes, POS are also molecules rather than particles. This can be proven categorically by experiment:

- POS have distinct molecular weights
- POS are spontaneously soluble in solvents

- POS can be recrystallized from the melt or from solution
- POS can be characterized by solution NMR (solids have no solution NMR spectrum)
- Soluble POS can be distinguished from POS agglomerated into domains by dielectric spectroscopy [13]
- Soluble POS can be distinguished from POS domains by wide-angle X-ray diffraction (WAXD) [14]
- Some POS types are liquids even in pure, undiluted form (particles are not liquid at 100% concentration)

When POS is added to transparent polymers, it is often observed that transparency is retained up to 5-10 wt % (or vol %) POS because the POS is soluble. Above that threshold however, turbidity sets in because additional POS is no longer soluble, forming domains that scatter light. This makes for an easy way to detect solubility, but caution must be exercised because the test is not infallible. Turbidity will only be detected if the POS domains have a different refractive index to the host polymer, and if the domains are of the right size to scatter visible light.

Although POS are clearly molecules, the POS cage conveys the visual impression that the substance is a particulate, due to the spherical / cubic shape. Indeed, some of the properties are reminiscent of particles and conceptually POS can be considered the smallest possible form of silica with a built in surface treatment, namely the organic moieties surrounding the central cage.

In physics parlance, light is referred to as a wavicle because it displays the properties of both a wave and of a particle. POS can therefore be considered "molicles" since they are molecules that display properties and appearance reminiscent of particles [5].

Often, siliceous fillers are surface treated with organosilanes to enhance their performance in polymers and coatings. Trialkoxysilanes, such as *iso*-octyltrimethoxy silane, react with the surface silanols on silica, whereby the *iso*-octyl groups protrude away from the surface, preventing agglomeration by sterically hindering close approach of adjacent particles. Thus, the non-reactive POS types such as octa-*iso*-butyl POS, octa-*iso*-octyl POS and octaphenyl POS can be considered minute silica particles with a perfectly applied layer of dispersant. Viewed from that perspective, it is no wonder that the POS are so easy to "disperse" in solvents, coating and polymers.

The other type of organosilane surface treatment applied to siliceous fillers such as talc, mica, wollastonite and silica, is the reactive sort commonly referred to as a coupling agent. These also contain a trialkoxy function which can react with the filler surface but they also contain a second functional group which can react chemically with the matrix coating or polymer. Typical functionalities include virtually the complete spectrum of organic moieties from amines, through alcohols, carboxylates, epoxides, urethanes, thiols and so on. Such coupling agents help with filler dispersion, and by increasing filler-matrix adhesion, yield strength in particular is enhanced, along with retention of matrix adhesion after exposure to heat and humidity.

When used in their conventional role as coupling agents, they react with the filler surface, and at the same time they self-condense in a non-ordered manner to produce an amorphous resin. In contrast, under POS synthesis conditions, these same coupling agents are directed to self-condense into ordered cage structures. Rather remarkably, high yields of single cage types are now possible, making purification easier and lowering manufacturing cost. However, early syntheses were fraught with problems including low yields, product mixtures that were hard to purify and multi-month reaction times [15] (see also Frank Feher's Foreword). POS varieties made from these coupling agent-type organosilanes include acrylate, methacrylate, vinyl, epoxy, thiol and amino compounds [6]. It should come as no surprise that these compounds, which are conceptually composed of molecular silica surrounded with a shell of coupling agent, are able to react and bond with polymers and coatings.

The properties of POS such as density, modulus, refractive index and so on fall between the properties of silica and those of organic compounds. Organic polymers such as polyethylene have Young's modulii in the 1-2 GPa range, whereas silica has a value closer to 70 GPa. Octacyclopentyl POS has been calculated to have an intermediate value of 11.8 GPa [16]. POS with shorter alkyl groups are expected to have higher modulii, whereas POS with long organic groups such as octa *iso*-octyl POS and PEG POS are found to be liquids at room temperature, implying a rather low modulus (Table **5.2**).

Table 5.2 POS spans the spectrum from inorganic to organic properties

Material	Weight% cage	Density (g cm^{-3})	Refractive index
Quartz (crystalline silica)	100	2.60	1.54
Octa hydrido POS	98	1.82	1.51
Octa methyl POS	78	1.50	1.51
Octa ethyl POS	64	1.33	1.50
Octa iso-butyl POS	48	1.13	1.48
Octa iso-octyl POS	32	1.01	1.45
Iso-octane	0	0.69	1.39

Interestingly, cubane, a carbon-based cage structured molecule, has the highest density of any hydrocarbon (1.29 g cm^{-3}). It seems then that cage structures such as POS and cubane confer unusually high densities and other related properties.

5.3 POS as Plastics Additives

The success of plastics is due in large part to the use of an array of additives including stabilizers, plasticizers, mold-release agents, antistatics, anti-blocks, flow aids, dispersants and acid scavengers to name just a few. Usually, such additives are low molecular weight compounds which have the advantage that they can easily be solubilized in the polymer. However, being low molecular weight, they also suffer disadvantages including:

1. Volatility – leading to loss through evaporation (the additive may be lost during polymer processing at high temperature), odor and safety concerns
2. Extractability – giving rise to unwanted taste and safety concerns when the additives migrate into food or water supplies
3. The combination of 1 and 2 above leading to loss of the additives, and thus loss of efficacy, causing plastic embrittlement and failure
4. Low molecular weight additives acting as plasticizers with a corresponding and undesirable loss of modulus and yield strength

For example, it has been shown that in hot water pipe applications, over 99% of the antioxidant is lost via evaporation and extraction, and less than 1% is chemically consumed, i.e., doing its job protecting the polymer from degradation [17]. The costly antioxidant is wasted, and the lost additive ends up burdening the environment where it pollutes water and soil. In the case of plasticizers used to keep plastics pliable, loss of additive leads to brittle plastics that break in use. There has been great interest in polymeric additives such as antioxidants [18,19] and plasticizers because their high molecular weights are expected to prevent volatility or migration. However, the stumbling block has been that polymers do not usually mix with one another, so it is difficult to incorporate such additives into the plastics.

POS bridges the molecular weight gap, having sufficient molecular weight to retard evaporation and migration, but having low enough molecular weight to enable it to dissolve molecularly into polymer melts and in the final fabricated part. In order to assure solubility, one must be able to fine tune the polarity of the additive so that it is similar to, and thus soluble in, the plastic or coating. This is where the customizability of POS comes to the fore. Similar to dendritic and hyperbranched structures, which can give higher molecular weight antioxidants with enhanced stabilization [20], POS antioxidants have also been synthesized and found to be effective [21]. POS is less expensive than the dendritic options, and has other advantages such as stability to high temperature processing.

Typical solubility for a POS with a polarity matching that of the host polymer is approximately 5 wt %, which is more than adequate for many polymer additive applications including antioxidants, plasticizers, flow aids, dispersants and lubricants.

The effect of POS upon a polymer depends upon many factors.

1. Addition of up to ~5 wt % of a soluble POS often leads to improved melt flow, but with no loss in modulus or yield strength [22]. Indeed, in some

cases, the POS acts as an antiplasticizer [23,24] filling free volume and raising modulus. Because the POS is molecularly dissolved, such systems retain full transparency.

2. Addition of ≥ 5-10 wt % soluble POS (or less for an insoluble POS) results in the formation of POS domains. This results in a loss in transparency but the domains on the surface give significantly reduced friction [25] and change the contact angle [26].
3. When POS is chemically bound to a polymer, the glass transition temperature is raised, and therefore the heat distortion and Vicat temperatures, because the rigid, bulky POS cages restrict chain mobility [27].

From the preceding discussion, it is clear that solubility and reactivity of POS are key factors. Therefore, we will now consider those aspects in more detail.

5.4 POS Solubility

Solubility follows the rule that "like dissolves like", meaning that the solvent must be similar to the substance being dissolved (solute). Specifically, the polarity of the two substances must be similar. So, how does one quantify polarity? Well, some common indicators are contact angle (although often misleading), surface energy, or testing the solubility of the new substance in solvents of known solubility parameter. For substances that have not yet been made, or have not yet been measured, the Solubility Parameter concept [28] offers a convenient method to calculate the polarity of a substance and its propensity to dissolve in different media. There are established tables which can be used to calculate the polarity, Solubility Parameter and many other attributes with some degree of confidence. The method is based on group contributions whereby each organic group has a fixed tabulated value, and the contribution from each group is assumed to be independent of the other groups in the molecule. The contributions for each group are summed and adjusted for the density of the substance. As the polarities and Solubility Parameters of POS have not yet been measured, calculating Solubility Parameters offers a way to predict solubility, thus saving considerable time and money (when compared to trial and error experimental work such as compounding).

5.5 Effects of POS on Polymer Properties

The precise effect of POS depends on many factors including whether the POS is miscible with the polymer, what type of polymer is being modified, the properties of the POS being added, and whether or not the POS is chemically bonded to the polymer.

5.5.1 *POS Solubilized in the Polymer*

To begin with, let us consider the case of an unreactive POS added to a polymer of similar polarity such that the POS is soluble. In such instances, the solubility of the POS normally has an upper limit at around 5-10 wt % POS, beyond which precipitation occurs and cloudiness is observed. We consider the case where POS is added at 5 wt % so that it remains molecularly dissolved.

Table 5.3 Typical response to different additive types (note that exceptions do exist)

Property	Solvent or plasticizer	POS	Polymer	Spherical filler
Solubility	Yes	Yes	No	No
Young's Modulus	↘	↗	→	↗
Yield Strength	↓	→	→	↘
Heat Distortion Temperature	↓	→	→	↗
Impact resistance	↗	→	↓	↘
Transparency	→	→	↓	↓
Flow / processability	↗	↗	→	↘
Leachable	Yes	No	No	No

Key: ↑ Property increases strongly, ↓ Property decreases strongly, → Property unchanged, ↗ Property increases, ↘ Property decreases

It is interesting to note that at low addition levels, POS can actually increase modulus somewhat, since it acts as an antiplasticizer. The POS cages are molecularly dissolved and fill in the free volume in the polymer, thereby increasing stiffness. Above the solubility limit, the POS forms domains of insoluble material, and since these have low cohesive strength, the POS acts as a plasticizer, thus softening the polymer.

Similarly, the effect of POS on transparency is different above and below the solubility limit. At low concentrations, where the POS is soluble, there is no loss in transparency even in cases where the refractive index of the POS and the polymer differ. This is because the POS molecules are too small to scatter visible light, being only 1-2 nm in diameter (and visible light having wavelengths in the range ~380-750 nm). As the POS concentration is increased above the solubility limit, turbidity sets in and one can often see at a glance that the POS is no longer soluble because of this haziness.

In other cases, POS acts as a plasticizer. It has been shown that POS is just as effective as conventional plasticizers like dioctyl phthalate on a per molecule (molar) basis in PVC [29] and PMMA [14]. Hence POS could find application as a non-volatile plasticizer in specialty applications that can bear the cost of using POS instead of standard plasticizers.

Dissolved POS was reported at a process aid, enhancing the melt flow in poly(phenylene oxide). Normally PPO is blended with polystyrene (PS) in order to improve the melt flow of the former, but an unwanted consequence is that the PS lowers the HDT and Vicat temperatures of the PPO. Any lower molecular weight additive would be expected to reduce the melt viscosity, but normally there would be a reduction in the modulus and yield strength of the polymer, in this case PPO.

However, it was found that POS, while improving the melt flow, did not have any detrimental effect on mechanical properties, and even gave some additional property enhancements. For instance, the color of the PPO was significantly reduced and the flame retardance was enhanced [30]. This initial proof of concept on PPO lead to subsequent investigations whereby it was shown that POS can also improve the melt flow of high performance polymers such as polyetheretherketone (PEEK), polyetherimide (PEI), polyphenylenesulfide (PPS) and cyclic olefin copolymer (COC) with no loss in mechanical properties [6,22] or degradation of molecular weight.

5.5.2 POS Insoluble Present at Concentrations Above the Solubility Limit

It would be natural to think that solubility of the POS is always desirable. However, it is found that some properties are only affected above the solubility limit. For example, addition of POS to PP or nylon 6 has little effect on the coefficient of friction until around 5 wt %, at which point additional POS is no longer soluble and blooms to the surface to give a strong lubricant effect [25,26,31] (see also Section 7.8, Chapter 7). The friction reduction has been reported in the literature with much fanfare, which is rather surprising given that oil filled nylon (for reduced friction) has been available commercially for decades, and at prices far more attractive than one can expect for POS-modified nylon.

Similarly, substantial work has been reported whereby fluorinated POS has been added to polymers, resulting in increased contact angles [32-35] (see also Chapter 6). Addition of POS above the solubility limit gives rise to high contact angles with water, although again, the same would be found for addition of any perfluorocarbon. Observation of a high contact angle should be interpreted with caution; it is assumed that contact angle is directly related to hydrophobicity of the bulk material, but that is only true under a strict set of conditions [36,37].

1. The surface must be completely smooth
2. The surface must be homogeneous with no domains of varying polarity

3. The surface must represent the bulk material (i.e., be chemically similar to the bulk material)
4. The test liquid must not swell or penetrate into the polymer under examination

It has been shown that condition 1 is not met because POS forms crystalline domains on the surface, thereby increasing roughness and resulting in a higher contact angle. This is the same phenomenon as seen in so-called ultrahydrophobic surfaces where the roughness boosts the contact angle to values higher than those arising from the hydrophobicity of the material itself.

In the case of POS in polymers, it is known that condition 2 is not fulfilled because POS above the solubility limit has been shown to form domains at the surface [25] which completely alter the contact angle measurements. Condition 3 is also unlikely to be satisfied, since the insoluble POS is expected to bloom to the surface as the system acts to achieve the lowest overall energy configuration; for example, the fluoro POS, having the lowest surface energy, will accumulate at the polymer surface. Thus it is possible to observe a high contact angle, but still find that the bulk polymer is hydrophilic and adsorbs water.

The interest in ultrahydrophobic and ultraoleophobic surfaces is fueled by the observation that Lotus plants have ultrahydrophobic surfaces, and are self-cleaning. Rainwater runs off them, taking dirt particles with it. There has been much effort expended to reproduce and commercialize this effect in artificial materials, but attempts to use such ultrahydrophobic surfaces in everyday life have failed for fundamental reasons. First, the effect depends upon the fine structure at the surface, and is easily destroyed by contact. Furthermore, the effect only works in the case of pure water without surfactants. So what works well with rainwater will not work well, for example, in a dishwasher.

5.5.3 POS Chemically Attached to the Polymer

Reactive POS types may be either polymerized into the polymer backbone or post-grafted onto a preexisting polymer. For instance, polymerization of a POS monomethacrylate with methyl methacrylate has been reported [38]. As expected, incorporation of such a bulky group into the polymer chain resulted in a significant increase in heat distortion temperature because the bulky groups restrict chain mobility, thus increasing the glass transition temperature (T_g) (see also discussion in Chapter 4). The incorporation of stiff, bulky groups is used commercially to increase HDT/Vicat, for example copolymerization of poly(styrene-*stat*-acrylonitrile) (commonly known as SAN) with a third monomer, either alpha-methyl styrene, or *N*-phenyl maleimide gives a high T_g SAN used to make specialty high heat ABS. Although POS is far larger and stiffer than those monomers, it is not cost-competitive with them. Use of POS to increase T_g would only be worth considering in the case of ultra-high temperature polymers such as PEEK, PAI (polyamide imide, e.g., Torlon®) and the like where the cost of POS might be tolerated.

5.5.4 *POS Network Thermosets*

Having looked at unbound POS and then POS bound chemically to the polymer backbone or pendant to it, we now consider the effect of making a network from POS. In this case, a highly functional POS is cross-linked to form a three dimensional structure. A conventional thermoset, for example an epoxy, will typically consist of monomers that are tetra-functional. POS epoxies, either the glycidyl type or the cycloaliphatic kind (Fig. 5.2), consist of POS cage mixtures whereby T8, T10 and T12 cages are intermingled, typically in the ratio 15:65:30, arising from the POS synthesis conditions (see also Section 4.6.1, Chapter 4).

Thus, the average number of epoxy groups is near to 10 and one can form an extraordinarily strong network. There are several advantages to POS epoxy systems.

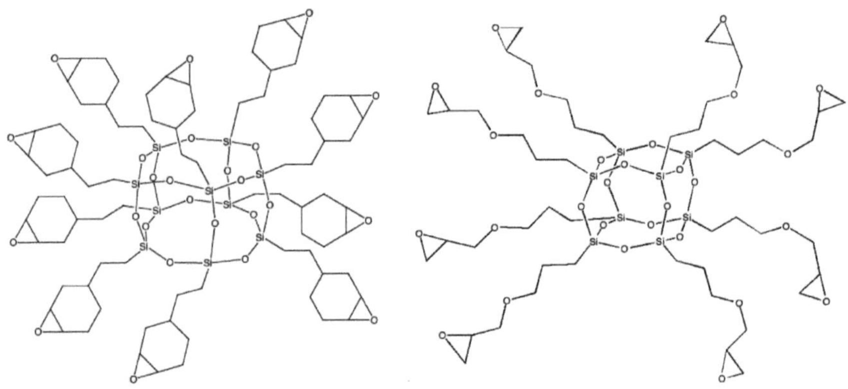

Fig. 5.2 Epoxycyclohexyl POS (left) and glycidyl POS (right)

1. They display low cure shrinkage, as low as 1% for the cycloaliphatic POS epoxy
2. Reduction in coefficient of thermal expansion (CTE) owing to the immutable structure of the POS cage
3. Retention of high modulus up to 300°C (cycloaliphatic type)
4. Resistant to solvent swelling due to high cross-link density and also because the POS cage acts like an unswellable molecular filler

As they share polarity similar to conventional epoxy monomers, the POS epoxies are miscible with them. Thus formulation is rather easy, as mixing is simple. Because the POS epoxies are viscous liquids, they can be blended with other epoxies at room temperature.

Curing of POS epoxies can be performed using the normal routes such as amine cure, anhydride cure or the use of imidazoles as appropriate for the epoxy chemistry, either glycidyl (aliphatic and aromatic amines) or cycloaliphatic (anhydrides and amines).

Gel time is the time at which an essentially infinite network is formed and that is critically dependent upon the average functionality of the system [39]. Adding just a small amount of highly functional monomer significantly reduces gelation time, thus giving a faster cure and increased productivity. As mentioned, POS epoxies have extremely high functionality and are therefore ideal for reducing cure time and thereby improving productivity.

The incorporation of POS epoxies often lowers the viscosity of the formulation, which is clearly advantageous in many instances, especially when there is a need to wet out a fiber tow or fabric. Another benefit is that the POS increases the high temperature modulus. The room temperature modulus remains unaltered, but at elevated temperatures the POS cage is so effective at restricting motion that there is often a complete absence of the glass transition temperature as measured by DMA [40].

5.6 POS Dispersants

Dispersants are an important class of surfactant because achieving good dispersion is often critical in commercial materials. Pigments when fully dispersed give higher tinting strength. Thus, for a given pigment concentration, more intense colors can be achieved. Alternatively for a given desired color, less pigment is required, thereby saving money on pigment, which tends to be expensive. Thus both plastic and coating formulations often contain some type of dispersant.

Agglomeration of pigments and fillers leads to problems. Specifically, the impact resistance of thermoplastics and thermoset materials is dramatically lowered by the presence of agglomerated particles. The other mechanical property most affected by agglomerates is elongation to break. Modulus and yield strength are not significantly altered however.

It is not just the mechanical properties that are affected. Electrical and thermal conductivity are both very much dependent upon the degree of dispersion. Agglomerated conductive particles can form a continuous pathway through the material, at which point (the so-called percolation threshold), the conductivity increases by several orders of magnitude.

The processability of materials can be as important as their final properties. Increasing the particulate concentration, be that fillers, pigments or other additives, increases the viscosity of the system. Progressively smaller particles have a stronger and stronger tendency to agglomerate and thereby raise viscosity. Thus, with the recent attention on nanoparticles, there is more need than ever for effective dispersants.

At low particle concentrations, the viscosity increase may be negligible, but some materials contain very high filler loadings. For instance, flame retardant fillers such as aluminum hydroxide and magnesium hydroxide are often used at 60 wt % or more. PVC tile contains 80 wt % of calcium carbonate, and polymer magnets can contain over 85 wt % filler. At such high loadings the viscosity of the material becomes an issue. Extruder throughput and mold filling suffer appreciably. High gloss is difficult or impossible to attain. Sometimes one desires to add more filler, but the viscosity limits the filler amount that can practicably be used. All of these issues can be mitigated, or avoided outright, by the use of an optimal dispersant. So, how do dispersants work and what constitutes an ideal dispersant for a given material? Dispersants coat particles and prevent them from adhering to one another.

The ubiquitous van der Waals forces attract particles together. Once they approach closely, those forces become very strong and the particles come into contact. Once agglomerated, they are hard to separate. By coating the particles, a dispersant prevents close approach and lessens the attractive forces. In non-aqueous systems, the primary mechanism for dispersant action is steric stabilization. The criteria [41] for effective steric stabilization are:

1. The surfactant must adhere well to the particle surface so that it is not displaced when two particles approach each other
2. The portion of the surfactant not adsorbed must be soluble in the surrounding medium, i.e., the solvent, coating or polymer as the case may be
3. The surfactant "tail" must be long enough to prevent the particles getting too close where van der Waals attractive forces dominate and cause strong agglomeration

AB and ABA type block copolymers make excellent dispersants whereby an insoluble block (B) adsorbs onto the particle and the other, soluble block (A), provides the particle separation. However, polymeric block copolymer stabilizers have disadvantages such as being relatively expensive. Furthermore, they are not ideal for nanoparticle dispersion because the size of the polymeric stabilizer is significant compared to the particles; hence one needs a lot of stabilizer and the effective particle size (the particle plus the layer of adsorbed polymer) is large, resulting in an increase in the effective filler loading and thus a concomitant increase in viscosity.

Small molecule surfactants have also been used as dispersants to good effect. Stearic acid is the standard dispersant for calcium carbonate because the carboxylic acid group adsorbs strongly to calcium carbonate and the aliphatic tail is compatible with low polarity polymers such as polyethylene and polypropylene. While stearic acid is highly efficacious on calcium carbonate, it is known to be ineffective on silica and other siliceous fillers (Fig. 5.2) because the acid group does not adsorb onto the surface [42].

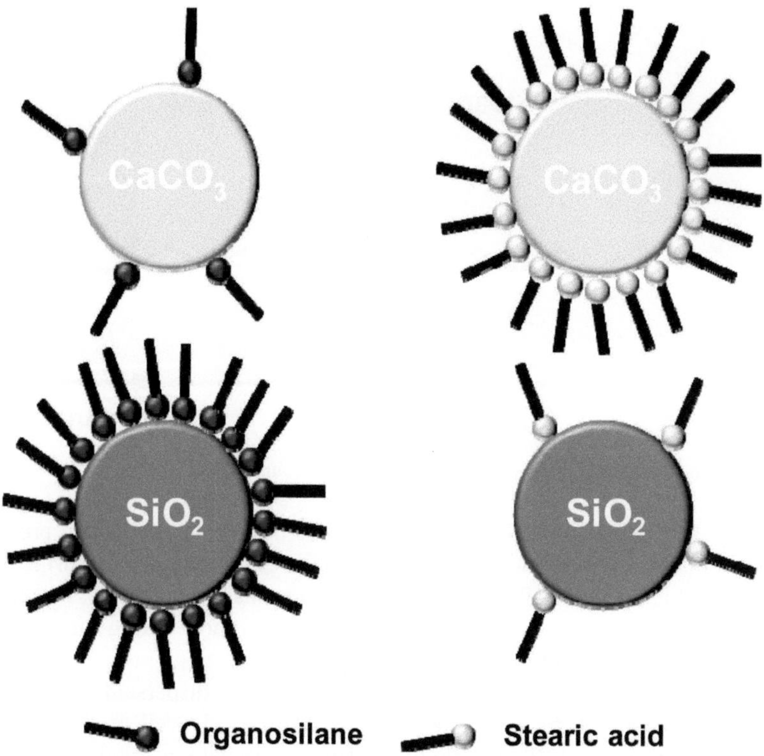

Fig. 5.3 Dispersants show specificity when adsorbing onto mineral surfaces

In some cases, there are no places for the dispersant head to "dock". For fillers such as silica and related fillers, organosilanes (Fig. 5.3) are a good choice because the head group can bond well to the filler. Thus, one must pick the appropriate dispersant in each case where the head group bonds well and the tail group is soluble in the medium. One large study investigated nine different mineral fillers and pigments to determine the optimal dispersant for each [43]. It was found that the surface chemistry, that is, the affinity of the surface for dispersants, was unique for every type of particle tested. Contrary to popular perception, it was shown that minerals do not behave as either acids (e.g., silica) or bases (e.g., calcium carbonate) but each were amphoteric, adsorbing both acid and basic dispersants. The unique surface chemistry of the minerals means that the quest for a universal dispersant is challenging.

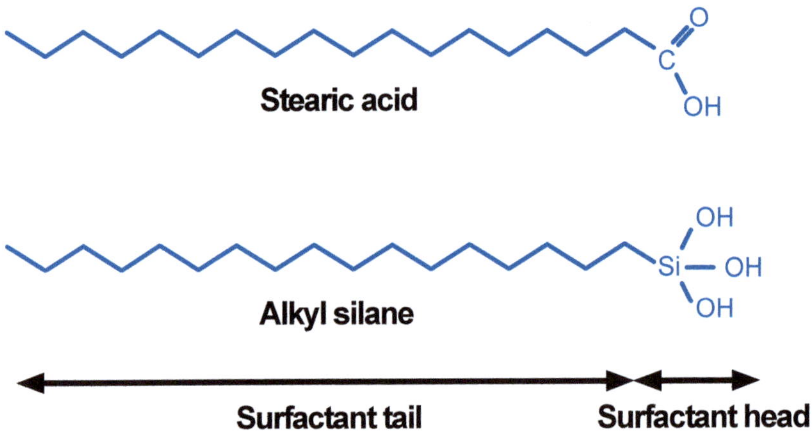

Fig. 5.4 Dispersant structure showing the "head" and "tail" components

Most POS are symmetrical and therefore do not display surfactant properties. However, there are exceptions. In particular, the POS trisilanols are surface active due to the highly polar silanol groups on the one hand and the less polar alkyl groups on the other (Fig. 5.4). Other POS expected to show strong surface activity are the new POS trisulfonic acids.

Indeed, POS trisilanols have proven to be effective dispersants for fillers and pigments [6,44,45]. Unusually, the POS trisilanols work well as dispersants for many different types of fillers including calcium carbonate, bismuth subcarbonate, yttrium oxide, titanium dioxide, hexagonal boron nitride, titanium carbide, magnetite and silica. As mentioned previously, a universal dispersant that functions on fillers as different as calcium carbonate and silica has hereto proven elusive. The broad applicability of POS trisilanols is probably due to their unique nature. They are highly acidic and therefore effective on basic fillers including carbonates, but also contain silanol functional groups that can bond to free silanol groups on the surface of silica, other siliceous fillers and metal oxides. Furthermore, because there are three reactive groups per molecule (Fig. 5.5), the POS trisilanols can chelate to surfaces forming bonds that survive even under extreme temperatures. Thus, particles can be dispersed using POS trisilanols in an extruder even in the case of PEEK which is extruded at rather high temperatures.

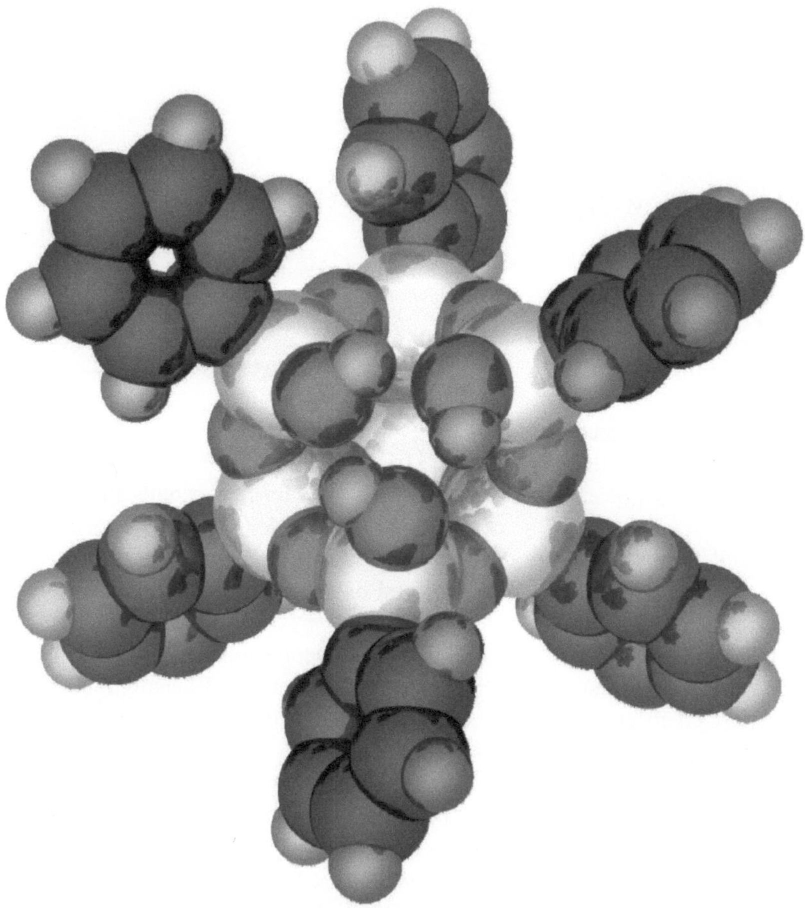

Fig. 5.5 POS Phenyl trisilanol showing the triad of central reactive Si-OH groups

5.7 POS Metal Deactivators

A related application, relying upon the ability of POS to adsorb strongly to surfaces, is the use of POS trisilanols as metal deactivators [6]. Transition metals attack organic substances including coatings and polymers so for example, the copper in a copper cable tends to attack the polyethylene insulation causing oxidation and premature failure. Metal deactivators adsorb strongly to the metal, thus preventing it from coming into contact with and degrading the surrounding material. POS trisilanols were tested as metal deactivators for copper and it was found that they were able to reduce the detrimental effect of the copper even at high temperature approaching 200°C under pure oxygen. Metal deactivators are rather expensive and are only needed in monolayer amounts so using POS may be commercially viable from a cost perspective.

5.8 New Applications and the Future

The utility of POS relies upon its special properties, namely:

- A rigid, bulky cage
- High degree of functionality
- Defined and intermediate molecular weight (between solvents and polymers)
- Defined molecular size
- Unique chemistry, i.e. the POS trisilanols
- Their unique chemistry offers possibilities for new patents and marketing

As well as those applications of POS already alluded to, there are other assorted applications either already commercialized or being explored. Some of those are mentioned here.

POS has been suggested for novel cosmetic formulations [46,47] where it could serve as a dispersant, emollient or viscosity modifier. Recently POS has been copolymerized to create a new, colorless and transparent polyimide [48] which can be solution cast or applied as a spray. The POS serves two vital roles in the polyimide which was developed to be a solar cell cover for space applications. Firstly, normal polyimide is colored, blocking some portion of the light and thereby reducing the efficiency of the solar panels. The copolymerization with POS results in a colorless film with improved solar cell efficiency. Presumably, the POS changes the intermolecular spacing, thereby disrupting the electronic interactions that normally result in color. Secondly, polyimides are attacked by atomic oxygen in Low Earth Orbit. Testing has shown that where a standard polyimide is first etched, and then destroyed, a POS enhanced polyimide is much more robust. Attack by atomic oxygen vitrifies the POS to form a protective glassy film, dramatically improving survivability of the material in space [48] (see Also Chapter 8).

POS molecules reside in an intermediate molecular weight range between solvents and plasticizers on the one hand, and polymers on the other. This has led to interest in using POS as calibration standards for mass spectroscopy. POS are also intermediate in size and are ideally suited to use as absolute calibration standards for gel permeation chromatography (GPC or SEC). GPC calibrations are normally performed using polystyrene standards but that method has serious drawbacks. Firstly, such calibration standards and expensive and are not truly monodisperse whereas POS is truly monomodal in size and also cheaper than polystyrene standards. Furthermore, solvated polystyrene coils change dimensions with temperature and with solvent type. POS on the other hand is rigid and does not change size with temperature or solvent polarity. This gives a tremendous advantage compared to traditional standards.

5.9 Conclusions

The POS family of molecules extends into the hundreds, but they are all based upon a rigid, inorganic, central core composed of silicon and oxygen. This characteristic structure makes them visually intriguing, but more importantly, it imparts a unique property set distinct from that of small molecules, polymers or particles. POS are immensely tunable, from non-reactive types to highly functionalized versions, and from hydrophobic to highly hydrophilic and water-soluble examples.

The scale-up of POS manufacture has brought prices down dramatically, thus opening up the possibility of using POS in new applications. However, POS is still relatively expensive compared to other chemicals and additives, so commercialization will remain limited to niche applications where POS can deliver what no other additive is able to.

5.10 References

1. Pielichowski K, Njuguna J, Janowski B, Pielichowski J (2006) Adv Polym Sci 201:225-296.

2. Joshi M, Butola BS (2004) J Macromol Sci Part C Polym Rev 44(4):389-410.

3. Pan G (2007) In Mark JE (ed) Physical Properties of Polymers Handbook Part VI, Springer, New York, USA.

4. Phillips SH, Haddad TS, Tomczak SJ (2004) Curr Opin Solid StateMater Sci 8:21.

5. DeArmitt C (2010) Polyhedral Oligomeric Silsequioxane (POSS®) Enhanced Plastics. In Xanthos M (ed), Functional Fillers for Plastics, 2nd Ed, Wiley-VCH, Ch 23.

6. DeArmitt C, POSS® User's Guide, Hybrid Plastics, USA.

7. Hartmann-Thompson C, Keeley D, Dvornic PR, Keinath SE, McCrea K. (2007) J Appl Polym Sci 104(5):3171-3182.

8. Hartmann-Thompson C, Keeley D, Pollock K, Keinath SE, Dvornic PR, Dantus M, Gunaratne T, Lecaptain D (2008) Chem Mater 20(8):2829-2838.

9. Hartmann-Thompson C (2009) US Pat. 0263287 A1.

10. Nowak R, Hartmann-Thompson C, Bruza K, Thomas L, Meier D (2008) World Pat. WO 2008127645 A1.

11. Hartmann-Thompson C, Merrington A, Carver PI, Keeley DL, Rousseau JL, Hucul D, Bruza KJ, Thomas LS, Keinath SE, Nowak RM, Katona DM, Santurri PR (2008) J Appl Polym Sci 110(2):958-974.

12. Decker B, Hartmann-Thompson C, Carver PI, Keinath SE, Santurri PR (2010) Chem Mater 22(3):942-948.

13. Hao N, Böhning M, Wohlrab S, Schönhals A (2008) Macromol Symp 267:69–73.

14. Kopesky ET, Haddad TS, McKinley GH, Cohere RE (2005) Polymer 46(13):4743-4752.

15. Feher FJ, Newman DA, Walzer JF (1989) J Am Chem Soc 111:1741-1748.

16. Capaldi FM, Boyce MC, Rutledge GC (2006) J Chem Phys 124:214709.

17. Viebke J (1996) PhD Thesis, Theoretical Aspects and Experimental Data on the Deterioration of Polyolefin Hot-Water Pipes, KTH, Stockholm, Sweden.

18. Al-Malaika S, Golovoy A, Wilkie CA, Chemistry and Technology of Polymer Additives, Blackwell Science, Oxford, UK, 1999.

19. Al-Malaika S, Reactive Modifiers for Polymers, Chapman and Hall, London, UK, 1997.

20. Bergenudd H, Eriksson P, DeArmitt C, Stenberg B, Malmström Jonsson E (2002) Polym Degrad Stability 76:503-509.

21. DeArmitt C, Wheeler P, Hait S (2009) unpublished work, Hybrid Plastics, Hattiesburg, MS, USA.

22. DeArmitt C, Wheeler P (2008) Plastics Addit Compound 10(4):36-39.

23. Iyer S, Schiraldi DA (2007) Macromolecules, 40(14):4942–4952.

24. Soong SY, Cohen RE, Boyce MC, Chen W (2008) Polymer 49(6):1440-1443.

25. Misra R, Fu BX, Morgan SE (2007) 65th Annual Technical Conference Society of Plastics Engineers, 62-66.

26. Misra R, Fu BX, Morgan SE (2007) J Polym Sci Part B Polym Phys 45:2441-2455.

27. Haddad TS, Lichtenhan JD (1996) Macromolecules 29(22):7302-7304.

28. Van Krevelen DW, te Nijenhuis K, Properties of Polymers, 4th Ed, Elsevier, Oxford, UK 2009.

29. Soong SY, Cohen RE, Boyce MC (2007) Polymer 48:1410-1418.

30. Kamo H, Kuga S, Ono T, Ikeda M (2002) US Pat. 193533 A1.

31. Misra R, Fu BX, Plagge A, Morgan SE (2009) J Polym Sci Part B Polym Phys 47(11):1088-1102.

32. Rios PF, Dodiuk H, Kenig S, McCarthy S, Dotan J (2007) J Adhesion Sci Technol 21(5-6):399-408.

33. Mabry JM, Vij A, Iacono ST (2007) Polym Prepr 48(2):970.

34. Xu J, Li X, Cho CM, Toh CL, Shen L, Mya KY, Lu X, He C (2009) J Mater Chem 19(27):4740-4745.

35. Iacono ST, Vij A, Grabow W, Smith DW, Mabry JM (2007) Chem Commun:4992–4994.

36. Adamson AA, Physical Chemistry of Surfaces, 5th Ed, John Wiley & Sons Inc., New York, USA, 1990.

37. Chan CM, Polymer Surface Modification and Characterization, Carl Hanser Verlag, Munich, Germany 1993.

38. Escudé NC, Chen EYX (2009) Chem Mater 21(24):5743-5753.

39. Flory PJ (1953) Principles of Polymer Chemistry, Cornell University Press, Ithaca, NY, USA.

40. Mya KY, He C, Huang J, Xiao Y, Dai J, Siow Y-P (2004) J Polym Sci Part A Polym Chem 43(14):3490-3503.

41. Shaw D (1992) Introduction to Colloid and Surface Chemistry 4th Edition, Butterworth-Heinemann, Oxford, UK.

42. Karger-Kocsis J (1995) Polypropylene, Structure, blends and composites, Volume 3, Composites, Chapman and Hall, London, UK.

43. Improved Thermoplastic Composites by Optimized Surface Treatment of the Mineral Fillers, http://www.phantomplastics.com/attachments/File/Optimized_Dispersants_for_Fillers_and_Pigments_Report.pdf

44. Wheeler PA, Misra R, Cook RD, Morgan SE (2008) J Appl Polym Sci 108:2503-2508.

45. Lichtenhan JD, Fu X, Blue M, Wheeler P, Misra R, Morgan S (2008) World Pat. 054418 A2.

46. DeArmitt C (2008) Cosmetics & Toiletries (Allured Publishing) 123(8):51-56.

47. Quadir M (2004) World Pat. 082611.

48. Poe GD, Farmer BS (2009) US Pat. 0069508 A1.

Chapter 6
Fluorinated Polyhedral Oligosilsesquioxane Surfaces and Superhydrophobicity

Scott T. Iacono, Andrew J. Peloquin, Dennis W. Smith, Jr. and Joseph M. Mabry

6.1 Introduction

Fluorinated compounds are a logical choice for hydrophobic applications owing to their generally low surface energy. Polyhedral molecules may also improve hydrophobicity by increasing material surface roughness. There have been many recent attempts to synthesize and characterize various types of fluorinated polyhedra. These reports include the fluorination or fluoroalkylation of C_{60} [1,2]. Unfortunately, $C_{60}F_{48}$ (fluorinated buckminsterfullerene) cannot be used as a hydrophobic material, since it is metastable and is hydrolyzed by water [3]. However, the perfluorocarborane species, perfluoro-deca-β-methyl-*para*-carborane, shows remarkable hydrolytic and oxidative stability [4]. Fluorinated carbon nanotubes and nanofibers have also been produced [5]. Many of these fluorinated polyhedral compounds may be useful in hydrophobic applications, but they are generally hazardous to prepare, require air and moisture sensitive manipulations, and have limited economies of scale. For these reasons, alternative fluorinated polyhedra, such as Polyhedral Oligomeric SilSesquioxanes (POS) are highly desired (Figure 6.1).

Scott T. Iacono
U.S. Air Force Academy, Department of Chemistry, USAF Academy, Colorado Springs, CO 80840, USA

Andrew J. Peloquin
U.S. Air Force, Patrick AFB, FL 32925, USA

Dennis W. Smith, Jr.
Department of Chemistry & Center for Optical Materials Science & Engineering Technologies, Clemson University, Clemson, SC 29634, USA

Joseph M. Mabry
Air Force Research Laboratory, Space & Missile Propulsion Division, Edwards AFB, CA 93524, USA
E-Mail: joseph.mabry@edwards.af.mil

C. Hartmann-Thompson (ed.), *Applications of Polyhedral Oligomeric Silsesquioxanes*, Advances in Silicon Science 3, DOI 10.1007/978-90-481-3787-9_6,

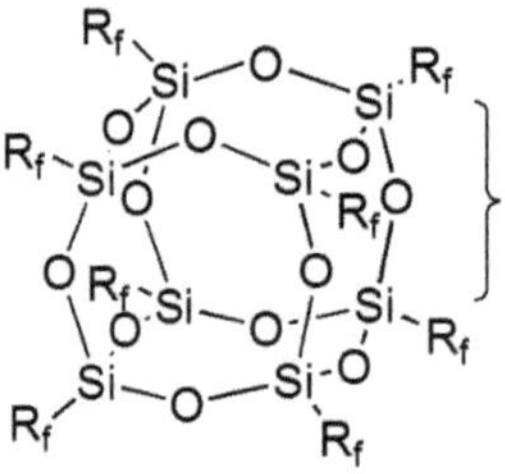

Fig. 6.1 General structure of fluorinated polyhedral oligomeric silsesquioxanes

The addition of fillers to polymeric matrices is of extreme technological importance (see also Chapter 5). Filler addition can impart enhanced scratch resistance, increase thermal or mechanical properties, and improve processing parameters. Silicate and carbon black based fillers are quite common. They are often inexpensive and their incorporation into many polymer systems is fairly straightforward. When miscibility is a problem, surface modification of the fillers to further enhance their compatibility is widespread. The silylation of surface silanol groups on silica fillers is a good example. Processing is another factor that has been optimized. The use of high shear to break up large agglomerates or aggregates of nanoscopic particles is common. These approaches yield nanoscopic species with large surface areas, which should favor physisorption and/or chemisorption between the polymer chain and the filler.

Polyhedral Oligomeric SilSesquioxanes (POS) are thermally robust cages consisting of a silicon-oxygen framework, most commonly possessing alkyl functionality on the periphery. POS have been used in the development of high performance materials for commercial aerospace and medical applications [6,7] (see also Chapters 8 and 9 respectively). POS molecules can be functionally tuned, are easily synthesized with inherent functionality, and are often commercially available [8]. POS compounds may possess a high degree of compatibility in blended polymers and can easily be covalently linked into a polymer backbone [9]. The incorporation of POS into polymers often produces nanocomposites with more desirable properties, such as higher glass transition temperature, mechanical strength, thermal and chemical resistance, and ease of processing. Applications include space-survivable coatings [10,11] and ablative and fire-resistant materials [12–14]. POS can be produced as either completely condensed cages or incompletely condensed cages with silanol groups that allow further modification. Owing to its physical size, the incorporation of POS into polymers generally serves to reduce polymer chain mobility, often improving both thermal and mechanical properties.

A number of recent reports have detailed POS materials as reinforcing fillers (or reinforcing co-monomers) in a number of composite systems with nanometer-sized domains [15–17]. The results reported herein are somewhat different, in that the monodisperse POS building blocks seem to be rather non-interacting. Specifically, the organic functionality surrounding the silsesquioxane core is composed of fluorinated alkyl groups. Fluoroalkyl compounds are known to be basically inert. This is largely because they are non-polarizable and have low surface free energies.

Fluoroalkyl chains are often rigid, due to steric and electronic repulsion. This large corona of the fluorinated POS compounds should help retard the van der Waals attraction between POS cores. These POS materials are monodisperse and crystalline. The melting point of the POS is lower than the processing conditions of various fluoropolymers, so one can safely assume that hard filler effects should not be an issue during melt blending. In this regard, one may expect that these materials could exhibit small molecule, solvent like, characteristics. This paper discusses many of the parameters and physical properties of simple blends of fluorinated POS materials in fluoropolymer matrices. It is demonstrated that these simple blends do not rigorously correspond to idealized filled polymer models, or to solvent swollen systems. Herein, we describe the properties resulting from the blending of these fluorinated POS compounds into various fluoropolymers.

6.2 Experimental

6.2.1 Materials

Fluorinated POS compounds **FH**, **FO**, **FD** (Fig. 6.2) and the corner-capped series **2–9** (Fig. 6.3) were prepared and fully characterized from previously published procedures [18,19]. **6F-BP** PFCB aryl ether polymer (Fig. 6.7, M_n = 22,000–25,000) was donated and is also commercially available from Tetramer Technologies, L.L.C. and distributed through Oakwood Chemicals, Inc. Polychlorotrifluoroethylene (PCTFE or Neoflon M 400-H) was obtained from Daikin. Hexadecane, and hexafluorobenzene were purchased through Aldrich and used without further purification.

6.2.2 Single Crystal X-Ray Structural Characterization

Crystal data for **FH** and **FD** POS were collected at T = 103.0(2) K using Bruker 3-circle, SMART APEX CCD with χ-axis fixed at 54.74°, running on SMART V 5.625 program (Bruker AXS: Madison, WI, 2001). Crystallographic data for all structures have been deposited in the Cambridge Cyrstallographic Data Center (CCDC). Assigned CCDC numbers for **FH** and **FD** are 608207 and 608209, respectively. Copies of the crystallography data can be obtained, free of charge, from CCDC, 12 Union Road, Cambridge CB2 1EZ, UK (e-mail: deposit@ccdc.cam.ac.uk).

6.2.3 *Fluorinated POS Coating and Composite Preparation*

6.2.3.1 Spin Cast Fluorinated POS Coating

POS powder surfaces were prepared by dissolving the fluoroalkyl POS in a minimal amount of hexafluorobenzene followed by mechanical agitation. The surfaces were spin cast at 2500 RPM onto borosilicate glass plates producing a well-adhered coating.

6.2.3.2 Fluorinated POS Solvent Blended Composites with 6F-BP PFCB Aryl Ether Polymer

Two methods were used to prepare polymer (or blend) films. For spin cast films, the dried polymer is initially applied in a minimal amount of THF onto a glass substrate, and then spin coated at 2500–3000 RPM using a Chemat KW-4A spin coater. The polymer-coated substrate is dried in a vacuum oven at 60 °C for 24 h. For drop-cast films, the polymer dissolved in a minimal amount of THF was dispensed onto a glass plate to uniformly coat the surface. The polymer solution was allowed to evaporate in a glass enclosure for 48 h and then finally dried in an oven at 60 °C for an additional 24 h. Spin and drop-cast films were approximately 1–2 µm thick and were measured by the Zygo NewView 6300 3D white light optical profiling system.

6.2.3.3 Fluorinated POS Melt Blended PCTFE

Fluoropolymer composites are prepared by adding the designated wt % of **FD** and **FO** POS melt blended into PCTFE. Melt processing and blending of POS/fluoropolymers was conducted on a DSM Micro 5 Compounder having a chamber free volume of 5 cm^3. Powders were pre-mixed in their appropriate ratios and charged to the mixer, imposing a residence time of 5 min under an inert nitrogen atmosphere with the modular heating profile set at a flat 280 °C. Blend extrudates were transferred to a DSM micro-injection molding machine for the fabrication of disks. Samples for dynamic mechanical thermal analysis (DMTA), contact angle measurements, and atomic force microscopy (AFM) were thin films. The films were made by compression molding two grams of the polymer-blend extrudate utilizing a Tetrahedron compression molder. The bars for SEM were also compression molded with the same molding times and temperatures, but utilizing only ½ ton of force.

6.2.4 *Thermo-Mechanical Analysis*

The compression molded films were cut into 3 × 20 mm rectangular samples for dynamic mechanical thermal analysis utilizing a DMTA V from TA Instruments. All fluoropolymer samples were analyzed using a 5 °C temperature ramp and a tensile geometry. The PCTFE samples were each tested from 30 °C to 150 °C. Stress/strain tests were performed on all samples to identify the largest force at which the material exhibited an elastic deformation, thus limiting the pre-tension force used to test the samples. Strain sweeps were also performed to ensure that the testing strain was within the linear viscoelastic region. The Fluorinated POS effect on the thermal stability of the fluoropolymers was examined using a 2050 TGA from Rheometric Scientific in the presence of nitrogen.

6.2.5 *Microscopy*

The dispersion of the fluorinated POS was investigated using atomic force microscopy (AFM) and scanning electron microscopy (SEM).

6.2.5.1 Atomic Force Microscopy (AFM)

The AFM measurements were performed in tapping mode using a Digital Instruments Dimension 3100 Scanning Probe Microscope (SPM), which utilizes automated atomic force microscopy (AFM) and scanning tunneling microscopy (STM) techniques. Atomic Force Microscopy was conducted on a Nanoscope IV controller (3100 SPM Head) in tapping mode. Etched Silicon probes of nominal spring resonance 300 kHz (spring constant approx. 0.3 mN m^{-1}) were used for light tapping (driving amplitude ca 1.1 V) of varying section size at 1-2 Hz collection times (512 points/line). For melt-processed samples, the AFM samples were cut from the compression molded films discussed in the results and discussion section. It was noted that most of the surfaces did not have a significant phase image, which was interpreted as being indicative of relatively uniform surfaces that differed only in surface topography. The surface roughness was dominated by the processing effects, e.g., gross striations indicative of film buckling upon removal. However, fine scale features were resolvable and will be discussed. In the AFM micrograph images, height is shown on the left side of the figure and a simultaneously collected phase image is shown on the right.

6.2.5.2 Scanning Electron Microscopy (SEM)

Scanning electron microscopy was performed using an ISI CL6 operating at 15 keV equipped with a Kevex X-ray detector. Elemental mapping was performed using energy dissipation X-ray analysis with IXRF Systems analysis software. Samples were cut from compression molded bars and the cross-section was imaged. Furthermore, an elemental mapping of the surface is discussed.

6.2.6 *Static and Dynamic Contact Angle*

Contact angle analyses were performed on a FDS Dataphysics Contact Analyzer System using a syringe metering pump. Deionized water (18 MΩ·cm, Barnstead) was used as the interrogating liquid. Small drops of water (approximately 2-5 μl) were accurately metered onto a flat surface, and the full screen image of the drop was captured with the frame grabbing software coupled to a CCD camera operating at the optimized zoom and contrast. The contact angles were determined via the software suite or via graphical fitting of the contact tangents in the captured image. Both approaches gave the same nominal value within ± 2 degrees. Static water contact angle values reported were an average of three values measured on various areas of the surface.

6.3 Results and Discussion

6.3.1 *Fluorinated POS Synthesis*

Fluorinated POS compounds fluorodecyl (**FD**), fluorooctyl (**FO**), and fluorohexyl (**FD**) were produced by the base-catalyzed hydrolysis of trialkoxy silanes (Figure 6.2). These compounds tend to condense into T_8 cages, rather than cage mixtures, as has been previously observed in the base-catalyzed synthesis. The yields for these reactions are often nearly quantitative. This is significant because the usual method to produce T_8 cages is the acid-catalyzed hydrolysis of trichlorosilanes, which produces an undesirable acidic by-product.

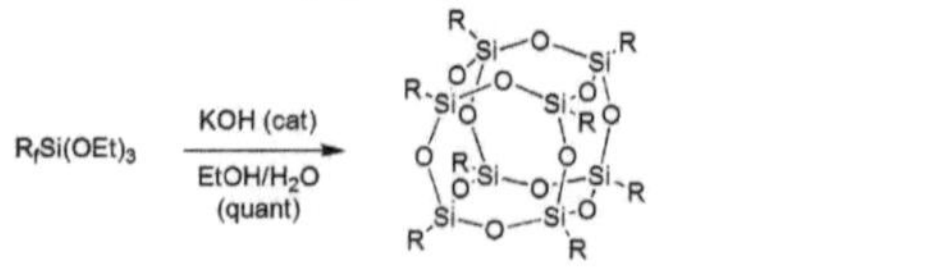

FH R = $CH_2CH_2CF_2CF_2CF_2CF_3$
FO $CH_2CH_2CF_2CF_2CF_2CF_2CF_2CF_3$
FD $CH_2CH_2CF_2CF_2CF_2CF_2CF_2CF_2CF_2CF_3$

Fig. 6.2 Direct synthesis of octahedral fluorinated POS compounds

Hepta(3,3,3-trifluoropropyl)tricycloheptasiloxane trisodium silanolate **1** was used as an interemediate for the prepartion of fluorinated POS compounds **2–7** by "corner-capping" with fluoroalkyltrichlorosilanes (Figure 6.3). The intermediate salt (**1**) is stable in air, but decomposes to silsesquioxane resin upon exposure to moisture. The intermediate adduct is the result of a pathway to fully

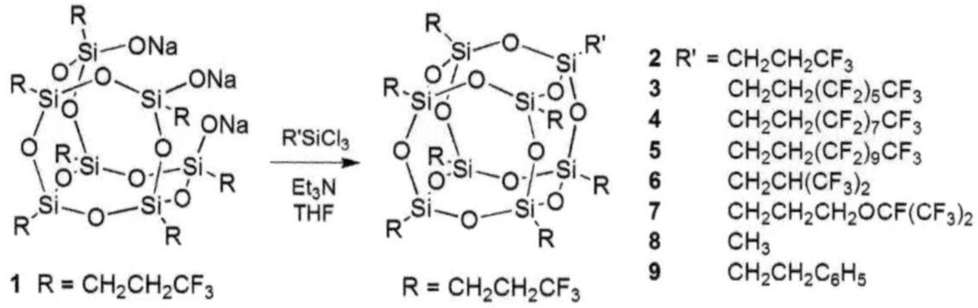

Fig. 6.3 Synthesis of corner-capped fluorinated POS compounds

condensed cage structures by simply controlling feedstock stoichiometry.

Corner-capping with commercially available functionalized fluoroalkyltrichlorosilanes afforded diverse structures with linear fluoroalkyl chains (**2**–**5**), a branched hexafluoroisopropyl (**6**), and a heptafluoroisopropoxypropyl ether (**7**). Fluorinated POS **2** (synonymously denoted **FP** for fluoropropyl) possesses octahedral symmetry like those fluorinated POS series **FH**, **FO**, and **FD**. This corner-capping methodology was employed in the incorporation of methyl (**8**) and phenylethyl (**9**) hydrocarbon groups in the predominantly fluorinated environment.

6.3.2 *Fluorinated POS Properties*

Selected properties of fluorinated POS compounds are illustrated in Table 6.1. These Fluorinated POS compounds are only soluble in fluorinated solvents such as AK-225G (Asahi Glass) and hexafluorobenzene. Unlike most non-fluorinated POS compounds, thermogravimetric analysis (TGA) indicates fluorinated POS volatilize rather than decompose. No residue remains after heating under either nitrogen or dry air. **FD** POS is the most stable compound, subliming at over 300 °C. Fluorinated POS are also very dense, high molecular weight materials. For example, **FD** POS has a molecular weight of 3993.54 g mol^{-1} and a density of 2.067 g cm^{-3}.

The structure–property relationships for fluorinated POS compounds were studied from experimental melting points (T_m) obtained from the amorphous powders precipitated from methanol. As a general trend, melting points become depressed as fluoroalkyl chain lengths increase due to weaker intermolecular wan der Waals attractions. All fluorinated POS compounds were observed to posess narrow melting points as high molecular weight compounds, indiciative of high purtiy (Table 6.1). In all cases, no decomposition was observed for these compounds at the melting point. It is clear from the data in Table 6.1 that melting point properties are primarily influenced by susbtrate structure and not solely by wt % F content. In this case, the low melting points of extended fluorinated POS compounds **3–5** would be valuable in the low temperature melt processing of polymers, affording blended composites.

The fluorinated POS compounds were spin-cast from hexafluorobenzene onto a glass substrate. The spin-casting produced well-adhered, white powder-like films, and did not visually expose any of the glass substrate. Their static (or advancing) contact angles were measured using deionized water and hexadecane as test fluids. Hexadecane is a standard test fluid for determining the oil repellency (or oleophobicity) of a material. The fluorinated POS series measured water contact angles are on average 50° higher than for hexadecane contact angles. As a consequence of the highest wt % F content, **FD** produced the highest water and hexadecane static contact angles, of 154° and 87° respectively. Since its water contact angle was greater than 150°, the surface of **FD** as a spin cast powder was formally classified as being ultrahydrophobic.

For the fluorinated POS octamer series of **FP**, **FH**, **FO** and **FD**, there is a near linear progression of water and hexadecane contact angle in relation to the increasing fluorine content of the fluorinated POS compounds. This is clearly shown by a 13% increase in water contact angle from **FP** (38.2% fluorine content) to the **FH** (57.2% fluorine content) compound. A leveling-off effect is anticipated for the series for fluoroalkyl substitution beyond the **FO**, since a negligible increase of 5% was observed from **FH** (57.2% fluorine content) to **FO** (61.9% fluorine content). However, the **FD** (64.7% fluorine content) showed an abrupt increase of 12% relative to **FO** POS. A similar trend was observed for hexadecane contact angles, but the standard deviation would indicate that the oleophobicities of **FH**, **FO**, and **FD** are the same.

Table 6.1 Selected properties of fluorinated POS compounds

Compound	Fluoroalkyl group	$\theta_{water}/°$	$\theta_{hex}/°$	$T_m/°C$[a]	wt % F[b]
FP (2)	$R = R' = CH_2CH_2CF_3$	116	69	234-237	38.5
FH	$R = CH_2CH_2(CF_2)_3CF_3$	131	80	121-123	57.2
FO	$R = CH_2CH_2(CF_2)_5CF_3$	138	82	120[c]	61.9
FD	$R = CH_2CH_2(CF2)_7CF_3$	154	87	150[c]	64.7
3	$R = CH_2CH_2CF_3$ $R' = CH_2CH_2(CF_2)_5CF_3$	109	77	104-106	45.0
4	$R = CH_2CH_2CF_3$ $R' = CH_2CH_2(CF2)_7CF_3$	112	76	88-90	50.5
5	$R = CH_2CH_2CF_3$ $R' = CH_2CH_2(CF_2)_9CF_3$	112	80	105-107	51.8
6	$R = CH_2CH_2CF_3$ $R' = CHCH(CF_3)_2$	122	74	234-236	40.9

Compound	Fluoroalkyl group	$\theta_{water}/°$	$\theta_{hex}/°$	$T_m/°C^a$	wt % F[b]
7	$R = CH_2CH_2CF_3$ $R' = CH_2CH_2CH_2OCF(CF_3)_2$	108	73	70-71	40.4
8	$R = CH_2CH_2CF_3$ $R' = CH_3$	102	51	168-170	35.9
9	$R = CH_2CH_2CF_3$ $R' = CH_2CH_2C_6H_5$	106	45	109-111	33.0

[a]Melting point capillary. [b]Calculated from molecular formula. [c]DSC analysis at midpoint.

Generation of potential energy surfaces for a packed lattice of **FH** (Figure 6.4) and **FD** (Figure 6.5) provided a better understanding of the increase in water contact angle observed for **FD** POS. **FD** POS appears to exhibit a molecular roughening arranged in a corrugated fashion. **FP** and **FO** POS appear similar to the flatter **FH** POS. This increased molecular-scale, surface roughness presumably contributes to the dramatic increase in the water contact angle of **FD** to 154°, which is approximately 23° higher than **FH**.

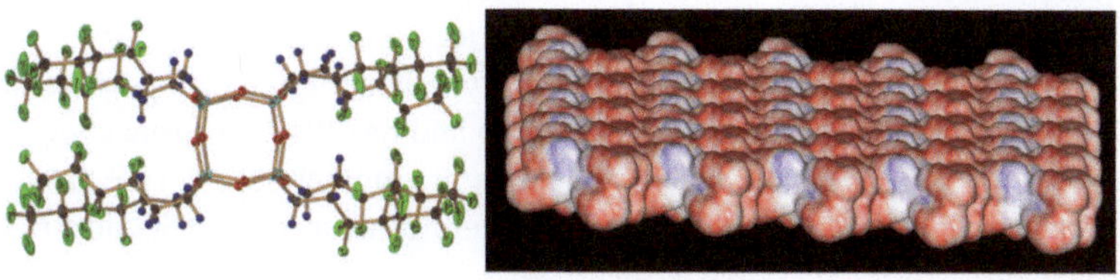

Fig. 6.4 X-ray crystal structure (left) and electrostatic potential diagram of FH POS (right) exhibiting a relatively flat surface

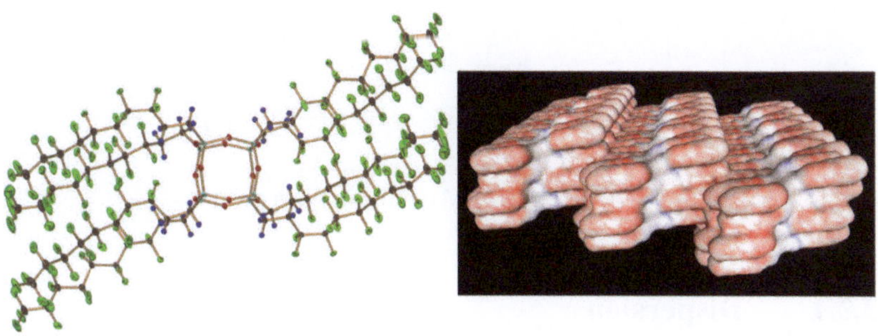

Fig. 6.5 X-ray crystal structure (left) and electrostatic potential diagram of FD POS (right) exhibiting a corrugated type surface

The potential energy surfaces shown in Figure 6.4 and Figure 6.5 represent crystalline surfaces obtained by single crystal X-ray analysis. It should be noted that the water contact angles for the observed trend were analyzed from spin cast, semicrystalline powder films. AFM analysis revealed the spin cast **FD** POS surface produced a root mean square (rms) roughness of approximately 4 μm (Figure 6.6).

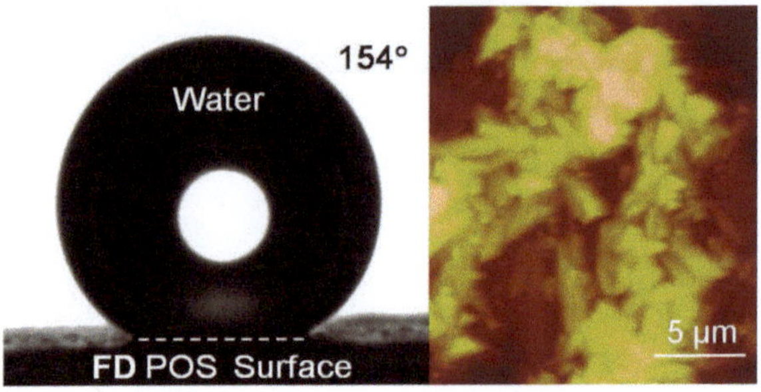

Fig. 6.6 Goniometer image of water drop on spin cast film (SCF) of FD POS with a measured contact angle of 154° (left) and AFM height image (right) of SCF of FD POS showing micrometer-size crystalline aggregates

6.3.3 POS Fluoropolymers

Selected fluorinated POS compounds possessing the highest hydrophobicity and oleophobicity were blended into several fluoropolymers in a preliminary study of fluorinated POS compatibility in polymer matrices. For the purposes of this study, **FD** POS **6F-BP** perfluorocyclobutyl (PFCB) aryl ether (Figure 6.7) polymer blends are used to describe fluorinated POS dispersion and its effect on surface wettability (Section 6.3.3.1), and **FO** and **FD** POS blends in commercial PCTFE are used to describe melt processing (Section 6.3.3.2), thermo-mechanical (Section 6.3.3.3), and surface properties (Section 6.3.3.4).

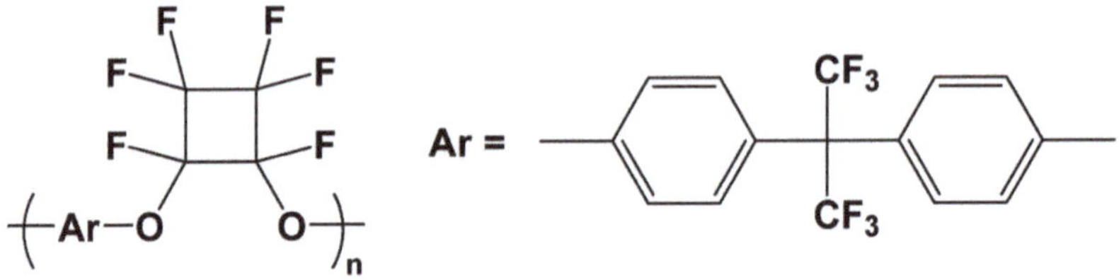

Fig. 6.7 6F-Biphenyl (6F-BP) perfluorocyclobutyl (PFCB) aryl ether polymer

6.3.3.1 Dispersion

The level of dispersion of fluorinated POS compounds into polymer systems is largely dependent on surface chemistry. PFCB aryl ether polymers are of interest in a multitude of materials applications, especially for their ability to produce optically transparent and processable films [20]. Increasing **FD** POS wt % loadings blended with **6F-BP** PFCB aryl ether polymer showed a gradual increase in water contact angle, but a significant increase in hexadecane contact angle (Figure 6.8). The **6F-BP** PFCB aryl ether polymer is intrinsically hydrophobic and produced water and hexadecane contact angles of 95° and 27°, respectively. **FD** POS loadings up to 15 wt % developed a plateau from static water contact angle; the blend showed an overall 32% increase in water contact angle (124°) at this loading compared with unblended **6F-BP**. At optimized **FD** POS loadings of 10 wt %, a maximum hexadecane contact angle of 80° was observed, increasing hexadecane repellency by 158%. While films prepared with 15 wt % **FD** POS loading still appeared transparent and homogenous, 20 wt % **FD** POS produced slight phase separation. At 30 wt % **FD** POS, significant incompatibility was observed, producing brittle opaque films with crystalline aggregates on the film surface.

It was notable that when blended surfaces were tilted beyond 90° or even inverted (180°), water remained pinned to the film; this is not the case with powdered surfaces solely prepared from spin cast fluorinated POS. A surface with a contact angle of 90° or higher is considered a "non-wetting" surface, while a surface with a contact angle below 90° is considered "wetting." Addition of fluorinated POS yielded "non-wetting" surfaces.

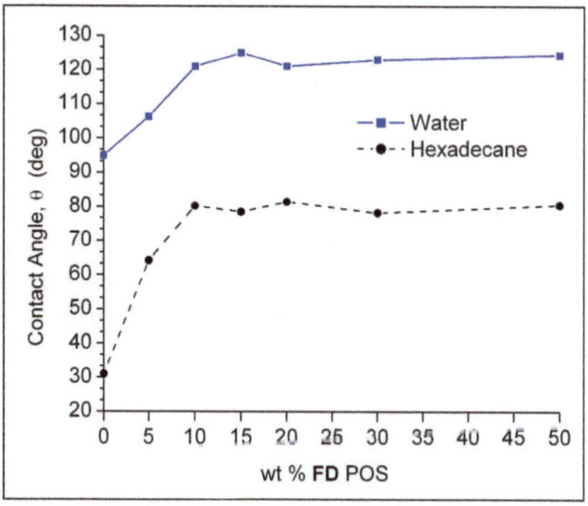

Fig. 6.8 Water and hexadecane contact angles of various wt % of FD POS blended into 6F-BP PFCB aryl ether polymer

Dynamic water and hexadecane contact angles were evaluated to determine the degree of hysteresis of **FD** POS blended into **6F-BP** PFCB aryl ether polymer. The angles of advancing (θ_a) and receding (θ_r) of water and hexadecane were obtained by placing a liquid drop on the surface and tilting the stage of the goniometer. The results of the measurements for water and hexadecane are shown in Figure 6.9. At all wt % of **FD** POS blended into the polymer, the water and hexadecane drops remained pinned on the surface, even when the stage was tilted 90°. Initially, the advancing and receding angles were taken at the onset of liquid drop perturbation on the uphill and downhill side as the stage was tilted. However, these measurements were difficult to assess "by eye" and produced a high deviation in recorded values. Therefore, the dynamic angles were recorded at a 90° tilt in order to ensure consistency. These results indicate a condition of high surface hysteresis where the surface energy (γ_{SV}) exceeds the surface tension (γ_{SL}) of the liquid drop.

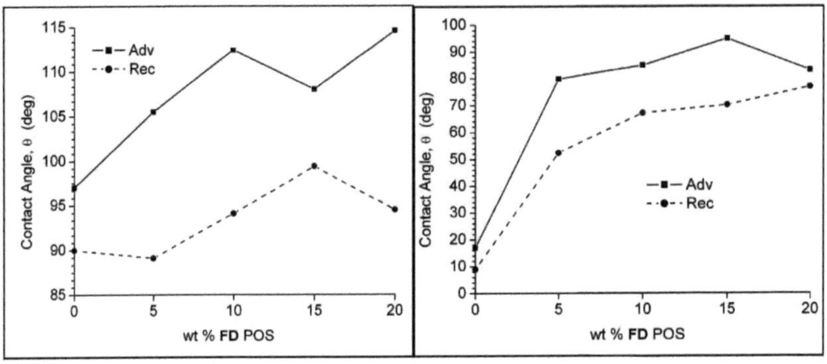

Fig. 6.9 Dynamic water (left) and hexadecane (right) contact angles of various wt % of FD POS blended into 6F-BP PFCB aryl ether polymer

Blending **FD** POS into the **BP-6F** PFCB aryl ether polymer introduced additional fluorine content and increased surface roughness. The relationship between contact angle and surface energy is governed by Young's equation, which relates surface interfacial tensions to the liquid and gas phases of water [21]. Furthermore, surface roughness imparts increased hydrophobicity to a material, as demonstrated by Cassie and Baxter, as well as Wenzel [22,23].

Atomic force microscopy (AFM) analysis of 15 wt % **FD** blend (Figure 6.10) compared with the virgin **6F-BP** PFCB aryl ether polymer (Figure 6.11) showed a marked increase in surface roughness. From AFM analysis, unblended **6F-BP** polymer and 15 wt % **FD** POS composite blend gave measured surface roughnesses of 0.527 nm and 1.478 nm, respectively. The incorporation of the fluorinated **FD** POS structures produced this threefold increase in surface roughness, possibly caused by blooming and aggregation of these structures on the surface during the spin-casting process. The increase in surface roughness was nearly 1 nm (the difference between 1.278 nm and 0.527) owing to the inclusion of **FD** POS. Such a slight increase in surface roughness is usually not enough to influence the overall macroscopic properties such as the contact angle. It is generally accepted that average surface roughness (R_a) < 100 nm has little effect on contact angles and hysteresis. Therefore, it is presumed that the low surface energy fluorine content contributed by fluorinated POS has the most influence on the surface contact angle, whereas the surface roughness is an important, but minor contributing parameter. On-going surface characterization is being investigated in order to determine the concentration gradient of the fluorinated POS structures on the surface compared to those entrained in the bulk material.

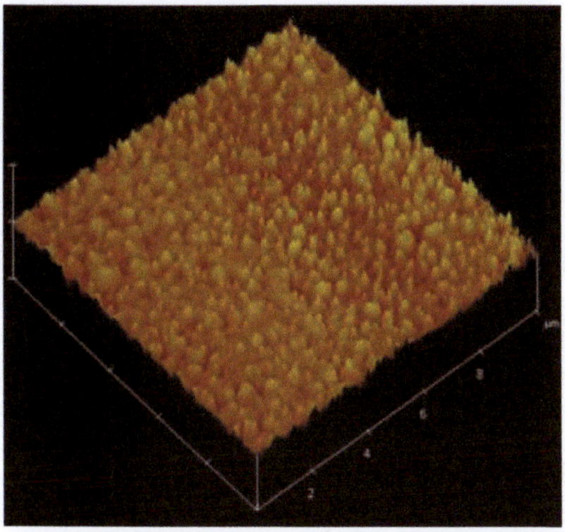

Fig. 6.10 AFM height image of **6F-BP** PFCB aryl ether polymer with 15 wt % **FD** POS

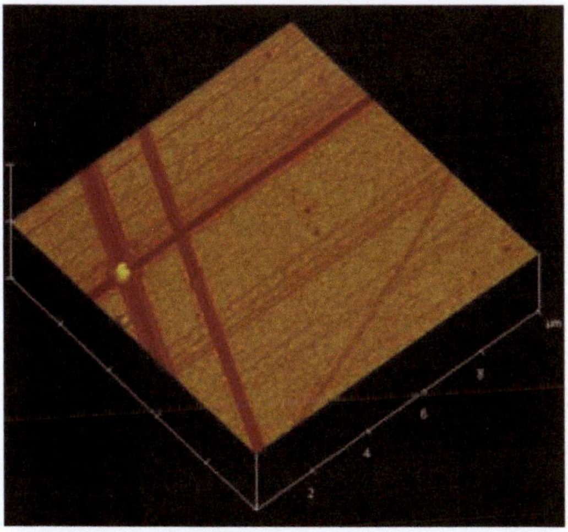

Fig. 6.11 AFM height image of **6F-BP** PFCB aryl ether polymer

6.3.3.2 Melt Processability

The processability of the samples was compared using torque and load of the compounder motor, a measure of the pressure generated in the mixer, where the pressure is generated as a result of the conical design of the mixer. For a constant volume of material compounded, and a fixed screw speed, the pressure generated is proportional to the viscosity of the material. The lower the pressure, the lower the viscosity, and the easier the material is to process. The second measure of processability is the torque output by the motor. This gives an indication of the mechanical energy put into the system, and is proportional to the current used by the motor. Hong et al. utilized a similar measure to characterize the processability of polyethylene and hyperbranched polymer blends [24]. The lower the torque output the more processable the polymer (for a given screw speed).

These two measures of the processability of the polymer blends were recorded at 30 s intervals during processing. It was found that within a 95% confidence interval, the load and torque values were constant for the duration of processing, excluding the first 30 s. Therefore, an average value for both torque and load is assigned to each processing run. In order to investigate the effect of the addition of fluorinated POS, relative torque and relative load values were computed utilizing the average values in comparison to the average values found for the unfilled resins. Figure 6.12 shows the relative torque and load values with respect to the wt % of POS added for the PCTFE blends. The solid symbols represent the relative torque values, and the relative load results are illustrated by the open symbols. The square symbols denote the results of the **FD** POS blends, and the circular symbols denote the **FO** POS blend results. One will note that PCTFE processability is improved by more than 30%.

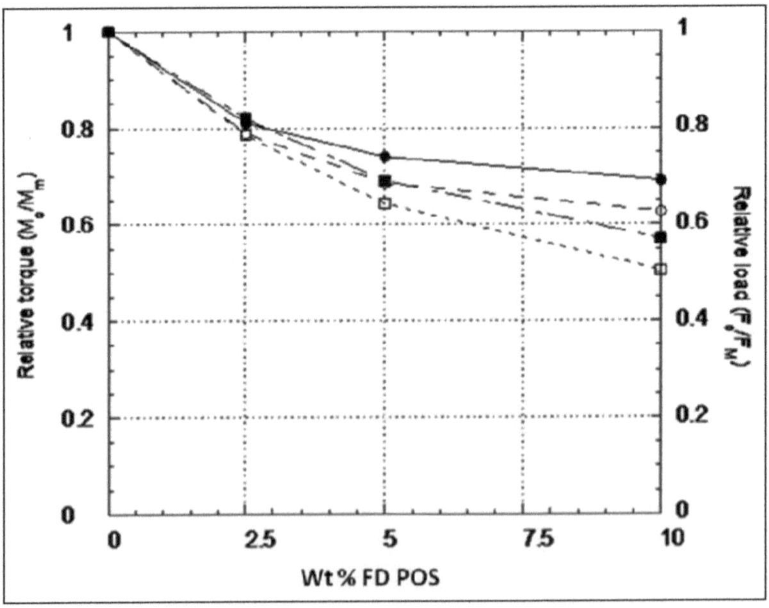

Fig. 6.12 Effect of FD POS on processing variables torque and load for PCTFE blends

6.3.3.3 Thermo-Mechanical Analysis

In order to determine the effect of the fluorinated POS on the mechanical properties of PCTFE, dynamic mechanical analysis (DMA) was performed. Figure 6.13 illustrates the storage and loss modulii for the PCTFE blends. Notice only a small decrease in the modulus values between the filled and unfilled samples. The loss modulii of the PCTFE blends are very similar in value to the unfilled PCTFE, with only a slight decrease. The mechanical properties of the fluoropolymer are only slightly altered by the addition of fluorinated POS. The possible exception to this is the composite with 10 wt % **FD** POS. This sample was very difficult to mold into quality films for testing. The poor quality of the films tested may lead to the scatter seen in the data, and possibly the lower temperature for the peak in the loss modulus curve. For this study, the glass transition temperature of the polymer is defined as the peak in the loss modulus curves. The addition of fluorinated POS to PCTFE decreases the glass transition temperature by approximately 2 °C. The variation in glass transition temperature seen with the addition of fluorinated POS is small enough to be statistically insignificant.

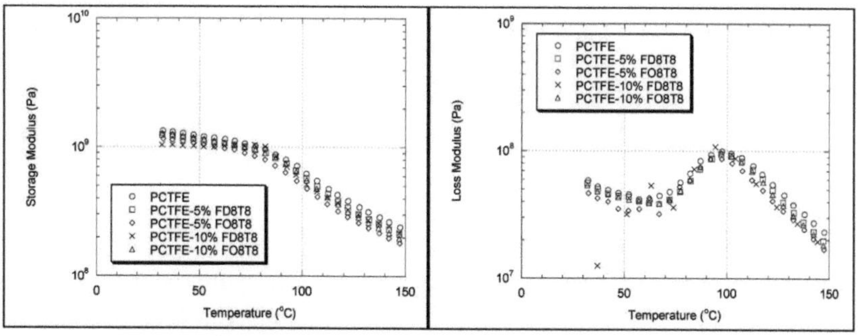

Fig. 6.13 Storage and loss modulus data for FD POS/PCTFE films

6.3.3.4 Surface Properties

While various fluoropolymers are known for their desirable properties, which include hydrophobicity and a low coefficient of friction, incorporation of fluorinated POS may help to improve these properties even further. Contact angles have been obtained on PCTFE nanocomposites containing **FO** and **FD** POS.

Technologies that may benefit from the blending of fluorinated POS into fluoropolymers include abrasion resistance, lubricity, anti-icing, and non-wetting applications. Figure 6.14 (left) shows a drop of water on the surface of a PCTFE film with a contact angle of 88°. A drop of water on the surface of a PCTFE blend containing 10 wt % **FD** POS (Figure 6.14, right) produced a contact angle of 128°. Hence there is a 40° increase in contact angle with a modest 10 wt % added **FD** POS.

Fig. 6.14 Water contact angles of 88° on PCTFE film (left) and 128° on PCTFE film containing 10 wt % FD POS (right), respectively

Similar to PFCB aryl ether polymer solvent blended coatings carrying **FD** POS, application of hexadecane on the PCTFE melt blended compositions with **FD** POS resulted in improvement of oleophobicity, with hexadecane contact angles increasing from approximately 32° to 58°.

6.4 Conclusions

Various fluorinated polyhedral oligomeric silsesquioxanes (POS) have been solvent or melt blended into a variety of fluoropolymers. These fluoropolymers include perfluorocyclobutyl (PFCB) aryl ether polymers and poly(chlorotrifluoroethylene) (PCTFE). The composite blends produce well-dispersed fluorinated POS based on microscopy analysis. The fluoroalkyl groups on the fluorinated POS cages demonstrate good miscibility in selected fluoropolymer matrices. These fluorinated POS fluoropolymer composites may be useful as low friction surfaces either as bulk components or coatings. Contact angle measurements of the POS fluoropolymers show an improvement of water and hexadecane contact angles over the unfilled materials. The low surface energy POS compounds also appear to act as a processing aid during fluoropolymer processing, significantly reducing both the torque and load measurements in the extruder. Thermal and mechanical properties of the blended fluoropolymers do not compromise the integrity of the unfilled polymers. Work continues to encompass fluorinated POS as simple drop-in modifiers for fluoropolymers. It is also worth noting that fluorinated Si_8O_{12} compounds have recently been used to develop non-wetting and stain resistant fabrics [25] and fabrics with tunable oleophobicity [26].

6.5 Acknowledgments

We gratefully acknowledge the Air Force Research Laboratory, Propulsion Directorate and the Air Force Office of Scientific Research for their financial support. The authors also wish to thank Brian Moore at AFRL for his assistance with AFM images and Marietta Fernandez at AFRL for help with SEM images.

6.6 References

1. Gakh AA, Tuinman AA, Adcock JL, Sachleben RA, Compton RN (1994) J Am Chem Soc 116:819.

2. Fagan PJ, Krusic PJ, McEwen CN, Lazar J, Holmes-Parker D, Herron N, Wasserman E (1993) Science 262:404.

3. Taylor R, Avent AG, Dennis TJ, Hare JP, Kroto HW, Walton DRM (1992) Nature 355:275.

4. Herzog A, Callahan RP, Macdonald CLB, Lynch VM, Hawthorne MF, Lagow RJ (2001) Angew Chem Int Ed 40:2121.

5. Hayashi T, Terrones M, Scheu C, Kim YA, Rühle M, Nakajima T, Endo M (2002) Nano Lett 2:491.

6. Gonzalez RI, Phillips SH, Hoflund GB (2000) J Spacecraft and Rockets 37:463.

7. Phillips SH, Haddad TS, Tomczak SJ (2004) Curr Opin Solid State Mater Sci 8:21.

8. POSS® is a registered trademark of Hybrid Plastics Inc., Hattiesburg, MS 39401.

9. Li G, Wang L, Hanli N, Pittman CU Jr. (2001) J Inorg Organomet Polym 11:123.

10. Hoflund GB, Gonzalez RI, Phillips SH (2001) J Adhesion Sci Technol 15:1199.

11. Gilman JW, Schlitzer DS, Lichtenhan JD (1996) J Appl Polym Sci 60:591.

12. Lu S-Y, Hamerton I (2002) Prog Polym Sci 27:1661.

13. Lee GZ, Wang L, Toghiani H, Daulton TL, Pittman CU Jr. (2002) Polymer 43:4167.

14. Zhang W, Fu BX, Schrag E, Hsiao B, Mather PT, Yang N-L, Xu D, Ade H, Rafailovich M, Sokolov J (2002) Macromolecules 35:8029.

15. Fu BX, Gelfer MY, Hsiao BS, Phillips S, Viers B, Blanski R, Ruth P (2003) Polymer 44:1499.

16. Deng J, Polidan JT, Hottle JR, Farmer-Creely CE, Viers BD, Esker A (2002) J Am Chem Soc 124:15194.

17. Fu BX, Hsiao BS, White H, Rafailovich M, Mather PT, Jeon HG, Phillips S, Lichtenhan J, Schwab J (2000) Polym Int 49:437.

18. Mabry JM, Vij A, Iacono ST, Viers BD (2008) Angew Chem Int Ed 47:4137.

19. Iacono ST, Grabow W, Vij A, Smith DW Jr., Mabry JM (2007) Chem Commun 47:4992.

20. Iacono ST, Budy SM, Jin J, Smith DW Jr. (2007) J Polym Sci Part A Polym Chem 45:5707.

21. Young T (1805) Phil Trans Roy Soc 95:65.

22. Wenzel RN (1936) Ind Eng Chem 28:988.

23. Cassie ABD, Baxter S (1944) Trans Faraday Soc 40:546.

24. Hong Y, Cooper-White JJ, Mackay ME, Hawker CJ, Malmstrom E, Rehnberg N (1999) J Rheol 43:781.

25. Misra R, Cook RD, Morgan SE (2010) J Appl Polym Sci 115(4):2322-2331.

26. Choi W, Tuteja A, Chatre S, Mabry JM, Cohen RE, McKinley GH (2009) Adv Mater 21(21):2190-2195.

Chapter 7
Polyhedral Oligomeric Silsesquioxanes in Electronics and Energy Applications

Claire Hartmann-Thompson

Introduction

This chapter is of broad scope and and covers the use of hybrid polyhedral oligomeric silsesquioxane (POS) materials to enhance performance in various electronics, optical and energy-related applications. It reviews the liquid crystal phase behavior of Si_8O_{12} compounds and their use in LC devices, their use as electroluminescent (EL) materials in light-emitting devices (with a particular focus upon the problematic area of blue emission), their use as lithographic resists in the fabrication of electronic and optical devices, their use in diverse sensor systems, and their use in fuel cell membranes, battery electrolytes and lubricants. Si_8O_{12} materials have also been successfully employed as coatings and adhesives in space photovoltaic solar cells, and this topic is reviewed in Chapter 8, along with various other space material-related applications.

7.1 Polyhedral Oligomeric Silsesquioxanes in Liquid Crystal Systems

Liquid crystalline (LC) phases [1] are phases intermediate between the disordered liquid state and the highly ordered crystalline solid state, and are an important

Claire Hartmann-Thompson
Michigan Molecular Institute, Midland, MI USA

C. Hartmann-Thompson (ed.), *Applications of Polyhedral Oligomeric Silsesquioxanes*, Advances in Silicon Science 3, DOI 10.1007/978-90-481-3787-9_7,

example of the phenomenon of self-assembly. LC phases may be either nematic (N), chiral nematic (N*, also termed cholesteric) or smectic (S_m), Fig. 7.1. Nematic phases have long-range orientational order and short-range positional order. Chiral nematic phases are periodically twisted about an axis perpendicular to the direction of molecular orientation (termed the director), and are associated with mesogens containing chiral centers. Smectic phases have layered structures, and possess long-range orientational order and a degree of long-range positional order. In the smectic A phase (S_mA), the director is perpendicular to the layers, and in the smectic C phase (S_mC), the director is inclined at an angle to the perpendicular (Fig. 7.1). Chiral groups (denoted by *) cause nematic and tilted smectic phases to form twisted helical structures. Nematic or chiral tilted smectic phases in materials containing rod-like mesogens can display ferroelectric effects. The nematic phase is of the greatest interest for commercial display technologies. The crystalline phase is denoted as C and the isotropic phase is denoted as I.

Polarized optical microscopy (POM) is used to identify textures and phase transition temperatures associated with the various mesophases. Different mesophases have distinctive textures, e.g., a Schlieren (threaded marble-type) texture is characteristic of nematic phases, a streaky oily texture is characteristic of cholesteric phases, a focal-conic fan texture is characteristic of smectic A phases, and a granulated texture is characteristic of smectic C phases. Differential scanning calorimetry (DSC) may also be used to determine transition temperatures, and X-ray diffraction may be used to characterize liquid crystalline phases. At the isotropization temperature (also referred to as the clearing temperature) the final transition from a mesophase to a normal liquid phase with no positional or orientational order occurs.

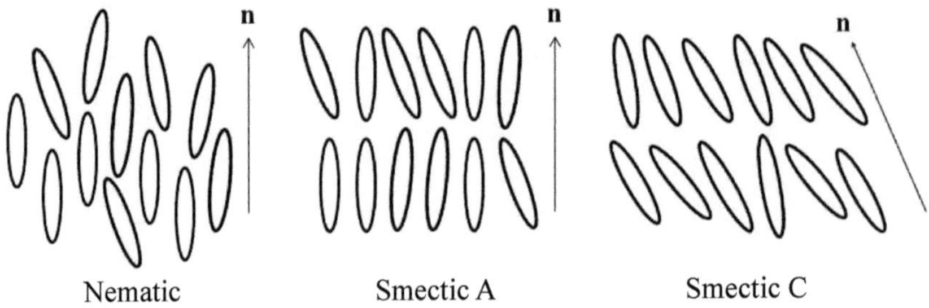

Fig. 7.1 Illustration of various liquid crystalline phases, where n denotes the direction of molecular orientation (i.e., the director)

Main chain LC polymers have rigid aromatic groups in their backbones, and form LC phases in solution or in the melt. Examples include aromatic polyamide fibers such as Dupont Kevlar® (the first commercial LC polymer, developed in the 1960s), and aromatic polyesters such as Ticona Vectra®, Dupont Zenite® and Solvay Xydar®. Side chain LC polymers are comprised of rigid rod mesogenic side-chains attached to polymer backbones. The first liquid crystal displays (LCDs), based on cyanobiphenyl materials, were developed in the early 1970s, and in order to obtain processible materials for commercial LCD applications, side-chain LC polymers were developed. It was found that a spacer between the backbone and the mesogen was necessary in order to decouple the mesogen from the backbone, and to enable the mesogens to align to form mesophases. The majority of electroluminescent (EL) polymers are also liquid crystalline (see Section 7.2), and LC POS materials and EL POS materials have been reviewed together [2]. Any rigid anisotropically-shaped molecule (e.g., rods or discs) has the potential to form mesophases. Discotic liquid crystals form either nematic phases (denoted N_D) or columnar phases (denoted D, the discotic analog of the smectic phase). In non-emissive thermotropic LC displays (attractive for large area display technologies), scattering occurs either from nematic liquid crystals in a continuous polymer phase, or from solid particles in a continuous nematic LC phase.

In this section, various categories of LC Si_8O_{12} materials are reviewed. Si_8O_{12} cores partially or fully functionalized with a range of mesogens (standard cyanobiphenyl, metallomesogen, end-on, side-on lateral, chiral or bent-core) constitute the largest body of work, but Si_8O_{12}-mesogen dyads, main chain LC polymers end-capped with Si_8O_{12}, Si_8O_{12} cores functionalized with side-chain jacketed LC polymers, and linear polymers carrying both pendant Si_8O_{12} and pendant mesogens are also reviewed. In addition, the use of Si_8O_{12} compounds at interfaces in LC devices in order to induce spontaneous vertical alignment and to improve device performance is discussed.

In the 1990s, hydrosilylation of polymethylhydridosiloxanes, $(SiMeHO)_n$, by vinyl-functionalized mesogens was the method of choice for creating a vast range of side-chain liquid crystalline polysiloxanes [3], although it is interesting to note that some polysiloxanes are able to form LC phases without carrying mesogens (e.g., di-*n*-alkylpolysiloxanes). The flexibility of the siloxane backbone was an advantage in enabling the mesogens to align. The structure-property relationships of these compounds were studied intensively; structure variables included calamitic versus discotic mesogens, the length and composition of the spacer, the fraction of siloxane repeat units carrying a mesogen and the distribution of the mesogen along the backbone, and property variables included the category of mesophase and the transition temperature. This approach was then extended beyond linear siloxanes [4,5,6] to include cyclic siloxanes, polysilsesquioxanes [7,8] and polyhedral oligomeric silsesquioxanes [9] as variables, where some or all of the silicon atoms in the Si_8O_{12} cage were substituted with mesogenic LC moieties. The attachment of a mesogen to an Si_8O_{12} core is also of interest because the mesogen is immobilized on a substrate, but is still free to align in an externally applied field [10]. Because most liquid crystalline side-chain polysiloxanes were synthesized by hydrosilylation, this also became the synthetic route of choice for silsesquioxane analogues [11] (although metal-containing mesogens [12] can create difficulties in conventional metal-catalyzed hydrosilylation chemistry). $Si_8O_{12}(H)_8$ and $Si_8O_{12}(OSiMe_2H)_8$ undergo clean α-hydrosilylation with most alkenes, while oxyallyl compounds give some β-silylated side-products. Since platinum catalyst residues can affect LC transition temperatures [13], deactivation of the catalyst (e.g., by using a phosphorus compound) and removal of the catalyst (e.g., by use of a silica column) is important [9].

The first study of LC Si_8O_{12} compounds was carried out in 1991 by Kreuzer and co-workers at Wacker-Chemie GmbH in Germany [9]. Si_8O_{12} $(OSiMe_2H)_8$ was reacted with an alkene-functionalized mesogen in a dicyclopentadienyl platinum dichloride-catalyzed hydrosilylation, and platinum residues were removed using a short silica gel column. All of the products displayed smectic phases (Table 7.1) with the associated focal conic textures. When a mesogen was linked laterally (side-on) to a T_8 core (Fig. 7.2), a nematic texture was observed at 23 °C. The lateral attachment of mesogens had previously been demonstrated for side-chain LC polysiloxanes [14].

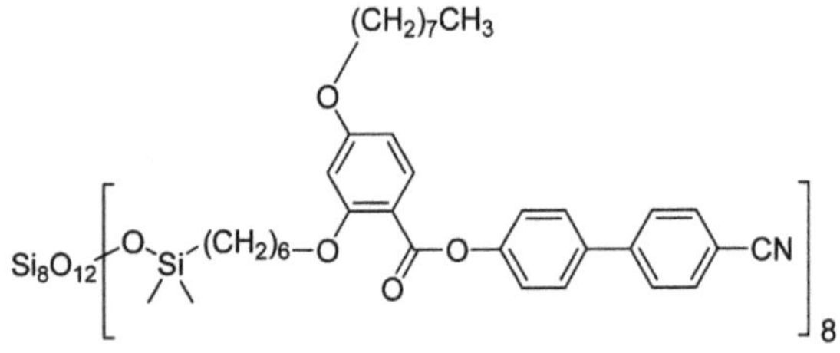

Fig 7.2 Cyanobiphenyl mesogen laterally attached to an Si_8O_{12} core [9]

Table 7.1 Phase transitions for various mesogen-functionalized polyhedral oligomeric silsesquioxanes where C denotes crystalline, G denotes glassy, I denotes isotropic, S_mA denotes smectic A and Chol denotes cholesteryl [9]

Core	Substituent	Phase transitions (°C)
T_8	$[(CH_2)_{10}COO\text{-}Chol]_8$	C 128 (G 46) S_mA 157 I
T_6	$[OSiMe_2(CH_2)_{10}COO\text{-}Chol]_6$	G 30 S_X 37 S_mA 142 I
T_8	$[OSiMe_2(CH_2)_{10}COO\text{-}Chol]_8$	C 108 S_mA 160 I
T_8	$[OSiMe_2(CH_2)_3O\text{-}Ph\text{-}COO\text{-}Ph\text{-}CN]_8$	C 119 S_mA 146 I
T_8	$[OSiMe_2(CH_2)_3O\text{-}Ph\text{-}COO\text{-}Ph\text{-}Ph\text{-}CN]_8$	G 102 S_X 127 S_mA >300 I
T_{10}	$[OSiMe_2(CH_2)_{10}COO\text{-}Chol]_{10}$	C 122 (G 51) S_mA 164 I
T_{10}	$[OSiMe_2(CH_2)_3O\text{-}Ph\text{-}COO\text{-}Chol]_{10}$	C 115 (G 87) S_mA 262 I
T_8	$[OSiMe_2(CH_2)_{10}COO\text{-}Chol]_4[Me]_4$	G 35 S_X 63 S_mA 148 I
T_8	$[OSiMe_2(CH_2)_{10}COO\text{-}Chol]_4[OSiMe_2(CH_2)_{10}\text{-}COO\text{-}Ph\text{-}Ph]_4$	G 17 S_X 60 S_mA 107 I
T_8	$[OSiMe_2(CH_2)_3O\text{-}Ph\text{-}COO\text{-}Chol]_5[OSiMe_2(CH_2)_3O\text{-}Ph\text{-}COO\text{-}Ph\text{-}Ph]_3$	S_X 114 S_mA 213 I

Three years later, the Laine group at the University of Michigan synthesized various allyl-terminated aromatic diester compounds [15] using an earlier reported method [16]. Some were isotropic and some showed nematic phases and C-LC onset temperatures in the range 160 to 188 °C. Temperature cycling studies were not possible because the compounds underwent Claisen rearrangements upon heating. When one of LC allyl compounds, 4-(4-allyloxybenzoyloxy)biphenyl, was reacted with octahydrido-Si_8O_{12}, both the ratio of the reagents and the ^{29}Si NMR analysis suggested that a symmetrical tetra-substituted compound may have been formed. This was based on the observation of SiC signals at -67 ppm and -69 ppm, but only one SiH signal at -84 ppm. The size exclusion chromatography (SEC) data were broadly consistent with this. Upon heating, the compound flowed at 90 °C, transitioned to a coarse nematic texture with inversion walls at 133 °C , and showed a clearing point at 184 °C. Upon cooling, the compound transitioned to a nematic Schlieren texture at 178 °C, and transitioned to an inversion wall texture at 134 °C. The remainder of this study concerned model hydrosilylation reactions between silanes and allyl-functionalized small molecule aromatic LC esters, and the various side products formed when different platinum or palladium catalysts were used (see also Section 1.6.2, Chapter 1). Platinum 1,3-divinyltetramethyldisil oxane (Karstedt's catalyst) gave the fewest side-products under the mildest reaction conditions, and an $Si_8O_{12}(H)_8$ >> Et_3SiH > $HMe_2SiOSiMe_2H$ > Ph_2SiH_2 order of reactivity was observed. The use of palladium dibenzylideneacetone-Ph_3P catalyst resulted in oxysilylation with loss of propene.

The same group [17] reported a liquid crystalline methacrylate $Si_8O_{12}(H)_4[(CH_2)_4OPhCOO\text{-biphenyl}]_4$ in 1994 (Fig. 7.3). This material was prepared with the aim of combining the properties of Si_8O_{12} compounds (i.e., good adhesion to surfaces, hardness and thermal stability) with those of rigid rod liquid crystalline (LC) polymers (i.e., toughness, low coefficient of thermal expansion (CTE), good oxidative and photochemical stability) for application as a curable dental composite, where abrasion resistance is achieved at lower filler loadings, and in more easily processed and less viscous materials (see also Section 9.4.2, Chapter 9).

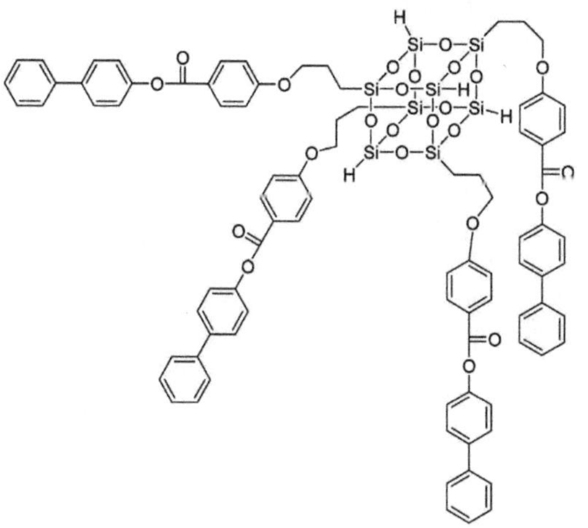

Fig. 7.3 A tetra-substituted LC Si_8O_{12} species [17]

The first Si_8O_{12} species functionalized with a metallomesogen [10] was synthesized by Saez and co-workers at the University of Hull (UK) in 1996. Most metallomesogens are bifunctional and carry two identical functionalized ligands, and are therefore potential cross-linkers; however, a high degree of cross-linking can cause a loss of liquid crystallinity in polymeric systems. In an attempt to avoid this, a mono-functional (asymmetric) square planar diamagnetic nickel(II) complex (Fig. 7.4) was prepared and reacted with $Si_8O_{12}(OSiMe_2H)_8$ in order to synthesize a free oligomeric LC compound. This hydrosilylation was catalyzed by Karstedt's catalyst and the product was confirmed by elemental analysis. The same compound was also reacted with 1,3,5,7-tetramethyl-cyclotetrahydridosiloxane (D_4H_4). The free Ni(II) complex (Fig. 7.4) melt-transitions to a smectic A phase at 181 °C and clears to the isotropic liquid at 190 °C, and the D_4 cyclic Ni(II) compound melt-transitions to a smectic A phase at 196 °C and clears to the isotropic liquid at 296 °C with some decomposition. Unfortunately, the Ni(II)-functionalized Si_8O_{12} compound was found to be non-mesomorphic, and simply melted with decomposition at 219 °C. A year later, the same group [18] reported a copper analog of the monofunctional nickel mesogen shown in Fig. 7.4.

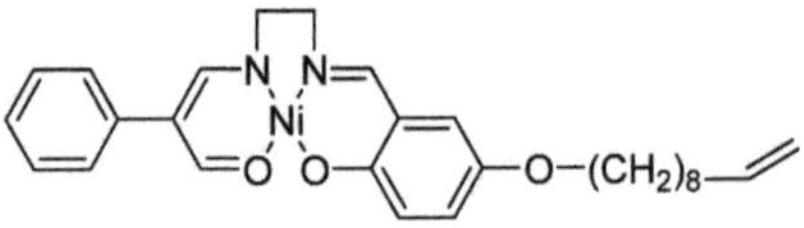

Fig. 7.4 Monofunctional nickel(II) metallomesogen [10]

In 1996, Mehl and Goodby at Hull [19] attached eight cyanobiphenyl units to an Si_8O_{12} core to form the compound Si_8O_{12} $[OSiMe_2(CH_2)_nO\text{-biphenyl-CN}]_8$ where n = 4, 6 or 11 for the spacer (Fig. 7.5, middle). A lamellar smectic A phase formed, and transition temperatures were observed at 94, 117 and 129 °C for n = 4, 6 and 11, respectively. X-ray diffraction and molecular simulation suggested that the mesogens were packing in two groups of four (Fig. 7.5, top), were forming a rod-like (rather than star-like) molecular shape, and that the rods were packed in disordered layers. Such ABCCBA arrangements (where A is a mesogen, B is a spacer and C is an Si_8O_{12}core) are very common for mesogen-functionalized Si_8O_{12} compounds [19,20,21]. In a later study by Saez and Goodby at Hull in 1999 [22], an analogous n = 11 Si_8O_{12} compound functionalized with 16 cyanobiphenyl units was prepared (Fig. 7.5, bottom) in a convergent synthesis that required five separate hydrosilylation, Grignard or SiCl substitution steps. Interestingly, despite the dense packing of the mesogens around the Si_8O_{12} core, a lamellar mesophase was still preferred over a cubic or a columnar phase, and enantiotropic smectic A and smectic C phases were observed (G -18 °C, S_mC 63 °C, S_mA 92 °C). This compound had a lower clearing point and a lower S_mC to S_mA transition temperature than its linear polysiloxane analog.

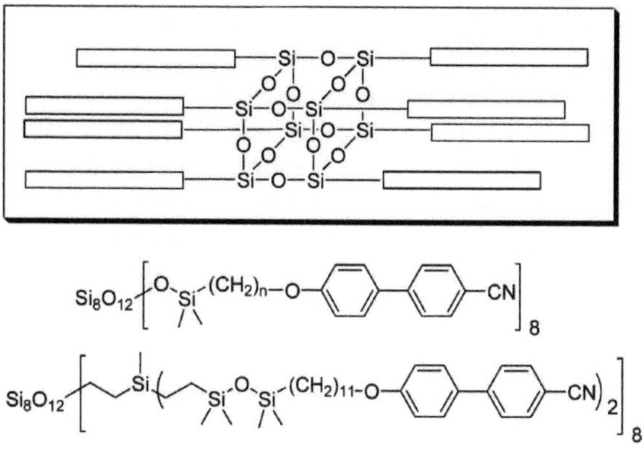

Fig. 7.5 LC Si_8O_{12} compounds carrying cyanobiphenyl mesogens [19,22]

The LC Si_8O_{12} systems discussed above (Fig. 7.5) are useful model compounds because they are monodisperse [11]. Hence they can be used as scaffolds to study the behavior of mesogens without having to consider the variables of polydispersity and tacticity that are unavoidable in studies of side-chain LC polymers. For side-chain LC polymers carrying a given mesogen, phase transition temperatures increase and then plateau out with increasing numbers of repeat units, with the plateau commonly occurring between 12 and 30 repeat units [13]. An Si_8O_{12} core carrying cyanobiphenyl mesogens was compared with tetrakis(dimethylsiloxane) ($Si(OSiMe_2)_4$) star cores, D_4 and D_5 cyclic cores, and a T_{10} polyhedral oligomeric silsesquioxane core. For the standard cyanobiphenyl mesogen on Si_8O_{12}, four or more methylene spacers between mesogen and Si_8O_{12} are required for LC behavior to occur, and a smectic A phase is observed at high temperatures. As the number of methylene spacers increases from four to eleven, the smectic A to isotropization transition temperature increases [19,20]. Analogous compounds with polysiloxane backbones, tetrahedral $Si(OSiMe_2)_4$ star cores and D_4 cyclic cores show the same trend in the transition temperature from smectic A to isotropic. The glass transition temperature of the mesogen-functionalized Si_8O_{12} compounds decreases with increasing spacer length owing to the plasticizing effect of the methylene segments [11]. Isotropization enthalpy and entropy values suggest that the systems become increasingly ordered as spacer length increases. The transition temperatures to smectic A, and also from smectic A to the isotropic phase, increase with the stiffness or the mesogen [9]. Transitioning from smaller to larger polyhedral oligomeric silsesquioxane cores (T_6, T_8, T_{10}) results in an increase in isotropization temperature [9].

Following on from these structure-property studies [9,11], in a much later 2007 study by Kawakami and co-workers at the Japan Advanced Institute of Science and Technology [23], the range of core architectures was extended to include the structures in Fig. 7.6. These were functionalized with 4-allyloxy-4'cyanobiphenyl (somewhat misleadingly classified by the authors as a 'nematic mesogen' C 52 °C,

N 70 °C, I) or cholesteryl-10-undecanoate (classified as a 'cholesteric mesogen' C 52 °C, S 65 °C, cholesteric phase 82 °C, I) by Karstedt-catalyzed hydrosilylation.

The compounds were fully characterized by elemental analysis, MALDI-TOF MS and ^{29}Si NMR. The LC phase behavior was found to be largely independent of the structure of the siloxane core, although the authors did not cite the earlier structure-property studies where core architecture was varied [9,11]. The double-decker compound tetra(dimethylsilanyloxy)octaphenyltetracyclooctasilsesquioxane (DDODMS, Fig. 7.6, bottom right) was an interesting choice of core, since its structure was intermediate between a cyclic siloxane and an open-cage polyhedral oligomeric silsesquioxane. Little work on open-cage cores functionalized with mesogens has been carried out, and the vast majority of work reviewed in this section concerns closed-cage Si_8O_{12} cores. When carrying the cyanobiphenyl mesogen, the DDODMS core gave a T_g of 10 °C, but no mesophases were observed, possibly because of the short spacer, and because of steric hindrance from the phenyl groups interfering with mesogen alignment. When carrying the cholesteryl mesogen, DDODMS showed a cholesteric to isotropic transition at 94 °C, whereas all of the other cholesteryl-carrying cores in Fig. 7.6 showed smectic to isotropic transitions instead. Interestingly D_4 (Fig. 7.6, top right) carrying cyanobiphenyl was reported to have a nematic to isotropic transition at 118 °C, in contrast to earlier studies that had reported smectic mesophases for cyanobiphenyl-functionalized cyclic siloxanes (albeit with longer spacers) [11].

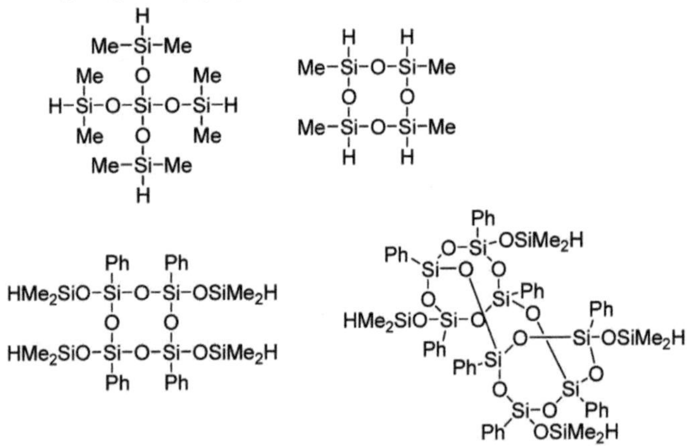

Fig. 7.6 Structures of various cores used to carry cyanobiphenyl or cholesteryl mesogens [23]

In a 2000 University of Hull study [24], two mesogen-functionalized Si_8O_{12} compounds with siloxane (rather than methylene) spacer groups were synthesized, and compared with the starting mesogens, and with related model compounds. (Fig. 7.7) They had lateral rather than end-on functional groups (Fig. 7.7) and were reacted with octavinyl- Si_8O_{12}. These mesogens required multi-step syntheses and were successfully purified by flash column chromatography. The Si_8O_{12} core functionalized with the upper mesogen (Fig. 7.7) showed C 59.5 °C [S_mX 33.1 °C] S_mC 111.3 °C, N 145.8, I, notable for the thermal stability of its mesomorphic state, with an isotropization temperature ~20 °C higher than that of the free mesogen. The Si_8O_{12} core functionalized with the lower mesogen (Fig. 7.7) showed T_g -19.3 °C, S_mX 37.6 °C, N 50.5 °C, I. In a later 2007 collaboration between the University of Patras (Greece) and the group at Hull University [25], Si_8O_{12} was functionalized with the vinyl analogue of the mesogen [24] shown in the top part of Fig. 7.7 and characterized by XRD, an extension of the earlier POM study [24] of the mesophases. The pure mesogen, the pure mesogen-substituted Si_8O_{12} compound, and mixtures of these two materials at 1:1, 3:1 and 7:1 w/w were studied. The free mesogen formed a nematic phase while the Si_8O_{12} compound and the free mesogen/ Si_8O_{12} mixtures transitioned from a columnar rectangular phase, to a columnar hexagonal phase, to a nematic phase as temperature increased. The columnar phases were destabilized as the amount of free mesogen in the mixture increased. The XRD data suggested that for the columnar hexagonal phase, some free mesogens were located in the inter-columnar space and others were located within the columnar structure (causing an increase in both lattice constant and stacking periodicity), whereas for the columnar rectangular phase all of the mesogens were withinin the columnar structure (causing no significant change in the lattice constant but an increase in stacking periodicity).

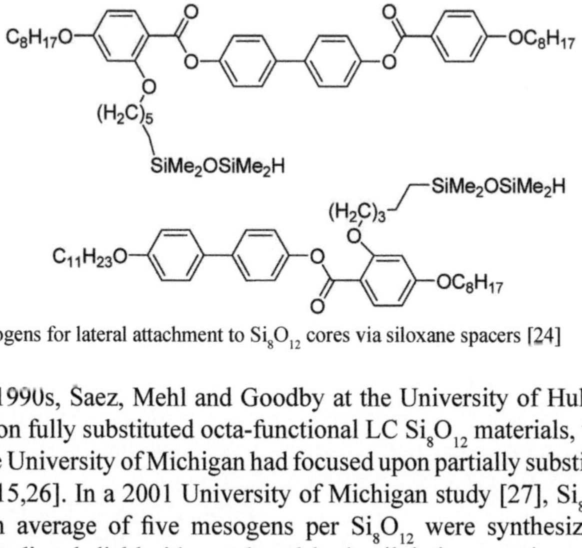

Fig. 7.7 Mesogens for lateral attachment to Si_8O_{12} cores via siloxane spacers [24]

In the 1990s, Saez, Mehl and Goodby at the University of Hull had generally focused upon fully substituted octa-functional LC Si_8O_{12} materials, while the Laine group at the University of Michigan had focused upon partially substituted LC Si_8O_{12} materials [15,26]. In a 2001 University of Michigan study [27], Si_8O_{12} compounds carrying an average of five mesogens per Si_8O_{12} were synthesized in platinum dicyclopentadienyl dichloride-catalyzed hydrosilylation reactions between $Si_8O_{12}(OSiMe_2H)_8$ and allyloxy-functionalized mesogens, following on from an earlier study where a tetra-substituted LC Si_8O_{12} compound had been prepared [15]. The aim of this work was to lower the phase transition temperatures to levels suitable for biological applications (e.g., dental composites, see also Chapter 9). Allyloxy-R mesogens with various spacers were synthesized, where R = $(CH_2)_3$O-Ph-COO-biphenyl (no mesophase), R = $(CH_2)_3$O-Ph-COO-biphenyl-OCH_3 (nematic), $(CH_2)_2O(CH_2)_2$O-Ph-COO-biphenyl (no mesophase), or R= $(CH_2)_2O(CH_2)_2$O-Ph-COO-biphenyl-OCH_3 (no mesophase). Elemental analysis was used to demonstrate the presence of five mesogens per Si_8O_{12}. Interestingly when the latter three mesogens were attached to Si_8O_{12} cores, the resulting compounds all showed S_mA and N phases, despite the non-mesomorphic nature of some of the starting mesogens. Side-chain LC polysiloxanes show similar behavior, and this has been rationalized in terms of rigid groups being more densely packed and more easily aligned when immobilized on a linear polymer than when free. The Si_8O_{12} materials carrying mesogens with flexible CH_2CH_2O spacers had lower melting and clearing temperatures than the Si_8O_{12} materials carrying mesogens with $CH_2CH_2CH_2O$ spacers. The introduction of OCH_3 end-groups resulted in a general increase in the various phase transition temperatures.

In a 2000 University of Kyoto study [28], the following three mesogens were attached to Si_8O_{12} in hexachloroplatinic acid-catalyzed hydrosilylations between the vinyl-functionalized mesogen and octahydrido- Si_8O_{12}: $(CH_2)_5$O-biphenyl-OCH_3, $(CH_2)_5$O-biphenyl-CN and $(CH_2)_5$-Ph-N=N-Ph-CN. Reaction stoichiometry was set to achieve full mesogen substitution. No mass or elemental analysis data were given for the products; however, the absence of an SiH peak in the ^{1}H NMR spectrum suggested that highly substituted products had probably been obtained. The three starting mesogens and the three mesogen-functionalized Si_8O_{12} compounds all showed nematic phases by DSC and POM (where the Schlieren texture was observed), but no smectic phase. The nematic phase existed over a much wider temperature range in the Si_8O_{12} compounds than in the starting mesogens. These compounds (synthesized from $Si_8O_{12}(H)_8$) are very closely related to the University of Michigan and University of Hull compounds discussed above (albeit synthesized from Si_8O_{12} $(OSiMe_2H)_8$), although these earlier studies are not cited. Si_8O_{12} $[OSiMe_2(CH_2)_nO\text{-Biphenyl-CN}]_8$ where n= 4, 6 or 11 shows a smectic phase [19] and $Si_8O_{12}[(CH_2)_2O(CH_2)_2O\text{-Ph-COO-biphenyl-}OCH_3]_5$ shows both smectic and nematic phases [27]. Nematic phases for mesogen-functionalized POS compounds had only previously been observed at partial substitution [27] or for laterally attached mesogens [9,24] so it is possible that full substitution may not have been achieved in this study.

In 2001, workers at the University of Hull functionalized Si_8O_{12} cores with chiral mesogens to achieve enantiotropic chiral nematic phases that were stable over a wide temperature range [29]. The two chiral alkene-functionalized mesogens shown in Fig. 7.8 (C 111.3 °C, N* 153.5 °C, I and C 101.1 °C, N* 109.2 °C, I respectively) were reacted with Si_8O_{12} $(OSiMe_2H)_8$ in a Karstedt-catalyzed hydrosilylation to give compounds showing G 23.7 °C, N* 116.9 °C, I, and G 11.2°C, N* 72.0 °C, I respectively. POM showed that the chiral nematic phase of the first compound was blue and iridescent in texture (the blue color suggesting a helix with a 0.2 to 0.3 μm pitch). The second compound showed a non-iridescent Schlieren texture. In contrast to the majority of the mesogens reviewed above (where the reactive group was at the 'end' of the rod), in these mesogens 'side-on' or lateral attachment was achieved. A Si_8O_{12} core carrying sixteen laterally attached chiral mesogens [30] was synthesized in a Karstedt-catlyzed hydrosilylation between Si_8O_{12} $[CH_2CH_2SiMe(vinyl)_2]_8$ and the SiH-laterally functionalized mesogen shown in Fig. 7.8. This compound may also be considered as a first generation dendrimer with an Si_8O_{12} core. Upon cooling, DSC and POM showed an I to N* transition at 107.7 °C, and an N* to Col_{hd} (hexagonal columnar) transition at 102.3 °C. The hexagonal columnar phase had a characteristic fan-like texture. Upon further cooling, POM showed a texture change that was tentatively assigned as a rectangular columnar phase, although no associated thermal transition was observed by DSC. The columnar phases suggested the molecules had a disc-like shape, and the structural organization of the mesogens was further investigated by molecular modeling. In the earlier study discussed above [22], sixteen 'end-on' cyanobiphenyl groups had been attached to an Si_8O_{12} core (Fig. 7.5) and had formed S_mA and S_mC phases. Columnar phases for mesogens laterally attached to Si_8O_{12} cores were also observed in later studies [25].

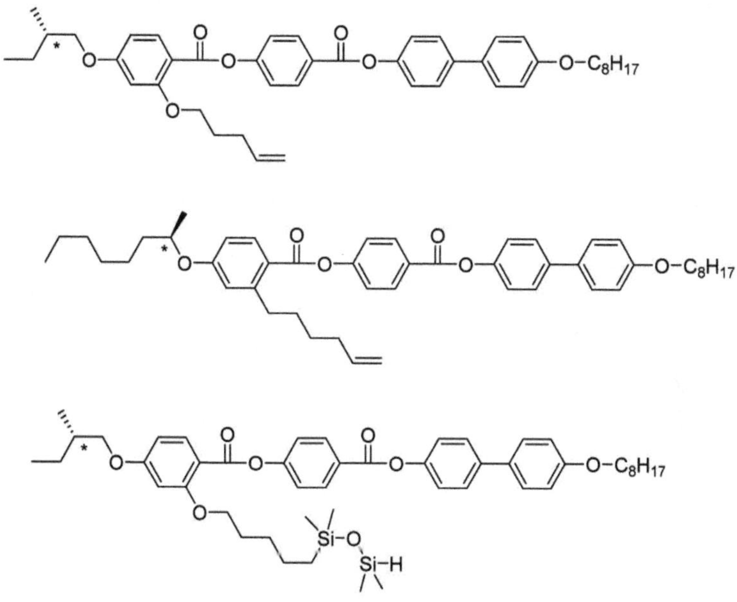

Fig. 7.8 Chiral mesogens used to functionalize Si_8O_{12} cores and create chiral nematic phases [29] and hexagonal columnar phases [30]

The effect of attaching a bent-core mesogen (Fig. 7.9) to an Si_8O_{12} core by hydrosilylation (Karstedt catalyst) at various substitution levels (3.5, 5 and 8 mesogens per Si_8O_{12}) was studied [31] by Fan, Shen, Zhou and co-workers at Beijing National Laboratory in 2008, with the aim of tailoring ferroelectric and antiferroelectric behavior. In bent-core systems, the symmetry is broken using an achiral molecule with a bent-core structure instead of using a mesogen with a chiral center (as in Fig. 7.8) [29,30,32]. Characterization by DSC and one-dimensional WAXD showed that the product carrying 8 mesogens formed two bilayer smectic C phases, while the products carrying 3.5 mesogens and 5 mesogens formed a single monolayer antiferroelectric smectic C phase.

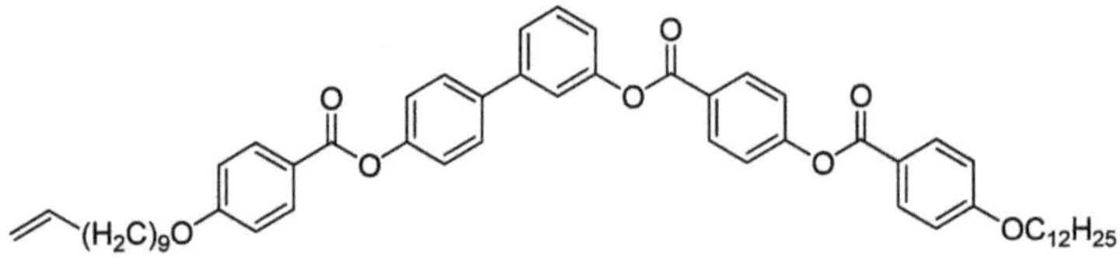

Fig. 7.9 Bent-core mesogen [31]

The discussion above covers Si_8O_{12} cores either partially or fully substituted by mesogens. When the substitution level is very low and the Si_8O_{12} core carries only one mesogen (Fig. 7.10), it may also be considered as an asymmetric molecular dyad, i.e., a dumbbell-shaped molecule in which two different entities are linked by a spacer. Dyads are of particular interest when the two components have complementary properties, e.g., they may be electron-donor acceptor pairs. In 2006 and 2007 University of Connecticut studies [33,34], the behavior of an Si_8O_{12}-discotic mesogen dyad was investigated (Fig. 7.10). This was synthesized by reaction between Si_8O_{12} $(i\text{-Bu})_7(CH_2)_3OH$ and a carboxylic acid-functionalized triphenylene (in an interesting departure from hydrosilylation chemistry), and the pure material and its 1:1 mol/mol mixture with POS were studied by DSC, two-dimensional XRD and TEM. Both compositions were nanophase-separated. The discotic triphenylene groups formed a bilayer LC lamella sandwiched between Si_8O_{12} lamellar crystals in the pure dyad, and in the 1:1 blend, the discotic triphenylene groups were interdigitated. No evidence of columnar mesophases was observed. The isotropization of the LCs and the melting of the Si_8O_{12} both occurred at 66 °C. In a more recent University of Connecticut study [35], analogous Si_8O_{12} compounds carrying eight triphenylene groups were synthesized via amidation. A number of variants were studied, with triphenylene alkyl groups of either 5 or 12 carbons, and with spacers of either 2, 6 or 10 carbons, and various columnar phases were characterized by XRD and TEM.

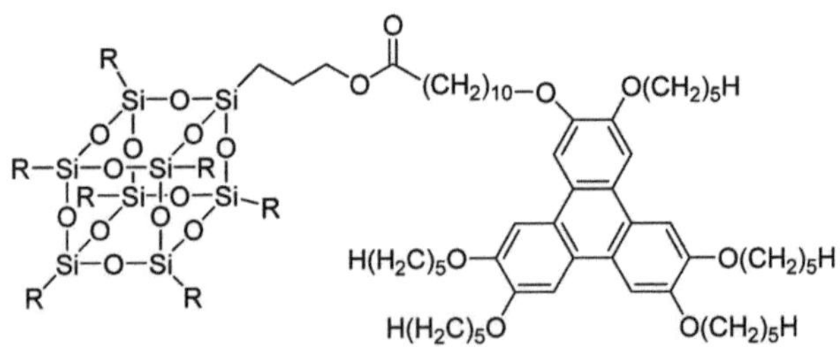

Fig. 7.10 An Si_8O_{12}-discotic mesogen dyad [33]

The materials reviewed above all have architectures where an Si_8O_{12} core carried one or more mesogens, but in an alternate approach, an Si_8O_{12} core carrying LC polymer arms was studied [36]. Atom transfer radical polymerization (ATRP) was used to grow eight arm star poly{2,5-bis[(4-methoxyphenyl)oxycarbonyl] styrene} (PMPCS) from an octafunctional Si_8O_{12} core, and the liquid crystalline phase behavior of the star polymers was studied. Si_8O_{12} $(OSiMe_2(CH_2)_3O(C{=}O)CMe_2Br)_8$ was used as an initiator, and 2,5-bis-4-methoxyphenyloxycarbonylstyrene (MPCS) was used as monomer in the presence of copper(I) bromide, an amine and chlorobenzene. This monomer was the precursor to a lateral jacketed LC polymer (Fig. 7.11). Products with PMPCS arms of various lengths (and various total molecular masses) were prepared and characterized by GPC, DSC, one-dimensional WAXD and POM. In addition, the Si_8O_{12} core was destroyed using hydrofluoric acid in order to release and characterize the arms, and to gain insight into what had occurred during the ATRP reaction (e.g., intramolecular coupling of arms and/or intermolecular coupling of stars). The T_g of the star increased with molecular mass, and plateaued at 116 °C when M_n exceeded 38,000. At M_n < 45,000, only amorphous phases were observed, but above this mass, columnar hexagonally-packed nematic phases with star-like textures were observed. Relative to their linear analogs, the star polymers required fewer repeat units per arm to stabilize the LC phase.

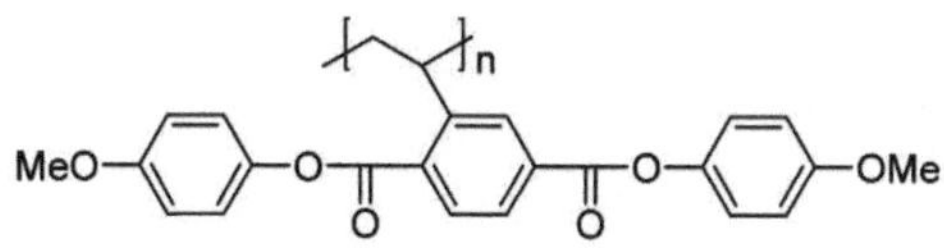

Fig. 7.11 Structure of lateral jacketed LC polystyrene derivative [36]

In contrast to the various Si_8O_{12} core species discussed above, an alternate class of LC Si_8O_{12} materials [37] are based on an architecture where conventional linear main-chain LC polymers are end-capped with Si_8O_{12} (Fig. 7.12). These materials were studied in an attempt to overcome the poor adhesion of LC polymers to metal surfaces, to other polymers and to themselves. In a 1996 study [37], a US Air Force Research Laboratory (AFRL) group reacted an $Si_8O_{12}(cyclohexyl)_7Cl$ end-capper with carboxylic acid-terminated LC polymer to give the structure shown in Fig. 7.12. DSC showed that both the starting LC polymer and the Si_8O_{12}-terminated LC variant had a T_g of ~95 °C and a nematic to isotropic transition at ~235 °C. In a 2004 Case Western Reserve University (USA) study by Schiraldi and co-workers, side-chain and main-chain LC species were combined by formulating 2.5 wt % cyanobiphenyl-functionalized Si_8O_{12} nano-additive into several main-chain LC polyesters [38]. The aim of this work was to improve the performance of the polyesters in fiber and high barrier packaging applications. Main-chain LC polymers with Si_8O_{12} fillers have also been patented as improved dielectric materials for electronic circuits [39].

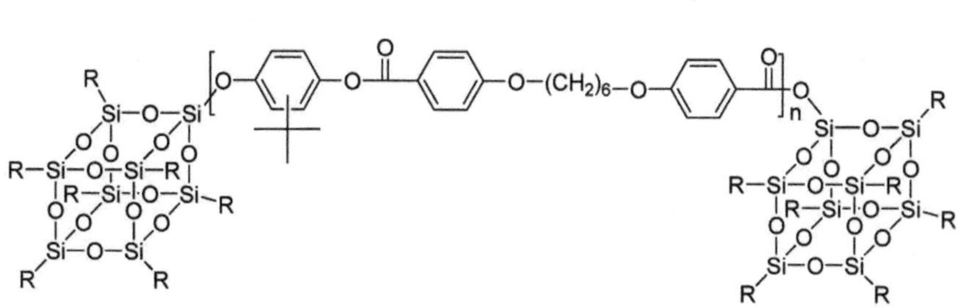

Fig. 7.12 An Si_8O_{12} end-capped main chain LC polymer [37]

In a further example of an unusual Si_8O_{12} LC architecture [40], a Kyoto University group prepared linear polymethacrylates carrying various levels of pendant Si_8O_{12} and pendant cyanobiphenyl mesogen. These polymers were prepared from mixtures of Si_8O_{12}-functionalized methacrylate monomers and cyanobiphenyl-functionalized monomers in an azobisisobutyronitrile (AIBN)-initiated free radical polymerization. The materials did not show any LC behavior until a composition of 90% cyanobiphenyl /10% POS was reached, and a smectic phase was observed by POM and DSC. Relative to the 100% pendant cyanobiphenyl polymethacrylate, the 90% cyanobiphenyl material showed improved thermal stability, and an LC phase that was stable over a wide temperature range. For cyanobiphenyl levels below 90%, no glass transitions or LC transitions were observed.

Most recently Si_8O_{12} species have been used to induce spontaneous vertical alignment in liquid crystal cells [41], thus eliminating the need for a conventional vertical alignment layer [42,43,44]. Many kinds of commercial LC display require that a nematic LC be aligned at an intermediate pre-tilt angle. A desirable decrease in switching time occurs as pre-tilt alignment angle increases from 0° to near 90°, and extremely fast response times of less than 1 ms have been achieved for some displays (e.g., no bias bend (NBB) displays). Cells were made from indium tin oxide (ITO)-coated substrates with a polyimide layer that could induce a near-horizontal pre-tilt angle of 2° for nematic LC molecules. When aminoalkyl-functionalized Si_8O_{12} was added to the LC molecule and injected into the cell, it was found that the pre-tilt angle could be controlled from 0° to 90° by varying the loading of Si_8O_{12} compound from 0 to 2 wt % [41]. This resulted from the contributions from the polyimide (imposing a pretilt angle of 2°) and from the Si_8O_{12} compound (imposing spontaneous vertical alignment after its adsorption on the inner surface of the ITO substrate). Alignment can also be achieved by doping the LC material with an azo-dye (e.g., methyl red) that adsorbs to the substrate surface when irradiated with light of an appropriate wavelength and intensity. When this approach was combined with the Si_8O_{12} approach, and both Si_8O_{12} compound and methyl red were added to the LC compound, homogenous alignments were achieved [45].

Si_8O_{12} compounds were also used to improve the response time of a holographic polymer dispersed liquid crystal (HPDLC) system [46], of importance in applications such as switchable windows, shutters and displays. HPDLCs have a periodic structure of alternating LC-rich and polymer-rich layers analogous to a diffraction grating, and can thus display volume holographic properties. They are fabricated by irradiating a mixture of LC and photosensitive monomer. A periodic composition is achieved because monomers are consumed in the irradiated areas while LCs are driven into the dark areas. Diffraction efficiency increases as the difference between the refractive index (RI) of the polymer and the LC increases, and also as the phase separation between the LC and the polymer improves. Octavinyl- Si_8O_{12} was added to a HPDLC system to obtain a fast response time, a low driving voltage, a high diffraction efficiency and a low shrinkage. SEM and AFM showed that Si_8O_{12} moieties were located at the LC-polymer interfaces.

7.2 Polyhedral Oligomeric Silsesquioxanes in Electroluminescent (EL) Materials and Light Emitting Devices (LEDs)

In 1962, blue light emission was demonstrated when anthracene was subjected to high voltages (>400 V) [47], and in 1990, a group at Cambridge [48] reported π-conjugated green light-emitting poly-*p*-phenylene-vinylene (PPV) polymers (Fig. 7.13). Some conjugated polymers were found to have electroluminescent properties [49], i.e., they emit light in the presence of an applied electric field or current, and LEDs based on low molecular mass organics (OLEDs) and polymers (PLEDs) have been widely studied [50]. Such devices comprise a cathode, an electron transport layer (ETL), an emissive layer, a hole injection layer (HIL) and an anode. Typical cathodic materials are aluminum or calcium, having suitably low work functions and low susceptiblities to oxidation. Typical anodic materials are ITO-coated glass with an ultra-high quality surface, PANI (polyaniline) and PEDOT (poly-2,3-ethylenedioxythiophene), Fig. 7.13. ITO is an important material because it is both transparent and electrically conducting. In addition, 2-(4'-*t*-butylphenyl)-5-(4'-diphenyl)-1,3,4-oxadiazole (PBD) is often used as an electron transport material and poly-N-vinylcarbazole (PVK) is often used to raise the highest occupied molecular orbital (HOMO) level in order to achieve a closer match to the HOMO level of PEDOT. Electrons come from the cathode, holes come from the anode, and electron-hole pairs form (excitons) in the emissive layer. Excitons create excited states that decay by emission of a photon. The emissive layer is of key interest, and

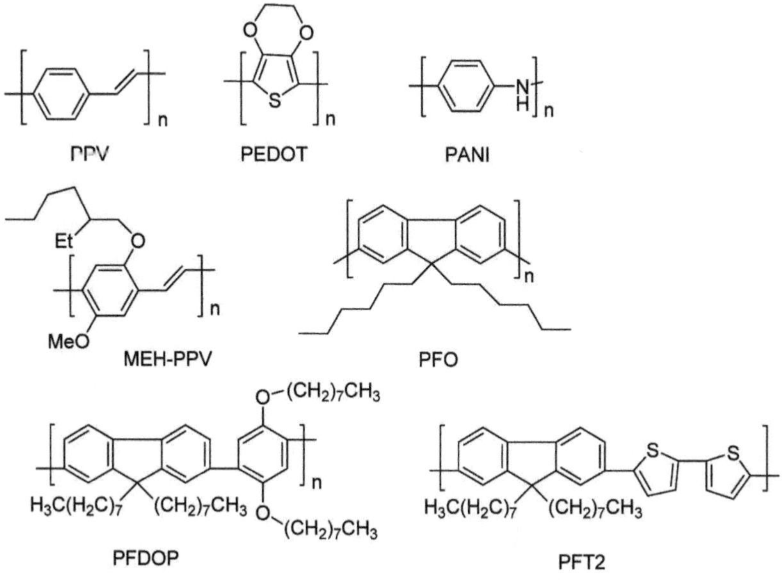

Fig. 7.13 Polymers commonly used in light emitting devices

the use of Si_8O_{12} hybrid materials to improve its performance, lifetime, stability and color balance is covered in this section [51]. Common EL polymers used in the emissive layer include poly(9,9'-dihexylfluorenyl-2,7-diyl (polyfluorene, PFO, a blue emitter [52] and poly-2-methoxy-5-[2-ethylhexyloxy]-1,4-phenylenevinylene (MEH-PPV, a yellow emitter) (Fig. 7.13). EL polymer architectures may include Si_8O_{12} as an endcapping group, as a pendant group and as a star core, and Si_8O_{12} compounds have also been introduced into these polymers as physical additives.

LEDs are currently being applied in flat display panels, and in flexible and rollable screens, leading to concepts such as 'electronic paper' (since LEDs do not require backlighting); however, coating large areas without defects in the fabrication of larger displays remains a challenge. In contrast to LCD devices, LEDs have far higher image quality, resolution and brightness (>2000 cd m^{-2}), plus better viewing angles (approaching 180°) and faster switching speeds (0.01 ms and 10 kHz refresh rate). LEDs are also used as low-power white light sources, where the emissions of two or more materials that emit at different wavelengths are combined to produce white light. White light emitting diodes are of interest for full-color displays (via color filters), as backlights for liquid crystal displays, and as light-emitting sources. Some white light emitting devices require complex fabrication one layer (i.e., one color) at a time. OLED performance parameters include wavelength (color of light emitted), luminance (a brightness greater than 10,000 cd m^{-2} is desirable), EQE (external quantum efficiency, probability that one electron will result in the emission of one photon, greater than 10% is desirable), Commision International de l'Eclairage (CIE) coordinates (relative intensities of red, green and blue emission), turn-on voltage, drive voltage and luminescence half life (greater than 10,000 hours is desirable).

Semi-conducting polymers have been widely studied for use in electroluminescent displays, solar cells, photovoltaics, sensors, thin film organic transistors, lasers and light-emitting electrochemical cells. However, the formation of aggregates, excimers and polaron pairs can degrade performance in these applications. Strategies to address this include increasing the molecular mass of the polymer, decreasing its polydispersity, introducing asymmetry (e.g., *cis*- or *meta*-linkages), or end-capping with a bulky group such as Si_8O_{12}. The introduction of Si_8O_{12} is known to stabilize color at high temperatures [53,54,55]. This is important since device temperatures can go up to 90 °C. Blue light emission has been the most problematic area in conjugated polymer organic light emitting device (OLED) research. PFO (Fig. 7.13) and its derivatives are widely used as blue light emitters because they have high solid-state quantum yields, good solubility, and good chemical and thermal stability. However PFO also has low energy emission bands resulting in the emission of undesirable blue-green light, and in quenching. Blue-green emission is caused by chain aggregation and formation of lower wavelength emitting excimers, and also by fluorenone defects. Many attempts have been made to stabilize PFO, and one approach has been to introduce Si_8O_{12} into the structure, where the enhanced thermal stability conferred by the presence of Si_8O_{12} reduces the tendency of fluorenone defects to form at higher temperatures.

7.2.1 *Polyhedral Oligomeric Silsesquioxane End-capped EL Polymers*

Heeger and co-workers [56] at the University of California (Santa Barbara) first introduced Si_8O_{12} onto the chain ends of PFO and MEH-PPV, two well-established EL polymers (Fig. 7.13), in 2003. Si_8O_{12}-endcapped MEH-PPV (DP=2000) and Si_8O_{12}-endcapped PFO (DP=500) were prepared with molecular masses in the region of 100,000, determined by both elemental analysis for silicon and by size exclusion chromatography (SEC). The Si_8O_{12}-MEH-PPV polymer was prepared by reaction between aryl chloride monomer and aryl chloride-functionalized Si_8O_{12} in the presence of potassium *t*-butoxide, and the Si_8O_{12}-PFO polymer was prepared in a Yamamoto coupling reaction between aryl bromide PFO monomer and aryl chloride functionalized Si_8O_{12} in the presence of bis(1,5-cyclooctadiene)-nickel(0) catalyst. Thermogravimetric analysis (TGA) showed that the presence of Si_8O_{12} end groups improved thermal stability by 50 °C. Solubility in organic solvents was also improved. The Si_8O_{12} end-capped materials and their Si_8O_{12}-free analogs had identical EL spectra. At 100 mA cm^{-2} in an ITO/PEDOT/polymer/Ca/Ag device, MEH-PPV- Si_8O_{12} gave a luminance of 1320 cd m^{-2} while MEH-PPV alone gave a luminance of 230 cd m^{-2}. These performances were rationalized in terms of better injection and transport of charge carriers. The Si_8O_{12} materials had improved coating quality, fewer pinholes and showed better adhesion to the substrate. The PFO - Si_8O_{12} polymer gave reduced excimer emission (i.e., the undesired blue-green color at 525 nm) relative to PFO itself. The same group filed a 2003 American Dye Sources, Inc. patent of broad scope that covered conjugated optoelectronic polymers (e.g., polypyrroles, polyacetylenes, polyphenylenes, polyphenylenevinylenes, polythiophenes, polyfluorenes and polyanilines) with bulky groups (organic or inorganic) in the polymer backbone or pendant to the backbone, with at least one silsesquioxane moiety serving as an anchor for the polymer [57].

Blue-emitting devices based on an Si_8O_{12} end-capped PFO (poly-9,9'-dioctylfluorene variant) were also reported by another group four years later [58]. The devices had an ITO/PEDOT:PSS/PVK/PFO-Si_8O_{12}/LiF/Al structure. The effect of plasma treatment and/or heat treatment upon the ITO surface was studied by atomic force microscopy (AFM). The best luminance and current density values were obtained when both plasma treatment and heat treatment were carried out (486 cd m^{-2} and 0.55 cd A^{-1}), and plasma treatment was shown to contribute more to performance than heat treatment. The presence of a PVK hole injection and transport layer resulted in a reduction of the intensity of undesirable PFO excimer emission. The ITO substrate is commonly treated with hydrogen peroxide-ammonium hydroxide-water (1:1:5 v/v) followed by ultrasonic cleaning with isopropanol and water, and polymer layers are applied by spin-coating and inorganic layers are applied by chemical vapor deposition. Further work [59] was reported on Si_8O_{12}-endcapped poly(9,9'-dioctylfluorenyl-2,7-diyl) in a study of current-voltage and luminance current characteristics of devices in which aluminum(III) tris(8-hydroxyquinoline) (Alq3) was assessed as an electron transport layer directly below an uppermost Mg:Ag/Ag layer.

7.2.2 EL Polymers with Pendant Polyhedral Oligomeric Silsesquioxane Groups

The first example of a conjugated polymer carrying a pendant Si_8O_{12} group [60] was reported in 2004 by Shim and co-workers at the Korea Advanced Institute of Science and Technology. A PFO carrying pendant Si_8O_{12} groups was synthesized in a nickel(0)-mediated Yamamoto coupling reaction using a Si_8O_{12}-functionalized monomer (Fig. 7.14, monomer **1**). The Si_8O_{12}-functionalized monomer was itself prepared in a hydrosilylation reaction between Si_8O_{12}(cyclopentyl)$_7$H and a diallyl-functionalized aryl bromide monomer. Poly-9,9'-dihexylfluorene-poly-9,9'-di- Si_8O_{12}-fluorene copolymers of various compositions (1, 2, 5, 10 and 20% Si_8O_{12} units) were then prepared. They have good solubility in organic solvents and could be readily spin-coated into high quality thin films. Photoluminescene spectroscopy on polymer-coated quartz substrates showed reduced aggregation and excimer formation, resulting from the large Si_8O_{12} group causing steric hindrance and reducing inter-chain interactions. Higher quantum yields and improved stability after thermal annealing (up to 150 °C) relative to PFO were also observed. A stable blue-emitting device was fabricated with the layer structure: ITO/PEDOT-PSS (40 nm)/pendant Si_8O_{12}-polyfluorene 20% (80 nm)/Ca(500 nm)/Al(800 nm). No green emission (500-600 nm) was observed. The HOMO and LUMO energy levels of the polymers were determined by cyclic voltammetry, and it was shown that the presence of Si_8O_{12} did not affect the redox properties of the polymers. In this study [60] each C9 position carried two Si_8O_{12} groups (Fig. 7.14). In a study by the same group published the following year, PFOs with one Si_8O_{12} group and one hexyl group at the C9 position were synthesized using an analogous synthetic route, and copolymers containing 5, 10 and 20% Si_8O_{12}-carrying repeat units were prepared [50]. In another 2005 study [61], the Shim group attached pendant Si_8O_{12} groups to photoluminescent blue-emitting poly-9,9'-dioctyl-2,7-fluorene-*alt*-2,5-bis(octyloxy)-1,4-phenylene (PFDOP) and poly-9,9'-dioctylfluorene-*alt*-bithiophene (PFT2) polymers (Fig. 7.13) in palladium-catalyzed Suzuki coupling reactions using Si_8O_{12}-functionalized aryl bromide monomers [61]. Once again, chain aggregation decreased, quantum yields were enhanced, thermal color stability increased and brighter devices were fabricated.

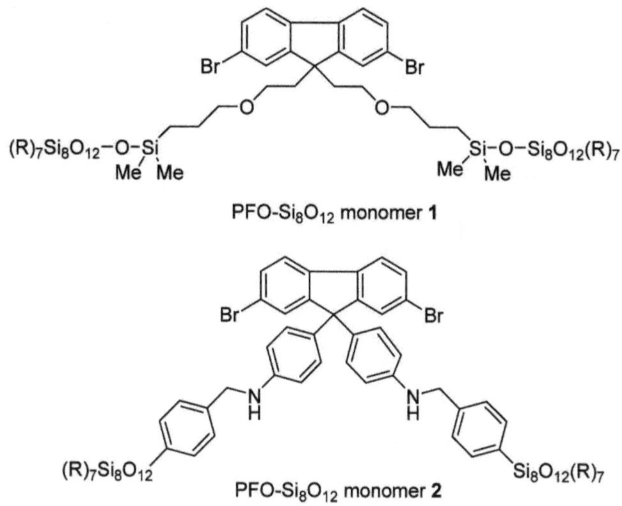

Fig. 7.14 PFO- Si_8O_{12} monomer **1** [60] and PFO- Si_8O_{12} monomer **2** [55]

Six months after the 2004 Shim study discussed above, an alternative PFO with C9-pendant Si_8O_{12} groups (Fig. 7.14, monomer **2**) was reported by Wei and co-workers [55] at National Chiao Tung University (Taiwan); hence both the Shim and Wei materials have been claimed as the first pendant- Si_8O_{12} PFOs. The Wei polymer was prepared in a palladium-catalyzed Suzuki coupling, where $Pd(PPh_3)_4$ and Si_8O_{12}-functionalized aryl bromide monomers were used. Once again, the presence of Si_8O_{12} reduced inter-chain aggregation and prevented the formation of keto defects. In PFO, the C9 position is most readily modified to enhance solubility in organic solvents. In this structure (Fig. 7.14), undesirable C9-benzyl- Si_8O_{12} links prone to photoxidation are avoided. Molecular mass decreased as the proportion of POS-functionalized monomer in the polymer increased (M_n = 27,000 for PFO versus M_n=12,000 at 10% pendant- Si_8O_{12} PFO copolymer). This was attributed to steric hindrance during polymerization. PFO had a T_g of 62 °C while no T_g was observed for the 10% pendant- Si_8O_{12} copolymer. The photoluminescence maxima for PFO and 10% pendant- Si_8O_{12} PFO copolymer were the same, but the quantum yield for 10% Si_8O_{12} copolymer was 54% higher than that of pure PFO.

The following year, Wei and co-workers carried out a second study [54] comparing a pendant- Si_8O_{12} EL polymer with the Si_8O_{12}-free EL polymer. In this case, pure MEH-PPV (Fig. 7.13) was compared with pendant-POS-PPV-*co*-MEH-PPV, a copolymer of poly-*p*-phenylenevinylene with pendant Si_8O_{12} units and MEH-PPV. The former had a quantum yield of 0.19 while the latter had a quantum yield of 0.85. With 10% Si_8O_{12}-PPV content, an ITO/PEDOT/polymer/Ca/Al device was five times brighter than the MEH-PPV control device (2196 cd m^{-2} versus 473 cd m^{-2}). As before, Si_8O_{12} was introduced by using an Si_8O_{12}-functionalized monomer, but on this occasion a Gilch polymerization route was followed, using benzyl bromide monomers and potassium *t*-butoxide. Introduction of 10% Si_8O_{12} units lowered M_n from 61,000 (pure MEH-PPV) to 39,000. Both MEH-PPV and the 10% pendant- Si_8O_{12} copolymer resulted in devices with a yellow EL emission peak at 590 nm.

In 2005, Takagi and co-workers (Nagoya Institute of Technology, Japan) reported another PFO variant carrying pendant Si_8O_{12} groups [62], where the C9 position carried two Ph-O-Ph-CH_2CH_2- Si_8O_{12}(cyclopentyl)$_7$ groups, and the polymer was prepared in another palladium-catalyzed Suzuki coupling reaction. These polymers were compared to PFO variants carrying C9 trimethylsilyl groups. The Si_8O_{12} polymers maintained strong blue emission even after thermal annealing, whereas the trimethylsilyl polymers showed an undesirable green emission that became stronger after annealing. In 2006, Fan, Huang and co-workers (Fudan University, Shanghai) made hybrid Si_8O_{12}-tethered polyfluorenylene-ethynylene (PFE) materials [63] in Sonagashira coupling reactions using di-bromo-fluorene monomers carrying two pendant Si_8O_{12} groups (as per Fig. 7.14). WAXD showed that introduction of Si_8O_{12} increased the interchain distance, and quantum yields and thermal stability were improved relative to the Si_8O_{12}-free PFE analog.

Recently, the introduction of Si_8O_{12} into a hyperbranched EL polymer architecture [64] has also been reported to improve electroluminescence, photoluminescence quantum yield and thermal stability relative to a linear Si_8O_{12}-free EL polymer analog. Iron(III) chloride oxidative polymerization was used to combine 4,7-bis(3-ethylhexyl-2-thienyl)-2,1,3-benzothiadiazole and octa(3-ethylhexyl-2-thienyl-phenyl)- Si_8O_{12} monomers to give a hyperbranched architecture, and both the linear and HB polymers in this study gave an EL maxiumum at 660 nm.

In many of the nickel (Yamamoto) and palladium (Heck or Suzuki) coupling chemistries above, it is worth noting the importance of removing any residual metal catalyst, since metal residues are detrimental to OLED performance; in this respect, small organics that can be purified by flash column chromatography are advantageous (see later sections).

7.2.3 EL Star Architectures with Polyhedral Oligomeric Silsesquioxane Cores

A number of electroluminescent star species with Si_8O_{12} cores have been described in the literature, and the use of this architceutre in EL, optoelectronic and photovoltaic applications has recently been reviewed [65]. In 2001 the Laine group at the University of Michigan filed a patent claiming hybrid LED materials where hole transport groups, electron transport groups or emissive moieties were attached to an Si_8O_{12} core [66], and in 2004 Jabbour (Arizona State University) and co-inventors filed a patent [67] claiming one or more luminophores (e.g., 2,7-bis-(2,2-diphenylvinyl)fluorene for blue emission or pyran-4-ylidene malononitrile for red emission) attached to a nanoparticle core (e.g., Si_8O_{12}) for the purpose of achieving white light-emitting devices. In this section, Si_8O_{12} cores functionalized with EL polymers and Si_8O_{12} cores functionalized with small aromatics are reviewed.

In 2004, Chen and co-workers at National Taiwan University (Taipei) synthesized a star polymer with an Si_8O_{12} core and PFO arms in an attempt to produce an EL material with higher Si_8O_{12} content than a Si_8O_{12} end-capped linear EL polymer [53]. High molecular mass is desirable in a linear EL polymer, but such polymers only have 1 wt % POS content [56]. However, when PFO is attached to the eight corners of POS in a star architecture, 3.8 wt % content is possible. Another Yamamoto bis-1,5-cyclooctadiene-nickel(0)-catalyzed coupling reaction was used, this time between octa(2-(4-bromophenyl)ethyl)octasilsesquioxane ($POS(CH_2CH_2PhBr)_8$) and aryl bromide-terminated polydioctylfluorene [53]. When used in a device, the star polymer had a quantum yield and a maximum luminescence intensity twice as large as the pure PFO control (5430 cd m^{-2} versus 2510 cd m^{-2}). Since both the linear polyfluorene and the octa-functional Si_8O_{12} carried aryl bromide functionality, the structure of the resulting material was ambiguous (i.e., the number of arms per Si_8O_{12} and the molecular mass of the arms were undefined); however approximately 75% of the crude product was soluble, making it possible for devices to be fabricated by spin-coating the EL layer from xylene.

In 2008, a University of Nantes (France) group made green-emitting hybrid Si_8O_{12}-DP-PPV materials [68]. An ITO/PEDOT:PSS/Si_8O_{12}-DP-PPV/Ca/Al device was fabricated [69], where Si_8O_{12}-DP-PPV was poly-2,3-diphenyl-1,4-phenylenevinylene (Fig. 7.13) grown from an Si_8O_{12} core. This was then studied by charge-based deep-level transient spectroscopy (Q-DLTS), a technique for observing defect states and traps in LED devices. Traps associated with the inorganic or the organic parts of the hybrid Si_8O_{12}-DP-PPV material were assigned on the basis of their capture cross sections.

However, the majority of Si_8O_{12} star species are comprised of an Si_8O_{12} core functionalized with small organic emitters rather than polymeric arms [53]. Since free aromatics have a tendency to assemble into π-stacks (causing excimer emission, and impacting electronic and photonic performance), the original rationale for producing Si_8O_{12}-aromatic stars was to keep the aromatics apart, and to prevent aggregation [70,71,72]. In 2004, He and co-workers (Institute of Materials Research and Engineering, Singapore) [70] claimed the synthesis of the first luminescent organic clusters displaying the same properties as inorganic quantum dots (Fig. 7.15, top left and top right), but without the disadvantage of dot size distribution resulting in line-broadening. These were synthesized from $Si_8O_{12}(PhBr)_8$ and the appropriate brominated aromatics in palladium-catalyzed coupling reactions, and thin films of the Si_8O_{12} compounds showed enhanced photoluminescence (PL) quantum yields relative to the analogous free aromatics. Si_8O_{12} was selected as a scaffold because of its thermal stability, low dielectric constant (~2.7) and its high band gap (6.0 eV absorption and 4.2 eV emission). Later quantum dot compounds [73] incorporating longer sequences of phenyl groups (Fig. 7.15, bottom left) also showed enhanced PL quantum efficiency and improved film forming properties. In 2005, the Laine group at University of Michigan reacted bromophenyl- Si_8O_{12} with various arylboronic acids, $(HO)_2B$-Ar, in $Pd(PPh_3)_4$-catalyzed coupling reactions [71]. In this way, phenyl, biphenyl, naphthyl, thiophene and 9,9-dimethylfluorene were attached to Si_8O_{12} cores. MALDI-TOF spectra showed that these compounds had a distrubtion of compositions corresponding to the distribution (varying numbers of bromine atoms) in the bromophenyl- Si_8O_{12} starting materials. The absorption and emission spectra of these materials showed them to behave as conventional organic aromatics rather than quantum dots [70], and this was attributed to the distribution of compositions, although the distribution of compositions of $Si_8O_{12}(PhBr)_8$, or possible lack thereof, was not reported in He's 2004 study [70]. The effect of an Si_8O_{12} substituent on the absorption and emission wavelengths of the aromatic was unclear in the 2005 University of Michigan study [72], with a red shift being reported for the aromatics above, but a blue shift being reported for stilbene. Recently a Korean group synthesized an EL Si_8O_{12} material carrying eight anthracenenaphthyl groups via hydrosilylation between $Si_8O_{12}(OSiMe_2H)_8$ and the allyl-functionalized aromatic, and used the material to fabricate a blue-emitting device [74].

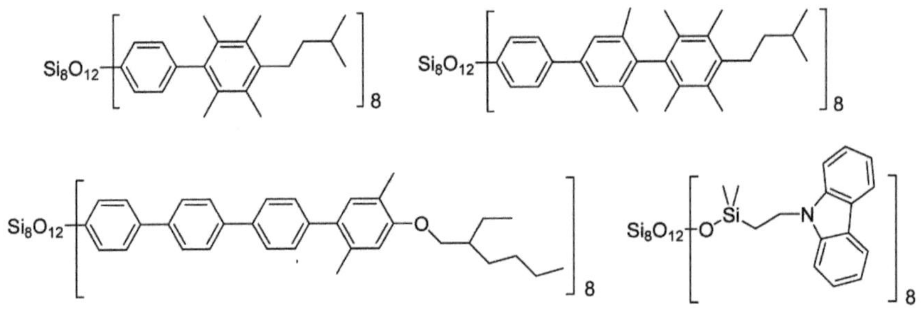

Fig. 7.15 First Si_8O_{12} 'quantum dot' compounds (top) [70], a later dot (bottom left) [73], and a carbazole-functionalized Si_8O_{12} (bottom right) [75]

Another early example of a small molecule star species with an Si_8O_{12} core [76] was synthesized by the Laine group at the University of Michigan in 2005. In contrast to the Si_8O_{12}-PFO star species above [53] that carried oligomeric PFO arms, in this material a small organic aromatic hole transport compound, FL03 (*N*4-(9,9-dimethyl-9*H*-fluoren-7-yl)-*N*4'-(9,9-dimethyl-9*H*-fluoren-2-yl)-*N*4,*N*4'-diphenylbiphenyl-4,4'-diamine), was coupled to octavinyl- Si_8O_{12} in a Heck reaction with a monobromo-FL03 derivative and a $Pd(P(t\text{-}Bu)_3)$ catalyst (Fig. 7.16, top). Small organics can be obtained with greater purity than polymers, but require vacuum deposition techniques to fabricate displays. Vacuum deposition is expensive and is only practical for small displays. Polymers can be deposited from solution using efficient and industrially scalable techniques such as ink-jet printing to form large surface area displays, but are less pure than small organics. Hence the rationale of this study was to achieve a solution-processable but high purity (monodisperse) OLED material by attaching a small organic OLED entity to an Si_8O_{12} core via a rigid link. Flexible links have been used to attach mesogens to Si_8O_{12} cores in liquid crystalline materials [15,27,30] (as discussed in the previous section) but crystalline domains are undesirable in OLED materials because they result in charge trapping.

The resulting Si_8O_{12}-FL03 star carried 3 to 10 FL03 units per Si_8O_{12} (i.e., some double Heck reactions took place), and exhibited intense blue photoluminescent emission at ~430 nm. When devices fabricated using Si_8O_{12}-FL03 were compared to devices fabricated using pure FL03, the Si_8O_{12}-FL03 devices were brighter and had higher external quantum efficiencies. In another 2005 study [75] by Kawakami and Imae (Japan Advanced Institute of Science and Technology, Ishikawa), Si_8O_{12} was functionalized with carbazole (a photoactive and electroactive π-chromophore, Fig. 7.15, bottom right) in a hydrosilylation reaction between octahydrido- Si_8O_{12} and *N*-vinyl carbazole. Elemental analysis showed it to be a pure octa-substituted compound, in contrast to the mixed compositions above [76]. The Si_8O_{12}-carbazole compound showed a single strong emission peak in the solid state, and no excimer emission.

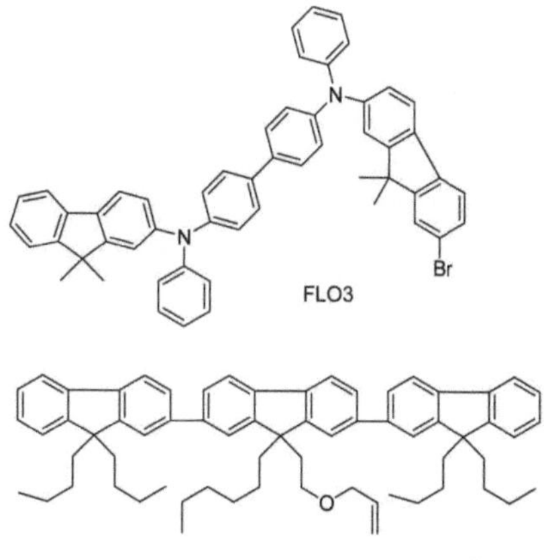

Fig. 7.16 FL03 reagent [76] and terfluorene chromophore [77] used to prepare POS-cored star species

In 2006, Shim and co-workers [77] at the Korea Advanced Institute of Science and Technology synthesized a blue-emitting Si_8O_{12} with terfluorene (FL3) chromophores on each of the eight corners (Fig. 7.16, bottom). This Si_8O_{12}-FL3 star was then doped into PFO. Si_8O_{12}-FL3 was prepared in a platinum-catalyzed hydrosilylation reaction between octakis(dimethylsiloxy)- Si_8O_{12} and allyl-functionalized terfluorene (Fig. 7.16, bottom) which itself required a six step synthesis [77]. This compound also showed good solubility and thin film-forming abilities. The maximum blue emission wavelength of Si_8O_{12}-FL3 (394 nm in THF solution) overlaps well with the maximum absorption wavelength of PFO (391 nm in THF solution), making it suitable for use as a dopant for blue light emitting polymers, and increasing quantum efficiency via energy transfer. This also applies in the solid state. EL devices (analogous to those described above) fabricated from Si_8O_{12}-FL3-doped polyfluorenes showed a four- to eight-fold increase in quantum efficiency relative to the non-doped PFO EL device. This was explained in terms of Si_8O_{12} disruption of terfluorene aggregation in this system. It is notable that an energy transfer from a shorter-wavelength blue emissive material to a longer-wavelength blue emissive material is being reported.

By 2007, a light-emitting pyrene-functionalized Si_8O_{12} had been reported with an external quantum efficiency (EQE) of 2.6% [78]. In addition an Si_8O_{12} core had been functionalized with two different emitters in a controlled ratio, and multiple-wavelength emitting Si_8O_{12} materials had been synthesized [79]. Si_8O_{12} cores functionalized with a combination of hole transport groups, electron transport groups and at least one cross-linkable group, and light-emitting devices incorporating these Si_8O_{12} compounds were also claimed in a 2007 patent [80].

7.2.4 Polyhedral Oligomeric Silsesquioxane Iridium Complexes

An important subset of EL Si_8O_{12}-cored star species are those functionalized with iridium complex emitters. Heavy metal ions such as osmium, iridium and platinum are used in phosphorescent LEDs because they facilitate singlet to triplet intersystem crossing (ISC) via spin-orbit coupling, and can harvest singlet and triplet excited states. Cyclometalated iridium complexes have been used to produce high efficiency systems where wavelength tunability across the visible spectrum has been achieved by varying the cyclometalating ligand.

In 2009, Jabbour (Arizona State University), Mochizuki (Nitto Denko Corporation) and co-workers reported a system where carbazole hole transporters and iridium complexes (Fig. 7.17) were attached to the same Si_8O_{12} scaffold in hydrosilylation reactions between vinyl-functionalized iridium complexes and SiH-functionalized Si_8O_{12} species [81]. Two series of materials were prepared. In the first series, monochromatic emitters were based on an Si_8O_{12} core carrying seven cyclopentyl groups and one iridium complex (Fig. 7.17). White-emitting devices were then fabricated by combining three Si_8O_{12} (cyclopentyl)$_7$(iridium) compounds of this type, carrying the three different iridium complexes, Ir-(fppy)$_2$- Si_8O_{12} (blue, 470 nm), Ir-(ppy)$_2$- Si_8O_{12} (green, 500 nm) and Ir-(pq)$_2$- Si_8O_{12} (red, 587 nm). In the second series, emitters based on an Si_8O_{12} core carrying one iridium complex and seven carbazole groups were prepared by stepwise stoichiometric hydrosilylation with a one equivalent of vinyl-functionalized iridium complex followed by seven equivalents of *N*-allylcarbazole. A distribution of products was obtained, but these could be separated chromatographically. The mononchromatic OLEDs had EQEs of 5 to 9% and the white light-emitting OLEDs had EQE's of 8%. In 2009, the same group [82] synthesized an electroluminescent Si_8O_{12} star compound carrying an iridium complex (Fig. 7.18) and used it in an ink-jet fabrication of an LED device that had a peak luminance of 10,000 cd m^{-2} and a peak quantum efficiency of 2.5%.

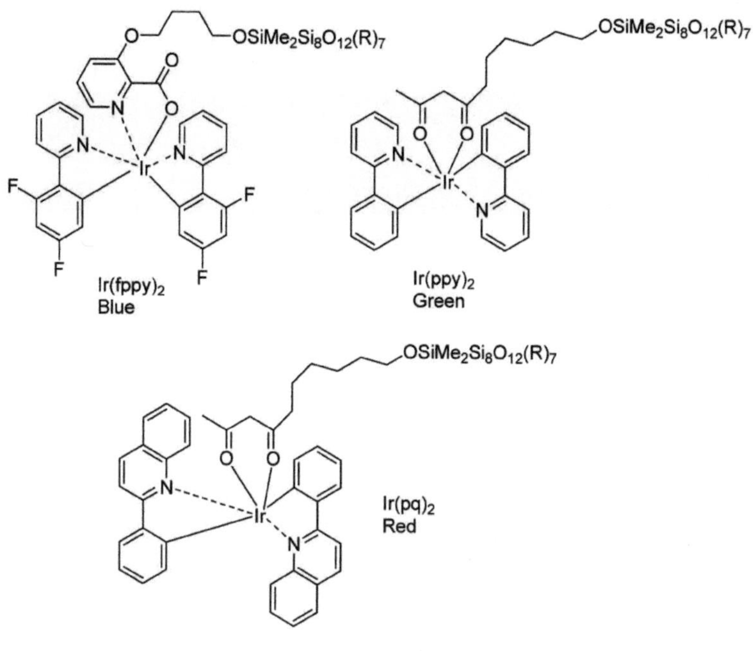

Fig. 7.17 Si_8O_{12} functionalized with various iridium complex emitters [81]

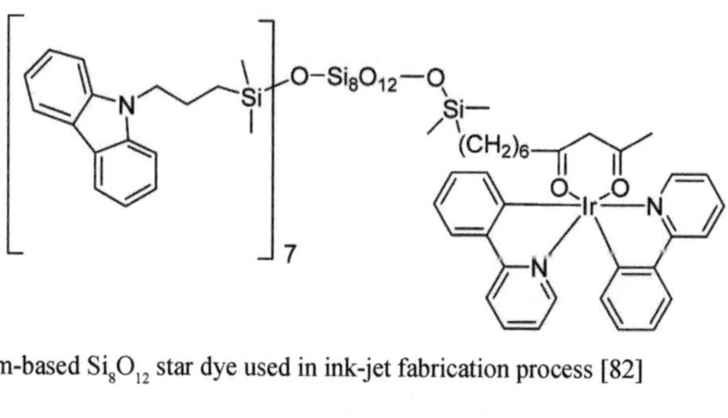

Fig. 7.18 Iridium-based Si_8O_{12} star dye used in ink-jet fabrication process [82]

In an earlier 2006 study, Hsu and co-workers (National Chiao Tung University, Taiwan) reported an Si_8O_{12} core functionalized with iridium(III) bis(2-phenylpyridine-C2-N')(13-terdecenylacetonate) had been added to CBP (4,4-*N,N'*-dicarbazole-biphenyl) and TPBI (1,3,5-tris-2-*N*-phenylbenzimidazoylbenzene), and used in an OLED device [83]. Three Si_8O_{12}-cored star light-emitting materials were prepared by reacting octasilane-POS with three allyl-functionalized conjugated light-emitting moieties (Fig. 7.19) in hydrosilylation reactions. SEC data suggested that five to eight chromophores were attached to the Si_8O_{12} core. All three materials had good solubility in organic solvents and could be spin-coated to form high quality films, and to fabricate devices (Table 7.2). The triplet Si_8O_{12}-**3** material (Fig. 7.19) was blended with a 4,4-*N,N'*-dicarbazole-biphenyl (CBP) host to prevent quenching.

Table 7.2 Characteristics of various devices incorporating Si_8O_{12}-star EL species [83]

Device	Color / λ^{EL}_{max} (nm)	Maximum brightness (cd m^{-2})	Maximum current yield (cd A^{-1})
ITO/PEDOT/ Si_8O_{12}-**1**/Ca/Al	496 (Green)	115	0.07
ITO/PEDOT/ Si_8O_{12}-**1**, 0.8 wt % PBD/Ca/Al	496	1469	0.8
Si_8O_{12}-**2**	476 (Blue-green)	70	0.02
Si_8O_{12}-**2** + 0.8 wt % PBD	476	1102	0.88
Si_8O_{12}-**3**	524 (Green)	1008	1.04
Si_8O_{12}-**3** + 18 wt % CBP	524	1172	3.99

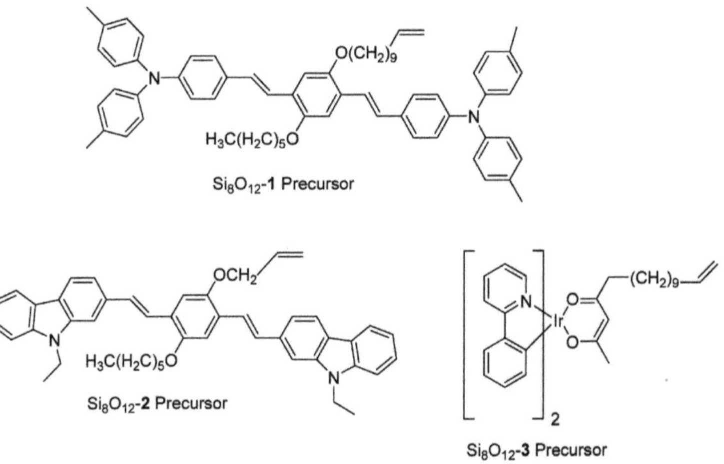

Fig. 7.19 Allyl precursors used to prepare Si_8O_{12}-cored stars [83]

In 2005, Peng and co-workers at South China University of Technology (Guangzhou) co-doped a blue emitting Si_8O_{12}-terminated PFO host (EL at 424 nm) with two phosphorescent iridium complex guests, green-emitting $Ir(Bu\text{-}ppy)_3$ (EL at 510 nm, where Bu-ppy denotes 2-4'-*t*-butylphenylpyridine) and red-emitting $(Piq)_2Ir(acaF)$ (EL at 610 nm, where acaF denotes 1-trifluoro acetylacetonate) in order to fabricate a white light-emitting diode [84]. The EL peaks appeared at slightly different wavelengths than those observed in the EL spectra of the individual emitters. In 2006, in another South China University of Technology study [85], the iridium complex $Ir(1\text{-}piq)_2pt$ was synthesized, where 1-piq denotes 1-phenylisoquinoline and pt denotes 3-(pyridin-2'-yl)-1*H*-1,2,4-triazole. This complex was added to an Si_8O_{12}-terminated PFO host polymer, and an ITO/PEDOT/

PVK/2% Ir complex-PFO- Si_8O_{12}-PBD/Ba/Al electrophosphorescent device was fabricated that had red emission at 605 nm, a quantum efficiency of 10.4% and a luminous efficiency of 9.4 cd A^{-1} at 10.8 mA cm^{-2}.

It is also interesting to note that luminescent iridium-Si_8O_{12} compounds have been used in oxygen sensing (see also discussion of sensor applications in Section 7.5). An Si_8O_{12} core carrying eight $Ir(ppy)_2$(cs-acac) groups was synthesized via reaction of the cs-acac (non-8-ene-2,4-dione) alkene groups, and different luminescent responses in the presence of nitrogen and oxygen were observed [86].

7.2.5 Physical Blending of Polyhedral Oligomeric Silsesquioxanes into EL Polymers

Si_8O_{12} has also been introduced into the EL layers of devices by physical blending (rather than by chemical attachment to EL materials, as per Sections 7.2.1 to 7.2.4), following work where silica or titanium dioxide nanoparticles were introduced into PPV films in OLEDs [87,88,89]. Physical blending has the advantage of being simple relative to the chemical introduction of Si_8O_{12}, where multistep synthetic and purification procedures are often required. In a 2006 study by Gong and co-workers at the University of California at Santa Barbara [90], Si_8O_{12} carrying one chloro group and seven cyclohexyl groups was physically blended into MEH-PPV at 0.5 wt %, and the resulting ITO/PEDOT:PSS/MEH-PPV:POS/Al device had a luminous efficiency of 1 cd A^{-1}. The improvement in performance was attributed to enhanced electron injection into the emissive layer associated with ionic conductivity contributed by Si_8O_{12}, although the identity of the ions and mechanism of conduction were not clearly defined. An ITO/PEDOT:PSS/MEH-PPV- Si_8O_{12}/Al device was used as a control, where Si_8O_{12} was present as a chemically bound end group on MEH-PPV rather than as a physical additive.

In a later 2007 study [91] by Lee and Lai at National Yunlin University (Taiwan), $Si_8O_{12}(OSiMe_2CH_2CH_2Ph)_8$ was physically blended with MEH-PPV at 5, 10 and 30 wt % Si_8O_{12} to fabricate composite films that were characterized by AFM and evaluated in devices (Table 7.3). It was shown that POS and MEH-PPV were highly compatible, and that surface roughness increased with increasing Si_8O_{12} content. Thus the interfacial area between the light-emitting layer and the cathode was increased, resulting in enhanced electron injection from the cathode, increased electron-current density and improved current efficiency. The presence of Si_8O_{12} also resulted in deceased hole-current density, showing that Si_8O_{12} compounds functioned as hole-trappers. The EL emission wavelength at 590 nm remained constant with varying Si_8O_{12} content, but intensity decreased with increasing Si_8O_{12} content. The intensity of excimer emission at 630 nm decreased with increasing Si_8O_{12} content. At high Si_8O_{12} contents thickness decreased and current leakage increased.

Table 7.3 Characteristics of various devices incorporating physically blended Si_8O_{12} species in the EL layer [91]

Device	Color / λ^{EL}_{max} (nm)	Maximum brightness (cd m^{-2})	Maximum current yield (cd A^{-1})
ITO/PEDOT/MEH-PPV 0% Si_8O_{12}/Ca/Al	593	2946	0.78
MEH-PPV 5% Si_8O_{12}	592	2376	0.83
MEH-PPV 10% Si_8O_{12}	592	1962	1.06
MEH-PPV-30% Si_8O_{12}	590	1693	1.10

Hence the use of Si_8O_{12} to improve the stability and electroluminescent properties of LEDs has become so well established that Si_8O_{12} materials are used as the default when other materials and variables associated with LED performance are being studied [59,69,85], and Si_8O_{12}-terminated PFO is now so successful as a blue EL material that it is commercially available from companies such as American Dye Sources, Inc. (Quebec, Canada).

7.3 Polyhedral Oligomeric Silsesquioxanes in Non-linear Optic (NLO), Optical Limiting (OL) and Laser Applications

Despite the extensive work done on Si_8O_{12} in LC and EL applications (Sections 7.1 and 7.2), very little work has been done in the related area of non-linear optical (NLO) materials. Sol-gel techniques have been used to prepare NLO hybrid materials comprised of an NLO component, e.g., an azo dye [92,93], a Disperse Orange 3 chromophore [94,95], nitroanilines [96] or *N*-4-nitrophenol-L-prolinol [97], and either a silsesquioxane or TEOS-derived silica component [98,99], with the aim of maintaining NLO properties while improving thermal post-poling orientation stability. Poling is carried out by exposing the material to an electric field (2-5 kV cm^{-1}), and the dipoles of the chromophores align in the direction of the field. In hybrid silsesquioxane-organic materials, refractive index and optical loss can be reduced with increasing silsesquioxane content [100]. A ladderlike polysilsesquioxane carrying stilbene chromophores (Me_2N-Ph-CH=CH-Ph-NO_2) was poled in-situ during film formation, and the film was found to have more stable poling-induced orientation than the analogous linear polysiloxane [101]. Stilbene chromophores were attached side-on (at a C=C group) or end-on by hydrosilylation with polyhydridosilsesquioxane. The less mobile the orientated NLO groups are, the greater the thermal stability of the second order non-linear response will be. This has been achieved by poling before or during cross-linking and then stabilizing the orientation after complete cross-linking has occurred. Hence sol-gel chemistry offers one route to cross-linking and 'fixing' chromophores in position.

Another area of increasing importance is that of optical limiting (OL) materials [102,103], a class of NLO materials that are transparent under normal conditions, but become opaque in response to high energy laser pulses. These materials are useful for protecting the human eye, and also various optical sensors, from laser damage, and exhibit a decrease in transmission when the energy of an incident laser pulse increases above a certain threshold. The ideal OL material responds rapidly over a wide range of wavelengths and has a high damage threshold. Structures with π-conjgation (e.g., carbon nanotubes, fullerenes, porphyrins, phthalocyanines, various organic dyes, chromophores and organometallics) may be used as optical limiting (OL) materials [104]. However many such materials have poor solubility, processability and low thermal stability, issues that are also encountered in the EL polymer field (as discussed in Section 7.2). A series of inorganic Si_8O_{12}-organic (azobenzene) hybrid materials was prepared in order to achieve good OL properties in combination with improved solubility in organic solvents and thermal stability [105]. Three different target material architectures were prepared in hydrosilylation reactions between the di-functional azobenzene monomer and octahydrido- Si_8O_{12}. The monomer was prepared in a four-step convergent synthesis. The first was a dumbbell molecule prepared by reaction between two equivalents of $Si_8O_{12}(H)_8$ and one equivalent of monomer. The second was a linear 'bead' polymer with POS units ('beads') within the backbone and pendant azobenzene groups, prepared using a one to one molar ratio of $Si_8O_{12}(H)_8$ and monomer. The third was a network of Si_8O_{12} units connected by spacers with pendant azobenzene groups, prepared using a two to one molar ratio of monomer to $Si_8O_{12}(H)_8$. Platinum dicyclopentadiene catalyst gave soluble products in moderate yields, while platinum divinyltetramethyldisiloxane catalyst (Karstedt's catalyst) gave cross-linked gels. For all products, mixtures of α- and β-adducts were obtained, and the highest proportions of β-adducts were observed for the highest monomer molar ratios. The products were characterized by GPC, IR, ^1H and ^{29}Si NMR. Incorporation of the chromophores into the various Si_8O_{12} architectures did not change their UV-visible spectra or impact their OL performance. TGA showed that the hybrids decomposed at temperatures 65 to 85 °C higher than the free chromophores (5 wt % loss). Interestingly the 'bead' materials had higher thermal stability than the dumbbells or the networks, and this was attributed to their more ordered structures. OL performance was studied using 13 ns optical pulses at 532 nm with a repetition of 1 Hz from a frequency doubled Q-switched mode locked ns/ps Nd:YAG laser, and incident and transmitted laser pulses from samples in solution in a quartz cell were monitored. Non-linear optical properties were measured using a Z-scan technique and the same laser system and sample solutions. This showed that the nitro materials had non-linear absorption and non-linear refraction, while the H and methoxy materials only had non-linear refraction. Recently the same Donghua University (Shanghai) group reported an Si_8O_{12} core functionalized with eight stilbene groups that was prepared in a hydrosilylation of octahydrido- Si_8O_{12} [106]. When compared with stilbene control OL materials, the hybrid material had comparable OL performance, but had improved thermal stability, with thermal decompositions temperatures 20 to 60 °C higher than the stilbenes.

Another example of a covalently bonded Si_8O_{12}-conjugated aromatic system used to make improved OL materials is Si_8O_{12}-metallophthalocyanine, with either cobalt(II), copper(II) or zinc(II) [107]. Two equivalents of mercaptopropyl-Si_8O_{12}(*i*-Bu)$_7$ were reacted with 4,5-dichloro-1,2-dicyanobenzene in the presence of potassium carbonate as base, in a nucleophilic aromatic substitution at the two aryl chloride sites. The product was used to synthesize the three metallophthalocyanines. Tetramerization to form the planar phthalocyanines resulted in structures carrying eight peripheral Si_8O_{12} units. The copper phthalocyanine showed significant non-linear absorption in a Z-scan measurement and promising OL performance, while the cobalt and zinc phthalocyanines showed no non-linear absorption.

Polyhedral oligomeric silsesquioxane compounds have also been used to improve the performance of laser dyes. When Si_8O_{12}(methylmethacryl)$_8$ was added to an ethyl acetate solution of the laser dye PM567, the laser action was significantly enhanced [108].This was attributed to weak optical scattering from the Si_8O_{12} compound (despite its nanometer dimensions), resulting in a phenomenon known as 'lasing with intensity feedback'. Hence the addition of the Si_8O_{12} compound at 1 wt % was reported to improve the dye laser efficiency by 65%. Similar enhancements in laser performance were seen when the same Si_8O_{12} structure was present in PM567-doped polymethyl methacrylate (PMMA) with no degradation in laser output after 100,000 pump pulses [109]. Si_8O_{12}(methylmethactyl)$_8$ and Si_8O_{12}(*i*-Bu)$_8$ were also found to enhance the laser action of other laser dyes such as Rhodamine 6G, Rhodamine 640, Sulforhodamine B and Perilene Red.

7.4 Polyhedral Oligomeric Silsesquioxanes in Lithographic Applications

Lithography has much in common with photography; a pattern is formed by exposing a chemically reactive surface to radiation, and then the pattern is fixed and stabilized in a series of additional processes [110]. The use of a polymeric resist is the most important method for the fabrication of patterns on the nano-scale, although other approaches besides nanolithography have been reviewed (e.g., imprinting and self-assembly) [111,112]. Lithography is also the method preferred by industrial fabricators of electronic and optical devices. Resists have played a key role in the steady decrease in feature size, and in the increase in memory capacity predicted by Moore's law. In a typical process, a substrate is coated with a continuous layer of resist material, and the resist is exposed to radiation through a mask having transparent and opaque regions that confer the desired pattern. The regions of the resist that are irradiated (corresponding to the transparent regions of the mask) undergo a chemical change, while the regions of the resist that are not irradiated (corresponding to the opaque regions of the mask) remain unchanged. The unexposed regions of the resist may then be removed (in a positive process) or the exposed regions may then be removed (in a negative process). The removal of the resist is termed the development stage, as in the processing of photographic film. The surface-patterned resist may be irradiated by exposure to UV-visible wavelengths in the range 13 to 436 nm, or by exposure to X-rays, electron beams, ion beams or plasma. Hence resists must be polymeric, in order to be spin-coated into films, although film-forming molecular glasses based on lower mass species have been reported [113]. Resists must also be thermally stable, and have suitable photochemical properties such that they change their solubility upon exposure to radiation. In the chemical amplification method, exposure generates the formation of a catalyst (normally a protic acid resulting from a photoacid generator, PAG, precursor) which then catalyzes the post-exposure bake (PEB) reaction resulting in the solubility change of the resist polymer. Important parameters associated with lithographic processes include sensitivity, that is inversely related to the mJ cm^{-2} dose of radiation required to achieve exposure without loss of resolution, and line edge roughness (LER), where low LERs are desirable and are associated with good quality pattern transfer.

The irradiation wavelength favored by the semiconductor industry has gradually decreased from 436 nm, to 365 nm, to 248 nm, to 193 nm, where 90 nm resolution could be achieved at 193 nm in 2006, and 65 nm resolution could be achieved in 2009. Each new wavelength required the development of new resists with appropriate transparency at the wavelength of interest. For example, conventional aromatic polymers are unsuitable for 193 nm applications despite their chemical and thermal stability, although polyaromatics have not been ruled out owing to the lower wavelength (red shift) of their absorbance. Currently devices with 65 nm critical dimensions are being produced, and new technologies with target critical dimensions of 45, 32 and 22 nm are in development. These targets will require thinner resist films (50-100 nm) in order to maintain reasonable aspect ratios; however, thinner films have higher defect densities, and different glass transition temperatures and LERs than bulk films of the same chemical composition [110]. 157 nm lithography and 193 nm immersion lithography were both considered as the successors of 193 nm lithography in the next step toward increased resolution and ever smaller features, but 193 nm has emerged as the front runner for 45 nm and 32 nm features because of technical obstacles encountered during the development of 157 nm resists. For 22 nm features, the current favored technology is EUV (13 nm) using polycarbocyclic organic resists. In immersion lithography, an immersion fluid is used between the lens and the photoresist layer, where the immersion fluid medium confers higher resolution than air. The drawback of immersion lithography is that components of the resist may leach out into the immersion fluid, or that the immersion fluid may contaminate the photoresist.

Lithographic resist materials used for the fabrication of sub-100 nm features (e.g., 193 nm, 157 nm, 13 nm for EUV) must meet very demanding performance requirements [110]. Fluoropolymers and siloxane polymers are useful in 157 nm systems because of their high transparency and low absorbance at this wavelength, although siloxane polymers are prone to fragmentation and to undesirable outgassing that can damage the lens. Hence polyhedral oligomeric silsesquioxanes are promising resist materials because of their transparency and increased chemical stability relative to polysiloxanes. An Si_8O_{12} cage architecture is also preferable to a polysilsesquioxane architecture because of the Si_8O_{12} monodispersity and precisely defined structure. In addition, polysilsesquioxanes carry residual SiOH functionality that can undergo condensation reactions, decrease shelf life and result in negative tone chemistry, whereas Si_8O_{12} does not. The Shipley Company (Marlborough, USA) studied the effect of polymer architecture on silicon outgassing in bilayer silicone resists in ArF and F_2 lithography [114], and it was found that an Si_8O_{12} moiety in the silicone backbone gave a high silicon content for superior etch resistance, no outgassing, and successful pattern resolution below 100 nm.

The use of Si_8O_{12} compounds as lithographic resists was originally suggested in a 1985 Russian report by Korchkov and Martynova [115], particularly as negative resists for electron and X-ray lithography. From 2000 onward [116], various resists comprised of Si_8O_{12}-acrylate copolymers (see for example Fig. 7.20) have been studied, primarily by groups at the University of Connecticut (USA) and at the Institute of Microelectronics (Athens, Greece) [116-125]. Structures and monomer ratios have been varied in order to achieve optimum properties for lithographic resists, and in some cases a combinatorial approach has been used [116]. Such copolymers have the potential disadvantage that incorporation of an Si_8O_{12} moiety changes the morphology (i.e., Si_8O_{12} domains form and phase separation occurs) relative to the Si_8O_{12}-free acrylate polymer (see Chapter 4), and thus the physical properties of the resist are affected [126]. An x-ray photoelectron spectroscopy (XPS) study showed that a 1.5 nm layer at the surface of an Si_8O_{12}-acrylate copolymer resist was rich in Si_8O_{12} relative to the bulk material below [127], and when an Si_8O_{12} copolymer was treated with oxygen plasma, ellipsometry and XPS showed that a silicon oxide layer formed at the surface [128]. This concept is further explored in Chapter 8, where Si_8O_{12} copolymers are used in space applications requiring resistance to atomic oxygen (instead of oxygen plasma).

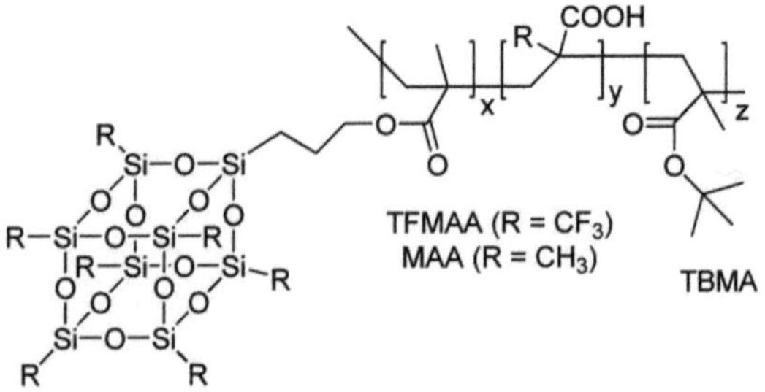

Fig. 7.20 An Si_8O_{12}-acrylate terpolymer evaluated as a resist for 193 nm bilayer lithography [129]

In a 2004 Greek-French collaboration by Argitis and co-workers [130], positive tone Si_8O_{12}-functionalized methacrylate photoresists were prepared from acrylate-Si_8O_{12} monomers copolymerized with various combinations of other acrylate monomers (methacrylic acid, MAA, *t*-butylmethacrylate, TBMA, *t*-butyl trifluoromethylacrylate, TBTFMA, itaconic anhydride, IA and 2-(trifluoromethyl) acrylic acid, TFMAA). TBMA and TBTFMA undergo acid catalyzed deprotection reactions (where acid comes from a perfluorooctylsulfonate photoacid generator PAG) to form the carboxylic acids that undergo positive tone lithography and dissolve in basic solution, while MA and IA are hydrophilic and enhance adhesion and solubility in base. TFMAA was also selected to decrease absorbance at 157 nm. Thin films of Si_8O_{12} copolymer were spin coated from MIBK (methyl isobutyl ketone) or PGME (propylene glycol methyl ether) onto a novolac epoxy resin substrate and etched by oxygen plasma. XPS showed that undesirable aggregation of Si_8O_{12} at the surface occurred most with the Si_8O_{12}-TBMA copolymers. The best copolymers (comprised of the multiple copolymers above) were found to have high sensitivity at 157 nm (less than 10 mJ cm^{-2}), no silicon outgassing, and sub-100 nm resolution.

In a 2006 study by the same group from Athens and collaborators at Interuniversity Microelectronic Center (Leuven, Belgium), similar Si_8O_{12}-methacrylate resists were assessed in 193 nm bilayer lithography [129]. These were terpolymers prepared from methacrylate POS, TFMAA or MAA (to control degree of fluorination), and TBMA (Fig. 7.20). IA was not used as a monomer because its tendency to hydrolyze was detrimental to shelf life. Although TFMAA was not required to reduce absorbance at 193 nm (in contrast to the 157 nm case above [130]), it was found that terpolymers with high TFMAA content gave resists with the best homogeneity. A partially fluorinated terpolymer was the first example of an Si_8O_{12} resist that could be directly incorporated into the standard industrial process using tetramethyl ammonium hydroxide developer. In some cases cyclopentyl groups were used as the other (blocking) groups on Si_8O_{12} [117,121] but in this study ethyl groups were used to confer higher etch resistance and lower surface roughness [120,131]. The best homogeneity was obtained when a hydrophobic PAG anion with a long fluorinated tail was used in order to achieve the best compatibility with the terpolymer matrix. The best-performing resists had a sensitivity of 10 mJ cm^{-2}, and the most suitable substrate was found to be a novolac resin.

In a 2006 Korean study by Kim and co-workers [132], an Si_8O_{12} molecular resist carrying diazoketo groups (Fig. 7.21) was reported for use in deep UV lithographic processes. This work was carried out to avoid the use of chemically amplified resists and the associated PAGs, since PAGs are environmentally unfriendly aromatic compounds, and have poor compatibility with photoresist polymers, resulting in poor resolution. Diazoketo groups were selected to achieve deep UV photobleaching and polarity change upon exposure (where a Wolff rearrangement occurs and carboxylic acid groups are generated), and once again Si_8O_{12} was selected for its thermal and chemical stability. The polycyclic cholate moiety (Fig. 7.21) was selected because its amphilicity conferred good film-forming ability. The Si_8O_{12} molecular resist was evaluated in a single layer (positive tone) process and in a bilayer process. Spin-coating of silicon wafers was carried out from propylene glycol methyl ether, and the exposed regions of the resist film were developed using industry standard 2.38 wt % tetramethylammonium hydroxide solution.

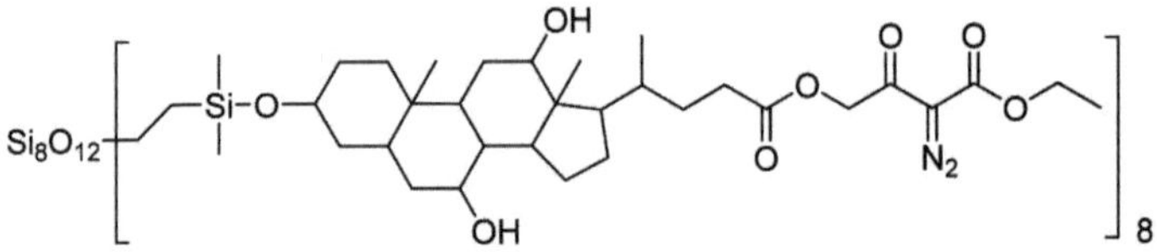

Fig. 7.21 Diazoketo Si_8O_{12} molecular resist for use in lithographic processes without photoacid generators (PAGs) [132]

In an attempt to avoid the post-exposure bake (PEB) step in conventional lithographic processes, a molecular resist comprised of Si_8O_{12} functionalized with trimethoxysilyl groups was designed [133], taking advantage of the fact that acid-catalyzed cross-linking of $Si(OMe)_3$ occurs readily at room temperature.

As well as the resist itself, Si_8O_{12} has also been used in topcoat materials for resists. A 2006 IBM Corporation patent claimed T_8, T_{10} and T_{12} polyhedral oligomeric silsesquioxane cores functionalized with groups soluble in aqueous bases (e.g., COOH, SO_3H, amides, ureas and so on) as topcoats for photoresists used in photolithography or immersion lithography [134]. The function of the topcoat was to prevent interactions between the photoresist material and the immersion fluid (see description of immersion lithography above), while being soluble in base developer and transparent at the irradiation wavelength. The materials were prepared in hydrosilylation reactions between hydrido- Si_8O_{12} species and various unsaturated cyclic aliphatics.

In a 2007 University of Texas (Austin) study [135], Si_8O_{12} functionalized with a mixture of photosensitive acrylate groups and thermally curable benzocyclobutane (BCB) groups (Fig. 7.22) was used in a step and flash imprint lithography reverse tone (SFIL-R) process. SFIL was developed in order to reduce the number of steps in the dual damascene process used to fabricate copper interconnects, where SFIL was an alternative to optical lithography [136]. In SFIL-R [137], a topcoat is applied on top of an organic imprinted layer. The topcoat must be a low viscosity and non-volatile liquid (for efficient 'filling' of the imprinted layer beneath) with a high silicon content for etch resistance, and must carry curable functional groups. In addition, in order to be compatible with a low-κ manufacturing process, the imprint material must have a low dielectric constant, thermal stability at 400 °C, high modulus, low shrinkage upon cure and low coefficient of thermal expansion (CTE).

In 1996, Laine and co-workers (University of Michigan) had reported that liquid Si_8O_{12} compounds functionalized with epoxy groups could be used as photochemically curable materials [138], and thus had potential for use in an SFIL process. In the 2007 Texas work [135], an Si_8O_{12} starting material was synthesized with long flexible arms that conferred liquid character, $Si_8O_{12}(OSiMe_2OSiMe_2H)_8$. This compound is an interesting variant of the more commonly used, $Si_8O_{12}(OSiMe_2H)_8$ mentioned elsewhere in this chapter, and is synthesized as shown in Fig. 7.22. The octasilanol-Si_8O_{12} intermediate (Fig. 7.22, top center) is a surprisingly stable white solid that can be stored long-term without self-condensation of the silanol groups, and that maintains its solubility over time [139]. In the final product (Fig. 7.22, bottom), a ratio of 5:3 BCB to acrylate was shown to give the optimum combination of properties for SFIL.

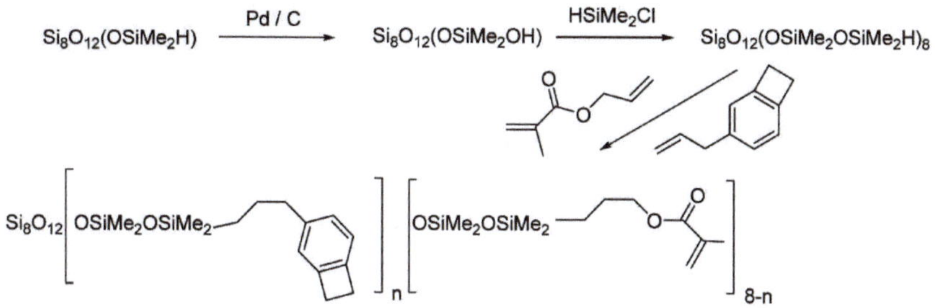

Fig. 7.22 Synthesis of a long-arm POS material for an SFIL process [135]

Recently four-beam interference lithography was used to fabricate three-dimensional porous Si_8O_{12} microstructures with high periodicity that could be used as templates for fabrication of various photonic crystals, phononic crystals, micro- and opto-fluidic devices and sensors [140]. Multibeam interference lithography is the most widely used technique for the fabrication of large-area 0.3 to 6 micron periodic structures, although this can only be used to create periodic structures from photosensitive materials. However, if a 3D periodic material is used as a template, it can be back-filled with an almost unlimited range of other (non-photosensitive) materials, and then the template can be removed by decomposition or dissolution to obtain an inverse 3D periodic structure. In some cases this structure is then used itself as a template in a second templating process. Si_8O_{12} is a good 3D template material because it is compatible with photolithographic processes (see earlier discussion), thermally and dimensionally stable above 400 °C and soluble (i.e., removable by dissolution) at room temperature. An epoxycyclohexyl polyhedral oligomeric silsesquioxane cage mix (T_8, T_{10}, T_{12}) was used as the negative-tone resist precursor [141], and exposed in the presence of 1 wt % PAG to form a 3D silica-like structure. This was then calcined, either by heating in argon to 500°C, or by exposure to oxygen plasma. Cracking was found to occur upon thermal treatment in oxygen, and crack resistance was associated with the presence of residual carbon in the calcined material. The 3D structure was stable when heated up to 1100 °C, and when oxygen plasma was used for calcinations, the volume fraction of pores could be controlled by varying the plasma etch time and power. Because Si_8O_{12} is transparent at UV-visible wavelengths, thicknesses of 30 micron were possible, in contrast to less transparent substrates used for multibeam lithography, where thicknesses of 5 micron are obtained [142]. The 3D structures were filled with elastomeric polydimethylsiloxane (PDMS) and then the Si_8O_{12} component was removed using hydrofluoric acid at room temperature. This resulted in a high fidelity PDMS replica of the periodic 3D structure that could be used as a flexible photonic crystal. Because PDMS has a low modulus and is easily deformed, it is possible to use it to fabricate tunable photonic crystals [142]. Hence the color of the PDMS photonic film in this study was a function of strain, and its behavior was reversible, i.e., the same color changes were observed as strain values were varied over multiple cycles between 0 and 23%.

7.5 Polyhedral Oligomeric Silsesquioxanes in Sensor Systems

Sensors have a transducer element that converts one stimulus or form of energy (input) into another form of energy (output signal). Inputs may be physical (e.g., temperature, pressure) or chemical (e.g., vapors, liquids). A sensor behaves reversibly, i.e., it should return to its original state after the sensing event (stimulus) has occurred. Sensors should not be confused with assays, which are irreversible, nor with actuators, which convert energy into motion (e.g., muscles). In this section, the use of Si_8O_{12} materials in sensor systems that change their fluorescent, optical, piezoelectric, resistive or electrochemical properties in response to an analyte in the vapor or liquid phase is reviewed. In some cases, single sensors are used, and in other cases arrays of sensors are used (Fig. 7.23). Sensor arrays for vapor detection are often termed 'electronic noses' and their liquid phase counterparts are termed 'electronic tongues'. Sensors may function according to a lock-and-key model, or according to a cross-reactive array model. Array systems are 'cross-reactive' when one component in the array responds to multiple analytes, and a particular analyte causes a number of components in the array to respond. This is in contrast to lock-and-key sensor arrays where each component is specific to a given analyte or to a class of analytes. The cross-reactive array-based vapor sensor systems incorporating Si_8O_{12} that are covered in this section include chemically responsive dyes (giving on-off fluoresence intensity changes, or wavelength changes in absorption or fluorescence emission), surface acoustic wave (SAW) sensors (giving changes in acoustic frequency) and carbon-polymer composite sensors (giving changes in volume and in resistivity).

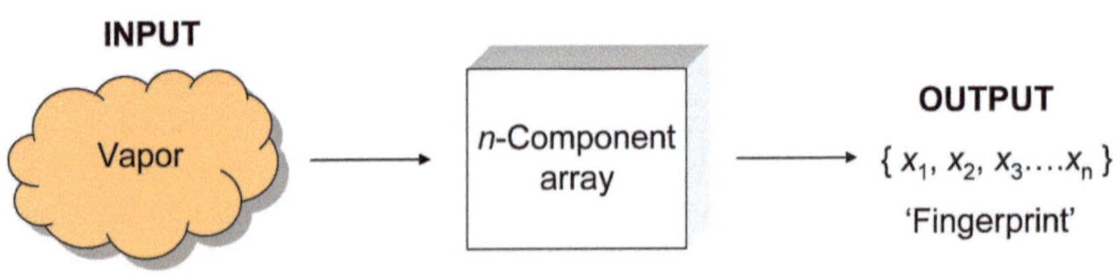

Fig. 7.23 Response of an *n*-component sensor array to a vapor input. Output *x* may be a set ('fingerprint') of *n* resistance, conductivity, frequency, wavelength or other values

Interestingly, in the light of the earlier discussion of predominantly aromatic liquid crystal Si_8O_{12} materials (Section 7.1) and electroluminescent Si_8O_{12} materials

(Section 7.2), many of the sensor materials reviewed in this section are also based on architectures where an Si_8O_{12} core is functionalized with aromatic groups. Most of these aromatic groups are fluorophoric. Apart from sensors, the other main application of fluorophore-functionalized Si_8O_{12} compounds is as fluorescent colorants for plastics or inks, e.g., a 2008 Ciba patent reports that POS has been functionalized with the *N*-(2'-ethylhexyl)-3,4,9,10-perylenetetracarboxylic monoimide monoanhydride chromophore to make fluorescent colorants for addition to thermoplastic or thermoset polymer formulations, or to formulate fluoresecent inks [143].

7.5.1 Fluorophore-Functionalized Polyhedral Oligomeric Silsesquioxanes as Sensors

In 2008, the use of fluorophore-functionalized Si_8O_{12} cores in various sensor applications was reported in three separate studies [144-146]. In the first 2008 study by Hartmann-Thompson and co-workers at Michigan Molecular Institute [144], a set of polyhedral oligosilsesquioxane nanosensors designed to change their wavelength of fluorescence emission upon interaction with analytes was synthesized, characterized and assessed in a remote detection application (Fig. 7.24). In remote detection, information may be obtained from a remote location either by passive detection of waves emitted from remote objects (e.g., passive FTIR), by actively sending a wave to a remote object, and obtaining information by analyzing the wave(s) returning from the object (e.g., Laser Detection and Ranging, LIDAR) or by introducing a sensor at the remote location that can send back information (e.g., use of metal nanoparticles on a remote surface to send back enhanced Raman data). In this study, the nanosensors deployed at the remote location were based upon Si_8O_{12} cores carrying various wavelength-shifting fluorophores more commonly used in the life sciences field (e.g., for multiphoton excitation microscopy and for fluorescence labeling of cells and tissues), where the wavelength of fluorescence emission is a function of the polarity of the chemical environment around the fluorophore. The Si_8O_{12} nanosensors were easily synthesized from commercially available fluorescent labels that had been designed to react readily with amino groups in amino acids; hence the labels were also able to react with amino- Si_8O_{12} compounds under mild conditions. A four-component cross-reactive array of nanosensors was used to generate fluorescence data sets (fingerprints) for a number of analytes including chemical warfare agent (CWA) simulants and toxic industrial chemicals (TICs) by measuring the one-photon fluorescence spectra of nanosensor-analyte pairs in solution (Fig. 7.24). The feasibility of using the nanosensor array for the remote detection of analytes in clouds and on surfaces was then demonstrated in collaboration with the Dantus laser research group at Michigan State University. A femtosecond laser was used to interrogate the array and induce two-photon fluorescence in the nanosensors, and a remote fluorescence probe was used to record

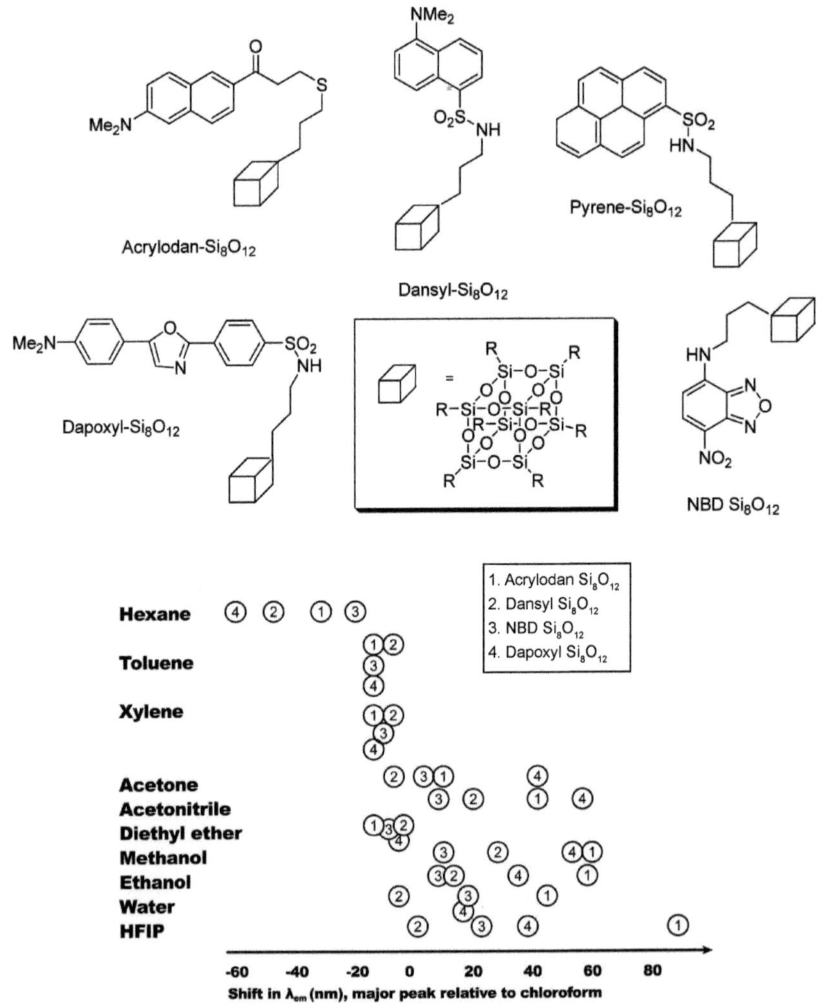

Fig. 7.24 An array of fluorophore-functionalized Si_8O_{12} compounds and the response of the array to various analytes [144]

the responses of the nanosensors [144].

It is interesting to note that the environmentally sensitive NBD (7-nitrobenz-2-oxa-1,3-diazole-4-yl) fluorophore (Fig. 7.24) has also been used in bridged silsesquioxane sensors, as well as in the discrete Si_8O_{12} nanosensors discussed above. These bridged silsesquioxane sensors were described in a 2006 study by Edmiston and co-workers at the College of Wooster (Ohio), and were based on molecularly imprinted sol-gels originating from bis(trimethoxysilylethyl)benzene and an additional fluorene-functionalized trimethoxysilyl compound (Fig. 7.25). The fluorene groups were removed from the sol-gel matrix via carbamate cleavage leaving fluorene-shaped voids (i.e., fluorene-specific binding sites) and an NBD fluorophore was placed in the void by reacting a free amino group with NBD chloride [147]. The aim of this study was to fabricate a chemical fluorescence sensor specific to fluorene. Molecular imprinting of polymeric substrates is an established way to achieve such specificity. Sol-gel films on glass slides were tested by immersion in 0.5 to 100 ppb aqueous fluorene solutions or by exposure to solid fluorene (and resultant vapor). Non-imprinted sol-gels were used as controls [148]. In this study changes in fluorescence intensity at a given wavelength (540 nm) were monitored rather than changes in fluorescence wavelength [144]. For the NBD fluorophore, both wavelength and intensity of fluorescence are highly solvent-dependent, and quantum yield is particularly low in water. Fluorescence intensity was found to increase with concentration of fluorene test solutions, and this was rationalized in terms of displacement of water by fluorene near NBD rather than direct interaction between fluorene and NBD fluorophore. Response times were less than 1 minute and the limit of detection was estimated to be 1 ppt, but the responses were not reversible, and rinsing with water or non-polar solvents (or flushing with air or nitrogen) failed to remove fluorene from the matrix, or to decrease the fluorescence intensity to its baseline value. The system successfully distinguished fluorene from naphthalene, fluoranthene and anthracene.

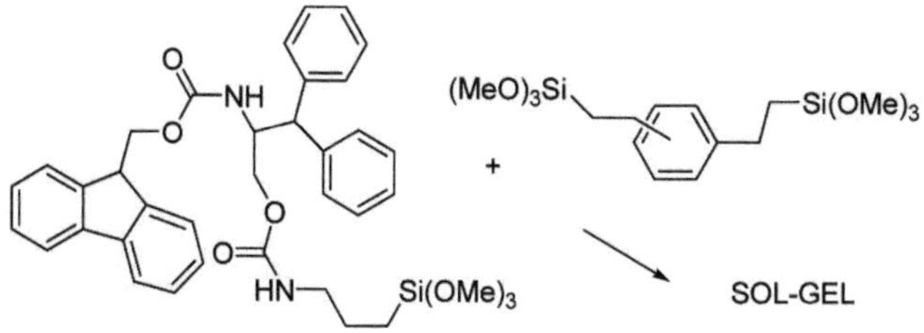

Fig. 7.25 Synthesis of fluorene-functionalized sol-gel [147]

In a second 2008 study [145] by Shi and co-workers (Tsinghua University, Beijing), a fluorescent pyrene- $Si_8O_{12}(i\text{-propyl})_7$ vapor sensor was synthezied in an esterification reaction between 1-pyrenebutyric acid (PBA) and a hydroxyl-functionalized Si_8O_{12} reagent (Fig. 7.26). Pyrene has a 380 nm emission in the UV region and a visible excimer emission at 450 to 500 nm, and is another example of an environmentally sensitive fluorophore having an emission wavelength dependent upon its immediate chemical environment. It can give either an on-off fluorescence sensor response (fluorescing or quenched), or a wavelength shift sensor response. The excimer emission results from π-stacking of the planar pyrene aromatic rings and is visible to the naked eye, making it suitable for sensing applications. Crystalline pyrene has strong excimer emission owing to its highly ordered structure, but spin-coated pyrene films show weak excimer emission because of their more random structure. Crystalline structures are disadvantageous in vapor sensing applications because it is difficult for the vapor to diffuse into the film and elicit a sensor response, especially in the case of low vapor pressure vapors such as nitroaromatics (explosives) and phosphonates (nerve agents and pesticides). In order to overcome this problem, Si_8O_{12} was used to carry the pyrene fluorophores (Fig. 7.26). This resulted in a spin-coated thin film with a highly ordered structure for strong excimer emission, but also with some degree of porosity for good vapor diffusion. A strong blue pyrene excimer emission at 475 nm was observed that was rapidly quenched in the presence of trinitrotoluene (TNT), 2,4-dinitrotoluene (2,4-DNT) and nitrobenzene vapors, with a response time of one minute, faster than that of a film of pure pyrene. In a physical blend of PBA and the pyrene-Si_8O_{12} compound, pyrene monomer emission dominated excimer emission, showing a disordered distribution of pyrene in the thin film. Response time increased as film thickness increased from 4.5 nm to 21 nm, as measured by ellipsometry. The 4.5 nm film was estimated to consist of four layers of pyrene-Si_8O_{12} molecules, and had a response time of 10 seconds. Interestingly, pyrene was evaluated in the Michigan Molecular Institute study [144] in which Si_8O_{12} was functionalized with a range of fluorophores (Fig. 7.24), but was excluded from the final sensor array precisely because of its pair of emission bands, making it unsuitable for a system where it was necessary that each array component generate a single output at a single emission wavelength.

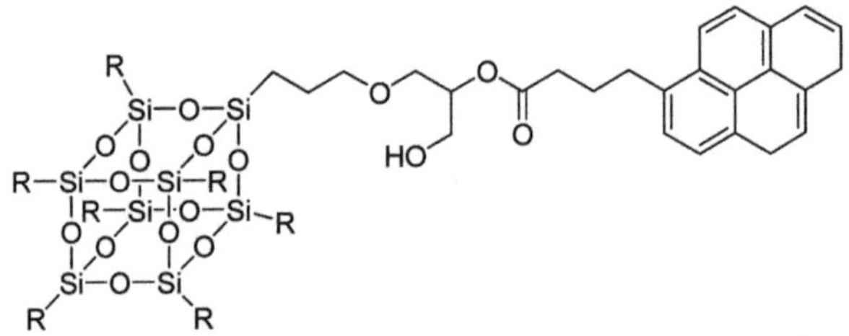

Fig. 7.26 On-off fluorescence sensor based on pyrene-functionalized Si_8O_{12} [145]

In a third 2008 study [146] by Chujo and co-workers (Kyoto University, Japan), Si_8O_{12} cores carrying multiple functional groups were used as ratiometric multimodal chemical sensors for monitoring solvent polarity via fluorescence spectroscopy and ^{19}F NMR spectroscopy. A multimodal imaging agent is a single probe that is able to generate two or more different kinds of information. The combination of fluorescence and ^{19}F magnetic resonance imaging (MRI) information is particularly useful for in-vivo small animal studies, and requires a probe that carries a fluorophore and a fluorinated contrast agent. Probes that respond to a stimulus by generating one signal of increased intensity, and a second corresponding signal of decreased intensity are termed ratiometric probes. Si_8O_{12} was selected as a scaffold for a ratiometric fluorescence/^{19}F MRI multimodal chemosensor because it is multifunctional, and has good biocompatibility and low toxicity (see Chapter 9). An Si_8O_{12} core carrying five trifluoroacetyl groups (for ^{19}F - MRI), two phosphoryl groups for enzyme interaction (e.g., with alkaline phosphatase, AP) and one CCVJ fluorophore (for fluorescence) was synthesized in several steps from octa-ammonium POS, and the desired molecular ions were observed by MALDI-TOF MS after each synthetic step. (Fig. 7.27, top). CCVJ (9-(2-carboxy-2-cyanovinyl) julolidine) is a molecular rotor fluorophore that increases its emission intensity in a high viscosity environment, and decreases its emission intensity in a low viscosity environment, where internal molecular rotations occur more easily and suppress emission [149]. In polar conditions, the Si_8O_{12} probes formed a cluster resulting in high fluorescence intensity (500 nm in methanol, inhibited molecular rotation) but a low intensity ^{19}F NMR signal (little tumbling). In non-polar conditions, the probes were widely dispersed, resulting in suppressed fluorescence (unhindered molecular rotation) and enhanced ^{19}F NMR signals (tumbling possible). The probe was used to monitor solvent polarity changes occurring during enzymatic reactions.

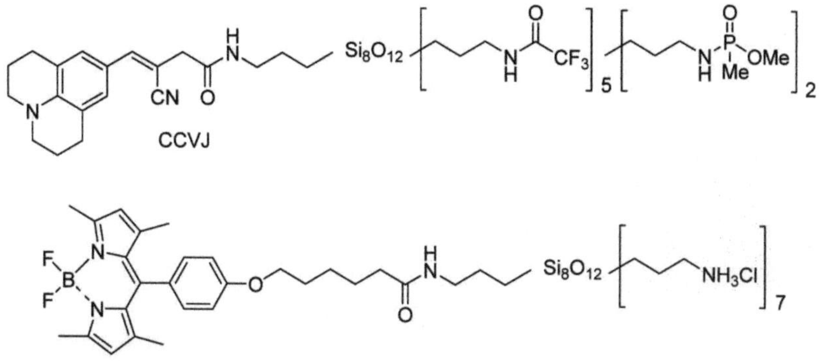

Fig. 7.27 Si_8O_{12} carrying a molecular rotor fluorophore (CCVJ), trifluoroacetyl groups and phosphoryl groups (top) [146], and Si_8O_{12} carrying a BODIPY fluorophore and ammonium groups (bottom) [150]

In another earlier example [150] of a fluorophore-carrying Si_8O_{12} core in a life sciences study (Rotello and co-workers, University of Massachusetts Amherst, 2005), Si_8O_{12} was functionalized with ammonium groups (~90%) and a BODIPY fluorophore (4,4-difluoro-4-bora-3a,4a-diaza-s-indacene, ~10%), see Fig. 7.27, bottom. It was evaluated as a drug carrier and drug delivery agent (see also Section 9.4.1, Chapter 9). Si_8O_{12} is biocompatible, stable, highly water-soluble with suitable functionalization, and has high cellular uptake and low toxicity (see Chapter 9). Water solubility means that the drug can be taken orally, and the small size and high charge density of octa-ammonium-Si_8O_{12} means that the unit may easily be transferred through vascular pores. The monodisperse nature of Si_8O_{12} compounds also eliminates the problems associated with polydisperse linear polymers and the complex syntheses associated with dendrimers. Octa-ammonium-Si_8O_{12} was neutralized with triethylamine and then reacted with the succinimidyl ester of BODIPY in order to attach the fluorophore. An amine terminated BODIPY compound (i.e., with no Si_8O_{12} content) was used as a control. Cos-1 cell cultures were exposed to Si_8O_{12}-BODIPY and control BODIPY. Live fluorescence confocal microscopy showed cell uptake of Si_8O_{12}-BODIPY in the cytosol (rather than the nucleus), but no uptake of the control. Hence it was shown that an Si_8O_{12} carrier has the potential to deliver drugs either with low water solubility or with poor tissue uptake. The BODIPY family of fluorophores [151] are widely used long-wavelength fluorescent labels that are comparatively non-polar and a have a high fluorescence intensity and a wide multiphoton excitation cross section.

In a 2007 study by Wu and co-workers, Fudan University, Shanghai [152] another catioinc Si_8O_{12} species, (octa-aminophenyl hydrochloride salt) was used as a probe for DNA via resonance light scattering (RLS), see also Section 9.4.3, Chapter 9. In this application, cationic Si_8O_{12} had the advantage of water solubility and stability across a wide pH range. Electrostatic interaction and complexation between cationic POS and DNA polyanion in aqueous solution enhanced the RLS signal, and the enhancement was proportional to nucleic acid concentration. Foreign species such as metal ions and amino acids did not affect signal enhancement. Complexation was shown by various band shifts in the FTIR spectrum, and cationic Si_8O_{12} was prepared from octaphenyl Si_8O_{12} by nitration with nitric acid followed by reduction to the amine over palladium-carbon.

7.5.2 *Polyhedral Oligomeric Silsesquioxane Sensors for Gas and Vapor Detection*

Si_8O_{12}-based materials have been used to improve the performance of optical waveguides, optical fibers and surface acoustic wave sensors for the detection of gases and vapors, and the use of luminescent compounds for oxygen sensing [86] has already been alluded to in Section 7.2.4.

Optical transducers are attractive for the sensing of flammables such as jet fuel and hydrogen because the electronics are at a remote location, reducing the risk of electrical sparking. An ideal coating for such a transducer must be robust, and must have high surface area and allow fast diffusion of vapors for rapid sensor response. In a 2008 study by Sirbuly, Ratto and co-workers (Lawrence Livermore National Laboratory, California), single crystalline tin dioxide (SnO_2) nanowire subwavelength optical waveguides decorated with octapropylammonium- Si_8O_{12}-stabilized palladium nanoparticles were used to detect hydrogen [153]. Palladium has an ultra-high affinity for hydrogen gas, and its optical properties change when it adsorbs hydrogen, making it a suitable coating for a fiber optic hydrogen sensor. Si_8O_{12}-stabilized nanoparticles were found to have a number of advantages over conventional palladium nanocrystal coatings including higher affinity for the tin dioxide surface, a better match between the palladium species absorbance spectrum and the tin dioxide waveguide fluorescence spectrum, shorter path lengths (50 micron vs. 2 mm for other optical fiber gas detectors) that could achieve reasonable sensitivity (0.5%), and higher porosity resulting in response times as fast as 1-2 seconds for hydrogen in solution. The Si_8O_{12}-Pd nanoparticles were attached to the waveguide by injecting a POS-Pd colloid in methanol into a sensing microchannel. This was subsequently dried with air, and then hydrogen-argon mixtures were introduced via the same microchannel. This system also has the potential to be extended into an array or electronic nose by coating a number of waveguides with a variety of chemo-responsive nanoparticles. The Si_8O_{12}-Pd nanoparticles could be washed off the waveguide using aqua regia, while conventional palladium coatings cannot be removed from the wave guide without damaging it and preventing its re-use.

In a related application, a 2007 Ocean Optics, Inc. patent [154] described a system in which an optical sensing molecule (capable of a change in absorbance or emission in response to an analyte) is encapsulated in a transparent silsesquioxane matrix, coated onto an optical fiber and used to detect oxygen in a jet fuel tank environment. One benefit is that the matrix is resistant to hydrocarbons (particularly jet fuels) but still permeable to oxygen. A mixture of fluorescent organoruthenium molecular oxygen sensor and 3,3,3-trifluoropropyl triethoxysilane and methyltrimethoxysilane sol-gel monomer was coated onto the tip of the fiber, which was designed to monitor dissolved oxygen in fuels, and headspace oxygen in fuel tanks. Tris(4,7-diphenyl-1,10-phenanthroline)ruthenium(II) perchlorate absorbs at 460 nm (blue) and emits at 620 nm (orange-red), and the fluorescence is quenched as a function of the concentration of oxygen (a triplet species capable of dynamic fluorescence quenching). Similar compounds are used as temperature sensors, but in these cases the luminophores must be sealed in order to prevent exposure to oxygen in the environment.

Surface acoustic wave (SAW) sensors are piezoelectric quartz systems where the frequency of the surface acoustic wave is a function of mass [155]. Hence if the sensors are coated with a material that can interact reversibly with a given vapor, a mass change and a response to that vapor may be recorded. SAW sensors are normally used in cross-reactive arrays [156], where each SAW in the array carries a different coating, and each SAW in the array gives a different mass change and frequency value in response to a given vapor (see Fig. 7.23), thus generating a 'fingerprint' of frequencies for that vapor. The collection of coatings must cover the full range of solubility interactions (dispersion, dipole-dipole and hydrogen-bonding), and needs to include non-polar, polarizable, dipolar, hydrogen-bond basic and hydrogen-bond acidic polymers. Almost all polymers exhibiting these types of interactions are commercially available, with the key exception of hydrogen-bond acidic polymers carrying phenol or fluorinated alcohol functionality, which are of particular importance in the detection of hydrogen-bond basic entities such as nerve agents (Fig. 7.28) and nitroaromatic explosives in security and defense applications [157,158]. Given the hazards of chemical warfare agents, hydrogen-bond acidic SAW polymers are normally tested against nerve agent simulants such as dimethyl methylphosphonate (DMMP, Fig. 7.28) [157-160]. In addition to carrying the desired hydrogen-bond acidic sensor groups and having good initial sensitivity to species such as DMMP and nitroaromatics, hydrogen-bond acidic SAW sensor polymers must be robust enough to withstand multiple vapor challenges and maintain their sensitivity over time. However, it is non-trivial to achieve a combination of good sensor properties and robust coating properties in a single polymeric material, although some attempt has been made to address this issue by cross-linking polysiloxane SAW coatings [161].

An alternative approach was taken by Hartmann-Thompson and co-workers at Michigan Molecular Institute in 2007, where a series of Si_8O_{12} nanofiller compounds functionalized with hydrogen-bond acidic sensor groups was prepared, characterized by IR, ^{1}H, ^{13}C and ^{29}Si NMR and MALDI-TOF MS, and formulated into polycarbosilane coatings for 500 MHz SAW sensor platforms [162,163]. Sensor responses to the explosives simulant dinitrotoluene (DNT) and to the nerve agent simulant DMMP were studied, and the performances of the Si_8O_{12} formulations were compared to those of conventional hydrogen-bond acidic linear SAW sensor polymers carrying the same sensor groups (Fig. 7.28). The Si_8O_{12} formulations gave good initial responses to the simulants, maintained 40 to 65% of their original response over a period of six months (Fig. 7.29) and maintained their sensitivity down to a simulant vapor concentration of 1 ppb v. The surface compositions of the SAW sensor coatings were characterized by sum frequency generation (SFG) spectroscopy. SFG is a comparatively novel technique for obtaining surface-specific vibrational spectra, and can be used to study chemical groups at interfaces to a depth of 10 nm or less. Unlike XPS, it does not require high vacuum conditions, and is thus of particular interest in life sciences studies. SFG spectra are generated by a non-linear optical phenomenon that occurs when two pulsed laser beams (one of fixed visible wavelength and one of variable IR wavelength) hit a point on an interface at the same time [164]. Highly ordered isobutyl surfaces dominated in SXFA Si_8O_{12} (Fig. 7.28) while highly ordered polycarbosilane surfaces (with contributions from silicon-methyl and silicon-phenyl groups) dominated in the FPOL Si_8O_{12} (Fig. 7.28). Another technique that has been used to look at the surface of Si_8O_{12}-polymer composites is time-of-flight secondary ion mass spectrometry (ToF-SIMS), which has been used to study the surfaces of octabenzyl- Si_8O_{12}-polycarbonate systems [165].

In another example of a hydrogen-bond acidic Si_8O_{12} species [166], Esker and co-workers at Virginia Polytechnic Institute (USA, 2005) described Langmuir-Blodgett (LB) films of open-cage trisilanolphenyl polyhedral oligomeric silsesquioxane that were found to bind strongly with the nerve agent simulant DMMP, probably owing to its acidic character when compared to the open-cage trisilanolisobutyl analogue, and also to conventional untreated silica (SiOH) surfaces. DMMP did not adsorb to silica, but heating to 150 °C was required to desorb DMMP from the trisilanolphenyl polyhedral oligomeric silsesquioxane surface. The ability of a material to interact selectively, rapidly and reversibly with a given vapor can also be important in the preconcentrator component of a sensor device (as well as in the transducer component of the device that actually performs the 'sensing'). In cases where a vapor has a very low concentration or a very low vapor pressure, it is necessary to pre-concentrate it before it reaches the transducer. This enables the transducer to encounter a high enough concentration of analyte to register a response. Insufficiently high concentrations of analyte are a problem across a range of sensing techniques including SAW sensors and ion mobility spectroscopy (IMS), especially for low vapor pressure species such as nitroaromatics (i.e., explosives). Molecularly imprinted nanoporous organosilicas have been used as preconcentrators in systems for the detection of TNT in an electrochemical cell by square-wave voltammetry [167]. Benzene and diethylbenzene-bridged organic-inorganic species were used, templating was achieved using decylaminetrinitrobenzene during the bistrialkoxyorganosilane sol-gel reaction, and pore size was controlled through use of surfactants.

The idea of dispersing a sensing silsesquioxane agent in a non-sensing matrix [162,163] is unusual in the sensor field, and it is far more common for a silsesquioxane to be used as a non-sensing matrix to carry some other sensor species. In one case sodium iodide or potassium iodide was used as the ozone sensitive material in a continuous polymethylsilsesquioxane matrix [168]. In another example, a silicone ladder (silsesquioxane) polymer was mixed with a quaternary ammonium salt, cast onto the gate of a field-effect transistor and polymerized, and the ion-sensitive field-effect transistor showed a linear response to nitrate ions [169]. In a third example bromocresol green was adsorbed onto a polymethylsilsesquioxane surface and its colorimteric indicator responses to ammonia and to sulfur dioxide in air were studied [170]. In a final example [171], a polysilsesquioxane matrix was used to carry nanostructured silicon powder luminophores. These materials were used to make pressure sensitive paints for use in wind tunnel testing. Pressure sensitive paints are generally based upon oxygen permeable polymers carrying photoluminescent fillers (e.g., ruthenium(II) complexes or various metal porphyrins) whose photoluminescence is quenched by oxygen, and quenching is a function of pressure, enabling pressure mapping of surfaces.

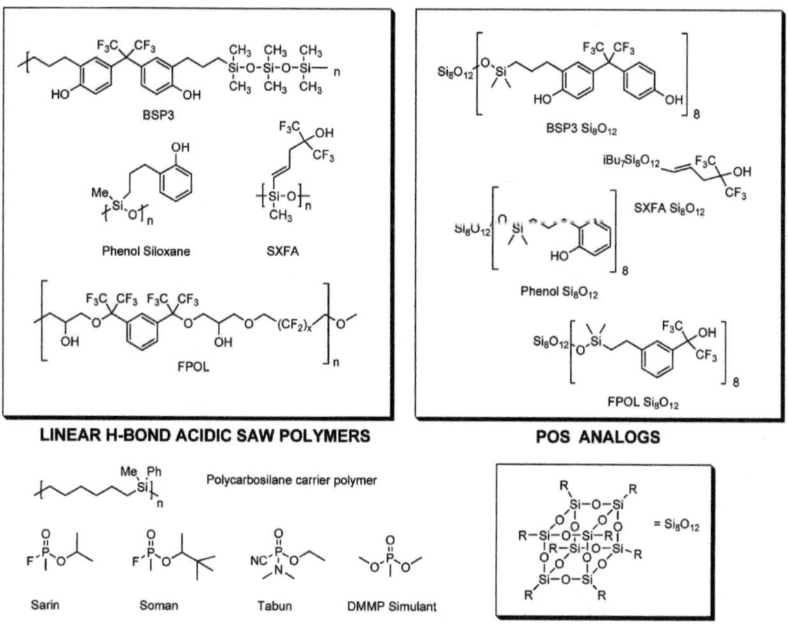

Fig. 7.28 Conventional linear hydrogen bond acidic polymers, hydrogen-bond acidic Si_8O_{12} analogues, the polycarbosilane polymer used to carry the Si_8O_{12} sensors, and hydrogen bond-basic nerve agents and DMMP simulant [162]

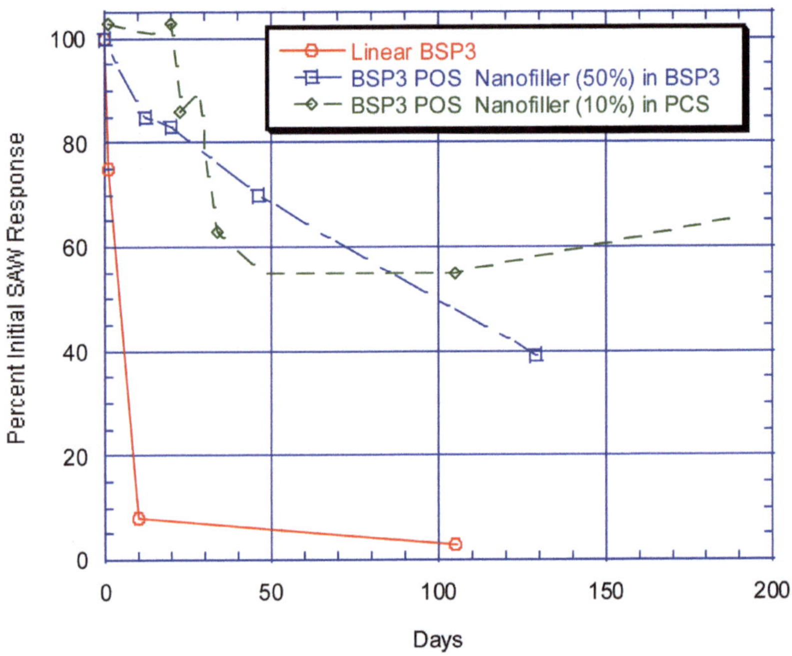

Fig. 7.29 Changes in SAW DMMP responses with time for systems containing BSP3-type sensor groups [162]

7.5.3 Polyhedral Oligomeric Silsesquioxanes in Conducting Composite and Electrochemical Sensors

One variety of electronic nose [172] may be fabricated by assembling many different polymers, each carrying carbon black, into an array of conducting carbon-polymer composites [173,174]. Each component of the array has a different gas-polymer partition coefficient for a given vapor, and gives a different degree of swelling and a different change in resistivity in response to that vapor (since resistivity increases as the conducting carbon structures move apart upon swelling). Thus the vapor generates a 'fingerprint' comprised of a unique set of *n* resistance changes for an *n* component array (Fig. 7.23). These polymer composite arrays have the advantage of operating at room temperature, in contrast to metal oxide systems that require high temperatures [175]. In several studies by Castaldo, Massera and co-workers (ENEA Centro Di Ricerche Di Portici, Naples, Italy, 2007-2008), a polymer carrying pendant Si_8O_{12} groups, poly[(propylmethacryl-heptaisobutyl- Si_8O_{12})-*co*-(*n*-butylmethacrylate)] (Fig. 7.30, left), plus 20 wt % additional filler (carbon, silicon, zinc, copper and carbon alloys of the three latter components) was evaluated in an array of this type [176-178]. The alkyl components of this copolymer impart weak dipolar character, and the methacryl groups impart hydrogen bond-basic character [179].

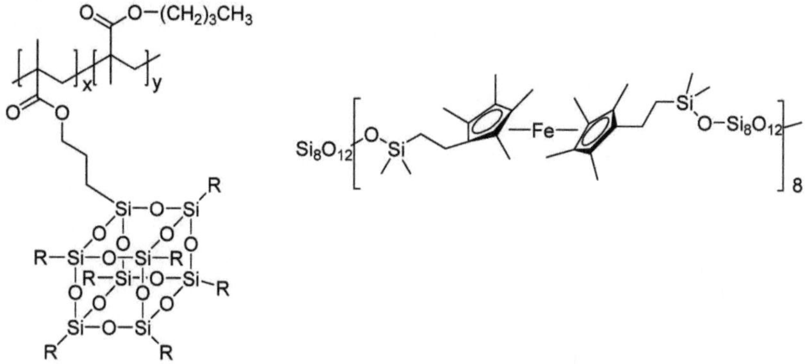

Fig. 7.30 Si_8O_{12} copolymer used as matrix in conducting composite sensor (left) [178], and as a component of three-dimensional Si_8O_{12}-ferrocenyl network used as a mediator on a glucose detection electrode (right) [180]

Instead of an array of different polymer-carbon black composites (where the polymer was varied and the filler did not change), this array was created by combining the single pendant Si_8O_{12} polymer (Fig. 7.30, left) with a range of fillers, where the filler was varied and the polymer did not change. Since noise varies with polymer in a conventional electronic nose, the authors speculated that using the same polymer throughout the array should enable the question of noise to be treated as a simple systematic error. A relative humidity sensor with a rapid response and complete reversibility was fabricated. However, the response could not be explained by the conventional polymer matrix-carbon swelling model. The sensor had an unexpectedly high sensitivity, and its conductivity varied by several orders of magnitude over the 0 to 100% relative humidity (RH) range. When the composite was tested with water and with hexane, the material did not swell, and conductivity increased (rather than decreased as would be expected in a polymer-carbon composite that was swelling in the presence of an analyte). Furthermore, a control film of pendant Si_8O_{12} polymer without additional filler behaved in the same way. The fillers were prepared using a planetary ball-milling technique and had particle sizes of 500 nm or less, and Raman spectroscopy was used to show that the two-component systems were true alloys, and that no chemical reaction had taken place between carbon and either silicon, zinc or copper. Composite films with a thickness of ~500 nm were prepared by spin-coating from particle suspensions in THF- Si_8O_{12} copolymer solutions. Gold interdigitated contacts were deposited on the films by electron-gun evaporation. Poly[(propylmethacryl-heptaisobutyl-Si_8O_{12})-*co*-(n-butylmethacrylate)] with 15 wt % Si_8O_{12} and with 25 wt % Si_8O_{12} contents were compared in the absence of additional filler, and their ac electrical frequency response properties were investigated over a range of temperatures in order to understand the sensing mechanism [178]. Glass transition temperature was measured in-situ in films and devices using this ac electrical technique rather than the usual techniques for measuring T_g (e.g., DSC), and the T_g values were found to increase with increasing ac field frequency. The authors suggest a porosity model in which water molecules create a hydrogen bond network that allows charge transfer between the Si_8O_{12} cages, although this does not fully explain the earlier observation of increased conductivity in the presence of hexane, albeit a smaller increase than that observed for water [176]. The 15 wt % Si_8O_{12} system performed better than the 25 wt % Si_8O_{12} system, and the authors attributed this to differences in Si_8O_{12} cage mobility and differences in water content in the pore structure [177]. Capacitance-temperature measurements (for permittivity) and conductivity-temperature measurements suggest that water is unable to crystallize at 273 K within the confinement of the pores, and that a super-cooled water phase is present below 230 K. The 15% copolymer was studied with and without additional filler at 20 wt %, and the responses to water, hexane and ammonia were studied [177]. The filled systems showed a 35% improvement in sensitivity relative to the unfilled system, and the filled systems all behaved similarly, further confirming that the fillers themselves did not directly contribute to conductivity, but did influence the pore structure, thus contributing to the conductivity and sensing mechanism.

In another array-based 'electronic nose' system, micromachined capacitors were used to detect changes in the dielectric constants of an array of chemoselective polymers in the presence of various analytes [181] (in contrast to the changes in resistivity in composites discussed above [176-178]), and the changes in frequency in the SAW sensors discussed in Section 7.5.2). Aminopropyl-polysilsesquioxane was a vital component in an array designed for the selective detection of hypergolic fuels, e.g., the nitrogen tetroxide oxidizer and various hydrazine compounds, associated with ballistic missiles.

Most of the earlier discussion has covered the detection of species in the gas or vapor phase, but some work has also been done on the detection of species in the liquid phase. In this area, electrodes carrying various smart coatings are used widely. In another example of silsesquioxane composites with conducting fillers, trimethoxysilanes with additional amino functionality were used to make silsesquioxanes with graphite present during the sol-gel process, where the amino groups acted as basic catalysts [182]. The resulting electrically conducting silsesquioxane-graphite composites were used as electrochemical sensors for the potentiometric detection of various anions in aqueous solution via amino acid-base chemistry at the electrode surface. In a second example, Losada, Armada and co-workers at Universidad Politécnica de Madrid (Spain, 2004) used Si_8O_{12}-ferrocenyl and Si_8O_{12}-permethylferrocenyl polymers (Fig. 7.30, right) as mediators in amperometric electrodes for the detection of glucose [180]. Such electrodes carry glucose oxidase enzyme, and generate information either by monitoring hydrogen peroxide generation (a mediator-less system) or by monitoring the oxidation state of a mediator affected by the enzyme reaction (a mediated system). Electron accepting materials such as ferrocene make excellent mediators, and hence much work on immobilizing ferrocene on electrode surfaces has been carried out (e.g., by addition of ferrocene to polymers, or by binding of ferrocene to carbon or platinum). Electron-accepting polycationic polymers are also good mediators, adsorb well to graphite, and readily bind enzymes. Methylated ferrocenes are known to have different redox properties and different enzyme interactions to non-methylated ferrocenes, and were found to be better mediators (requiring lower working potentials) for glucose detection. Analogous three-dimensional materials based on cyclotetrasiloxanes instead of Si_8O_{12} were also studied. The networks were synthesized in hydrosilylation reactions between octahydrido Si_8O_{12} and divinyl-ferrocene species. A platinum electrode carried both the Si_8O_{12}-ferrocene material (applied by solution casting) and the glucose oxidase (applied by electrostatic immobilization), and the performance of the electrodes was studied by cyclic voltammetry.

7.6 Polyhedral Oligomeric Silsesquioxanes in Fuel Cell Applications

Fuel cells are electrochemical cells that convert hydrogen, methanol or related fuels into electrical energy, and are a widely researched area of the alternative energy field. The two most important categories of fuel cells are solid oxide (SOFC) and polymer electrolyte membrane (PEMFC, Fig. 7.31). The former tend to be used in stationary power and co-generation applications, while the latter tend to be used in automotive and portable electronic power applications [183]. The high operating temperatures of SOFCs (>600 °C) rule out the use of polymers, but many polymer material challenges may be found in the PEMFC area. Fuel cell proton exchange membranes (PEMs) need to have high proton conductivity across a wide temperature and humidity range, low fuel permeability (to prevent fuel crossover), good mechanical strength and dimensional stability, good resistance to oxidative and acidic conditions (i.e., peroxides at the cathode, and protons respectively), good catalyst compatibility, and the ability to operate above 80 °C (to increase catalyst efficiency and to minimize the size of automotive radiators). The current industry standard material for hydrogen or methanol fuel cell PEM is Dupont's Nafion® sulfonated fluoropolymer (Fig. 7.32). Nafion® has a hydrophilic phase and a hydrophobic phase, and proton conduction occurs in the hydrophilic channels. Sulfonated aromatic polymers of various compositions and architectures have also been extensively studied as proton exchange membranes [184]. There have been many attempts to improve fuel cell performance by adding microscale or nanoscale particles to these PEM polymers, or by forming particle-like domains in the membranes in-situ. Examples include zirconium phosphate [185-187], calcium phosphate [188], titanium dioxide nanoparticles [189], silica [190-197] and nanosilica [198-200]. Hydrophilic particles are used to improve membrane water retention, leading to improved conductivity in low humidity conditions, but water-soluble hydrophilic particles can be at risk of being leached out of the membrane.

Particles also improve the physical properties of the membrane and reduce methanol permeability, but often decrease the conductivity because they interfere with the paths and mechanisms of proton conduction. One way to circumvent this trade-off is to use particles functionalized with proton conducting groups, e.g., silica carrying sulfonic acid groups [201-209] or zeolites carrying sulfonic acid groups [210]. Such groups also improve compatibility between the reinforcing filler and the sulfonated matrix that carries the filler. The use of sulfonated Si_8O_{12} (S-Si_8O_{12}) fillers and related silsesquioxane materials will be reviewed in this section.

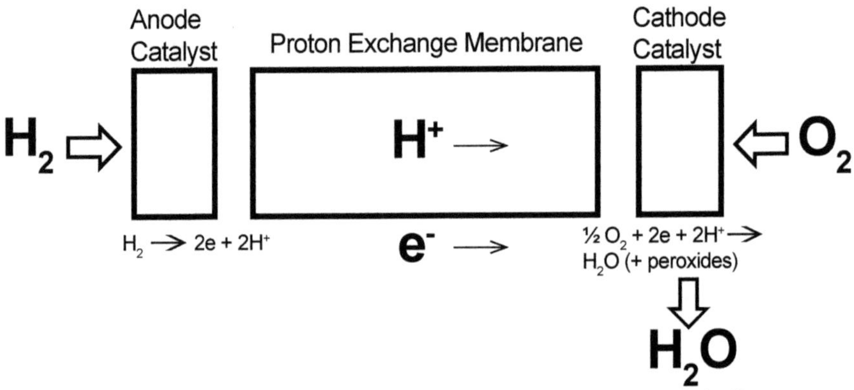

Fig. 7.31 Processes occurring within a polymer electrolyte membrane hydrogen fuel cell

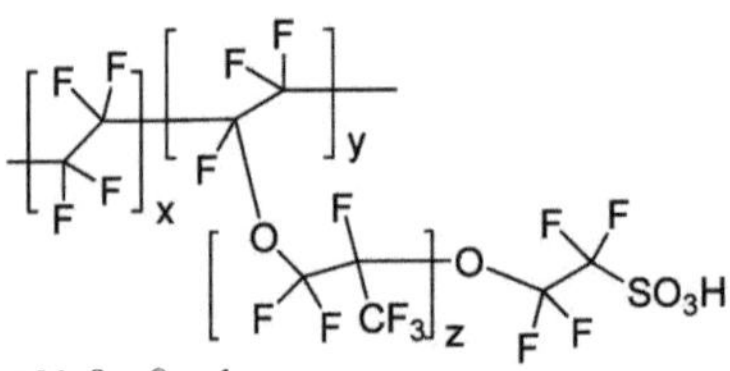

Fig. 7.32 Structure of Dupont Nafion® polymer

A 1992 French patent described a silsesquioxane electrolyte for ion exchange resin, battery or fuel cell applications [211,212] where benzyltriethoxysilane ($PhCH_2Si(OEt)_3$) was used to prepare a silsesquioxane that was subsequently sulfonated to give a material with a conductivity in the range 2 x 10^{-3} S cm^{-1} to 7 x 10^{-3} S cm^{-1} at room temperature. As the membrane was heated, conductivity decreased as water content decreased. The membrane was evaluated as an electrolyte in a Zn/MnO_2 battery system, and also in a hydrogen fuel cell system. The silsesquioxane was prepared using triflic acid in water at room temperature, and an M_n of 3600 was determined by SEC. The primary composition of matter claim covered polyorganosiloxane of the form $O_{3/2}$Si-R in which R was a sulfonated benzyl group (-CH_2PhSO_3H) imparting conductivity, or in which R was some other organic group deployed to tailor the mechanical properties of the membrane, and in which at least 10% of the R groups were sulfonated benzyls. The preferred sulfonating agent was chlorosulfonic acid ($ClSO_3H$), and the sulfonation was carried out in carbon tetrachloride solution for 30 minutes at room temperature. The sulfonated product precipitated out and was characterized by IR, where a band at 1730 cm^{-1} was assigned to $SO_3H.H_2O$. In the two decades since this patent was published, the use of silsesquioxanes in fuel cells has expanded dramatically.

The sulfonation of octaphenyl- Si_8O_{12} to make S- Si_8O_{12} S-POS (Fig. 7.33) has two non-trivial aspects, and it is suggestive that sulfonated phenyl groups bonded directly to silicon (in which R = $PhSO_3H$, and the phenyl group is not separated from silicon by a methylene spacer) were not claimed in the 1992 patent [211,212]. The first is that a phenyl group directly bonded to silicon (SiPh) is less activated to electrophilic aromatic substitution than a benzyl group bonded to silicon ($SiCH_2Ph$). However, methylene units such as those found in the benzyl group (CH_2Ph) are undesirable in fuel cell membranes where chemical stability in both acid and oxidizing conditions is required. The second is that *ipso* electrophilic attack at silicon (PhSi) competes with the desired electrophilic attack at hydrogen (PhH) positions [213], and in a phenyl-Si_8O_{12} system this would result in cleavage of the phenyl groups from the Si_8O_{12} cage. Hence S- Si_8O_{12} materials require full characterization by IR, ^{1}H NMR, ^{13}C NMR, ^{29}Si NMR and ion exchange capacity (IEC). IR will show that sulfonic acid groups are present, with four bands corresponding to SO_3 (symmetric and asymmetric) and SO_2 (symmetric and asymmetric) [214,215], and IEC will determine the number of sulfonic acid groups per unit mass (mmol g^{-1}). ^{1}H NMR will give information on the ratio of *meta*-substituted to unsubstituted phenyl rings, but neither IEC nor ^{1}H NMR will reveal whether the phenyl groups are attached to the Si_8O_{12} cage [215,216]. Hence ^{29}Si NMR is essential to determine which phenyl groups are still attached to Si_8O_{12} (O_3SiPh, -60 to -70 ppm), and which Si_8O_{12} silicon atoms have lost their phenyl groups in an *ipso* attack and been converted to O_3SiOH groups (-120 ppm). A 2006 study [217] mentions the use of sulfonated Si_8O_{12} and alkyltrimethylammonium bromide surfactant to make lamellar organic-inorganic nanocomposites, but the octaphenyl-Si_8O_{12}/ chlorosulfonic acid reaction conditions (Fig. 7.33) are not described, and no characterization data are given for the S- Si_8O_{12} product.

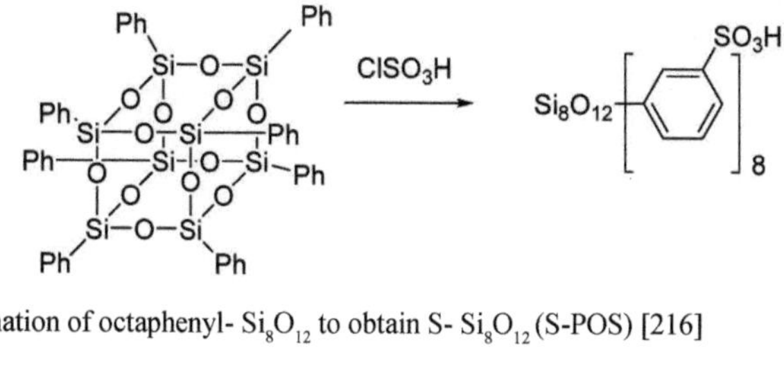

Fig. 7.33 Sulfonation of octaphenyl- Si_8O_{12} to obtain S- Si_8O_{12} (S-POS) [216]

In a 2008 Michigan Molecular Institute study, an almost fully-sulfonated Si_8O_{12} was prepared by reaction of octaphenyl-Si_8O_{12} with neat chlorosulfonic acid at room temperature (Fig. 7.33) and fully characterized [216,218]. This material had an IEC value of 3.5 mmol g^{-1} SO_3H, corresponding to mono-sulfonation of 74% of the phenyl rings, and ^{29}Si NMR showed that an aryl-substituted POS structure had been retained [219]. S-Si_8O_{12} was formulated into single layer [216] and multilayer [214] sulfonated polyphenylsulfone (S-PPSU) membranes (Fig. 7.34) by Chemsultants, Inc. (Mentor, Ohio, USA). S-PPSU was prepared by sulfonating Solvay Radel R5000® to a level of 1.5 mmol g^{-1} SO_3H. Multilayer membranes were fabricated with the aim of improving water retention and reducing brittleness. S-Si_8O_{12}/S-PPSU membranes cast from a mixed NMP/DMAc/DMSO solvent system had 100 nm Si_8O_{12} domains, and optimum conductivity was achieved with an S-Si_8O_{12} mass loading of 20 wt %. Loadings of S-Si_8O_{12} below 20 wt % resulted in lower overall IEC values for the membranes, and loadings of S-Si_8O_{12}above 20 wt % resulted in particle aggregation that impeded proton conductivity. Casting from NMP resulted in larger Si_8O_{12} domains. In low humidity conditions (25% RH) the three-layer membrane shown in Fig. 7.34 had superior proton conductivity to a Nafion® membrane, a pure S-PPSU membrane and a single layer 20% S-Si_8O_{12}/80% S-PPSU membrane (Fig. 7.35). Across all humidity conditions, both single-layer and multilayer S-Si_8O_{12}/S-PPSU membranes showed superior dimensional stability and less swelling when compared with Nafion®. No evidence of loss of additive was observed in water leaching experiments upon the single-layer S-Si_8O_{12}/S-PPSU membranes, and in the multilayer membranes, the water soluble S-Si_8O_{12} resided in a central layer protected from water leaching by the two pure S-PPSU outer layers (Fig. 7.34).

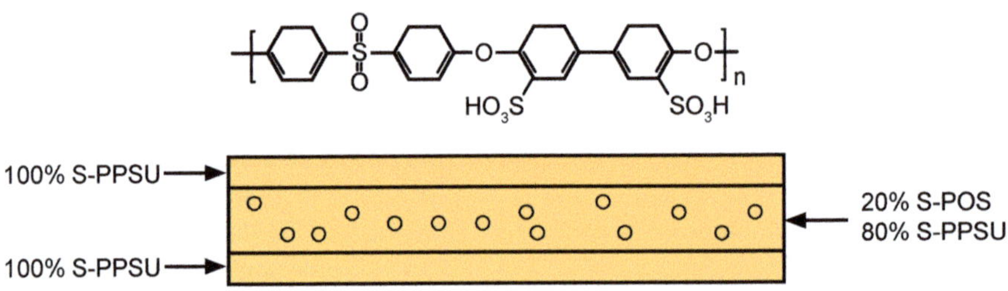

Fig. 7.34 Structures of sulfonated polyphenylsulfone (S-PPSU) repeat unit and of three-layer fuel cell membrane [214]

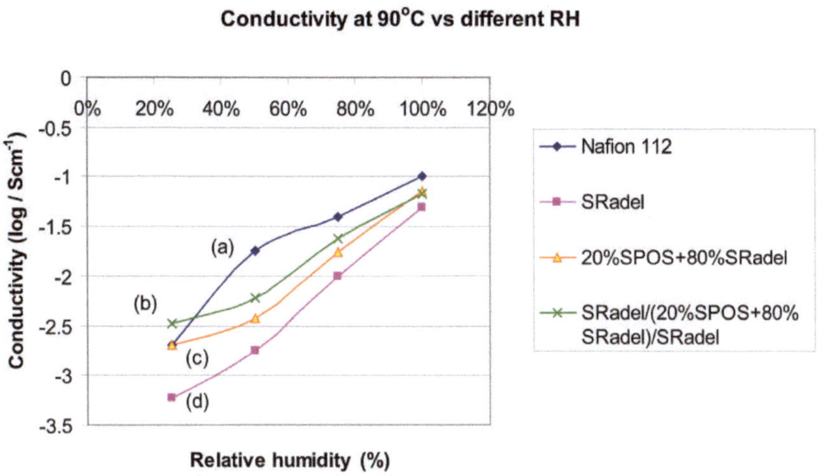

Fig. 7.35 Conductivity for different membranes. The conductivity data were collected at 90 °C at various % RH (a) Nafion® 112 (b) three-layered membrane (c) single-layered membrane with 20% S-POS and 80% S-PPSU (d) S-PPSU [214]

In this study [216], two other variants of octaphenyl-Si_8O_{12} carrying proton-conducting groups were also synthesized and evaluated in S-PPSU membranes. The first variant was an octaphenyl-Si_8O_{12} carrying both sulfonic acid groups and R_{18} alkyl groups, where the R groups were introduced in a Friedel Crafts alkylation of octaphenyl-Si_8O_{12}, and the product was subsequently sulfonated with sulfuric acid. Owing to the activating effect of the alkyl groups, mild sulfonation conditions could be used (in contrast to the more forcing chlorosulfonic acid sulfonation conditions necessary for octaphenyl-Si_8O_{12}). The second variant was an octaphenyl-Si_8O_{12} carrying phosphonic acid groups, where octaphenyl-POS was brominated and reacted with $(PhO)_2POCl$ in the presence of butyl lithium. The phosphonic ester was then hydrolyzed to the half ester.

In a later University of South Australia study [215], a partially sulfonated S-Si_8O_{12} with lower water solubility was prepared in order to reduce the likelihood of the additive being leached out of the membrane by water. A material with an IEC value in the range 2 to 2.5 mmol g^{-1}, corresponding to sulfonation of ~40% of the phenyl rings, was prepared by reaction of octaphenyl-Si_8O_{12} with chorosulfonic acid in chloroform at 0 °C for 4 hours. This S-Si_8O_{12} material was incorporated into a Nafion® 117 membrane, and also into a three-component membrane with Nafion® 117 and the ionic liquid 1-butyl-3-methylimidazolium bis(trifluoromethylsulfonyl) imide (BMI-BTSI). BMI-BTSI was introduced to act as a charge carrier in low humidity conditions, and S-Si_8O_{12} was used as a reinforcing filler to offset the plasticizing effect of BMI-BTSI upon the physical properties of the Nafion® 117 membrane. S-Si_8O_{12} was introduced at low levels (~1%) by immersing a swollen Nafion® 117 membrane into an alkaline aqueous solution of S-Si_8O_{12}, and BMI-BTSI was introduced by immersion of the membrane in a methanol solution of BMI-BTSI. S-Si_8O_{12} aggregates of 20 to 200 nm in size were observed by TEM throughout the membrane. At 100% RH and 80 °C the S-Si_8O_{12} / Nafion® membrane had a conductivity slightly higher than that of Nafion® (0.11 S cm^{-1} versus 0.09 S cm^{-1}), but a lower water uptake than Nafion®. In anhydrous conditions the conductivity of the S-Si_8O_{12} / Nafion® membranes increased as temperature increased, and was much higher than the Nafion® membrane. This was attributed to the ability of S-Si_8O_{12} to facilitate proton hopping between ionic Nafion® clusters. The addition of the ionic liquid to the S-Si_8O_{12} / Nafion® membrane increased conductivity under anhydrous and low humidity conditions.

In contrast to the closed-cage S-Si_8O_{12} species and to the composite membranes discussed above [214-218], a 2007 US-Korean collaboration [220] between Case Western Reserve University and Hanyang University describes a cross-linked membrane incorporating open-cage species (Fig. 7.36). An open-cage compound carrying three glycidyl epoxy groups was reacted with 4-hydroxybenzenesulfonic acid, and the resulting product (Fig. 7.36) was blended with polyvinylalcohol (PVA). As well as proton-conducting sulfonic acid groups, the compound also carried hydroxyl groups for PVA compatibility. The blend was cross-linked using ethylenediaminetetraacetic dianhydride (EDTAD). Membranes with various open cage to PVA ratios were prepared by mixing the two components with EDTAD in DMSO for 30 minutes at room temperature and then casting from DMSO solution. Unreacted (i.e., non-crosslinked) PVA and open cage compound were removed by immersion in DMSO. Proton conductivity increased with increasing sulfonated cage content, and was comparable with Nafion® at 20 wt % loading, and superior to Nafion® at 50 wt % loading. The membranes also had desirably low methanol permeability (an order of magnitude less than Nafion®) that decreased with increasing cage content. Related polysilsesquioxane composite membranes [221] based on (3-glycidoxypropyl)trimethoxysilane and oxidized (3-mercaptopropyl) trimethoxysilane have also been reported to have good conductivity (~0.1 S cm^{-1} at 70 °C in fully hydrated conditions).

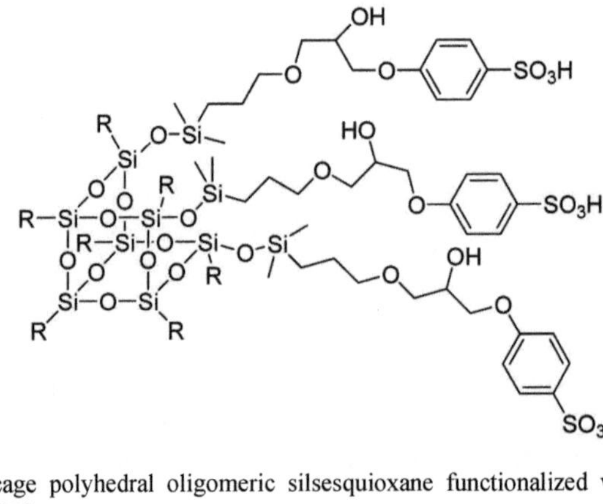

Fig. 7.36 Open-cage polyhedral oligomeric silsesquioxane functionalized with sulfonic acid and hydroxyl groups [220]

Instead of using perfluoro polymers carrying sulfonic acid groups (e.g., Nafion®) or sulfonated aromatic polymers [184], and attempting to manipulate their morphologies and create proton-conducting channels, e.g., block copolymers [223-225], or by fabricating composite membranes (such as the Si_8O_{12} composite membranes discussed earlier in this section), an entirely different approach [222,226] to fuel cell membrane design has been taken by Pintauro and co-workers at Vanderbilt and Case Western Reserve Universities in the US. In this approach proton-conducting channels were created by fabricating a three-dimensional network of interconnected polyelectrolyte nanofibers in an inert non-conducting matrix, where the nanofibers were made in an electrospinning process. Performance could be optimized by varying the composition of the nanofiber, the nanofiber diameter, the nanofiber volume fraction and the composition of the inert matrix. Performance was found to improve when the fully sulfonated variant of S-Si_8O_{12} (Fig. 7.33) [214,216,218] was added to the nanofibers [226]. The inert matrix, being entirely free of hydrophilic functionality (e.g., SO_3H) had no tendency to swell, and had excellent mechanical properties. In the initial 2008 study [222], a polyphenylene ethersulfone (PES) with 58% repeat units sulfonated was used as the nanofiber. A nanofiber mat was electrospun from DMAc solution, and then compressed to increase the density of nanofibers. Intersecting fibers were then interconnected to form proton-conducting paths by exposing the mat to DMF vapor. The mats were observed by SEM at various stages of fabrication and it was shown that the diameter of the nanofibers remained constant. The nanofiber mat was then impregnated with a UV-curable urethane adhesive to form the inert matrix. When in-plane conductivity *versus* nanofiber volume fraction was measured, no percolation threshold was observed, demonstrating the interconnectedness of the structure even at low nanofiber volume fractions. Conductivity increased linearly with nanofiber volume fraction as expected, and a 0.8 volume fraction nanofiber membrane had a conductivity of ~0.12 S cm^{-1} at 25 °C, versus 0.09 S cm^{-1} for Nafion® 117 under the same conditions. In a later study [226], conductivity was shown to increase with increasing IEC (SO_3H content) of the PES nanofiber material, and hence S-Si_8O_{12} (with its high density of SO_3H functionality) was introduced into the nanofiber in order to further increase its IEC. Nanofibers with 40 wt % S-Si_8O_{12} content were slightly thicker than those made from pure PES (400 nm versus 300 nm) and some precipitation of S-Si_8O_{12} outside the fiber was apparent. At 30 °C and 80% RH the 40 wt % S-Si_8O_{12} nanofiber membrane had a conductivity of 0.094 S cm^{-1} while the Nafion® 212 control membrane had a conductivity of 0.037 S cm^{-1}. The S-Si_8O_{12} nanofiber membrane retained its mechanical integrity at high relative humidity (>80%), while conventional composite membranes cast from PES/S-Si_8O_{12} mixtures lost their mechanical strength.

The introduction of polar nitrile groups [227] into sulfonated aromatic polymer PEMs is known to increase their affinity for inorganic fillers and to decrease their water uptake and swelling [228]. In a 2009 National University of Singapore study [227], polyacrylonitrile (PAN) chains were grafted to the corners of a mixture of T_8, T_{10} and T_{12} methacryl-substituted cages via atom transfer radical polymerization

(ATRP) (Fig. 7.37). The Si_8O_{12}-ATRP initiator was synthesized by reaction of the methacryl groups with bromine, and then polymerization with acrylonitrile was carried out using 2,2'-bipyridyl and copper(I) bromide. SEC showed that chain length was largely independent of reaction time (6 hours versus 20 hours) but M_n increased from 11,300 to 33,700 as the acrylonitrile-CuBr molar ratio increased from 200 to 600. No further mass increase was observed above the 600 ratio, but a decrease in the rate of polymerization was observed. This was rationalized in terms of steric hindrance occurring after comparatively short chains had grown from the vertices of the Si_8O_{12} cages [229]. The resulting starburst Si_8O_{12} product was added to Nafion® to fabricate composite membranes with low methanol permeability for direct methanol fuel cells (DMFC). Membranes carrying 5, 15 and 25 wt % Si_8O_{12} were cast from solutions of Nafion® 117 and Si_8O_{12} compound in DMAc. FTIR for hydrogen bonding between nitrile (CN) and sulfonic acid groups, intrinsic viscosity measurements and DSC studies all showed that the Si_8O_{12} additive had an affinity for the hydrophilic sulfonic acid regions of Nafion®, and was repelled from the hydrophobic perfluoro regions owing to the polar nitrile groups in PAN. This caused a severe disruption to the normal two-phase Nafion® morphology and its proton conducting channels. At 5 wt % loading, the Si_8O_{12} additive improved the power output of a DMFC by 122% at 80 °C. The 5% membrane left enough proton-conducting channels unperturbed for efficient conduction, but enough hydrophilic regions were perturbed by nitrile groups (repelling water and methanol) to give low methanol permeability. Performance also improved with the length of the PAN chains, although the nature of the Si_8O_{12}-initiated ATRP process limits the length of the chains, and keeps the overall size of the starburst structure below 20 nm in diameter.

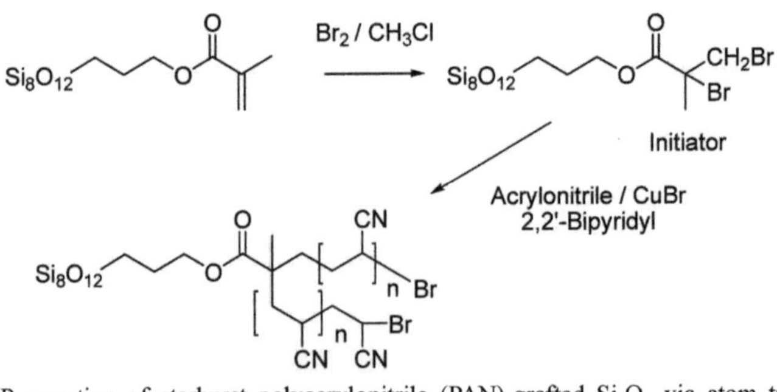

Fig. 7.37 Preparation of starburst polyacrylonitrile (PAN)-grafted Si_8O_{12} via atom transfer radical polymerization (ATRP) [227]

In an earlier study [230] carried out by the same group at the National University of Singapore, a hydrophobic octavinyl-Si_8O_{12} was added to Nafion® 117 at 5, 15 and 25 wt % loadings and used as a membrane in direct methanol fuel cells in order to reduce methanol crossover. The vinyl groups were polymerized in-situ in the membrane in the presence of UV light and a 2,2-dimethoxy-2-phenylacetophenone (DMPA) initiator, and nanosilica domains were generated and observed by SEM.

When DMAc-recast Nafion® membranes were compared with recast Nafion®-Si_8O_{12} membranes, the Nafion® membrane had a random matrix while the Nafion®-Si_8O_{12} membrane had a non-random matrix. The Nafion®-Si_8O_{12} membranes displayed lower methanol permeation, higher power density, lower activation energy to proton conduction and lower thermal stability than the pristine Nafion® membrane. A 2008 Chinese study by Zhao and Huang [231] describes the addition of Si_8O_{12} to a sulfonated polyphthalazinone ether ketone membrane that was solvent cast from a mixture of the two components. A conductivity comparable to that of Nafion® 117 was reported.

A number of studies of polysilsesquioxane fuel cell membranes [221,232-235], hybrid Nafion®-polysilsesquioxane membranes [236-238] and hybrid sulfonated polyetheretherketone (S-PEEK)-polysilsesquioxane [239,240] have also been carried out, especially in methanol fuel cells where reduction of methanol permeability is vital. In these hybrid membranes, polysilsesquioxanes are either formed in-situ [236,240] or else introduced as additives [237-239]. The structure, proton conductivity and chemical, thermal and mechanical stability can be optimized by controlling the sol-gel reaction conditions, e.g., pH, solvent, silicon to water ratio, catalyst and drying. Polysilsesquioxanes with either pendant propylsulfonic acid functionality or pendant ethylphosphonic acid functionality were prepared in sol-gel hydrolysis and condensation reactions using 3-(trihydroxysilyl)-1-propanesulfonic acid and trihydroxysilylethane phosphonic acid, respectively, and clear and colorless membranes were prepared by further reaction with various di-functional cross-linking agents [234]. The propylsulfonic acid membrane had a through-plane conductivity of 0.04 S cm^{-1} at 80 °C temperature and 100% RH, while the ethyl phosphonic acid membrane had lower conductivity (0.01 S cm^{-1} at 80° C and 100% RH) but higher thermal stability. Sulfonated polysilsesquioxane membranes prepared in nitric acid oxidations of disulfide-bridged polysilsesquioxane membranes had conductivities of 0.0062 S cm^{-1} under full humidification [232,233].

In another study, a Nafion® 117 membrane, pre-treated to generate the Nafion® sodium salt, was immersed in a TEOS-alkoxysilane mixture, and the sol-gel reaction proceeded inside the swollen Nafion® to give a hybrid Nafion®-ORMOSIL (organically modified silica) membrane [236]. The alkoxysilanes included vinyltriethoxysilane, diethoxydimethylsilane and diethoxydiphenylsilane. In another example of hybrid Nafion®-polysilsesquioxane membranes, a Nafion®/polyphenylmethylsilsesquioxane(PPSQ) / calcium hydroxyphosphate membrane and a Nafion®/PPSQ membrane were fabricated by dispersive mixing and sonication in DMF followed by solvent casting [237,238]. The membranes were then evaluated for application in direct dimethylether (DME) fuel cells above 100 °C. SEM showed that the PPSQ powder was 10 to 200 micron in size, but the domains in Nafion® were smaller than 1 micron in diameter and good dispersion was achieved. DME is a high energy density fuel with comparable performance to methanol, but in order to overcome slow oxidation kinetics at the electrode it is necessary to run DME fuel cells at temperatures above 100 °C. Both the crystallinity and the temperature of onset of thermal decomposition increase as the PPSQ content increases. The composite membranes appeared to have superior conductivity at high temperatures and high RH, and better single cell performance than Nafion® 115 at 100-120 °C and 1-2 atm pressure. A hybrid fuel cell membrane based on phosphonated polysilsesquioxane and S-PEEK has been reported, and a conductivity of 0.142 S cm^{-1} was measured at 120 °C, 100% RH and 40 wt % polysilsesquioxane loading [239]. An imidazole-functionalized polysilsesquioxane-S-PEEK membrane has also been reported [240]. Polysilsesquioxanes carrying sulfonic acid groups have also been used as hollow particle additives with holes on the surface [241] and have been added to membranes to decrease swelling. The work reviewed above concerns acidic membranes for proton conduction, but polysilsesquioxanes have also been used in alkaline membranes [242].

7.7 Polyhedral Oligomeric Silsesquioxanes in Battery Applications

In battery applications, polyhedral oligosilsesquioxanes have primarily been used in the electrolyte component of secondary (i.e., rechargeable) lithium ion batteries, although an early patent [211,212] suggested a sulfonated silsesquioxane electrolyte for a Zn/MnO_2 battery application as discussed in the previous section. Of all classes of battery, lithium has the highest energy density (Wh kg^{-1}) for electrochemical reasons associated with its high charge density in cationic form and its low atomic mass. Today's ubiquitous lithium ion battery is comprised of a carbon anode, a lithium salt cathode (e.g., $LiCoO_2$) and one of two major classes of electrolyte. The first class are liquids comprised of a lithium salt, e.g., $LiPF_6$, $LiAsF_6$, or $LiCF_3SO_3$, and an aprotic liquid, e.g., ethylene carbonate (EC), propylene carbonate (PC), ethyl methyl carbonate (EMC), dioxolane (DN), or fluorinated variants [243]. These systems have good conductivity (0.001 to 0.02 S cm^{-1} at 20 °C) and good lithium transport numbers, but do not function well at higher temperatures. In addition, liquid systems must be carefully sealed to prevent leaking, and may be flammable or explosive under certain conditions. The second class comprises the polymer electrolytes used in lithium-ion-polymer batteries, often referred to as 'lithium-polymer' batteries. These are safer and can be used at higher temperatures (60 to 100 °C). They can be used as both electrolyte and separator (enabling lightweight packaging and much greater flexibility in battery shape, design and aspect ratio), but they have low ionic conductivities and lithium transport numbers. Since electrolytes must have low reactivity with lithium, polymer electrolyte systems are usually based on polyethylene oxide (PEO, also known as polyethylene glycol, PEG), poly(methyl methacrylate) (PMMA), polyurethane, polyacrylonitrile (PAN) or fluoropolymers such as poly(vinylidene fluoride) (PVDF). Polymers may be used either alone, or as gelled polymer electrolytes (GPEs) in combination with a liquid and/or plasticizer. GPEs may be either homogenous and single phase, or else consist of a solvent phase in a porous polymer matrix. PEO alone with lithium salt has a very low conductivity (10^{-8} S cm^{-1} at 20 °C) whereas GPEs have conductivities as high as 0.009 S cm^{-1} at 20 °C [243]. Polymer-lithium salt electrolytes and GPEs have different mechanisms of lithium ion transport, and the GPE mechanism is dominated by diffusion of the liquid component [243].

Many strategies have been employed to improve conductivity without compromising the other beneficial properties of polymer electrolytes. High viscosity, high crystallinity, high glass transition temperature (T_g) and high cross-link density are all severely detrimental to ionic conductivity in linear polymer electrolytes. Hence, liquid plasticizers such as polypropylene carbonate have been added to improve conductivity, and gel electrolytes in which lithium salt/aprotic liquid electrolytes are trapped in polymer matrices have been developed [244]. However both of these approaches have some of the same drawbacks as the lithium salt/aprotic liquid electrolytes themselves. One way to obtain a system with a decreased glass transition temperature is to switch from a linear polymer architecture to a hyperbranched polymer architecture [245-252], and another approach has been to add micro- or nanooxide fillers (e.g., TiO_2, SiO_2 or Al_2O_3) in order to disrupt or suppress crystallinity, to lower T_g, and hence to improve conductivity [253,254].

Extensive studies of the effect of Si_8O_{12} filler addition upon the properties of lithium battery polymer electrolytes have been carried out by the Wunder research group at Temple University (Philadelphia, USA). In the first reports in 2002 and 2003, a study of PEO-grafted silicas for battery and pharmaceutical applications was extended to include Si_8O_{12} [255,256]. Two trimethoxy-PEO silanes (PEO-$Si(OMe)_3$, M_w 460 and 5000) were grafted onto untreated fumed silica, and several allyl-terminated PEO species were grafted onto octakis(hydridodimethylsiloxy)-Si_8O_{12} (Fig. 7.38) [255]. Allyl-terminated PEO variants with 2, 4, 8 and 12.5 repeat units were prepared in allylation reactions using allyl bromide and base. The glass transition temperatures of the starting allyl materials and the PEO-functionalized Si_8O_{12} materials (Fig. 7.38) were measured (Table 7.4). These results show that grafting to Si_8O_{12} has the potential to suppress the crystallinty of PEO, e.g., in the n = 4 case, the free PEO is completely crystalline while the Si_8O_{12}-grafted PEO is an amorphous liquid with a T_g of -85 °C. An elimination of crystallinity was also observed for PEO grafted to silica [257].

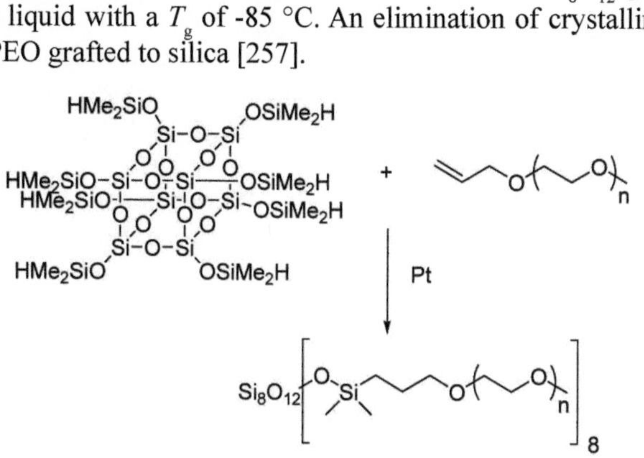

Fig. 7.38 Preparation of PEO-functionalized Si_8O_{12} [255]

Table 7.4 Glass transition temperatures (T_g) and melting points (T_m) measured by DSC for various free PEO(n = x) species and PEO(n = x) grafted onto Si_8O_{12} [255]

Material	T_g (°C)	T_m (°C)
Allyl PEO (n = 2)	-126	Completely amorphous
Allyl PEO (n = 4)	None	-28
Allyl PEO (n = 8)	None	-5
Allyl PEO (n = 12.5)	None	22
Si_8O_{12}-PEO (n = 2)	-89	None
Si_8O_{12}-PEO (n = 4)	-85	None
Si_8O_{12}-PEO (n = 8)	-80	-4
Si_8O_{12}-PEO (n = 12.5)	-69	19

Binary mixtures of lithium salts and Si_8O_{12}-PEO (n = x)$_8$ compounds of the type above (with the range of compositions extended to include n = 2, 4, 6, 8 and 12.5) were used to form viscous electrolyte solutions, and their conductivities were measured by ac impedance spectroscopy [258]. T_g increased with increasing chain length n, and the tendency of the chains to crystallize increased with increasing n. The lithium salts lithium chlorate ($LiClO_4$) and lithium bis(perfluoroethylsulfonyl) imide (LiBETI, $LiN(CF_3CF_2SO_2)_2$) were evaluated at oxygen to lithium (O/Li) ratios in the range 8/1 to 32/1. The conductivities of the Si_8O_{12}-PEO (n = x)$_8$ / lithium salt mixtures increased with decreasing T_g of the PEO component at low temperatures (close to the T_g). At high temperatures, conductivity increased with increasing PEO content, consistent with the lack of contribution of the Si_8O_{12} component to conductivity. Si_8O_{12}-PEO (n = 4)$_8$ with an O/Li ratio of 32/1 gave a conductivity of 2 x10^{-4} S cm^{-1} at 90 °C. Binary mixtures of Si_8O_{12}-PEO (n = x)$_8$ compounds and $LiClO_4$ had conductivities of 10^{-3} S cm^{-1} at 50 °C and 10^{-4} S cm^{-1} at room temperature, comparable to oligomeric PEO (M_w 300-500). The T_g values for these mixtures were between -60 and -70 °C [259]. In a later study [260], the same set of Si_8O_{12}-PEO (n = x)$_8$ compounds was formulated with a wider range of lithium salts including the original pair plus $LiN(CF_3SO_3)_2$, $LiPF_6$, $LiAsF_6$ and $LiBF_4$. The glass transition temperature increased with increasing salt concentration and PEO chain length. When comparing lithium salts at high temperatures and low viscosities, the less associative salts ($LiClO_4$, $LiN(CF_3CF_2SO_2)_2$, $LiN(CF_3SO_3)_2$, $LiPF_6$ and $LiAsF_6$) contributed more charge carriers and gave higher conductivities than the more associative salts ($LiBF_4$ and $LiCF_3SO_3$). When comparing lithium salts at low temperatures and high viscosities, the low T_g systems gave higher conductivity than the high T_g systems. Lithium salts either lower the T_g through the plasticizing effects of large organic anions (e.g., $LiN(CF_3CF_2SO_2)_2$) or affect the T_g by complexation with PEO, where the less associative salts have a greater tendency to complex with PEO and raise the T_g. The highest conductivity obtained in this system was 1.1 x 10^{-4} S cm^{-1} for Si_8O_{12}-PEO (n = 4)$_8$ and $LiN(CF_3CF_2SO_2)_2$ with an oxygen to lithium (O/Li) ratio of 16/1.

This work on viscous liquid electrolytes formed a basis for the design of solid polymer electrolytes that were prepared by blending Si_8O_{12}-PEO (n = 4)$_8$ with high molecular weight PEO (M_w 600,000) and $LiClO_4$ at various ratios [261]. The O/Li = 8/1 and 12/1 compositions had a tendency to microphase-separate into a high T_g amorphous phase (dominated by PEO) and a low T_g amorphous phase dominated by Si_8O_{12}-PEO (n = 4)$_8$. The O/Li = 16/1 compositions had the same two amorphous phases plus a third crystalline phase comprised entirely of PEO. As expected the conductivities were dictated by the nature of the continuous phase, where a continuous low T_g amorphous phase gave the highest values and a continuous crystalline phase gave the lowest values. The highest conductivity obtained was 8 x 10^{-6} S cm^{-1} at room temperature versus 2 x 10^{-6} S cm^{-1} for PEO itself at the same O/Li = 12/1 ratio.

Beyond closed-cage Si_8O_{12} itself, several other attempts have been made to use sol-gel derived materials to improve electrolyte conductivity. In one example, PEO chains of approximately 40 repeat units were grafted to a sol-gel derived siliceous framework via urea linkages, and a conductivity value of 6 x 10^{-6} S cm^{-1} at room temperature was recorded [262]. In a second, highly successful study, conductivity was increased by two orders of magnitude, and values in the region of 10^{-4} S cm^{-1} were obtained [263]. This was achieved by performing in situ sol-gel chemistry in PEO using 3-glycidopropyltrimethoxysilane (GLYMO) to produce a solid single-ion conducting (SIC) composite electrolyte with a three-dimensional network structure, where the glycidyl groups were selected to enhance compatibility with the PEO. Silicate particles with a diameter of 15 nm were prepared by reacting the three sol-gel reagents (GLYMO, tetramethoxysilane (TMOS) and the sodium salt of trihy droxysilylpropylmethylphosphonate) to give a system with immobile phosphonate anions and mobile sodium cations. Owing to the SIC nature of the system, near-unity cation transferance values were obtained, in contrast to conventional PEO electrolytes where ion mobility is associated with aggregates of ions; hence there is less of a contribution from single ions, and transfer numbers are significantly less than one.

Other examples of silsesquioxanes in battery technology include a Samsung patent [264] reporting that a hydrogen silsesquioxane was sintered and used in an anode for a lithium battery with improved charge and discharge performance, and a German patent [265] reporting that oligomeric silicon-oxygen clusters improved the chemical and thermal stability of lithium battery separators. The separator was made from a porous non-conducting polymer support coated with a ceramic, and $Si_8O_{12}(OSiR_3)_8$ was incorporated into the polymer support. The function of a battery porous separator is to separate cathode and anode, to be permeable to electrolyte under normal operating conditions, but to close its pores and shut down the battery in the event of a thermal runaway.

7.8 Polyhedral Oligomeric Silsesquioxanes as Lubricants

In 1982, a Russian group [266] first suggested that alkyl oligosilsesquioxanes with aliphatic chain lengths up to 6 carbons in length (i.e., hexyl) might be useful as lubricants, and more recently, Si_8O_{12} compounds have been patented [267] as lubricants and as additives to lubricant materials for control of viscosity, wear and thermal performance, where Si_8O_{12} chemical and thermal stability and tailorable properties have proved to be an advantage. These Si_8O_{12} compounds, functionalized with various octyl and hexyl groups, have been described as 'molecular ball-bearings'. In the patent examples they were formulated with a polyester lubricant, MIL-L-7808, and enabled the lubricant to avoid mass loss for longer at a given temperature, and to reduce the wear of metal components in a three-ball disc test. This patent also covered the use of Si_8O_{12} compounds in place of conventional mold-release agents (i.e., mineral oils, silicones or fluorinated polysiloxanes, as also alluded to in Chapter 5). Since mechanical systems must function from low temperatures up to extremely high temperatures (e.g., 500 °C and above in jet engines), the key materials challenges are to achieve stability at high temperatures, and to achieve lubricating properties across a wide temperature range. In addition, lubricants should be non-volatile, non-corrosive and should not decompose to form solid deposits that can impede mechanical functions. Next-generation automotive engines are expected to be smaller, and to function at higher temperatures in order to maximize fuel efficiency; hence this could be beyond the abilities of existing hydrocarbon mineral oil-based lubricants. Various high temperature lubricants based upon complex formulations of silicones, carboxylic acids or polyhydroxyl esters with antioxidants have been proposed. When used as an additive, the size of the Si_8O_{12} compound (0.7 to 3.0 nm) enables it to occupy sites within the lubricant, and to alter its free volume. The additive may be physically mixed or reactively blended with the lubricant. Si_8O_{12}-based lubricants resemble conventional lubricants in that they behave in a Newtonian manner, and their viscosity decreases at higher temperatures.

A Nigerian study by Dare [268] showed that the hydraulic fluid properties of Si_8O_{12} compounds were comparable with the commercial hydraulic fluids Mobil® EAL 224H (vegetable oil-based), Super-V® (AP) (petroleum-based) and Puroil® SHO (synthetic siloxane-based). Several alkyl-functionalized Si_8O_{12} compounds were prepared in hexachloroplatinic acid-catalyzed hydrosilylation reactions of $Si_8O_{12}(H)_8$ with 7-bromo-1-heptene, 8-bromo-1-octene, allyltrichlorosilane and 2-methyl-3-butyn-2-ol, respectively. The three colorless liquids in Table 7.5 showed the greatest promise as lubricants, as did the compound $Si_8O_{12}(CH_2Ph)_8$ that had been prepared in a Friedel Crafts alkylation of octavinyl-Si_8O_{12} with benzene and aluminum chloride. Typical industrial lubricants have kinematic viscosities (kv) in the range 5 to 15 cSt at 100 °C, and pour points from -40 to +6 °C.

Table 7.5 Some properties of Si_8O_{12}-based lubricants versus commercial lubricants [268]

	kv (cSt) 40°C	kv (cSt) 100°C	Onset of solidification (°C)	Pour point (°C)
$Si_8O_{12}[(CH_2)_7Br]_8$	35	6	0	--
$Si_8O_{12}[(CH_2)_8Br]_8$	38	6	8	--
$Si_8O_{12}[(CH_2)_6Cl]_8$	40	7	-5	-38
$Si_8O_{12}(CH_2Ph)_8$	48	11	-20	-25
Mobil® EAL	35	8	--	--
Super-V®	22	15	--	--
Puroil® SHO	42	7	--	--

In addition, methylsilsesquioxane particles (0.3 micron) prepared by emulsion polymerization were proposed as anti-wear lubricants for magnetic memory devices [269]. The use of Si_8O_{12} materials as lubricants during the processing of polypropylene [270-272] or nylon 6 [273] has also been studied in depth (see also Section 5.5.2, Chapter 5).

7.9 References

1. Donald A, Windle A, Hanna S (2005) Liquid Crystalline Polymers, Cambridge University Press, 2nd Ed. Cambridge, UK.
2. Pan Q, Fan X, Chen X, Zhou Q (2006) Huaxue Jinzhan 18(5):616-621.
3. Teyssie D, Boileau S (2000) Liquid Crystalline Silicon-containing Polymers. In Jones RG, Ando W, Chojnowski J (eds) The Science and Technology of their Synthesis and Applications, Kluwer, Dordrecht, Netherlands, Ch 22, pp 593-613.
4. Gray GW (1987) Thermotropic Liquid Crystals, John Wiley and Sons, Chapter 6.
5. Mehl GH, Goodby JW (1997) Mol Cryst Liq Cryst 303:15-21.
6. Mehl GH, Thornton AJ, Goodby JW (1999) Mol Cryst Liq Cryst 332:2965-2971.
7. Xie P, Guo JS, Dai DR, Jin SZ, Liu DS, Li Z, Zhang RB (1996) Polym Adv Technol 7:98.
8. Xie P, Wan Y, Zhou B, Hou J, Dai D, Li Z, Liu D, Zhang R (1996)

Macromol Chem Phys 197:745.

9. Kreuzer FH, Maurer R, Spes P (1991) Macromol Symp 50:215-228.

10. Saez IM, Styring P (1996) Adv Mater 8(12):1001-1005.

11. Mehl GH, Saez IM (1999) Appl Organomet Chem 13(4):261-272.

12. Serrano JL (1996) Metallomesogens, VCH, Weinheim.

13. McArdle CB (1989), Side Chain Liquid Crystals, Blackie, Glasgow, UK.

14. Lewthwaite AR, Gray GW, Toyne K (1992) J Mater Chem 2:119.

15. Sellinger A, Laine RM, Chu V, Viney C (1994) J Polym Sci Part A Polym Chem 32:3069-3089.

16. Aquilera C, Bartulin J, Hisgen B, Ringsdorf H (1983) Makromol Chem 184:253.

17. Sellinger A, Laine RW (1994) Polym Prepr (Am Chem Soc Div Polym Chem) 35(2):665-666.

18. Saez IM, Styring P (1997) Mol Cryst Liq Cryst 299:163-168.

19. Mehl GH, Goodby JW (1996) Angew Chem Int Ed 35:2641.

20. Mehl GH, Goodby JW (1996) Chem Ber 129:521.

21. Ibn Elhaj M, Skoulios A, Gouillon D, Newton J, Hodge P, Coles HJ (1995) Liq Cryst 19:373.

22. Saez IM, Goodby JW (1999) Liq Cryst 26(7):1101-1105.

23. Ahsan MS, Sasaki S, Kawakami Y (2007) React Funct Polym 67(11): 1200-1210.

24. Elsasser R, Mehl GH, Goodby JW, Photinos DJ (2000) Chem Commun 10:851-852.

25. Karahaliou PK, Kouwer PHJ, Meyer T, Mehl GH, Photinos DJ (2007) Soft Matter 3(7):857-865.

26. Laine RM, Zhang C, Sellinger A, Viculis L (1998) Appl Organomet Chem 12:715.

27. Zhang CX, Bunning TJ, Laine RM (2001) Chem. Mater 13:3653-3662.

28. Kim K-M, Chujo Y (2001) Polym Bull 46(1):15-21.

29. Saez IM, Goodby JW (2001) J Mater Chem 11(11):2845-2851.

30. Saez IM, Goodby JW, Richardson RM (2001) Chem- Eur J 7:2758-2764.

31. Pan Q, Chen X, Fan X, Shen Z, Zhou Q (2008) J Mater Chem 18(29):3481-3488.

32. Link DR, Natale G, Shao R, Maclennan JE, Clark NA, Korblova E, Walba DM (1997) Science 278:1924.

33. Cui L, Collet JP, Zhu L (2007) Polym Mater Sci Eng Prep 97:408-409.

34. Cui L, Collet JP, Xu G, Zhu L (2006) Chem Mater 18(15):3503-351.

35. Miao J, Lei (2010) J Phys Chem B 114(5):1879-1887.

36. Pan Q, Gao L, Chen X, Fan X, Zhou Q (2007) Macromolecules 40(14):4887-4894.

37. Mather PT, Chaffee KP, Haddad TS, Lichtenhahn JD (1996) Polym Prepr Am Chem Soc Div Polym Chem 37(1):765-766.

38. Somlai AP, Iyer S, Schiraldi DA (2004) Polym Prepr (Am Chem Soc Div Polym Chem) 45(1):650-651.

39. Sethumadhaven M, Kennedy SD, Barton CL (2004) World Pat. 2004114732 A1.

40. Kim KM, Chujo Y (2001) J Polym Sci 39(22):4035-4043.

41. Jeng S-C, Hwang S-J, Yang C-Y (2009) Opt Lett 34(4):455-457.

42. Jeng S-C, Kuo C-W, Wang H-L, Liao C-C (2007) Appl Phys Lett 91(6):061112/1-061112/3.

43. Teng WY, Jeng SC, Kuo CW, Lin YR, Liao CC, Chin WK (2008) Opt Lett 33:1663.

44. Hwang S-J, Jeng S-C, Yang C-Y, Kuo C-W, Liao C-C (2009) J Phys D: Appl Phys 42(2):025102/1-025102/6.

45. Chen W-Z, Tsai Y-T, Lin T-H, (2009)Appl Phys Lett 94(20):201114/1-201114/3.

46. Kim E-H, Myoung S-W, Lee W-R, Jung Y-G (2009) J Korean Chem

Soc 54(3):1180-1186.

47. Pope M, Kallmann H, Magnante PJ (1962) J Chem Phys 38:2042.

48. Burroughes JH, Bradley DDC, Brown AR, Marks RN, Mackey K, Friend RH, Burns PL, Holmes AB (1990) Nature (London) 347:539.

49. Friend RH, Gymer RW, Holmes AB, Burroughes JH, Marks RN, Taliani C, Bradley DDC, Dos Santos DA, Bredas JL, Logdlund M, Salaneck WR (1999) Nature (London) 394:121.

50. Lee J, Ch H-J, Cho NS, Hwang D-H, Shim H-K (2006) Synth Met 156(7-8):590-596.

51. Chan KL, Sonar P, Sellinger A (2010) J Mater Chem 19(48):9103-9120.

52. Ranger M, Rondeau D, Leclerc M (1997) Macromolecules 30:7686-7691.

53. Lin W-J, Chen W-C, Wu W-C, Niu Y-H, Jen AKY (2004) Macromolecules 37(7):2335-2341.

54. Chou C-H, Hsu S-L, Yeh S-W, Wang H-S, Wei K-H (2005) Macromolecules 38(22):9117-9123.

55. Cho C-H, Hsu S-L, Dinakaran K, Chiu M-Y, Wei K-H (2005) Macromolecules 38(3):745-751.

56. Xiao S, Nguyen M, Gong X, Cao Y, Wu H, Moses D, Heeger AJ (2003) Adv Funct Mater 13(1):25-29.

57. Xiao S, Nguyen MT (2003) US Pat. US 20030204038.

58. Shin S-B, Gong S-C, Jang J-K, Gong M-S, Chang Y-C, Sun Y-B, Chang H-J (2008) J Appl Polym Sci 110(6):3678-3682.

59. Fenenko L, Adachi C, Nakanishi Y, Smertenko P, Svechnikov S (2007) Mol Cryst Liq Cryst 467:303-309.

60. Lee J, Cho H-J, Jung B-J, Cho NS, Shim H-K (2004) Macromolecules 37(23):8523-8529.

61. Lee J, Cho H-J, Nam S, Hwang D-H, Kang J-M, Lim E, Lee J-I, Shim H-K (2006) J Polym Sci Part A Polym Chem 44(9):2943-2954.

62. Takagi K, Kunii S, Yuki Y (2005) J Polym Sci Part A Polym Chem 43(10):2119-2127.

63. Pu K-Y, Zhang B, Ma Z, Wang P, Qi X-Y, Chen R-F, Wang L-H, Fan Q-L, Huang W (2006) Polymer 47(6):1970-1978.

64. Xiao Y, Lu X, Tan L-W, Ong KS, He C (2009) J Polym Sci Part A Polym Chem 47(21):5661-5670.

65. Nguyen TP, Lee CW, Hassen S, Le HC (2009) Solid State Sciences 11(10):1810-1814.

66. Sellinger A, Laine RM (2002) World Pat. WO2002005971.

67. Cammack JK, Jabbour GE, Li S, Froehlich J (2005) US Pat. 0123760 A1.

68. Lee CW, Josse Y, Hsu CH, Nguyen TP (2008) Eur Phys J Appl Phys 42(3):213-218.

69. Renaud C, Josse Y, Lee C-W, Nguyen T-P (2008) J Mate Sci Mater Electron 19(1):S87-S91.

70. He C, Xiao Y, Huang J, Lin T, Mya KY, Zhang X (2004) J Am Chem Soc 126:7792-7793.

71. Brick CM, Tamaki R, Kim S-G, Asuncion MZ, Roll M, Nemoto T, Ouchi Y, Chujo Y, Laine RM (2005) Macromolecules 38(11):4655-4660

72. Brick CM, Ouchi Y, Chujo Y, Laine RM (2005) 38(11):4661-4665.

73. Xiao Y, Liu L, He C, Chin WS, Lin T, Mya KY, Huang J, Lu X (2006) J Mater Chem 16:829-836.

74. Eom J-H, Mi D, Park M-J, Cho H-J, Lee J, Lee J-I, Chu HY, Shim H-K, Hwang D-H, (2009) J Nanosci Nanotech 9(12):7029-7033.

75. Imae I, Kawakami Y (2005) J Mater Chem 15:4581-4583.

76. Sellinger A, Tamaki R, Laine RM (2005) Chem Commun 29:3700-3702.

77. Cho H-J, Hwang D-H, Lee J-I, Jung Y-K, Park J-H, Lee J, Lee S-K, Shim H-K (2006) Chem Mater 18(16):3780-3787.

78. Froelich D, Young R, Nakamura Y, Ohmori S, Li S, Mochizuki A, Lauters M, Jabbour GE (2007) Chem Mater 19:4991.

79. Lo MY, Zhen CG, Lauters M, Gabbour GE (2007) J Am Chem Soc 129:5808.

80. Yang X, Froehlich JD, Chae, HS, Li S, Mochizuki A, Jabbour GE (2009)

Adv Funct Mater 19(16):2623-2629.

81. Park J-L, Lee T-W, Kakimoto MA, Pu LS (2007) US Pat. US0045619 A1.

82. Singh M, Chae HS, Froehlich JD, Kondou T, Li S, Mochizuki A, Jabbour GE (2009) Soft Matter 5:3002-3005.

83. Chen K-B, Chang YP, Yang SH, Hsu C-S (2006) Thin Solid Films 514(1-2):103-109.

84. Xu Y, Peng J, Jiang J, Xu W, Yang W, Cao Y (2005) Appl Phys Lett 87(19):193502/1-193502/3.

85. Liang B, Jiang C, Chen Z, Zhang X, Shi H, Cao Y (2006) J Mater Chem 16(13):1281-1286.

86. Dukart CR, Ranzau SL, Pappenfus TM (2010) 239th ACS National Meeting, San Francisco, CA, USA, 21-25 March 2010, CHED-740.

87. Bozano L, Tuttle SE, Carter SA, Brock P (1998) J Appl Phys Lett 73:3911-3913.

88. Carter SA, Scott JC, Brock P (1997) J Appl Phys Lett 71(9):1145-1147.

89. Blom PWM, Schoo HFM, Matters M (1998) J Appl Phys Lett 75:3914.

90. Gong X, Soci C, Yang C, Heeger AJ, Xiao S (2006) J Phys D Appl Phys 39(10):2048-2052.

91. Lee R-H, Lai H-H (2007) Eur Polym J 43(3):715-724.

92. Cui Y, Chen L, Qian G, Wang M (2008) J Non-Crystall Solids, 543(12-13):1211-1215.

93. Xi H, Li Z, Liang Z (2001) Chinese J Polym Sci, 19(4):421-427.

94. Xie G, Sun D (2007) PMSE Prep 97:730-732.

95. Jeng RJ, Chen YM, Jain AK, Kumar J, Tripathy SK (1992) Chem Mater 4(5):972-975.

96. Oviatt HW, Shea KJ, Kalluri S, Shi Y, Steier WH, Dalton LR (1995) Chem Mater 7(3):493-498.

97. Han S, Li Z, Ji S, Dai D, Zhang R, Zhu C, Wang C (2000) J Sol-Gel Sci Tech, 18(2):137-144.

98. Choi KM, Shea KJ (1998) Plastics Engineering, Photonic Polymer Systems, Marcel Dekker, Inc., New York, pp. 437-480.

99. Luther-Davies B, Samoc M, Woodruff M (1996) Chem Mater 8(11): 2586-2594.

100. Su H-W, Chen W-C, Lee W-C, King J-S (2007) Macromol Mater Eng 292(5):666-673.

101. Xie P, Guo JS, Dai DR, Jin SZ, Liu DS, Li Z, Zhang RB (1996) Polym Adv Technol 7:98-103.

102. Prasad PN, Williams DJ (1991) Introduction to Nonlinear Optical Effects in Molecules and Polymers, John Wiley and Sons, Inc.

103. Sutherland RL (1996), Handbook of Nonlinear Optics, Marcel Dekker Inc., New York.

104. Sarkar A, Mirza S, Rahman S, Rayfield G (2011) Carbon Nanotubes For Optical Power Limiting Applications, In Wang ZM, Neogi A (Eds) Nanoscale Photonics and Electronics, Springer.

105. Su X, Guang S, Xu H, Liu X, Li S, Wang X, Deng Y, Wang P (2009) Macromol 42:8969-8976.

106. Su X, Guang S, Li C, Xu H, Liu X, Wang X, Song Y (2010) Macromolecules 43(6):2840-2845.

107. Ceyhan T, Yuksek M, Yaglioglu HG, Salih B, Erbil MK, Elmali A, Bekaroglu O (2008) Dalton Trans 2407-2413.

108. Costela A, Garcia-Moreno I, Cerdan L, Martin V, Garcia O, Sastre R (2009) Adv Mater 21:4163-4166.

109. Garcia O, Sastre R, Garcia-Moreno I, Martin V, Costela A (2008) J Phys Chem 112:710.

110. Argitis A, Niakoula D, Douvas AM, Gogolides E, Raptis I (2009) Int J Nanotech 6(1-2):71-87.

111. Xia Y, Rogers JA, Paul KE, Whitesides GM (1999) Chem Rev 99(7): 1823-1848.

112. Hirai T, Leolukman M, Liu CC, Han E, Kim YJ, Ishida Y, Hayakawa T, Kakimoto M, Nealey PF, Gopalan P (2009) Adv Mater 21(43):4334-4338.

113. Dai J, Chang SW, Hamad A, Yang D, Felix N, Ober CK (2006) Chem Mater 18:3404-3411.

114. Kanagasabapathy, S., Barclay, G.G., Cameron, J.F., Pohlers, G., Huby, F., Wiley, K. Abstracts of Papers, 227th ACS National Meeting, Anaheim, CA, USA, March 28-April 1, 2004, POLY-550.

115. Korchkov VP, Martynova TN (1985) Zhur Prikladnoi Khimii 58(9):2089-2096.

116. Gonsalves KE, Wang J, Wu H (2000) J Vac Sci Technol B Microelectron Nanometer Struct Process Meas Phenom 18(1):325-327.

117. Wu H, Hu Y, Gonzales KE, Yakaman MJ (2001) J Vac Sci Technol B 19:851.

118. Azam AM, Gonzales KE, Golovkina V, Cerrina F (2003) Microelectron Eng 65:454.

119. Wu H, Gonazles KE (2001) Adv Mater 13:670.

120. Bellas V, Tegou E, Raptis I, Gogolides E, Argitis P, Iatrou H, Hadjichristidis N, Sarantopoulou E, Cefalas AC (2002) J Vac Sci Technol B 20:2902.

121. Gonzales KE, Merhari L, Wu H, Hu Y (2001) Adv Mater 13:703.

122. Sarantopoulou E, Kollia Z, Kocevar K, Musevic I, Kobe S, Drazic G, Gogolides E, Argitis P, Cefalas AC (2003) Mater Sci Eng C23(6-8): 995-999.

123. Ali MA, Gonsalves KE, Batina N, Golovkina V, Cerrina F (2003) Proc SPIE Int Soc Opt Eng 5039:1173-1180.

124. Tegou E, Bellas V, Gogolides E, Argitis P, Dean KR, Eon D, Cartry G, Cardinaud C (2003) Proc SPIE Int Soc Opt Eng 5039:453-461.

125. Gogolides E, Argitis P, Bellas V, Tegou E (2003) World Patent WO2003102695.

126. Koh K, Sugiyama S, Morinaga T, Ohno K, Tsujii Y, Fukada T, Yamahiro M, Ijima T, Oikawa H, Watanabe K, Miyashita T (2005) Macromolecules 38:1264.

127. Eon D, Cartry G, Fernandez V, Cardinaud C, Tegou E, Bellas V, Argitis P, Gogolides E (2004) J Vac Sci Technol B Microelectron Nanometer Struct Process Meas Phenom 22(5):2526-2532.

128. Eon D, Raballand V, Cartry G, Cardinaud C, Vourdas N, Argitis P, Gogolides E (2006) J Vac Sci Technol B Microelectron Nanometer Struct Process Meas Phenom 24(6):2678-2688.

129. Douvas AM, Van Roey F, Goethals M, Papadokostaki KG, Yannakopoulou K, Niakoula D, Gogolides E, Argitis P (2006) Chem Mater 18:4040-4048.

130. Tegou E, Bellas V, Gogolides E, Argitis P, Eon D, Cartry G, Cardinaud C (2004) Chem Mater 16:2567-2577.

131. Zheng L, Waddon AJ, Farris RJ, Coughlin BE (2002) Macromolecules 35:2375.

132. Kim J-B, Ganesan R, Choi J-H, Yun H-J, Kwon Y-G, Kim K-S, Oh TH (2006) J Mater Chem 16:3448-3451.

133. Park JY, Kim MG, Kim J-B (2008) Macromol Rapid Commun 29(18):1532-1537.

134. Allen RD, Sooriyakumaran R, Sundberg LK (2006) US Pat 0189779 A1.

135. Hao J, Lin MW, Palmieri F, Nishimura Y, Chao H-L, Stweart MD, Collins A, Jen K, Wilson CG (2007) Proc SPIE Int Soc Opt Eng 6517: 651729-1-651729-9.

136. Palmieri F, Stewart MD, Wetzel J, Hao J, Nishimura Y, Jen K, Flannery C, Li B, Chao H-L, Young S, Kim WC, Ho PS, Wilson CG (2006) Proc SPIE Int Soc Opt Eng 6151:61510J/1-61510J/9.

137. Palmieri F, Stewart M, Jen K, Wilson CG, Schmid G (2007) Solid State Technol 50(9):42-45.

138. Sellinger A, Laine RM (1996) Chem Mater 8:1592-1593.

139. Hartmann-Thompson C, Hu J, Xu N (2007) unpublished results, Michigan Molecular Institute.

140. Xu Y, Zhu X, Yang S (2009) ACS Nano 3(10):3251-3259.

141. Moon JH, Seo JS, Xu Y, Yang S (2009) J Mater Chem 19:4687-4691.

142. Jang JH, Ullal CK, Gorishny T, Tsukruk VV, Thomas EL (2006) Nano Lett 6:740-743.

143. Jeganathan SG, Bramer D, Kote R, Maladkar GJ (2008) US 0029739.

144. Hartmann-Thompson C, Keeley D, Pollock K, Keinath SE, Dvornic PR, Dantus M, Gunaratne T, Lecaptain D (2008) Chem Mater 20(8): 2829-2838.

145. Bai H, Li, C, Shi G (2008) Chem Phys Chem 9(13):1908-1913.

146. Tanaka K, Inafuku K, Chujo Y (2008) Bioorg Med Chem 16(23):10029-10033.

147. Carlson CA, Lloyd JA, Dean SL, Walker NR, Edmiston PL (2006) Anal Chem 78(11):3537-3542.

148. Haupt K, Mosbach K (2000) Chem Rev 100:2495-2504.

149. Haidekker MA, Ling T, Anglo M, Stevens HY, Frangos JA, Theodorakis EA (2001) Chem Biol 8:123.

150. McCusker C, Carroll JB, Rotello VM (2005) Chem. Commun:996-998.

151. Haugland RP (2005) The Handbook. A Guide To Fluorescent Probes and Labelling Technologies, 10th edn. Invitrogen Corp, Eugene, Oregon.

152. Zou Q-C, Yan Q-J, Song G-W, Zhang S-L, Wu L-M (2007) Biosens Bioelectron 22(7):1461-1465.

153. Sirbuly DJ, Letant SE, Ratto TV (2008) Adv Mater 20(24):4724-4727.

154. Shahriari MR (2007) US Pat. 0122311 A1.

155. Ballantine DS, Rose SL, Grate JW, Wohljen H (1986) Anal Chem 58:3058.

156. Albert KJ, Lewis NS, Schauer CL, Sotzing GA, Sitzel SE, Vaid TP Walt DR (2000) Chem Rev 100:2595-2626.

157. Grate JW (2000) Chem Rev 100:2627-2648.

158. Grate JW (2008) Chem Rev 108:726-745.

159. Hartmann-Thompson C, Hu J, Dvornic PR, Kaganove SN, Keinath SE, Keeley D (2004) Chem Mater 16(24):5357-5364.

160. Hartmann-Thompson C, Keeley DL, Voit B, Eichhorn KJ, Mikhaylova M (2008) J Appl Polym Sci 107(3):1401-1406.

161. Barie N, Rapp M, Ache HJ (1998) Sens Actuators B 46:97-103.

162. Hartmann-Thompson C, Keeley D, Dvornic PR, Keinath SE, McCrea K (2007) J Appl Polym Sci 104(5): 3171-3182.

163. Hartmann-Thompson C (2009) US Pat. 0263287 A1.

164. Lambert AG, Davies PB, Neivandt DJ (2005) Appl Spectrosc 40:103-145.

165. Smentkowski VS, Duong HM, Tamaki R, Keenan MR, Ohlhausen JAT, Kotula PG (2006) Appl Surf Sci 253(2):1015-1022.

166. Esker AR, Dawson KJ, Huffer SM, Karabiyik,U, Deng J, Viers BD, Ferguson-McPherson MK, Morris JR, Mao M, Ducker WA, Satija SK (2005) Polym Prepr (Am Chem Soc Div Polym Chem) 46(1):76-77.

167. Trammell SA, Zeinali M, Melde BJ, Charles PT, Velez FL, Dinderman MA, Kusterbeck A, Markowitz MA (2008) Anal Chem 80(12):4627-4633.

168. Siemens AG (2004) US Pat. 0025568 A1.

169. Tsujimura Y, Yamane M, Wakida S-I (2001) Anal Sci 17(4):485-489.

170. Merzlyuk RM, Yashina NI, Shevchenko YN (1997) Zhurnal Prikladnoi Khimii (Sankt-Peterburg) 70(4):631-635.

171. Castaldo A, Massera E, Quercia L, Di Francia G (2006) Sens Actuators B 118:328-332.

172. Gardner JW, Bartlett PN (1999) Electronic Noses: Principles and Applications. Oxford University Press, New York.

173. Lonegran MC, Severin EJ, Doleman BJ, Beaber SA, Grubbs RH, Lewis NS (1996) Chem Mater 8:2298-2312.

174. Hartmann-Thompson C (2009) Carbon-Polymer Composite Sensors in Electronic Noses. In: Lechkov M, Prandzheva S (eds) Encyclopedia of Polymer Composites: Properties, Performance and Applications, Nova Science Publishers, New York.

175. Gardner JW, Shurmer HV, Corcoran P (1991) Sens Actuators B 4:117.

176. Castaldo A, Massera E, Quercia L, DiFranca G (2007) Macromol Symposia 247:350-356.

177. Massera E, Castaldo A, Quercia L, DiFranca G (2008) Sens Actuators

B129(1):487-490.

178. Castaldo A, Quercia . DiFrancia G, Cassinese A, D'Angelo P (2008) J Appl Phys 103(5):054511/1-054511/6.

179. Ryan MA, Shevade AV, Zhou H, Homer ML (2004) Polymer Carbon Black Composite Sensors in Electronic Noses for Air Quality Monitoring, NASA Bulletin,714-719.

180. Losada J, Garcia Armada MP, Cuadrado I, Alonso B, Gonzalez B, Casado CM, Zhang J (2004) J Organometal Chem 689(17):2799-2807.

181. Hobson ST, Cemavolic S, Patel SV, Warburton M, Mlsna TE (2006) ECS Trans 2(25):11-18.

182. Colilla M, Darder M, Aranda P, Ruiz-Hitzky EJ (2005) Mater Chem 15(35-36):3844-3851.

183. O'Hayre R, Cha S-W, Colella W, Prinz FB (2006) Fuel Cell Fundamentals, John Wiley and Sons, New York.

184. Hickner MA, Ghassami H, Kim YS, Einsla BB, McGrath JE (2004) Chem Rev104:4587-4612.

185. Costamagna P, Yang C, Bocarsly AB, Srinivasan S (2002) Electrochim Acta 47:1023.

186. Alberti G, Casciola M (2003) Annu Rev Mater Res 33:129-154.

187. Silva V, Weisshaar S, Reissner R, Ruffmann B, Vetter S, Mendes A, Madeira L, Nunes S (2005) J Power Sources 145:485-494.

188. Park YS, Yamazaki Y (2005) Solid State Ionics 176:1079-1089.

189. Prashantha K, Park SG (2005) J Appl Polym Sci 98:1875-1878.

190. Miyake N, Wainright JS, Savinell RP (2001) J Electrochem Soc 148:A905.

191. Shao Z-G, Joghee P, Hsing IM (2004) J Membrane Sci 229:43-51.

192. Lin CW, Thangmuthu R, Chang PH (2005) J Membrane Sci 254(1-2):197-205.

193. Antonucci PL, Ario AS, Creti P, Ramunni E, Antonucci V (1997) Solid State Ionics 125:431.

194. Adjemian KT, Lee SJ, Srinivasan S, Benziger J, Bocarsly AB (2002) J

Electrochem Soc 149:A256.

195. Ren S, Sun G, Li C, Liang Z, Wu Z, Jin W, Qin X, Yang X (2006) J Membrane Sci 282(1-2):450-455.

196. Hong L, Liu Z-L, Zhang X, Guo B (2008) US Pat. 0233451 A1.

197. Tay SW, Zhang X-H, Liu Z-L, Hong L, Chan SH (2008) J Membrane Sci 321:139-145.

198. Su YH, Liu YL, Sun YM, Lai JY, Guiver MD, Gao Y (2006) J Power Sources 155:111-117.

199. Chang HY, Lin CW (2003) J Membrane Sci 218:295-306.

200. Wilson BC, Jones CW (2004) Macromolecules 37:9709-9714.

201. Kim D, Scibioh A, Kwak S, Oh I-H, Ha H-Y (2004) Electrochem Commun 6:1069-1074.

202. Acosta JL,Gonzalez, L, Ojeda MC, del Rio C (2003) J Appl Polym Sci 90:2715-2720.

203. Wang H, Holmberg BA, Huang L, Wang Z, Mitra A, Norbeck JM, Yan Y (2002) J Mater Chem 12:834-837.

204. Kim DS, Park HB, Lee YM, Park YH, Rhim JW (2004) J Appl Polym Sci 93:209-218.

205. Gomes D, Buder I, Nunes SP (2006) Desalination 1999:274-276.

206. Ladewig BP, Knott RB, Hill AJ, Riches JD, White JW, Martin DJ, Diniz DA, Costa JC, Lu GQ (2007) Chem Mater 19(9):2372-2381.

207. Lin YF, Yen CY, Ma CCM, Liao SH, Lee CH, Hsiao YH, Lin HP (2007) J Power Sources 171(2):388-395.

208. Su YY, Liu YL, Sun YM, Lai JY, Wang DM, Gao Y, Liu B, Guiver MD (2007) J Membrane Sci 296(1+2):21-28.

209. Su YY, Wei TH, Hsu CH, Liu YL, Sun YM, Lai JY (2006) Desalination 200(1-3):656-657.

210. Holmberg BA, Wang H, Norbeck JM, Yan Y (2004) Polym Prepr (Am Chem Soc Div Polym Chem) 45(1):24-25.

211. Poinsignon C, Denoyelle A, Sanchez JY (1992) French Pat. FR 2670212 A1.

212. Poinsignon C, Denoyelle A, Sanchez JY, Armand M (1995) EP0560899 B1.

213. Carlier E, Revillon A, Guyot A, Baumgartner P (1993) React Polym 21(1-2):15-25.

214. Decker B, Hartmann-Thompson C, Carver PI, Keinath SE, Santurri PR (2010) Chem Mater 22(3):942-948.

215. Subianto S, Mistry MK, Choudhury NR, Dutta NK, Knott R (2009) Appl Mater Interfaces 1(16):1173-1182.

216. Hartmann-Thompson C, Merrington A, Carver PI, Keeley DL, Rousseau JL, Hucul D, Bruza KJ, Thomas LS, Keinath SE, Nowak RM, Katona DM, Santurri PR (2008) J Appl Polym Sci 110:958-974.

217. Liu L, Song L, Zhang G, Guo H, Hu Y, Fan W (2006) Mater Lett 60(15): 1823-1827.

218. Nowak R, Hartmann-Thompson C, Bruza K, Thomas L, Meier D (2008) World Pat. WO 2008127645 A1.

219. Unpublished results, Hartmann-Thompson C, Keeley D, Michigan Molecular Institute, 2009.

220. Chang Y-W, Wang E, Shin G, Han J-E, Mather PT (2007) Polym Adv Technol 18:535-543.

221. Park YI, Moon J, Kim HK (2005) Electrochem Solid-State Lett 8: A191-A194.

222. Choi J, Lee KM, Wycisk R, Pintauro PN, Mather PT (2008) Macromolecules 41:4569-4572.

223. Lee HS, Roy A, Badami AS, McGrath JE (2006) Polym Mater Sci Eng 51:210-211.

224. Wang H, Badami AS, Roy A, McGrath JE (2006) Polym Mater Sci Eng 51:202-203.

225. Serpico JM, Ehrenberg SG, Fontanella JJ, Jiao X, Perahia D, McGrady KA, Sanders EH, Kellogg GE, Wnek GE (2002) Macromolecules 3: 5916-5921.

226. Choi J, Lee KM, Wycisk R, Pintauro PN, Mather PT (2008) Electro Chem Soc Trans 16(2):1433-1442.

227. Zhang X, Tay SW, Liu Z, Hong L (2009) J Membrane Sci 329:228-

235.

228. Gao Y, Robertson GP, Guiver MD, Wang G, Jian X, Mikhailenko SD, Li X, Kaliaguine S (2006) J Membr Sci 278:26-34.

229. Costa ROR, Vasconcelos WL, Tamaki R, Laine RM (2001) Macromolecules 34:5398-5407.

230. Zhang X-H, Tay SW, Hong L, Liu Z-L (2008) J Membrane Sci 320:310-318.

231. Zhao L, Huang Y (2008) Huaxue Yu Nianhe 30(1):9-12, 33.

232. Khiterer M, Loy DA, Cornelius CJ,Fujimoto CH, Small JH, McIntire TM, Shea KJ (2006) Chem Mater 18(16):3665-3673.

233. Khiterer M, Loy DA, Small JH, Shea KJ (2004) Polym Prepr 45(1): 686-687.

234. Kalaw GJD, Yang Z, Musselman IH, Yang D-J, Balkus KJ, Ferraris JP (2008) Sep Sci Technol 43:3981-4008.

235. Honma I, Nakajima H,Nishikawa O, Sugimoto T, Nomura S (2003) J Electrochem Soc 150(5):A616-A619.

236. Kim YJ, Choi WC, Woo SI, Hong WH (2004) J Membrane Sci 238(1-2):213-222.

237. Nam S-E, Song S-A, Kim S-G, Park S-M, Kang Y, Lee JW, Lee KH (2006) Desalination 200(1-3):584-585.

238. Nam S-E, Lee K-H, Kang Y, Park S-M, Lee JW (2007) Sep Sci Technol 42(13):2927-2945.

239. Pezzin SH, Stock N, Shishatskiy S, Nunes SP (2008) J Membrane Sci 325(2):559-569.

240. Karthikeyan CS, Nunes SP, Schulte K (2006) Macromol Chem Phys 207(3):336-341.

241. Fujinami T, Mase T, Takami M (2009) US Pat. 0117439 A1.

242. Wu Y, Wu C, Xu T, Lin X, Fu Y (2009) J Membrane Sci 338(1-2):51-60.

243. van Schalkwijk WA, Scrosati B (2002) Advances in Lithium Ion Batteries, Kluwer Academic/Plenum Publishers, New York.

244. Linden D, Reddy TB (eds) (2002) Handbook of Batteries, McGraw-

Hill, New York.

245. Hawker CJ, Chu F, Pomery PJ, Hill DJT (1996) Macromolecules 29:3831-3838.

246. Tiglaar DM, Meador MAB, Bennett WR (2007) Macromolecules 40(12):4159-4164.

247. Tiglaar DM, Meador MAB, Kinder JD, Bennett WR (2006) Macromolecules 39(1):120-127.

248. Itoh T, Horii S, Hashimoto S, Uno T, Kubo M (2004) Ionics 10:450-457.

249. Tsujiko A (2007) World Pat. 024003.

250. Bai Y, Pan CH, Wu F, Wu C, Ye L, Feng ZG (2007) Gaodeng Xuexiao Huaxue Xuebao 28(9):1796-1800.

251. Pan CH, Bai Y, Wu F, Ye L, Wu C, Feng ZG (2007) Gongneng Cailiao 38(2):210-213.

252. Shi Q, Zhou X (2004) Gaofenzi Xuebao 1:114-120.

253. Liccocia S, Traversa E (2004) Nano-Micro Interface: 289-301.

254. Quartarone E, Mustarelli P, Magistris A (1998) Solid State Ionics 110:1.

255. Maitra P, Ding J, Wunder SL (2003) Polym Mater Sci Eng 88:568-569.

256. Maitra P, Wunder SL (2002) Chem Mater 14:4494-4497.

257. Maitra P, Ding J, Huang H, Wunder S (2003) Langmuir 19:8894-9004.

258. Zhang H, Kulkani S, Wunder SL (2004) Polym Mater Sci Eng 91:509-510.

259. Maitra P, Wunder SL (2003) Electrochem Solid-State Lett 7(4):A88-A92.

260. Zhang H, Kulkani S, Wunder SL (2006) J Electrochem Soc 153(2):A239-A248.

261. Zhang H, Kulkani S, Wunder SL (2007) J Phys Chem B 111(14):3583-3590.

262. Nunes SC, Bermudez VZ, Ostrovskii D, Silva MM, Barros S, Smith MJ, Carlos LD, Rocha J, Morales E (2005) J Electrochem Soc 152(2):A429-A438.

263. Bronstein LM, Karlinsey, RL, Stein B, Yi Z, Carini J, Zwanziger JW (2006) Chem Mater 18(3):708-715.

264. Kim H-S, Mah, S-K (2008) US Pat. 80166634 A1.

265. Hennige V, Hying C, Hoerpel G, Jost C, Kuehnle A (2004) German Pat. DE 10304734 A1.

266. Ismailov BA, Zhdanov AA, Zhurnal Obschchei Khimii, 1982, 52(3): 642-6.

267. Blanski RL, Phillips S, Rodgers SL, Lichtenhahn JD, Schwab JJ (2007) US Patent 7217683 B1.

268. Dare EO (2006) Turkish J Chem, 30:585-593.

269. Noda I, Isikawa M, Yamawaki M, Sasaki Y (1997) Inorg Chim Acta 263(1-2):149-152.

270. Misra R, Rollins K, Morgan S (2008) Polym Prepr 49(1):517-518.

271. Misra R, Fu BX, Morgan SE (2007) J Polym Sci Part B Polym Phys 45:2441-2445.

272. Misra R, Morgan SE (2006) Polym Prepr 47(1):410-411.

273. Misra R, Fu BX, Plagge A, Morgan SE (2009) J Polym Sci Part B Polym Phys 47(11):1088-1102.

Chapter 8
Polyhedral Oligomeric Silsesquioxanes in Space Applications

Henry W. Brandhorst, Jr.

8.1 The Space Environment

The space environment around the earth is exceptionally complex and has a significant impact on the lifetime and performance of any orbiting satellite. At least six factors influence the satellite: atomic oxygen (AO), electrons and protons trapped in the earth's magnetic field, solar flare protons, solar radiation, thermal cycling resulting from the passage of the spacecraft through the Earth's shadow, and debris. To deal with all these environmental factors, a space power system that uses solar cells is built to ameliorate these problems. First the solar cells are covered with a protective glass cover that contains a cerium additive. This absorbs ultraviolet light that would damage the silicone rubber adhesive that attaches the cover to the cell. The primary purpose of this cover is to limit the radiation damage from electrons and protons. The cells are usually then attached to a rigid composite substrate that can be oriented to track the sun as the satellite moves around the earth. The cover glass process is expensive, and there has long been a desire to find a material that would be resistant to these space environmental factors. POS additives have particular attributes that make them especially attractive for use in photovoltaic space power systems.

Atomic oxygen is the first hazard that a solar array must overcome. Atomic oxygen is created in earth orbit by the interaction of sunlight with residual oxygen molecules at the top of the atmosphere. Ultraviolet light transforms the oxygen molecule into a single oxygen atom, hence the term "atomic oxygen". It is the most prevalent atmospheric species in Low Earth Orbit (LEO), at about 300 km altitude. These atoms are highly reactive and are in high concentrations, therefore at a

Henry W. Brandhorst, Jr.
Director, Space Research Institute
Auburn University, AL 36849-5320

C. Hartmann-Thompson (ed.), *Applications of Polyhedral Oligomeric Silsesquioxanes*, Advances in Silicon Science 3, DOI 10.1007/978-90-481-3787-9_8,

satellite velocity of about 17,000 miles per hour (7.6 km s^{-1}), spacecraft materials are eaten away rapidly [1,2,3,4]. The actual atomic oxygen flux on a spacecraft varies with altitude, but the flux ranges from ~10^{12} – 10^{14} atoms cm^{-2} s^{-1}. The collision energy is about 5 eV, while the energy of a C-C bond is only ~4 eV. Therefore, polymeric materials such as Kapton® and polymer-matrix carbon-fiber composites are especially eroded by atomic oxygen, and this impacts spacecraft performance and/or structural failure. Kapton® is used as a substrate for solar arrays as well as part of thermal control blankets, so the problem of erosion can be severe. Thus, any solar array has to be designed to withstand attack by atomic oxygen. In general, the approach used to develop either protective coatings for polymers, or inherently durable polymers, has focused upon the use of metal atoms that, upon exposure to atomic oxygen, form stable non-volatile oxides that resist further attack. For example, silicones have been incorporated into polyimides, where exposure to atomic oxygen will cause the formation of sufficient silicon dioxide at the surface to protect the underlying polymers. For example, a polydimethylsiloxane-polyimide mixture has been developed [5]. However, the spatial variation and low surface concentration of the silicone constituents allowed gradual atomic oxygen attack of the bulk material when evaluated in ground laboratory testing.

Polydimethylsiloxanes, which contain one silicon atom per oxygen atom, are gradually converted to silica by atomic oxygen attack. In this process the loss of the methyl groups and conversion to SiO_2 results in the shrinkage of the polymer, with attendant cracks that can lead to attack of any underlying polymers [6,7]. However, the use of textured surfaces on the polydimethylsiloxanes has produced coatings that do not crack from the same atomic oxygen fluences that would cause the smooth surfaces on the same materials to crack [8]. Silsesquioxanes have shown promise over conventional polydimethylsiloxanes in that they contain 1.5 silicon atoms per oxygen atom, and do not show the shrinkage and cracking phenomena associated with polydimethylsilioxanes. These materials will be discussed later.

The Earth is encircled by electrons and protons trapped in the Earth's magnetic field. These radiation belts were discovered in 1958 by Dr. James Van Allen and carry his name. They are located in two primary regions above the Earth and along the Earth's magnetic plane as shown in Fig. 8.1. The inner radiation belt consists primarily of high concentrations of energetic protons (>100 MeV) and some energetic electrons (~100 keV). It extends from about 100 to 10,000 km (0.016 to 1.6 Earth radii) above the Earth's surface. The outer radiation belt consists primarily of high energy electrons, and extends from about 19,000 to 64,000 km (3 to 10 Earth radii) above the surface. This belt peaks at about 25,000 to 32,000 km (4-5 Earth radii). This belt contains electrons with energies from about 0.1 to 10 MeV. These energetic particles damage electronic devices such as solar cells by creating surface defects, and thus reducing output power. Thus protective cover glasses are used to limit the damage. The thicker the cover glass, the less will be the damage. However, any substitute material will also have to be resistant to these trapped electrons and protons. In addition, the amount of radiation a spacecraft receives depends upon its orbit. Satellites in LEO (<400 km) generally receive low doses of damaging radiation. Those satellites in geostationary space orbit (GEO) receive substantially

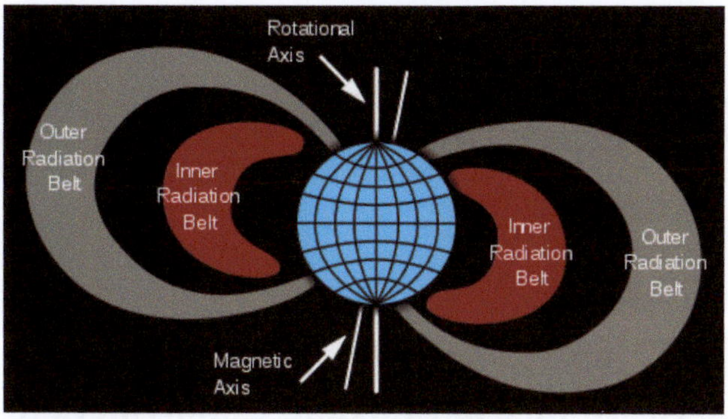

Fig. 8.1 Schematic drawing of the Earth's Van Allen radiation belts.

more, and those satellites in highly inclined orbits, or in mid-Earth orbits, receive the most radiation damage. In addition, the type of material in the solar cell also dictates the degree of damage.

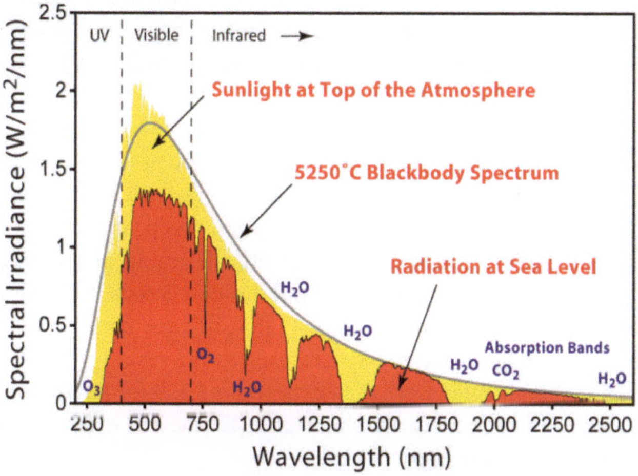

Fig. 8.2 Spectrum of solar radiation at the Earth

The next hazard for space vehicles traveling in Earth orbit is sunlight itself. The solar spectrum in space differs greatly from that of the sunlight that arrives at the Earth's surface. The sunlight that arrives at the top of the atmosphere is modified through absorption and scattering by atmospheric constituents such as water vapor, carbon dioxide, ozone, oxygen, nitrogen, dust and clouds. For a surface at sea level, the solar intensity reaching that surface is about 1,000 W m^{-2} when the sun is directly overhead. This represents a ~25% reduction in the sunlight intensity at the top of the atmosphere (~1360 W m^{-2}). The atmospheric absorption occurs in bands in the infrared region of the spectrum but almost uniformly across the visible and ultraviolet region of the spectrum, as shown in Fig. 8.2. The red curve shows the spectrum at the earth's surface while the yellow shows the outer space spectrum. In space, the major parts of the spectrum that affect the lifetime of a satellite are the ultraviolet and vacuum ultraviolet regions, from 150 to 300 nm. This high energy light breaks chemical bonds and thus leads to the darkening of the adhesives used to attach the cover glasses to the solar cells. Thus efforts were directed toward finding a polyhedral oligomeric silsesquioxanes (POS)-based replacement for the traditional cover glasses.

8.2 Resistance of Siloxane Copolymers to Atomic Oxygen in Low Earth Orbit

As noted in the initial paper on POS materials in space [9], materials with resistance to degradation by the harsh conditions (e.g., atomic oxygen and UV light) present in LEO are essential for extended use of lightweight, polymeric structures in space [10,11]. As noted above, the degradation of carbon-based materials in the LEO environment is due to the presence of ground-state atomic oxygen that has been created from the photo-dissociation of molecular oxygen [12] combined with solar UV radiation, and with the 8 km s^{-1} orbital velocity of vehicles in LEO [13]. A class of non-carbon-based materials with rates of erosion one to two orders of magnitude slower than organic polymers in LEO has been developed. These are the siloxane polymers which, when exposed to atomic oxygen (AO), convert to a protective silica-like coating. In an attempt to utilize this desirable property, the seminal work in this field was performed using polyimide-siloxane (PISX) copolymers and polyhedral oligomeric silsequioxane (POS)-siloxane copolymers as coatings and matrix materials. The copolymers were used as the resin for PISX-carbon fiber composites. These composites were then exposed in LEO for 50 hours as part of the "Effect of Oxygen Interactions with Materials" (EOIM-III) experiment aboard the STS-46 NASA space shuttle flight [14].

A composite made with a 37% loading of PISX with Celion 6000 carbon fibers had a density of 1.54 g cm^{-3}, a flexural strength of 194 Ksi, and a short beam shear strength of 8 Ksi. These properties were comparable to typical PMR-l5/graphite composites (having flexural strengths of ~235 Ksi), and hence were expected to be useful in similar applications. PMR-15 is a high temperature polyimide resin usually used with carbon fibers for excellent high temperature behavior, and Celion 6000 carbon fibers are commonly used in such composites. The fabrication conditions were chosen to demonstrate the feasibility of using PISX as the matrix for composites that would be suitable for space applications. The samples were flown on STS-46 in March 1992, and were exposed to the LEO environment in trays heated to 60 and 120 °C. The heated trays allowed investigation of the effect of temperature, atomic oxygen, and UV radiation on the composite samples. The samples were exposed to LEO for ~50 hours, the total A0 fluence was 2.2×10^{20} atom cm^{-2}, and the total time of exposure to the solar UV was 7.6 hours [9]. The investigation of the surface of this PISX composite using X-ray photoelectron spectroscopy (XPS) reveaed that the siloxane-rich composite surface was oxidized to the silica-like material as expected. The following conclusions may be drawn from the experiments above: (1) PISX is a suitable matrix for composites for space applications; (2) PISX composites are *one-to-two orders of magnitude more resistant* to the simulated LEO environment than are standard polyimide composites; (3) PISX composites *erode more slowly* when far-UV radiation is combined with the atomic oxygen, in contrast to other materials, where UV radiation accelerates the degradation process.

In another corollary experiment, a POS-siloxane copolymer was developed to use as an LEO-resistant coating. These polyhedral oligomeric silsesquioxanes ($RSiO_{1.5}$) contain a silicon-oxygen ratio that is intermediate between that found in silica (SiO_2) and that found in siloxanes ($-R_2SiO-$). The composition is very similar to that found on the surface of silicon-based materials after exposure to atomic oxygen in LEO. A family of tractable POS-siloxane copolymers was fabricated. Evaluation of the resistance of this POS-siloxane copolymer to a simulated LEO environment was conducted. The samples were exposed to atomic oxygen with and without far-UV radiation. XPS analysis revealed that the surface of the copolymer samples became more silicon- and oxygen-rich after exposure to atomic oxygen, both with and without far-UV radiation.

Comparison of these results showed that the POS-siloxane copolymer had a greater resistance to the simulated LEO environment than PISX. The POS-siloxane copolymer actually gained weight during the exposures. Combining far-UV radiation with atomic oxygen does not significantly change the results. The samples were exposed to a simulated atomic oxygen fluence of 4.3×10^{21} atoms cm^{-3} both with and without UV exposure at 15 mW cm^{-2} (about 7 Earth suns equivalent). The samples gained weight, and the initial microcracks on the surface were actually healed by the atomic oxygen exposure. This annealing phenomenon seen for the POS-siloxane copolymers may result from surface heating, both from the heat of reaction (ΔH), and from absorption of the kinetic energy (2-3 eV) of the atomic oxygen. This also could be a significant advantage for these materials, since micro-cracking exposes a coated surface to the LEO environment and to accelerated degradation, and contributes to loss of long-term durability for composites used in space. This finding led to a more detailed study of the POS composites and their use in space.

Several series of ground tests were run on a range of POS compositions [15] with a hyperthermal atomic oxygen source [16]. This source is ultrahigh vacuum (UHV)-compatible, operates at room temperature, and produces a high-purity, hyperthermal, AO flux with an AO:OC ratio of $\sim 10^8$. These sources are superior to plasma sources because they operate at UHV pressures ($\sim 1.33 \times 10^7$ Pa) and produce hyperthermal, ground-state O atoms and negligible quantities of other species (e.g., ions, contaminants and UV radiation). The AO produced by this source has an energy of 2-5 eV (although the energy distribution within this range was not measured). A POS-polyurethane (PU) copolymer was produced by solvent casting from THF solutions of 5 mg cm^{-3} of each polymer. They were dried, cleaned and put into the UHV chamber above.

XPS probes the near-surface region of a sample and yields a weighted average composition, with the atomic layers near the surface being weighted more heavily, since the photo-emitted electrons from these layers have a lower probability of inelastic scattering. The sampling depth is ~4–6 nm, and ~10% of the signal originates from the outermost atomic layer. This near-surface region is inhomogeneous because the AO reacts with the outermost few atomic layers. Therefore, the region that is affected to the greatest extent as a result of the reaction

Table 8.1 Near-surface composition determined from XPS data obtained from the as-entered, solvent cleaned, AO, and air-exposed 20-wt% POS-PU sample. Reprinted from [15] with permission from Journal of Adhesion Science and Technology.

Surface Sample Treatment	AO fluence O/cm^2	Composition (at. %)						Atom ratio
		O	Si	C	Sn	Na	N	O/Si
As-entered, solvent cleaned	—	18.5	8.1	72.5	0.9	—	—	2.28
2-h AO exposure	1.44 x 10^17	20.4	7.9	70.7	1.0	—	—	2.58
24-h AO exposure	1.77 x 10^18	21.8	9.5	61.7	1.0	3.0	3.0	2.29
63-h AO exposure	4.53 x 10^18	32.6	11.1	37.8	1.8	13.6	3.1	2.93
4-hr air exposure following 63-h AO exposure	4.53 x 10^18	38.9	13.7	43.4	2.0	2.0	—	2.83

with AO also makes the largest contribution to the XPS signal. This fact implies that XPS is an excellent technique for studying AO erosion of spacecraft materials. Even though both the distribution functions involving the depth of chemical reactions in the near-surface region, and the XPS determination of the weighted average composition of the near-surface region are complex, the compositional values provide a trend which is indicative of the chemical alterations occurring during AO exposure. The compositions determined using the homogeneous assumption are shown in Table 8.1 as a function of AO fluence. The O/Si atomic ratio is 2.28 for the as-entered sample; increases to 2.58, decreases to 2.29 and again increases to 2.93 after the 2, 24, and 63 hour AO exposures respectively. After the 4 hour air exposure, the O/Si atomic ratio is 2.83, but C also accumulates on the surface during the air exposure. The decrease in the O/Si ratio to 2.29 after 24 h of exposure could be attributed to the relative increase of Na and Sn on the surface. However, the overall increase in the O/Si atomic ratio resulting from exposure to the AO flux is a trend that has previously been observed in two other similar studies of a POS–PDMS (polydimethylsiloxane) and 60 wt % POS–PU copolymer.

It is attributed to the formation of SiO_2 and is consistent with the high-resolution spectra that follow. A significant reduction in the C 1s peak is observed as a result of the incremental exposures to the O-atom flux. The near-surface C concentration decreases from 72.5 atom % for the as-entered sample to 37.8 atom % after the 63 hour exposure. This decrease in C is most likely caused by the reaction of C in the near-surface region with O to form CO and/ or CO_2 which desorb. The increase in

C concentration to 43.4 atomic % observed after the air exposure, is most likely due to the adsorption of C-containing molecules from the air. Air exposure has been a contributor to erroneous results in the past. High-resolution XPS C 1s, O 1s and Si 2p obtained from the as-received, solvent-wiped POS-PU surface before and after the 2, 24 and 63 hour AO exposures are shown in spectra (a) to (d) of Fig. 8.3. Spectrum (e) was obtained after the 3.3 hour air exposure following the 63 hour AO exposure. Variations in peak shapes and positions are observed between the unexposed, AO-exposed, and air-exposed surfaces, indicating that the chemical species distribution is altered by exposure to the AO flux and then to air.

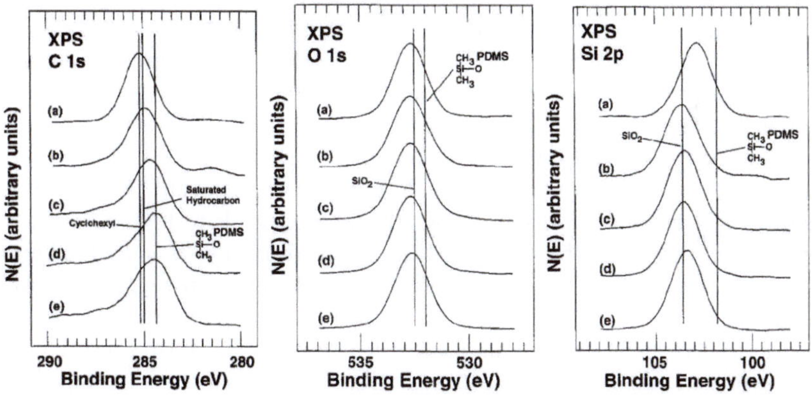

Fig. 8.3 XPS high Resolution C 1s (left), O 1s (center) and Si 2p (right) spectra from a 60 wt % POS-polyurethane (a) after insertion into the vacuum system, (b) after 2 hours, (c) after 24 hours, (d) after 63 hours exposure to the hyperthermal AO flux, and (e) 3.3 hour air exposure following the 63 hour AO exposure. Reprinted from [15] with permission from the Journal of Adhesion Science and Technology.

The C 1s peak, shown in Fig. 8.3 (left), is broad and centered at 285.0 eV, indicating that the predominant form of carbon present for the as-entered POS-PU sample is aliphatic, located on the hard and soft segments of the polymer chain and on the cyclopentyl groups of the POS cages. In spectra (b) to (d), the C 1s peak becomes broader and displays visible shoulders with increasing exposure to the O-atom flux. A shoulder associated with aromatic carbon is present in these spectra at 284.7 eV. Small shoulders are also visible on the high binding energy (BE) side of the C 1s peak in spectra (c) and (d). These are associated with species such as alcohols, formaldehydes (BE -286.0 to 287.7 eV) and organic acids (BE -287.5 eV), which form by reaction with the AO flux. These changes coincide with a decrease in the total carbon concentration in the near surface region, from 70.1 to 37.3 atom %. A dramatic reduction in the carbon concentration after AO exposure was also seen in the POS-PDMS sample studied previously. Exposure to air (spectrum e) produces an 11% increase in C near 285.0 eV, indicating adsorption of hydrocarbons from the air at reactive surface sites produced during the AO exposure. A similar observation was also noted for the POS-PDMS sample.

The O 1s spectra obtained from the as-entered POS-PU sample is shown in Fig. 8.3, center, spectrum (a). This peak is broad and centered at 531.9 eV. The predominant form of oxygen present for the as-entered sample corresponds to the carbonyl in the urethane segment (531.9 eV) of the polymer and the oxygen present in the POS cages (532.0 eV). Initially, the oxygen contribution is reduced from 18.2 to 17.5 atom % after a 2 hour exposure. Spectrum (b) shows this decrease to coincide with the O 1s peak shifting to a higher BE, and thus decreasing the contributions from the urethane segment and from the oxygen in the POS cages. This reduction is consistent with the production and subsequent desorption of CO_2. After the 24 hour exposure, the near surface O content increases to 23.7 atom %, and the predominant peak featured in spectrum (c) corresponds to silica. This feature becomes more prominent with the 63 hour exposure as the surface O content increases to 35.3 atom %, and is consistent with the formation of an SiO_2 surface. Exposure to air results in a small 3.7 atom % decrease in the surface O content as the reactive surface adsorbs hydrocarbons from the air.

The Si 2p peaks obtained from the POS-PU sample after the various treatments are shown in Figure 8.3 (right). The Si 2p peak for the as-entered sample, spectrum (a), is broad indicating the presence of several chemical states of silicon. This peak is centered at a BE of 102.7 eV, which corresponds to $RSiO_{1.5}$ in the POS cage. However, spectra (b), (c), and (d) reveal the formation of an SiO_2 layer with incremental exposures to the O-atom flux. The fact that little difference is observed in the spectra obtained after the 24 hour and 63 hour exposures indicates that this silica layer forms a protective barrier on the surface which prevents further degradation of the polymer with longer exposure to the O-atom flux. The significant compositional changes observed indicate that most of the near-surface region examined by XPS is altered by the AO exposure.

The chemical reactions that form CO_2 and H_2O are exothermic, so the local surface temperature may be relatively high. This, and the fact that the AO induces a chemically induced driving force, results in diffusion of sub-surface C and H to the surface where they react with the AO. This mechanism is responsible for the subsurface compositional alterations observed using XPS. Again, these findings are consistent with those observed for the POS-PDMS sample.

It was clear from these results that a major focus of the work should be upon making Kapton® polyimide (PI) resistant to AO attack, owing to its extensive use in the space environment. Kapton® polyimide is used widely on spacecraft in flexible substrates for lightweight high-power solar arrays because of its inherent strength, temperature stability, excellent insulation properties, UV stability and IR transparency. It is also used in conjunction with Teflon® FEP (fluorinated ethylene-propylene copolymer) in multilayer blankets for thermal control insulation because of its superior optical properties, including low solar absorbance. In these multilayer insulation blankets aluminum or gold is typically applied to Kapton® owing to these metals' low emissivity.

Over the last twenty-five years, it has been well established through space-based experiments and ground simulations that polymeric materials and films undergo severe degradation as a result of the aggressive environment encountered in LEO [14,17,18]. As noted in the introduction, in this high vacuum environment, materials are subjected to the full spectrum of solar radiation and must endure constant thermal cycling from -50 °C to 150 °C and bombardment by low- and high-energy charged particles, as well as high incident fluxes of AO. These harsh conditions, combined with the need for lighter weight and lower cost man-made orbiting bodies, necessitate the design of space-survivable materials. Thus Tomczak et al. published a major paper in 2005 that addressed this issue with persuasive results [19]. This study expanded the concept of space survivability by the addition of POS entities into the polymer, such that when high energy AO erosion inevitably occurs, the POS materials form a silica passivating layer that prevents further attack and erosion.

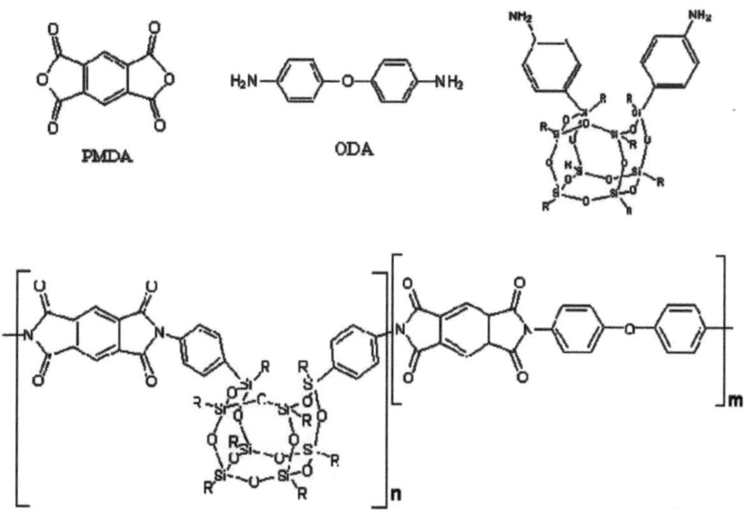

Fig. 8.4 POS Kapton polyimide synthesis. Reprinted from [19] with permission from the Materials Research Society.

Polyimides with the same chemical formula as Kapton® were synthesized via condensation polymerization of 4,4′-oxydianiline (ODA) and pyromellitic dianhydride (PMDA) in an N,N′-dimethylacetamide (DMAc) solvent [20]. A POS framework with two aniline pendant groups was then synthesized [21] as shown in Fig. 8.4. Using this monomer, POS-polyimide random copolymers were synthesized with POS loadings corresponding to 0, 5, 10, 20, and 25 wt %, and were exposed to AO in the pulsed beam hyperthermal facility described above. The mean translational energy of these AO atoms was 4.9 to 5.0 eV, matching the species encountered on orbit. The samples were evaluated before, during and after AO exposure using surface profilometry, atomic force microscopy (AFM) and X-ray photoelectron spectroscopy (XPS).

Fig. 8.5 shows the AFM surface roughness images. It is clear that the increasing concentration of POS in the Kapton® creates substantial resistance to erosion by atomic oxygen. Table 8.2 summarizes the AFM surface roughness data for 0, 10, and 20 wt % POS polyimide films.

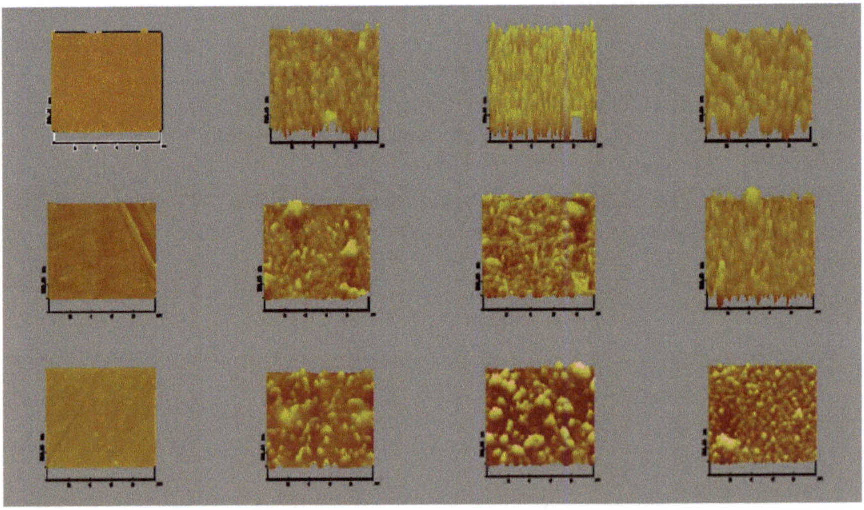

Figure 8.5 Atomic Force Microscopy images of POS polyimides with increasing atomic oxygen flux. (a) 0 wt %, (b) 10 wt % and (c) 20 wt % POS Polyimide surfaces after exposure to atomic oxygen fluences of 0.0, 8×10^{19}, 1.6×10^{20}, and 4.1×10^{20} O atoms cm^{-2}. Reprinted from [19] with permission from the Materials Research Society.

The erosion of Kapton® polyimide is substantially reduced by each successive addition of POS. These erosion rates have been converted into percentages relative to the untreated Kapton®. In addition, a sample of 25 wt % POS-polyimide was exposed to $2.03\text{x}10^{20}$ O atoms cm^{-2}. It showed continued resistance to erosion with this increased POS content. Owing to the very low erosion combined with sample roughness, the step height of erosion of 25 % POS polyimide was not measurable, but was less than the minimum measurable step height of 0.15 µm. The estimated erosion ratio of POS-polyimide to 0 % POS-polyimide for 20 wt % POS -polyimide at $2.03\text{x}10^{20}$ O atoms is 0.046, and that for 25 wt % POS is less than 0.025, showing the erosion decreased by approximately a factor of two with increased POS content. The surface atomic concentrations were measured by XPS, and these results further confirm the mechanisms involved (see Table 8.3).

The hyperthermal AO exposure is provided as the number of pulses from the facility described above. The Kapton® equivalent fluence was determined against a Kapton® H standard. The increasing Si and O concentrations with both POS addition and AO fluence (shown in red) provide solid evidence of the process by which the erosion rate of Kapton® is being reduced by to the formation of a silica-like layer.

A separate exposure of 25 wt % POS-polyimide at a fluence of 7.33 x 10^{20} O atoms cm^{-2} showed an erosion yield of about 1.1% that of Kapton®. For 0% POS-polyimide, the oxygen concentration approximately doubled, with a greater increase in the POS-polyimides. More importantly, in the POS-polyimides, the silicon concentration reached values of 25%, and at the highest AO fluence the Si:O ratio was approximately 1:2, consistent with the formation of a silica layer.

These surface changes are presumably responsible for the reduced erosion rates

Table 8.2 AFM surface roughness values. Reprinted from [19] with permission from the Materials Research Society.

	AFM RMS roughness values (nm)			
0 wt% POS polyimide	2.48	70	120	126
10 wt% POS polyimide	2.47	22.4	34.3	78.9
20 wt% POS polyimide	2.86	17.2	23.7	39.1
	0	3.8x10^19	1.6x10^20	4.1X10^20

and improved AO reaction efficiencies. Fig. 8.6 shows the effective erosion rates with increasing concentration of POS in polyimide. The fitted curve shows the relationship is exponential with increasing POS content. In the referenced 2005 study, Tomczak et al. explored the use of molecular dynamics simulations to give

Table 8.3 Surface atomic concentrations (%) from XPS scans. Reprinted from [19] with permission from the Materials Research Society.

Sample	AO exposure (beam pulses	Kapton-equivalent atomic oxygen fluence (10^20 O atoms/cm^2)	C	O	Si	N
0 wt% POS polyimide	0	0	72	19.5	1	7
	6	0.1	69	20	2	9
	100	1.63	69	24	1	6
	250	4.1	55	36	0	9
10 wt% POS polyimide	0	0	77	16	2	5
	6	0.1	73	18.5	5	3.5
	100	1.63	48	30	19	3
	250	4.1	20	56	23.5	0.5
20 wt% POS polyimide	0	0	70	20	6	4
	6	0.1	66	24	7	3
	100	1.63	20	54	25	0
	250	4.1	12	60	26	1

insight into the effect of atomic oxygen on POS-polyimides. They performed molecular dynamics simulations of 5 eV O atom collisions with a POS cage ($Si_8O_{12}H_8$), and with a POS-coated alkane thiol self-assembled monolayer (SAM) on a gold surface. In order to determine the force field for these calculations they used a semi-empirical electronic structure method known as MSINDO (modified symmetrically orthogonalized intermediate neglect of differential overlap). Troya and Schatz had calibrated this approach previously [22] for O atom reactions with hydrocarbon surfaces. In their studies of the reaction of O with the POS cage, they found three reaction pathways: (1) hydrogen abstraction to give OH plus a radical

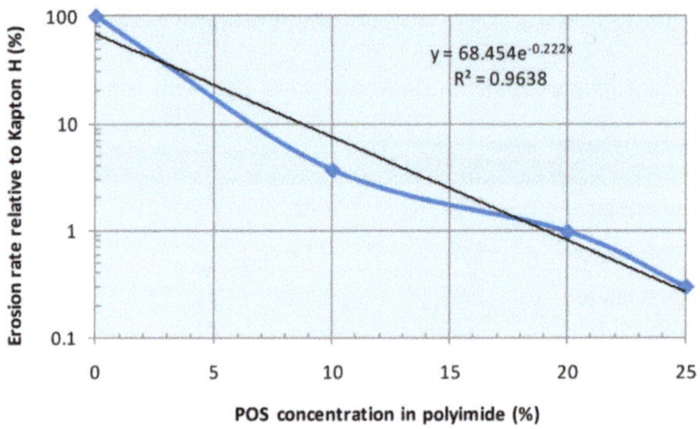

Figure 8.6 Effective erosion rates

cage species (a relatively minor channel), (2) O addition with H elimination to give H plus an oxygenated cage, and (3) O addition with cage opening. Their results show that H elimination and cage opening (by O atom addition to the Si-O bond followed by O-O scission) are the predominant reactions with the POS molecule. Similar mechanisms were found to apply to the POS-coated SAM. These results indicate 5 eV O atoms easily disrupt and oxidize the POS cage. Successive reactions with O atoms would presumably further oxidize the POS, ultimately to give silica.

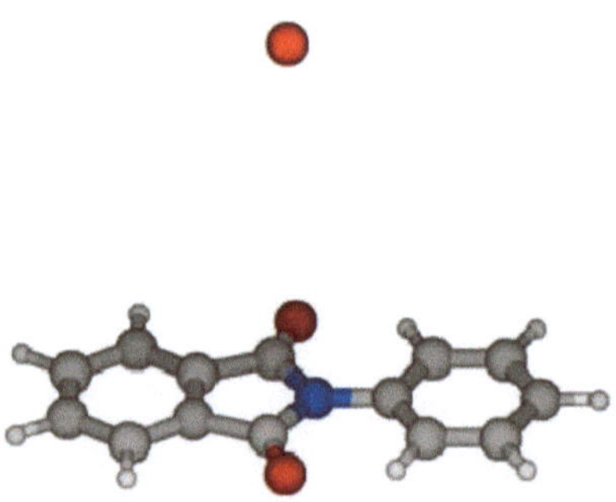

Figure 8.7 QM model of the primary reaction zone. red=oxygen, grey=carbon, blue=nitrogen and white=hydrogen atoms. Reprinted from [19] with permission from the Materials Research Society.

The erosion of polyimide by O atoms involves the breakage and formation of various chemical bonds such as C-C, C-N, C-H, C-O and O-H. Because analytical force fields for O + polyimide were not available, they used a direct dynamics classical trajectory method to simulate the reaction dynamics instead. In the direct dynamics method the classical trajectory is integrated with the forces on each atom obtained directly from quantum mechanical electronic structure calculations (MSINDO semi-empirical calculations in this case). Past studies have suggested that the focus should be upon the electronic ground state potential energy surface, since this makes a dominant contribution to reactivity. A simplified polyimide monomer model (Fig. 8.7) was used for the primary reaction zone and the direct dynamics calculation was carried out with this model.

Results for a thousand trajectories were generated for 5 eV O atoms colliding with the polyimide monomer is in its ground state with zero point vibrational energy. The unphysical flow of the zero point energy among different vibrational modes (which may cause artificial dynamics behavior especially for systems with large numbers of degrees of freedom) was ignored. Because this simulation was focused on the reactions that occur with the polyimide monomer on a very short time scale (less than 1 ps), the zero point energy flow did not significantly influence the dynamics. The calculations showed that polyimide is very reactive, and over 67% of collisions led to reaction. The classical trajectory simulations revealed a large number of reaction pathways, but O addition reaction was found to be the dominant process. The reaction pathways were grouped into four types: (1) complex formation in which O is added to the monomer without breaking C-C, C-N and C-H bonds (within 1 ps), (2) O addition associated with ring-opening (C-C and C-N bond cleavage); (3) H elimination; (4) OH abstraction. All these reactive channels were found to be exothermic, and hence were thermodynamically allowed. The branching ratios were 1:17:3:3 for reaction types (1) to (4) respectively. Other reaction types were also observed, however they were negligible compared to the above four types of reactions.

Thus in this landmark study it was shown that the incorporation of POS nanostructures into polyimides would likely substantially extend the lifetime of these materials in LEO applications where atomic oxygen erosion is the predominant damage mechanism. Their comparative erosion yields following exposure to a hyperthermal AO source provided clear evidence that POS polyimides exhibited significantly improved oxidation resistance. XPS data and molecular dynamics simulations suggested that this increased resistance was associated with the rapid formation of a self-passivating and self-healing silica layer upon exposure to high fluxes of AO. It was also suggested that this approach could easily be extended to other polymeric systems. Such systems include structural polymers such as polyethylene and polypropylene. Thus this study laid the foundation for a potential revolution in the field of space-survivable materials.

In a subsequent study [23], a series of main-chain POS polyimides (MC-POS-PI) and side-chain POS polyimides (SC-POS-PI) were synthesized with the tailorable side chain organic R-group being cyclopentyl. In this study, there were eight Si atoms, eleven O atoms and eight cyclopentyl groups. This is denoted as Si_8O_{11} because the twelfth oxygen atom has been replaced by two cyclopentyl groups. These materials were exposed to the hyperthermal AO source with results that showed increased AO tolerance with increasing Si_8O_{11} MC-POS-PI additions up to 8.75%. These results demonstrated that POS cages could be incorporated into many different polymeric materials simply by modifying the organic R groups attached to the POS cage.

Table 8.4 CTEs of POS-PI films

Sample	CTE (μm/m°C)
Kapton H®	30.25
0% POS-PI	33.11
7% MC-POS-PI	33.5
8.75% MC-POS-PI	35
7% SC-POS-PI	35.86
7% cage SC-POS-PI	33.64

Finally, because of the sudden temperature changes experienced by materials in the LEO environment, the coefficient of thermal expansion (CTE) is a highly important material property that can impact the usefulness of the POS-PI material. Mismatches in the CTEs between polymeric materials and their coatings lead to cracks, crazing, and mechanical failure. POS-PIs eliminate the need for AO-resistant coatings, but in order for POS-PIs to be a drop-in replacement for Kapton®, it is necessary that the new materials have predictable CTEs with values similar to Kapton®. The CTE values for several samples are shown in Table 8.4. The addition of POS gives a slight increase in the CTE relative to Kapton®. However, a slight decrease is seen after exposure of the SC-POS-PI to AO. The CTE of fused silica is about 0.55 μm m^{-1} $°C^{-1}$, and the CTE of the silica passivation layer formed on POS-PIs in the presence of AO is expected to be close in value. This mismatch between the silica passivation layer and the underlying POS-PI layer is likely to cause cracks in the silica passivation layer, which will again form silica in the POS-PI exposed areas. Thus this result confirms that the addition of POS to these polymeric materials does not radically change the CTE.

As a confirmatory test, samples of 0, 1.75 and 3.5 wt % Si_8O_{11} MC-POS-PI films were exposed to the LEO environment (AO, UV) for 3.9 years as part of the Materials International Space Station Experiment (MISSE-1), and then retrieved and studied. It was found that a 32.55 ± 0.87 μm thick polyimide film was completely eroded. The 1.75 % MC-POS-PI film showed some survival with the inner portion of a circular sample completely eroded and a step height of 5.79 ± 1.31 μm from the outer portion and the masked area, and the 3.5 % MC-POS-PI film maintained a step height of 2.12 ± 0.34 μm from the outer portion and the unexposed area. It was determined by XPS that the atomic percentages of the top 10 nm of the films were 34% Si, 59% O, and 7% C for both the 1.75 and 3.5 wt % Si_8O_{11} MC-POS-PI samples. These results confirm the existence of a silica-like layer formation (as also observed in the laboratory tests), and substantiate the model that had been developed.

Thus in this foundational effort, POS addition to a range of polymeric materials was shown to substantially increase their resistance to damage from atomic oxygen in LEO. Mechanical and physical changes of polyimide materials were characterized, and the mechanisms of damage (and of protection) were elucidated. However, the space environment presents two other hazards to spacecraft: UV/VUV light and particulate radiation (electrons and protons), and the sub-system on a spacecraft that is most affected by both of these hazards is the solar (photovoltaic) array that uses sunlight to produce power for the vehicle.

8.3 Polyhedral Oligomeric Silsesquioxanes in Space Solar Power Systems

A space photovoltaic power array consists of solar cells and their associated cover glasses that are attached together using a clear silicone adhesive. This combination is called a Cover Integrated Cell (CIC). The CICs are interconnected in series into a string of cells that produce the voltage required by the satellite. The strings are placed in parallel to produce the current needed by the spacecraft. The strings of series- and parallel–connected cells are then bonded to a substrate called a face sheet, which may itself be bonded onto an aluminum honeycomb material. This is called a panel. Several panels are connected together and attached to the spacecraft along with a deployment scheme. During launch, the panels are held tightly against the body of the spacecraft and once in space, the panels are deployed, usually on either side of the spacecraft. They are then oriented toward the sun and begin producing power for the vehicle and its energy storage system. Fig. 8.8 shows a typical communications satellite with the solar arrays extended.

Fig. 8.8 Wideband global SATCOM system (U.S. Air Force)

The solar cell cover glasses are fabricated from a special formulation that contains a small percentage of the element cerium. They may also have anti-reflection coatings on the outer surface to maximize the amount of sunlight absorbed. The cerium addition is very important because it absorbs ultraviolet/vacuum ultraviolet (UV/VUV) light in wavelength bands below about 0.35 μm. This absorption prevents the UV/VUV light from reaching the silicone adhesive, which darkens when exposed to those wavelengths. The cover glasses also protect against the AO in LEO. These cover glasses are very expensive and come in a range of thicknesses (from ~50 μm to 300 μm) to aid in protection against the particulate radiations in space. The thickness of the cover glass is chosen based upon the intended vehicle orbit. For a satellite in GEO, exposed to electrons, protons and solar flare protons, a thickness of 150 μm is sufficient. In LEO the cover glass can be very thin because very little particulate radiation is present there.

Hence because of the costs involved in the materials, as well as the costs from attaching the cover glasses to the solar cells with the silicone adhesive, it was logical to seek an alternative cover glass material that incorporated POS additives, both to provide the protection desired, and to have lower materials and attachment costs. The goal was a highly transparent POS-containing polymeric material that could be deposited by spraying or other low cost application method. It must be resistant to UV/VUV light and proton radiation. Electron radiation is of sufficiently high energy that it generally passes through the CIC. Addressing this would require excessively thick coatings which are not practical owing to high mass.

It was thought that a good first target for the development of this new polymeric cover would be thin film solar cells, because their flexibility may be incompatible with rigid cover glasses. There are several types of thin film solar cells. Some are made of cadmium telluride (CdTe), some of Copper Indium Gallium Selenide (CIGS), and others are made of amorphous silicon. The latter cells were selected for study because they are readily available, are thin and flexible, and are made with a "reel-to-reel" system for low cost. Success with this coating could then extend to other thin film solar cell systems, and could perhaps be transferred to the conventional crystalline GaAs-type multijunction solar cells.

The first use of POS materials was in conjunction with amorphous Si thin film solar cells [24]. The POSS® building blocks used in this study contained hybrid (organic-inorganic) compositions in which the internal frameworks were comprised primarily of inorganic silicon-oxygen bonds. As noted previously, the exterior of the nanostructure is covered by both reactive and non-reactive organic functionalities (–R) that ensure both compatibility and tailorability of the nanostructure with organic polymers. The resulting nanoscopic chemicals have low density, range in diameter from 0.5 to 3 nm, and can be tailored through variation in the R groups and in the size of the nanocage [9,25]. The molecularly dispersed POSS® readily forms a passivating silica layer when attacked by atomic oxygen. This layer in turn protects the virgin material from degradation. Furthermore, the silica-like composition of POS should provide enhanced UV and VUV resistance. The UV and VUV resistance provided by POS can even be enhanced further by replacing for the silicon atoms at the vertices in the nanocage with metals such as cerium. Metal atom addition to the POS nanocages leads to a new designation: polyhedral oligomeric metallic silsesquioxanes (POMS).

In addition, siloxanes (e.g., Dow Corning DC 93-500®, the industry standard cover glass adhesive) are generally considered stable under photo-aging or photo-oxidation conditions. Although they can protect underlying material from oxidation, they do not significantly absorb incident UV radiation, especially if they are alkyl-substituted. A related class of materials that are more highly oxidized and thereby more stable to oxidation and UV are silsesquioxanes. Previous research has shown that chemical manipulation of the organic groups surrounding the POS cage enables dispersion of POS nanostructures throughout the polymer matrix at high POS loadings via blending and co-polymerization techniques described previously.

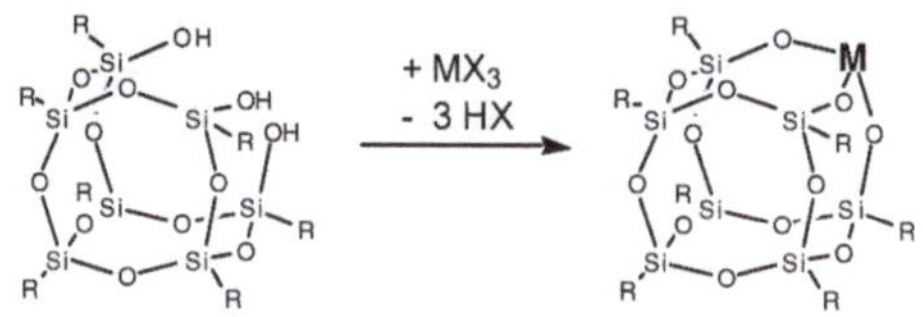

Fig. 8.9 Synthetic method utilized for the manufacture of POMS

For the thin film cell tests, the metalized POS coatings were prepared by adding 10 wt % metalized POS into PM1287 (a siloxane-POS resin). All the metalized POS could be readily dissolved into PM1287 under gentle heat. The incorporation of metals into a POS cage is easily accomplished through the reaction of a POS silanol with a metal halide or alkoxide as shown in Fig. 8.9. A total of five POS coatings: PM1287 control, 10 wt % Ce-POS, 10 wt % Tb-POS, 10 wt % Ti-POS and 10 wt % Gd-POS were tested. Table 8.5 lists the variety of polyhedral oligomeric metallasilsesquioxanes (POMS) prepared, their primary and secondary radiation absorbance targets, and potential component applications. The POMS systems were applied by dissolving them in a high performance conformal coating resin, PM1287. Each of the POMS was found to be soluble at 50-75 wt % levels in the PM1287 resin. Notable advantages of this resin are that it is nonflammable, contains no volatiles, is optically clear, and has a

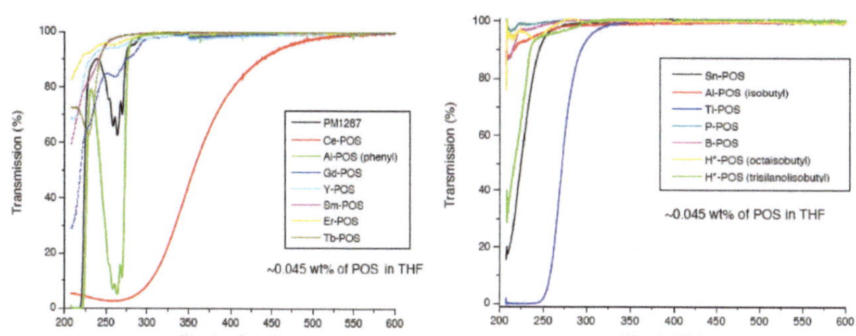

Fig. 8.10 Transmission spectra of various POMS formulations

continuous use temperature of 400 °C. The resin is designed for use as an optically clear coating, but can also be utilized as a composite resin in both glass and carbon fiber composites. Optical components on space vehicles are known to undergo material degradation, discoloration, loss of optical properties and adhesion due to chain cleavage, and oxidation upon exposure to photonic radiation and X-rays. Such degradation limits the service life and performance level of optical systems. A lower cost solution than deployment of metalized glass cover glasses and vapor-deposited coatings would be to coat sensitive optical components with a polymeric POMS conformal coating that absorbed the damaging incident radiation. The selectivity of POMS toward various wavelengths of VUV through visible radiation is shown in Fig 8.10. The two POMS absorbing the broadest spectrum of radiation are the cerium (Ce) and titanium (Ti) systems. The influence of the R group on the POMS absorption can be seen through examination of the two Al POMS systems.

The Al POMS containing R = Ph is strongly absorptive from 200 to 225 nm, and from 250 to 275 nm, while the Al POMS containing R = *i*-butyl only begins to absorb strongly at 225 nm. Similarly the Gd, Sn POMS also show strong absorptions at 225 nm, while POMS bearing other metals are only weakly absorbing above 200 nm, and absorb strongly below this level. In the light of these properties, the Er, Tb, Al POMS would likely be well suited as protective coatings for radiation below 200 nm. Note that space grade silicone is damaged by 150-160 nm VUV radiation, while polycarbonates are damaged by 243nm UV radiation.

Table 8.5 Radiation targets and applications for various POMS.

POMS additive	Primary targeted radiation	Secondary targeted radiation	Targeted component application
Gd	Neutrons	X-ray, e^-	ICs
Sm	Neutrons	X-ray, e^-	ICs
B	Neutrons	X-ray, e^-	ICs
W	X-ray	e^-	Sensors, PVs
Ta	X-ray	e^-	Sensors, PVs
Ti	VUV-UV	X-ray, e^-	Sensors, PVs
P	VUV-UV	Electrons	Sensors, PVs
Sn	VUV-UV	Electrons	Sensors, PVs
Si	VUV-UV	Electrons	Sensors, PVs
Al	VUV-UV	X-ray, e^-	Sensors, PVs
H	Protons	e^-, VUV-UV	IC and Optical
F-	Electrons	Protons	IC and Optical
Ce	VUV-UV	X-ray, e^-	Sensors, PVs
Tb	VUV-UV	X-ray, e^-	Sensors, PVs

Each of the POMS systems was found to be suitable for application as a coating on both thin film and triple junction solar cells, since they were optically transparent at 360 nm through longer wavelengths. The adhesion and suitability of Er and Tb POMS in PM1287 as VUV protective conformal coatings for solar cells was demonstrated by coating a series of active solar cells, and sending them to NASA Glenn Research Center for long term tests which confirmed their stability.

Interestingly, the Tb POMS was identified as being especially beneficial as a solar cell conformal coating, since it was found to absorb short wavelength radiation (200-300 nm) that the cell cannot use, and very efficiently re-emit it through luminescence in the useful 540 nm (green) range. Fig. 8.11 shows a solar cell coated with Tb POMS.

Fig. 8.11 Solar cell coated with Tb POMS

Ten 70 cm^2 amorphous silicon (α-Si) solar cells were measured before and after coating at the NASA Glenn Research Center under AM0 conditions. Five were coated with the POMS formulations listed above. The coatings were all between 125 and 150 nm in thickness. Fig. 8.12 shows a photograph of the five samples after coating compared to one uncoated sample. This color difference was not unexpected. From left to right the cells are coated with PM1287 control, 10% Ce-POS, 10% Tb-POS, 10% Gd-POS, 10% Ti-POS. An untreated cell is at the far right. From visual inspection, it can be seen that the anti-reflection coating on the cell has not been tailored for the refractive index of the POMS/PM1287, and hence the cells look dark grey rather than black.

Fig. 8.12 Coated solar cells

Examples of the pre-and post-coating current-voltage (I-V) curves are shown in Fig 8.13. The change in the curve is primarily due to a drop in short circuit current. No significant changes were seen in any of the other components. Table 8.6 provides a listing of the pre- and post-coating cell performance data. The average loss in short circuit current (I_{sc}) for the four cells with the POMS coatings was 6.1%, and the loss in I_{sc} for the PM1287 was 6%. Thus the loss in I_{sc} was the same for all cells, confirming the mismatch between the refractive index of the POMS (~1.45) and the cell anti-reflection coating. In this test level, no significant change was seen in the open circuit voltage (V_{oc}). Only one cell showed a loss in fill factor (FF), and that was the cell coated with PM 1287 without the POMS additions. Because PM 1287 was used in all the other formulations, this loss, which also impacted this cell's maximum power (P_{max}), is most likely caused by a handling problem, since amorphous Si thin film cells can be easily scratched.

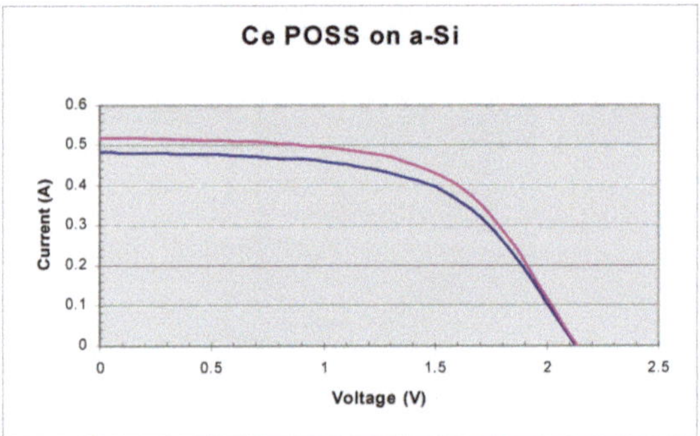

Fig. 8.13 Typical pre- and post-coating I-V curves

Table 8.6 Pre- and post- coating solar cell performance data

Cell coating	Voc (mV)	Isc (mA)	Pmax (mW)	FF (%)	Effic. (%)
Ce POMS pre-	2133	520	647	58.3	6.25
post-	2129	484	591	57.3	6.18
Ce POMS pre-	2143	521	658	58.9	6.87
post-	2131	492	616	58.7	6.43
Ce POMS pre-	2136	517	642	58.1	6.87
post-	2130	485	594	57.4	6.43
Ce POMS pre-	2129	522	652	58.2	6.81
post-	2133	493	613	58.2	6.4
Ce POMS pre-	2140	521	650	58.2	6.79
post-	1964	490	468	48.6	4.88

This was the first test of the application of a PM1287/POMS coating onto thin film solar cells. The coatings were adherent and thermally stable but the optical properties of the anti-reflection coating on the amorphous Si cells was not matched to the refractive index of the coating, hence no further work was done on these samples. It was deemed more important to confirm the radiation stability of these coatings before returning to solar cell applications.

In light of the results on thin film cells, the next steps not only involved looking at the radiation tolerance of the PM 1287, but also developing a new methacrylic POS. For all space applications the materials must be glass-like in their final state, resist AO, UV/VUV and be resistant to electron and proton radiation. Flexibility would be a benefit, but is not essential. In addition, they should be able to be deposited by spraying or other similar techniques, and, when ultimately applied to solar cells, have excellent adhesion and do not peel away under the extreme thermal cycling conditions that are experienced in space (-100 to +100 °C). The initial investigation of these new materials was directed toward determining their resistance to low energy proton irradiation. Many space missions are only possible by flying through the heart of the Van Allen radiation belts around the Earth. One mission of this type is a solar electric propulsion mission that moves a satellite from low earth orbit (LEO) to another location. The location may be geosynchronous earth orbit (GEO) as for a communications satellite, or a lunar orbit like ESA's Smart 1. While these missions take more time than using a chemical kick motor, the costs are substantially lower. Other missions of interest are those that would benefit from observing the Earth, and that fly either elliptical orbits that pass through the belts, or that stay within the belts (e.g., the Global Positioning System (GPS) of satellites). The initial study [26,27] looked at the resistance of these coatings to 2 MeV protons. These protons are absorbed in glass-like materials to a depth of about 75 μm, and these coatings were to be about 150 μm thick. Protons are exceptionally damaging to polymeric and other materials because most of the damage occurs at the end of their path, thus causing maximum damage in a very narrow region of the material. If the new materials can withstand this punishment, the next step is generally to assess their resistance to VUV/UV illumination.

Several different compositions of binder and POS additives were used. For most of the samples the binder was either PM1287 as before, or a new methacrylic POS called MA8000. Fig. 8.14 shows the chemical composition of these two binders. The octomeric Si-O cage is the basic building block in PM1287 and is usually introduced via hydrosilylation. To this structure, both metallic additives and organic ligands can be added. In this study, the PM1287 was usually prepared with a phenyl additive in concentrations up to 50%. These additions were generally increased in 10% increments. The MA8000 contained methacrylic additives up to 30%. Several different cure cycles were used from room temperature for long times, to 80 °C cures for a few hours. Metallic additives (POMS) included Gd and B. Both the PM1287 and MA8000 were found to be compatible with the POMS. As described above, the metal is inserted directly into the POS cage and is very stable. The addition of

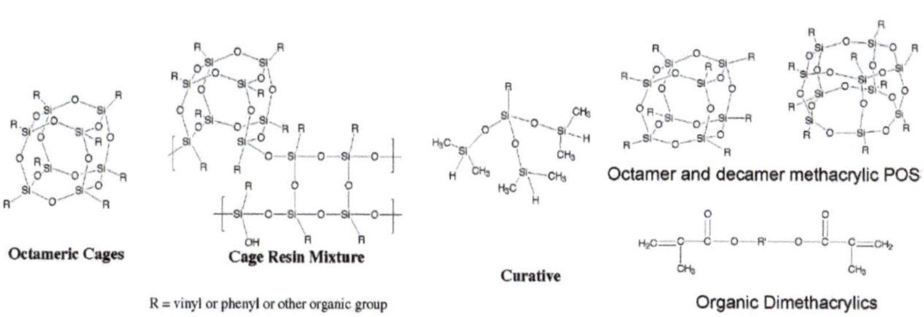

Fig. 8.14 PM1287 structure (left) and MA8000 structure (right)

POMS can further improve the overall performance of those POS based adhesives, especially if they emit light in the visible region of the spectrum.

The samples were coated onto standard 3.5 cm by 2.5 cm Thales CMG200 Ce-doped microsheet glass slides (the standard solar cell cover glass material). It was used in order to ensure adhesion, and to ensure that the spectral transmission through the layers would be the same as if the coating were used on a solar cell.

PM1287 samples were thermally cured using platinum catalyst. MA8000 samples were cured under UV light. The POS layer was deposited by brush. Optical transmission of the samples was measured between 200 and 1200 nm using an uncoated Ce-doped slide as the background reference. Because an uncoated slide was used as the reference, the data do not take into account transmittance differences associated with reflection at the interfaces.

A 2 MV Dual Source Tandem Accelerator shown is in Figure 8.15. This accelerator can provide a beam of protons from a SNICS ion source with energies from 100 keV to 4 MeV. It can also provide a beam of alpha particles ranging in

Fig. 8.15 NEC Pellatron 2 MV Dual Source Tandem Accelerator

energy from 100 keV to 6 MeV. In addition, a range of heavier atoms including nitrogen, aluminum, and phosphorus can be provided for ion implantation. The ion implantation energies exend up to 12 MeV. For this work, all irradiations were done in vacuum ($5x10^{-7}$ to $1x10^{-6}$ Torr) and at room temperature. The dose rate was kept constant for each irradiation, with the exception that for the highest dose, the rate was tripled. The initial irradiation was a dose of 10^{12} protons cm^{-2}. After each irradiation the samples were visually inspected for damage , and another transmission spectrum was acquired. The following irradiation total dose schedule was used: second irradiation 10^{13} protons cm^{-2}, third irradiation 10^{14} protons cm^{-2} and fourth irradiation 10^{15} protons cm^{-2}. Beam currents ranged from 55 nA at the lower doses to 190 nA at the highest dose. The proton beam was scanned over a 5 cm diameter area with the samples at room temperature. The scan rate was 517 Hz in the *x*-direction and 64 Hz in the *y*-direction. Two samples were irradiated at a time.

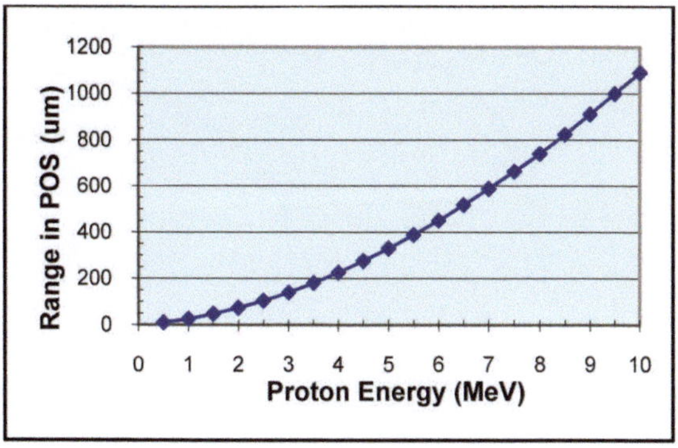

Fig. 8.16 Proton range-energy curve in POS having a density of 1.4 g cm^{-3}

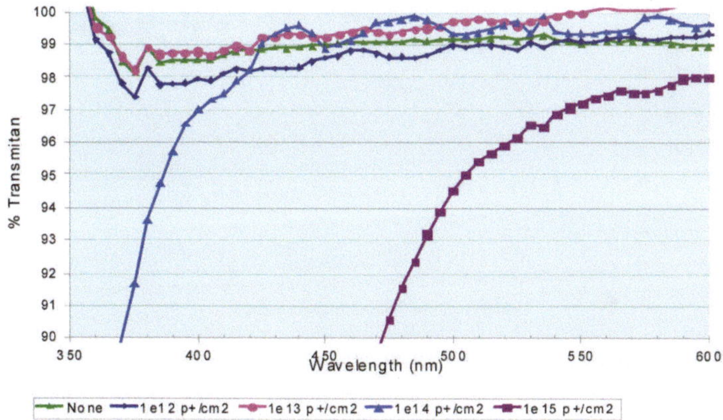

Fig. 8.17 Spectral transmission curves of PM1287 with 50% phenyl content

Fig. 8.18 Images of PM1287 with 50% phenyl content after 2 MeV proton irradiations

The range-energy calculation for the POS materials is shown in Fig. 8.16. These results were obtained using the measured density of the POS materials. At 2 MeV, the range of the protons in POS is 73 μm with a straggle of ~2 μm. Thus the entire proton beam is absorbed within the 150 μm thick POS layer, creating maximum damage within the layer.

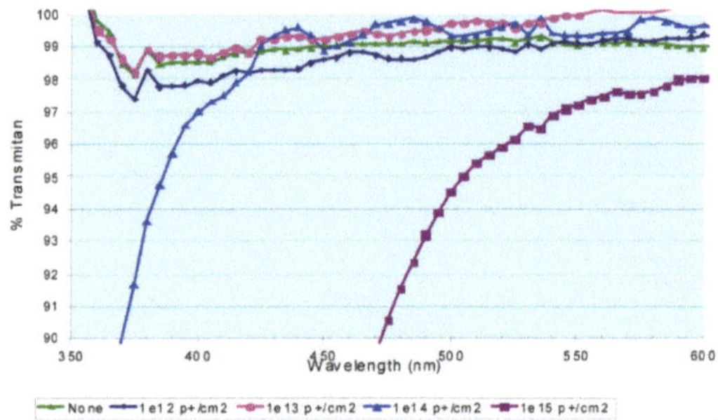

Fig. 8.19 Spectral transmission of PM 1287 with 10% to 50% phenyl content at 10^{12} p+ cm^{-2}

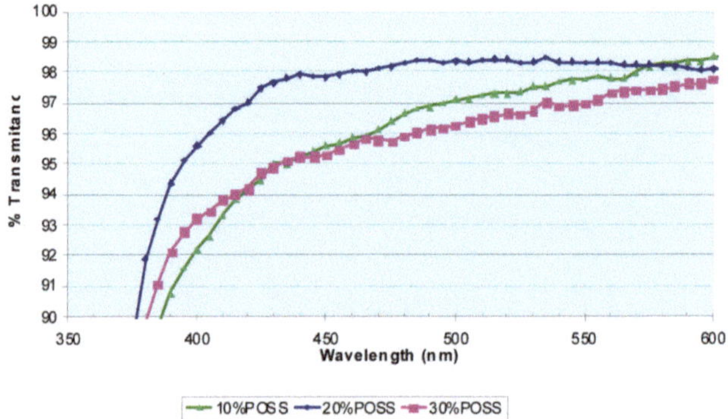

Fig. 8.20 Spectral transmission of MA8000 with various POS additions

Fig. 8.21 Images of MA8000.2 (20% POS) at several proton fluences

The PM1287 resin was tailored to have various possible phenyl contents. The phenyl percentage used herein indicates the mole percentage of the phenyl groups compared to the vinyl groups that are attached to the POS cage (as shown in Fig. 8.14). A total of four samples were made: 10% phenyl, 15% phenyl, 20% phenyl and 50% phenyl. The 50% phenyl PM1287 demonstrated the best proton radiation tolerance as shown in Fig 8.17. The data were limited to the region of interest between 300 and 600 nm. Below 300 nm the CMG200 slide has limited transparency, and above 600 nm the film transmittance does not vary with wavelength. For the sample with 50% phenyl concentration, the transmittance did not decrease until the 10^{14} p^+ cm^{-2} dose, and substantial darkening and cracking did not occur until the 10^{15} p^+

cm^{-2} dose as shown in Fig. 8.18. For the 10^{12} p^+ cm^{-2} dose, the high concentration of phenyl (50%) provided the best radiation tolerance and the lowest concentration (10%) the least tolerance as seen in Fig. 8.19. Because the intermediate 15% and 20% samples do not follow this trend there may be additional factors (i.e. catalyst loading) that affect the radiation tolerance. In addition to darkening, the sample films showed evidence of structural stress at large total dosages. The damage typically started as micro-fractures on the surface which then developed into long cracks, followed by delamination and separation from the cover glass substrate. The 10% and 15% phenyl films cracked during the 10^{13} p^+ cm^{-2} irradiation, the 20% cracked 72 hours after the 10^{13} p^+ cm^{-2} test, and the 50% phenyl film developed one long crack after 10^{15} p^+ cm^{-2}. These results suggested that the thermal expansion coefficient on the PM1287 was larger than the cerium-doped microsheet glass, and the additional shrinkage of the film caused by the proton irradiation was causing ultimate failure of the films.

The same procedures noted above were followed for the MA8000 samples. For MA8000 samples, the amount of functional POS molecules dissolved in the

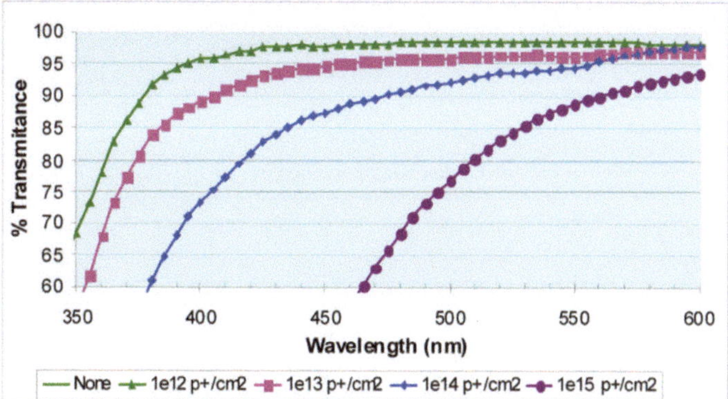

Fig. 8.22 Optical transmission curves of MA8000.2 (20% POS) with increasing proton dose

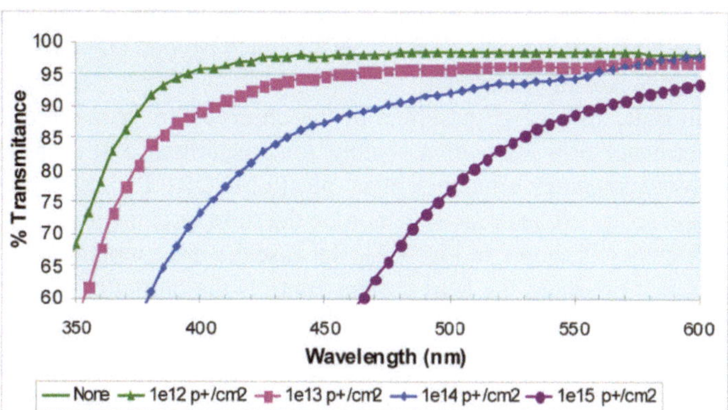

Fig. 8.23 External quantum efficiency of a typical ATJ space solar cell

polymer was directly related to the films' radiation tolerance. Three samples, 10 wt %, 20 wt %, and 30 wt % POS-reinforced MA8000 (MA8000.1, MA8000.2 and MA8000.3, respectively) were made. These samples were more opaque below 500 nm than were the PM1287 samples, as shown in Fig. 8.20. The sample with 20% POS in MA8000 was the most transparent, and showed the least radiation-induced darkening (see Fig. 8.21). Layers based on MA8000 also showed indications of stress within the film after irradiation, just as had been seen with the PM1287 films.

At 10^{14} p^+ cm^{-2}, both the 10% and 30% POS samples cracked, but the 20% POS sample did not crack until the 10^{15} p^+ cm^{-2} irradiation. Photographs of the MA8000 with 20% POS are shown in Fig 8.21 and the spectral transmission curves are shown in Fig. 8.22 (note the scale change). The darkening of the sample seems very great at the 10^{15} p^+ cm^{-2} dose level. Because spectral transmission curves by themselves do not give any indication of the performance of a solar cell beneath that layer, another study was performed to attempt to quantify this loss.

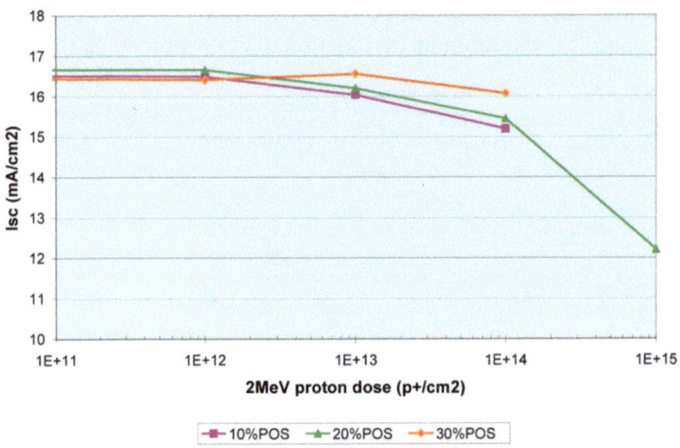

Fig. 8.24 Calculated I_{sc} variation with 2 MeV proton dose for PM1287-coated samples

This is a most informative approach because the internal quantum efficiency of the solar cell determines the ultimate performance beyond any darkening of the cover glass or encapsulant material.

Thus, in order to predict the effect of the observed cover material darkening on the performance of a solar cell assembly, the transmittance of each sample was multiplied by the quantum efficiency of a typical EMCORE Photovoltaics, Inc. ATJ triple junction solar cell, in order to determine the short circuit current, I_{SC}. The ATJ quantum efficiency is shown in Fig. 8.23. Because the three junctions are in series, the junction that produces the least current controls the output of the cell. The I_{SC} calculated for this typical ATJ cell is 17.1 mA cm^{-2}. This value agrees well with data provided by the company. As expected, the middle junction (indium-gallium-phosphide, InGaP) controls the short circuit current. The lowest junction captures much more current in the infrared region, but the InGaP junction limits the current.

Because the transmittance measurements were made using an uncoated cover glass for reference, the data do not incorporate the effect of reflective losses at the interfaces. A final solar cell design tailored for the POS encapsulant would include optimization of the anti-reflective coating between the cell and the cover material as noted for the thin film solar cell samples. Fig. 8.24 indicates that the PM1287 with 50% phenyl showed no decrease in I_{SC} for dosages up to 10^{14} p^+ cm^{-2}. Even at lower phenyl substitutions, the I_{SC} decrease was less than 5% for the sample containing 10% phenyl with a dose of 10^{13} p^+ cm^{-2}. Although the darkening of the MA8000 was more substantial, the decrease in I_{SC} for the 30% POS sample at a 10^{14} p^+ cm^{-2} dose is only 2%. This dose is roughly equivalent to the radiation dose received over a 20 year mission in GEO. Thus the stability of this POS coating is quite acceptable for that type of mission. However, it is important to note that there will be additional solar cell degradation that is not taken into account in this study.

The results show that the material compositions exhibit a regular progression in resistance to damage by 2 MeV protons as the phenyl content increases in the PM1287 resin, and as the POS content increases in the MA8000. It is important to note, that with the calculated I_{sc} of an ATJ cell for these coatings, the PM1287 with 50% phenyl substitutions decreased by only about 13% at 10^{15} p^+ cm^{-2}. This shows exceptional durability. Furthermore, the MA8000 with 20% POS showed a 17% drop under the same dose. It can be speculated that the chemical stability of the phenyl group may play a key role in limiting proton damage in these tests.

Fig. 8.25 Solar cell with POS conformal coating

While these results were very encouraging, many other space durability tests had to be taken into account. First among these is UV/VUV testing. The cracking seen in the coatings is also a serious concern. Because of the inherent adaptability of the material compositions of the POS, it may be possible to find solutions to the cracking, and ultimately to the darkening of these unique materials. While these coatings must be applied to current production cells to confirm that no inherent damage occurs to them, preliminary work with thin film solar cells uncovered no problems. These coatings are exceptionally clear, and are able to conformally coat solar cells (Fig. 8.25). This is a key factor in the protection of solar arrays from arcing in GEO orbit. The coating should prevent the charge build-up that is a precursor to an arc.

Recent work [28] on POS-silicone compositions was carried out in order to advance the understanding of these new types of flexible lightweight solar cell coatings through space environmental testing. These tests also aimed to provide information on solar cell coatings having the potential to significantly improve mission capability (reduced solar array weight and volume), to increase the solar array lifetime (increased radiation hardness), and to reduce the solar array cost (simplified manufacturing process). POS trisnorbornyl carrying additional *i*-butyl and/or *i*-octyl alkyl groups was used with DC 93-500, resulting in a 20 wt % reinforced DC 93-500® silicone material. Commercial cover glass materials coated with this resin were evaluated for their response to low energy proton radiation simulating the effects of the proton space environment. Exposures for multiple orbits including LEO (833 km, 98° inclination), Global Positioning System (GPS, 20,372 km, 55° inclination) and GEO (35,688 km, 0° inclination) were planned.

Table 8.7 Proton 10 year orbital simulations

Proton energy	Fluence (p^+/cm^2)	Fluence (p^+/cm^2)	Fluence (p^+/cm^2)
Energy (keV)	GEO	GPS	LEO
	10 Year	10 Year	10 Year
400	1.2E+13	1.84E+14	2.29E+11
300	2.74E+13	2.28E+14	1.55E+11
200	1.18E+14	6.19E+14	3.43E+11
100	3.58E+14	1.26E+15	1.00E+12
50	1.03E+15	1.80E+15	1.60E+13
20	2.54E+16	1.08E+16	1.00E+15

Radiation transport modeling was used to generate energy absorption relationships for the following systems: DC 93-500 coated bare Ce-doped cover glass, 20 wt % POS (DC 93-500 – NB1070) coated microscope slide, and an antireflection (AR) coated Ce-doped cover glass. The initial one year GPS orbit proton test was performed on these three systems. The minimum thickness of the POS-silicone coating was 8 μm. Table 8.7 provides the proton energy distributions and doses for 10 years in the three orbits. Optical transmission spectra were measured for all samples, and convolved with a typical triple junction solar cell quantum efficiency, and with the outer space (AM0) spectrum to obtain an estimate of potential current loss for one year in this high radiation orbit. These results, though preliminary and not done at the 10 year level, showed slightly greater loss for the 20 wt % POS (DC 93-500 – NB1070) compared to the DC 93-500 coated slide. Both showed 98.3% current remaining for the solar cell. The AR coated cover glass retained 99.7% of its initial current. These tests show promise for the POS system, but are not conclusive at this time. Much more work on specific compositions must be done to ensure that POS systems are useful for this application. It should be noted that the very low energy protons at 20 keV are exceptionally damaging because they are absorbed in the topmost layer of the material. The 400 keV protons are absorbed within the first

6 μm of the film.

Initial UV/VUV testing was done with four different samples: PM1287 with 50% phenyl content, MA8000.2, PM1287 with Gd POMS, and PM1287 with Tb POMS. The metallic additives were selected to determine if stability of the films was enhanced under these test conditions. The samples were tested for 1024 hours in the UV/VUV testing facility at the NASA Glenn Research Center. The exposures were made in a vacuum chamber using a deuterium lamp (Hamamatsu L7293) that provides radiation as low as 115 nm. The lamp is outside the chamber (to prolong its life) and the radiation is transmitted into the sample through a magnesium fluoride window. The transmission spectra were converted into a calculated J_{sc} loss for an ATJ cell. These results are shown in Fig. 8.26. It appears that the MA8000.2 and the PM1287 with the Tb POMS showed saturation of damage, but this is not conclusive.

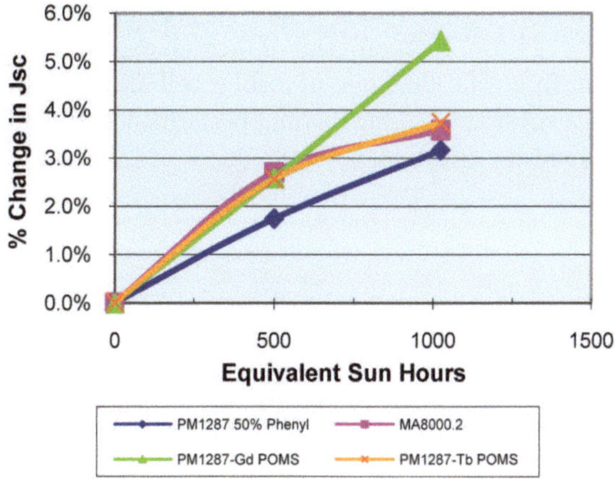

Fig. 8.26 UV/VUV preliminary test results

Interestingly, as noted in Fig. 8.11, the terbium POMS was identified as being especially beneficial as a solar cell conformal coating, since it was found to absorb 'non-useful' 200-300 nm radiation and very efficiently reemit it through luminescence at a 'useful' 540nm wavelength. This green wavelength is absorbed strongly by the top and middle junctions of the triple junction solar cell. As efficiency of the triple junction solar cell increases, the middle junction will play a more important role in radiation resistance and longevity. The application of such a coating would be beneficial, and provide additional photons for energy generation by the solar cell while protecting it against damage by unusable higher energy VUV wavelengths. The high atomic mass of terbium is also beneficial, since it provides shielding against proton, electron, and X-ray radiation. This work is still ongoing.

Because of the potential for a conformal solar cell coating, attention was devoted to producing a sprayable form of a clear, colorless POS-polyimide formulation. The POS-polyimide chain is depicted in Fig 8.27. This new material resulted from a collaborative effort between Hybrid Plastics, Inc. and Mantech International Corporation. This material received an IR100 award in 2008 and is marketed under

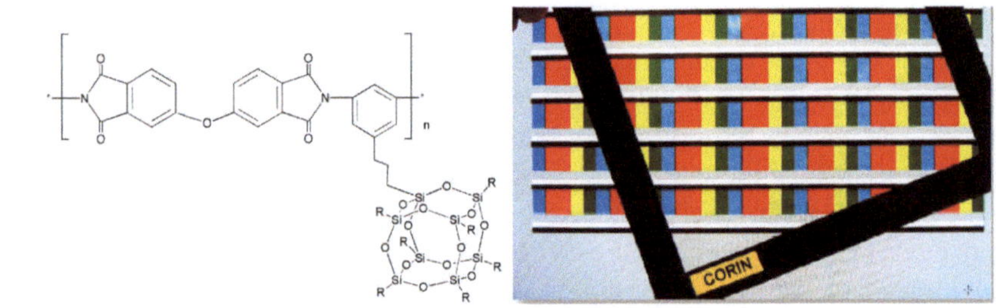

Fig. 8.27 Clear POS-polyimide formulation and its optical properties

the name Corin XLS at Mantech, and POSS® ImiClear at Hybrid. A 25 μm thick film has a 50% transmission cut-off wavelength of 392 nm, making it suitable as a coating for space solar cells. The refractive index is 1.54 and its solar absorptivity is 0.08 at a thickness of 25 μm. Its coefficient of thermal expansion is $68 \mu m\ m^{-1}\ {}^{\circ}C^{-1}$. Although this is much greater than that of a solar cell alone, space solar cells are mounted on various substrates, so the thermal performance of the package can be tailored to the required specification.

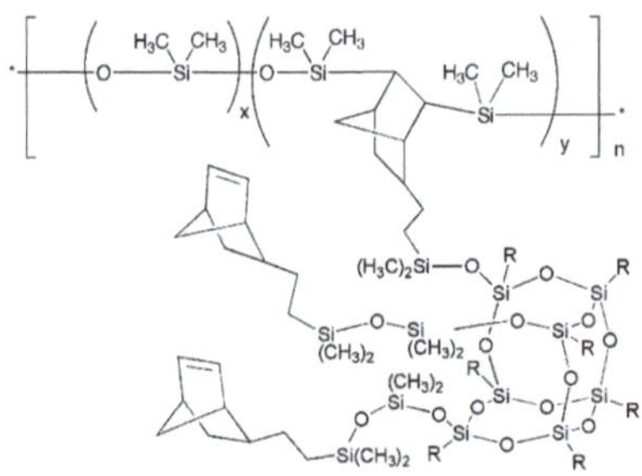

Fig. 8.28 Silicone-tris-norbenyl (isobutyl)$_7$POS network

Because of the need for a compliant POS solar cell covering material, Hybrid Plastics has developed two compliant adhesive POS silicone formulations. The primary difference is the POS loading (20% versus 40%). POSS® Coat SC4300 is a two-part adhesive coating that contains 20% POS loading and POSS® Coat SC4302 contains 40% of the trisnorbornyl(isobutyl)$_7$POS additive (Fig. 8.28). It can be used as an encapsulant, a protective coating or as an adhesive. It is important to note that the silicone used in this formulation is not the industry standard space silicone Dow Corning DC 93-500®. A 20% POS loading into DC 93-500 was demonstrated, and a flight experiment (TacSat-4) has a single multijunction solar cell covered with that material. However, because of the high cost of the space silicone, lower cost silicones are used for POSS® Coat SC4300 and POSS® Coat SC4302. These formulations cost three to six times less than DC 93-500.

The future of these coatings as replacements for cover glasses on solar cells is unclear at this time. It appears that the commercial adoption of a polymeric coating as a general replacement for solar cell cover glasses is unlikely. The reason for this is that the range of orbital radiation environments is extremely varied, and the need for decades of flight heritage to mitigate mission risk is essential. However, technically, such an effort is achievable. This work has shown that radiation shielding can be achieved using POS polymeric systems. Both radiation-resistant compliant adhesives and radiation-resistant polymer films for glass microsheets have been developed and tested. It appears that the best prospects for deployment of such materials would be on thin film solar cells, if or when they are used in space. While they appear to save mass and volume, their low efficiency requires larger areas that could interfere with the satellite sensors. At present, no similar drivers exist for crystalline solar cell arrays. However, if mission power requirements were to increase greatly, then radiation-resistant conformal coatings for high power arrays would be considered mission critical, since array reliability and mitigation of cell electrical arcing (short circuits) would likely become a critical path. Before any of those uses can occur, the results of space tests such as the MISSE exposure tests must provide persuasive evidence of success.

8.4 Summary

Hybrid inorganic-organic polymers containing POS monomers were synthesized, characterized and tested for atomic oxygen resistance. Samples were placed in a specialized AO source, with the ability to perform in-situ XPS analysis. For both the POS-polydimethylsiloxane and POS-polyurethane systems a passivating SiO_2 layer formed, resulting from the AO-induced degradation of the organic components and subsequent oxidation of the POS framework. Changing the composition of the backbone of the polymer system from entirely inorganic (POS-PDMS) to an inorganic-organic hybrid (POS-PU) does not adversely affect the formation of the SiO_2 layer. Furthermore, the resulting ceramic SiO_2 layer serves to rigidize the elastomeric materials and prevent further degradation of the virgin polymer. These promising results, combined with the numerous property enhancements previously reported for POS incorporation into traditional polymer systems, makes the use of POS an attractive alternative to filler or coating systems in space-based material applications.

The surface of a film of a POS-PU copolymer was characterized in situ using XPS before and after exposure to different fluences of AO produced by an ESD hyperthermal oxygen atom source. XPS data indicated that exposure to AO reduces the carbon content on the surface from 72.5 to 37.8 atom % after 63 hours exposure to an AO flux of 2.0×10^{13} atom cm^{-2} s^{-1}. The oxygen and silicon concentrations in the near-surface region increase with increasing exposure to the AO flux. The oxygen-to-silicon ratio increases from 2.28 for the as-entered sample to 2.93. High-resolution XPS data suggest that the AO initially attacks the cyclopentyl groups on the POS cage, most likely forming CO and/or CO_2 and water which then desorb.

Increased exposure to the AO flux results in the formation of a silica layer on the surface which acts as a protective barrier, preventing further degradation of the underlying polymer. These results are persuasive in showing that the addition of POS cages to many polymeric systems will enable those systems to resist the effects of AO in LEO, thereby representing a major breakthrough in materials survivability and usefulness in space.

Screening of materials that can be used for an easy-to-apply, space-durable, conformal encapsulant for solar arrays was conducted. The results showed that some of these coatings were exceptionally resistant to damage by 2 MeV protons, which were completely absorbed within the layers. More materials combinations and modifications will be tested over a range of other conditions, beyond proton exposure. Furthermore, the direct application of these coatings onto production solar cells is an essential next step. If this extensive environmental and radiation testing proves successful, a new approach to the total encapsulation of solar arrays will have been demonstrated. With that success, solar array costs should drop, and the development of high voltage arrays for high power space missions should proceed smoothly. However, despite positive results from diligent ground testing, the ultimate verification will only come through successful demonstration in space, owing to the conservative nature of the space business and the costs that failures entail.

We expect that the space market will be transformed by the various POS options, that the impact in performance will be high, and that the reductions in costs will be impressive. Thus the next decade will be exciting to watch!

8.5 References

1. Banks BA, Demko R (2002) Atomic Oxygen Protection of Materials in Low Earth Orbit, NASA/TM-2002-211360, February 2002.

2. Banks, BA (1997) The Use of Fluoropolymers in Space Applications. In Scheirs J (Ed) Modern Fluoropolymers: High Performance Polymers for Diverse Applications, John Wiley & Sons, Ltd, Ch. 4.

3. Banks BA, Rutledge S, Paulsen P, Stueber T, Simulation of Low Earth Orbital Atomic Oxygen Interaction with Materials by Means of an Oxygen Ion Beam, 18th Annual Symposium on Applied Vacuum Science and Technology, Clearwater Beach, Florida, February 6–8, 1989.

4. Banks BA, Rutledge S, Low Earth Orbital Atomic Oxygen Simulation for Materials Durability Evaluation, 4th European Symposium on Spacecraft Materials in Space Environment, Toulouse, France,

September 6–9, 1988.

5. Rutledge S, Mihelcic J, The Effect of Atomic Oxygen on Altered and Coated Kapton® Surfaces for Spacecraft Applications in Low Earth Orbit, Proceedings of the Materials Degradation in the Low Earth Orbit Symposium at the 119th annual meeting of the TMS, Anaheim, CA, Feb. 17–22, 1990.

6. Banks BA, deGroh K, Rutledge S, Haytas C (1999) Consequences of Atomic Oxygen Interaction with Silicone and Silicone Contamination of Surfaces in Low Earth Orbit, NASA/TM-1999-209179, July 1999.

7. Banks BA, Rutledge S, Sechkar E, Stueber T, Snyder A, Haytas C, Brinker D (2000) Issues and Effects of Atomic Oxygen Interactions with Silicone Contamination on Spacecraft in Low Earth Orbit, NASA/TM-2000-210056, June 2000.

8. Hung C, Cantrell G, Reaction and Protection of Electrical Wire Insulators in Atomic-Oxygen Environments. NASA TM–106767, AE-8D Wire and Cable Subcommittee Meeting sponsored by the Society of Automotive Engineers, Albuquerque, New Mexico, April 12–14, 1994.

9. Gilman JW, Schlitzer DS, Lichtenhan JD (1996) J Appl Polym Sci 60:591-596.

10. Leger JL, Visentine JT, Santos-Mason B, Low Earth Orbit Resistant Siloxane Copolymers, Proceedings of the 18th International SAMPE Technical Conference, October 7-9, 1986, pp. 1015-1026.

11. Munjal AK, Long Service Life Considerations of Composite Structures for Space Applications, Proceedings of the 23rd International SAMPE Technical Conference, Kiamesha Lake, NY, Oct. 21-24, 1991, Proceedings (A92-51501 21-23). Covina, CA, Society for the Advancement of Material and Process Engineering, 1991, p. 802-816.

12. Koontz SL, Albyn K, Leger LJ (1991) J Spacecraft Rockets 28:315-323.

13. Koontz SL, Leger LJ, Albyn K, Cross J (1990) J Spacecraft Rockets 27:346-348.

14. Koontz SL, Leger LJ, Visentine JT, Hunton DE, Cross JB, Hakes CL (1995) J Spacecraft Rockets 32(3):483-495.

15. Hoflund, GB, Gonzalez RI, Phillips SH (2001) J Adhesion Sci Technol

15(10):1199-1211.

16. Hoflund GB, Weaver JF (1994) J Measur Sci Technol (1994) 5(3): 201-205.

17. deGroh KK, Banks BA, Techniques for Measuring Low Earth Orbital Atomic Oxygen Erosion of Polymers, 2002 Symposium and Exhibition sponsored by the Society for the Advancement of Materials and Process Engineering, Long Beach, CA May 12-16, 2002.

18. deGroh KK, Banks BA (1994) J Spacecraft Rockets 31(4):656-664.

19. Tomczak SJ, Marchant D, Svejda S, Minton TK, Brunsvold AL, Gouzman I, Gonzalez RI (2005) Properties and Improved Space Survivability of POSS (Polyhedral Oligomeric Silsesquioxane) Polyimides, Material Research Society Symposium Proc Vol 851.

20. Mather PT, Jeon HG, Romo-Uribe A, Haddad TS, Lichtenhan JD (1999) Macromolecules 32(4):1194-1203.

21. Wright ME, Schorzman DA, Feher FJ, Jin RZ (2003) Chem Mater 15:264-268.

22. Troya D, Schatz GC (2004) J Chem Phys 120:7696-7707.

23. Tomczak SJ, Marchant D, Mabry J, Vij V, Minton T, Brunsvold AL, Guenthner A, Comparisons of Polyhedral Oligomeric Silsesquioxane (POSS) Polyimides as Space-Survivable Materials, 2005 International Chemical Congress of Pacific Basin Societies (Pacifichem), Honolulu, Hawaii, USA, December 15-20, 2005. (NOTE: this is also available as AFRL-ERS-PAS-2006-275. Air Force Research Laboratory (AFMC), AFRL/PRS, 5 Pollux Drive, Edwards AFB, CA 93524-7048).

24. Brandhorst HW, Lichtenhan JD, Fu BX, A POSS® Coating for Thin Film Solar Cells, 31st IEEE Photovoltaic Specialists Conference, Orlando, Florida, January 2005.

25. Gonzalez RI, Phillips SH, Holland GB (2000) J Spacecraft Rockets 37(4):463-467.

26. Brandhorst HW, Smith TIS, Wells BK, Lichtenhan JD, Fu BX, POSS® Coatings as Replacements for Solar Cell Cover Glasses, 19th Photovoltaic Research and Technology (SPRAT) Conference, NASA Glenn Research Center, Cleveland, OH September 20-22, 2005.

27. Brandhorst HW, Smith TIS, Wells BK, Lichtenhan JD, Fu BX, POSS® Coatings as Replacements for Solar Cell Cover Glasses, 4th World

Conference of Photovoltaic Energy Conversion, May 7-12, 2006, Hawaii.

28. Liu SH, Granata JE, Meshishnek MJ, Ciofalo MR, Simburger EJ, Space Radiation Environmental Testing on POSS Coated Solar Cell Coverglass, Space Photovoltaic Research and Technology (SPRAT) Conference, Cleveland, OH, 2007.

Chapter 9
Biomedical Application of Polyhedral Oligomeric Silsesquioxane Nanoparticles

Hossein Ghanbari, Sayed Mahdi Marashi, Yasmin Rafiei, Karla Chaloupka and Alexander M. Seifalian

9.1 Introduction

Recent years have been characterized by the emergence of novel biomaterials with improved properties for a variety of biomedical applications. In particular, the development of advanced polymeric materials, composites, and nano-structured materials has been at the centre of research attention in this era. Composite materials are combinations of at least two constituent materials with significantly improved physical or chemical properties. Matrix (host) and reinforcement (guest) are the two main constituents in the majority of composite materials. The interaction between components and the final structure of a composite vary according to the type of the constituent materials and the fabrication technique. When at least one of the components is present at the nanometre scale, the composite may be considered as a nanocomposite material [1]. Since these materials have improved physical, chemical, and mechanical properties, e.g., excellent compression properties combined with good tensile strength, they are sufficiently versatile to be used in a wide range of applications.

Nanocomposite materials are sometimes classified as either organic, inorganic, or hybrid in composition [2]. Bone tissue, an example of an organic composite [3], is a relatively hard and lightweight material formed from chemical arrangements of hydroxyapatite (providing the bone's rigidity and brittle properties), with a significant degree of elasticity supplied by collagen which is an essential requirement for bone (given its physiological function). Inspired by the enhanced strength of such natural nanocomposites, there has been much interest in preparing

Hossein Ghanbari, Sayed Mahdi Marashi, Yasmin Rafiei, Karla Chaloupka and Alexander M. Seifalian
University College London Division of Surgery and Interventional Science
Royal Free Hampstead NHS Trust Hospital
Hampstead Campus, London, U.K.

C. Hartmann-Thompson (ed.), *Applications of Polyhedral Oligomeric Silsesquioxanes*, Advances in Silicon Science 3, DOI 10.1007/978-90-481-3787-9_9,

synthetic nanocomposites and examining their properties for various potential applications. Synthetic nanocomposites may contain different organic or inorganic components in their composition in order to achieve the properties required for a specific application across a wide range of biomedical fields.

In this chapter, polyhedral oligomeric silsesquioxane (POS) nanoparticles and POS-based nanocomposite polymers and copolymers are briefly introduced, and then the biomedical application of POS-containing materials are discussed in detail.

9.2 Nanocomposites

Materials behave very differently at the nano-scale when compared to the macroscopic scale. The properties of the materials change considerably when their size becomes significantly small, i.e., in the 1 to 20 nm size range. Recent developments have aimed to combine a variety of improved properties from different nanomaterials together in a single highly advanced nanocomposite that has the capability to be used in diverse applications.

Nanocomposite refers to a multiphase solid material where at least one phase has a dimension in the size range of 1-20 nanometres (nm). These solid phases can either be amorphous, crystalline or a combination of both, with organic and inorganic characteristics which directly influence the mechanical, electrical, thermal, optical, electrochemical and catalytic properties of the nanocomposite. Nanoparticles have been used as bridges between molecules in different polymers [1,4].

As with standard composite materials, nanocomposites use a matrix to bind together dispersed nano-reinforcement agents in order to enhance the optical properties of the materials, to increase heat resistance, and to improve mechanical properties such as stiffness, strength and resistance to wear and damage. The nature of the nanocomposites makes them different from conventional composite materials [5]. As a material becomes smaller in scale, it mainly consists of surface with only a small amount of volume, and hence the surface to volume ratio increases. This is a distinctive feature of nanocomposites, making them relatively advanced compared to conventional composite materials, since larger surface areas create the opportunity for higher numbers of reactions to take place on the surface. Abalone shells, bones, and teeth are examples of nanocomposites in nature. Composite abalone shells consist of alternating calcium carbonate ($CaCO_3$) and nano-scale aragonite asperities, making the shells 3000 times stronger than monolithic $CaCO_3$ crystals [6].

Nanomaterials are very sensitive, and even a very small change in the synthetic method can alter their characteristics. The method by which nanomaterials are synthesized is a crucial factor behind the behaviour of nanocomposite materials. For instance, in the case of polymer nanocomposites, the synthetic method along with the nature of nanofiller, the method of nano-reinforcement, and the type of

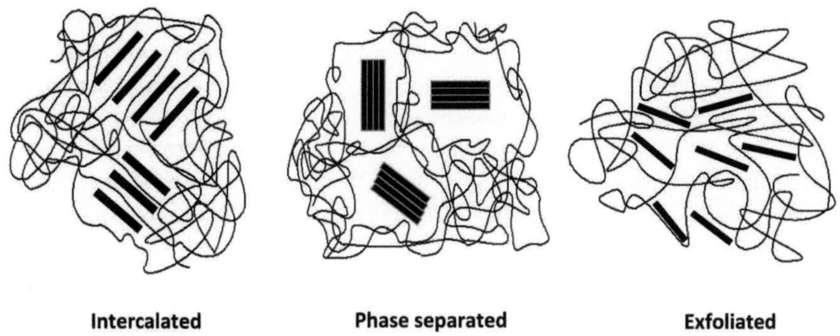

Fig. 9.1 Types of synthesized nanocomposites

interaction between the polymeric component and the reinforcing component regulate the behaviour of the polymer [7]. In intercalated nanocomposites, nanofillers are able to intercalate between the layers of matrix, or be further dispersed uniformly throughout the matrix to form exfoliated nanocomposites, thus maximizing the surface area available for component interaction. While intercalated nanocomposites have regular interlayers, exfoliated ones have larger interlayers (Fig. 1).

Currently available synthetic nanocomposites can be classified into clay-, carbon-, metal- and glass reinforced [8]. However silicon and metal oxides (e.g., ZnO and TiO_2) are commonly used as nanofillers owing to their beneficial mechanical properties. Clay-based nanocomposites are commonly formed into layers of montmorillonite, hectorite or saponite, while carbon-based nanocomposites consist of either single or multi-walled nanofillers [9].

The excellent properties of organic nanocomposites have encouraged scientists to develop synthetic nanocomposites for diverse areas of application [10,11,12] These advanced materials have the potential to overcome the main restrictions and shortcomings of the standard synthetic materials owing to their enhanced properties. However, for a material to be used in a specific application, some essential requirements should be fulfilled before actual use. In biomedical applications, for instance, the synthetic nanocomposite materials must have a substantial degree of biocompatibility, defined as the ability of the material to give the desired biological response in a certain biological environment.

Among the most commonly studied nanofillers in the development of nanocomposite materials is the silsesquioxane family.

9.3 Polyhedral Oligomeric Silsesquioxanes

The chemical structure of the silsesquioxane family of compounds is defined as $R_nSi_nO_{1.5n}$, generating structures consisting of an inner inorganic framework of silicon and oxygen atoms that is externally covered by organic groups (R), Fig. 9.2. R may be hydrogen, alkyl, alkene, aryl, arylene or almost any other functional group. Based on their molecular architecture, silsesquioxanes can be classified into two main categories of nanostructure: non-caged and caged. The non-caged class of silsesquioxane molecules can be sub-divided into random, ladder, and partial-cage nanostructure [13] (Fig. 9.2). Ladder-like polysilsesquioxanes include poly(phenylsilsesquioxane) [14], poly (methylsilsesquioxane) [15], and poly(hydridosilsesquioxane) [16].

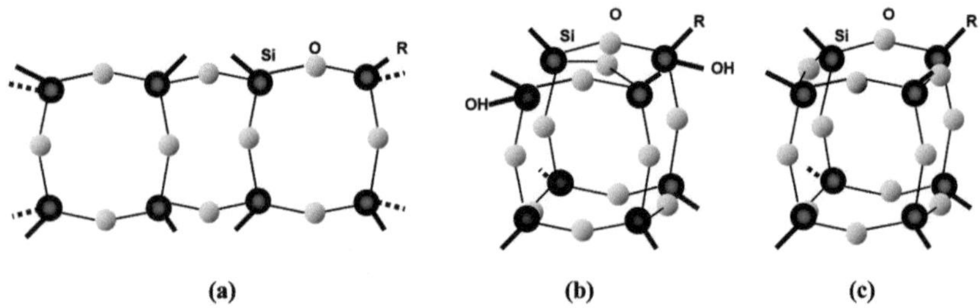

Fig. 9.2 Different structures of silsesquioxanes, (a) ladder structure, (b) partial cage structure, (c) cage structure

Conventional cage-like silsesquioxanes are designated polyhedral oligo-silsesquioxanes or polyhedral oligomeric silsesquioxanes [14]. Polyhedral oligomeric silsesquioxane (POS) is one of the molecules in the silsesquioxane family that has a regular three-dimensional shape (since polyhedra are multi-faceted three dimensional entities) formed by a few units (hence oligomeric), each containing silsesquioxane (SiO_3). The most common structure has a stoichiometric formula, and has an inner inorganic framework of molecular silica comprised of 8 silicon atoms, 12 oxygen atoms and 8 organic groups (R) that merged together form a molecule (Fig 9.2c). Hence, each silicon atom is bonded to three oxygen atoms and one organic group that may be either inert or reactive. These are highly symmetric molecules with a nanoscopic size of approximately 1.5 nm in diameter including the R groups, and can be considered as the smallest achievable silica particles [17].

Based on the type of R group (e.g., hydrogen, alkyl, alkylene, or aryl arylene) there is an almost unlimited number of POS types such as alkyls, olefins, alcohols, esters, anhydrides, acids, amines, imides, epoxies, thiols, sulfonates, fluoroalkyls, silanols, and siloxides (see also Chapter 1). Due to their hybrid organic-inorganic framework, POS molecules are sufficiently thermally and chemically robust for use in the space environment [18,19] (see Chapter 8). In addition, owing to their outstanding thermal stability, oxidative resistance, good insulating properties, and high gas permeability, ladder-like polysilsesquioxane polymers are used in a range of electronic applications, optical and semiconductor devices, gas separation membranes and drug compositions [14].

Research has shown that POS cubes can be incorporated as building blocks [20] into polymers to form hybrid inorganic–organic copolymers with improved thermal, mechanical and physical properties [21,22] (see also Chapters 4 and 5). Physical and thermal properties improved by incorporation of POS include decreased dielectric constants (κ) [23,24], increased glass transition temperature [23,25-28], decreased coefficients of thermal expansion, thermal stability, and heat evolution [17,27]. Improvements in the mechanical properties include tensile strength [24-26,29] and viscosity [17,29-31], and enhanced viscoelastic properties [4,22,32]. Further improvements have been reported as a result of the inclusion of POS into the structure of different polymers such as improved oxidation resistance, reduction of flammability, oxygen permeability, and reduced inflammatory reactions, showing the advantage of the use of these materials in various engineering applications [24-26,29].

POS can be incorporated into copolymers, either by grafting, or as a co-monomer, as dendrimer cores or as nanobridges (see also Chapter 4). Based on the number of side groups, POS may exist as pendant cages, be part of a polymeric backbone, or be cross-linked. For instance, POS molecules which are incorporated into the hard segment of a polyurethane-based polymer as a pendant side chain confer resistance to degradation, and enhance viscoelastic properties, biocompatibility and induce a smaller in vivo inflammatory response [4,22,32]. The incorporation of POS can also change the surface properties including hydrophobicity, surface energy and surface topography (Fig. 9.3), see also Chapter 6. Efficient surface coverage and stability to a variety of conditions are features that make POS derivatives attractive for use as surface modification agents or other modifiers [33]. Using POS as a non-covalent modifier, a variety of three-dimensional aggregates has been recently obtained from various nanoparticles such as palladium, magnetic, and gold nanoparticles [34-39].

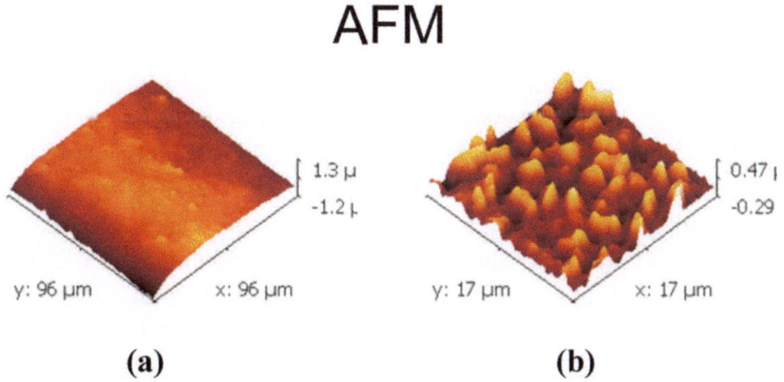

Fig. 9.3 AFM images of (a) PCU polymer and (b) POS-PCU nanocomposite surfaces. The nano-topographic pattern of the surface is attributed to the presence of POS molecules in the polymer.

9.4 Biomedical Applications of Polyhedral Oligomeric Silsesquioxane-Containing Polymers

The unique structure and superior properties of polyhedral oligomeric silsesquioxanes allow them to be introduced into different polymers and copolymers developed for various biomedical applications. As mentioned earlier, owing to their biocompatibility and ability to be easily incorporated into different polymers, POS nanostructures have been shown to have potential for various applications in diverse areas of the biomedical field such as drug delivery, dental applications, biomedical devices, and tissue engineering. Owing to their inert nature and low inflammatory response, the framework of POS, consisting of Si-O and Si-C, is very similar to that of silicone which has been a widely used biomaterial since the 1960s, when it was introduced in breast implants [40]. Biocompatibility is the main characteristic of POS, and results from the increased surface energy in the foci of the silicon-rich areas. Non-toxicity and cytocompatibility are other features of POS, making them suitable for biomedical application [41-43]. Nano-scale building blocks (POS cages in this case) incorporated into the polymers also improve their mechanical and viscoelastic properties. Hence, polymeric POS-based nanocomposites have been largely studied by material scientists with the aim of using them in biomedical and tissue engineering applications.

9.4.1 Drug Delivery

Generally, the systemic administration of drugs is associated with major clinical side effects. In contrast, newly proposed drug delivery systems aim to carry the intended drugs directly to the area where they are supposed to have therapeutic function. The basic concept behind the use of nanotechnology-based systems for drug delivery is related to the modulation of pharmacokinetics of incorporated molecules, i.e., the properties that govern drug absorption, distribution, and elimination in the human body are determined by the physico–chemical properties of the nano-system, particularly by surface exposed molecules, and by their electric charge and size [44]. For instance, recently developed silicone hydrogels can be utilized as a matrix for transdermal drug delivery [45], while silicone microspheres are being developed for pH-controlled drug delivery in the gastrointestinal tract [46].

Nanocomposites can also be used in drug delivery systems. The features of biodegradation, thermodynamic stability and biocompatibility, along with improved surface features, are all desired characteristics that allow nanocomposites to be considered as ideal nanocarriers with high distribution potential within biological systems. The application of POS nanoparticles in drug delivery systems also has the potential advantage of easy transfer via vascular pores, owing to their small size and high charge density that increase tissue uptake [47].

In order to assess the efficacy of a silsesquioxane nanocomposite to be used as a drug delivery agent, McCusker et al. [47] labelled octa-ammonium Si_8O_{12} units with a fluorescent dye (BODIPY) by neutralization of ammonium sites with triethylamine, and subsequent substitution with a succinimidyl ester derivative (see also Section 7.5.1, Chapter 7). BODIPY is a commonly employed fluorescent cellular membrane marker, which can be readily conjugated to various systems to track cellular migration patterns. It was found that the POS unit was the key factor that enabled the POS-BODIPY conjugate to disperse in the cytosol, as compared against a BODIPY control. Moreover, the POS-BODIPY did not influence the cellular morphology after uptake. The viability assay further proved that the cells with POS-BODIPY had the same activity level as the control, indicating that POS had extremely low toxicity. Furthermore, dispersion in the cytosol demonstrated that the POS-BODIPY conjugate entered into a cell via diffusion, and not via endocytosis. The lack of nuclear uptake demonstrated specific localization in the cell, and observations showed that the POS cages had potential as drug delivery carriers via direct conjugation with drug molecules that were insoluble in water, or exhibited a lower cellular uptake [47].

Recently Tanaka et al. demonstrated enhanced entrapment ability of dendrimers by incorporation of a polyhedral oligomeric silsesquioxane core [48]. Their results showed that the POS-core dendrimer could entrap a larger number of guest molecules without loss of affinity, and consequently, the water solubility of the entrapped guest molecules could be increased. In addition, they demonstrated that a fluorophore entrapped in the POS-core dendrimer was prevented from undergoing fluorescence photo-bleaching.

As well as the optimization of the chemical and physical properties of the nano-systems (as determinant factors in drug distribution and kinetics in the biosystems), the manufacturing techniques of mass production for commercialization and clinical translation purposes also need to be optimized and further investigated [49]. Simplification of the manufacturing techniques and optimization of targeted drug delivery vehicle systems by self–assembly of pre-functionalized biomaterials can facilitate high volume manufacturing processes [50,51]. The self assembly process results in the construction of vehicle shells that contain drug molecules in the centre. These shells are expected to prevent the occurrence of any drug leakage, and to avoid immune system recognition by impeding protein adhesion [49].

Owing to their strong framework and degradation resistance characteristics, resulting from the strong intermolecular forces between their constituent molecules and neighbours, POS molecules can be used as scaffolds for the production of these drug delivery shells [47]. Recently, self–assembled spherical amphiphilic nanoparticles (NPs) containing polyhedral oligosilsesquioxane as the inner hydrophobic core, and poly(vinylalcohol) (PVA) as the hydrophilic outer shell have been reported [52]. In vitro studies demonstrated that this nanosystem was able to release the model drug in a controlled manner. POS incorporation also improved the thermal stability and hydrophilicity of PVA, making it a potential carrier for sensitive drugs, proteins, peptides, and DNA [53,54]. In addition, owing to the size and specific nanostructure, this drug delivery system was able to travel easily within the body, and to carry the drug to the target regions [52,55].

In a recent study, a new nano-drug delivery system based on poly(L-glutamic acid) dendrimers with a polyhedral oligomeric silsesquioxane nanocubic core has been propsed [56]. Using pH-sensitive hydrazine bonds and a targeting moiety, the cellular internalization and anti-tumour potential of the OAS-G3-Glu dendrimer conjugated with doxorubicin was assessed in vitro. In this study the release rates of doxorubicin at different pH levels were investigated, and the cellular uptake of the biotin modified conjugates was analyzed. The results of this study demonstrated that the spherical morphology, along with the compact structure and multiple peripheral functional groups, made this nano-system suitable for drug delivery applications.

In line with other applications, it appears that increased attention is being paid towards the application of nanoparticles for chemotherapy, drug delivery, and imaging. For tumour diagnosis purposes, the physico-chemical features of the nanoparticles (such as size, surface coating, surface charge and stability) allow the qualitative or quantitative in vitro detection of tumour cells at the site of interest. Nanoparticles can act either by direct or indirect mechanisms. Taking advantage of tumour vasculature hypermeability can give these nanomaterials the flexibility required for the direct targeting of tumour tissues. However, in an indirect way, these particles can also target the tissue or cells near the tumour, and act as a drug reservoir to fight against the adjacent neoplastic cells. Controlled drug delivery at the site of interest, cell internalization, efficient drug protection, and protection of the drug from premature inactivation during its transport all seem to be feasible with appropriate design of the nanoparticle, perhaps by manipulating physico-chemical characteristics such as structure [57]. A schematic illustration of therapeutic drug delivery nanosystem based on the POS nanocage is shown in Fig. 9.4.

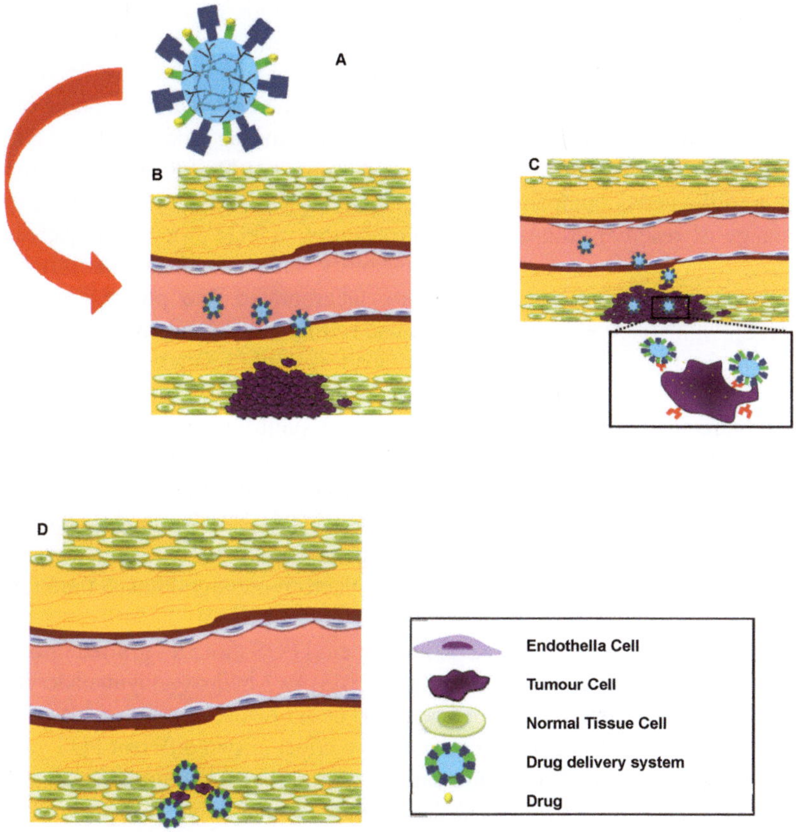

Fig. 9.4 Schematic illustration of therapeutic targeted drug delivery system, (A) POS incorporated spherical nano-drug delivery system (B) migration of the drug delivery system through capillary walls towards the targeted tumour cells (C) attachment of the nanosystem to the malignant cells via specific receptors (D) therapeutic effect of the delivered drug and significant reduction/elimination of malignant cell population

9.4.2 *Dental Nanocomposites*

A methacrylate-based resin matrix with glass or ceramic fillers is the main class of dental composite material. The resin matrix is usually cured (hardened) by photo-initiated free radical polymerization [58]. Polymeric dental composites have been used for over 40 years, and methacrylate-based polymers have been used extensively as dental implants alongside metallic and ceramic materials. In spite of intensive efforts to improve the mechanical properties of denture materials, there are still considerable shortcomings to be addressed. Shrinkage, wear resistance, biocompatibility, lack of strength, toxicity of polymeric materials, and modulus of elasticity are all areas that need to be improved. Additionally, inflammation and hypersensitivity reactions to denture materials, although rare, have been reported [42,59] and need to be addressed. Consistent with other areas, newly available POS nanomaterials are being considered as a potential solution to improve some or all of these properties. The potential for POS utilization for the modification of dental materials was first proposed by Sellinger and Laine in 1996 [60]. Small size (1.5 nm in core diameter) compared to other "nano" fillers and the ability to carry a wide range of reactive groups are unique features of POS. Moreover, increased glass transition temperature, oxygen permeability, and reduced flammability have been observed when POS derivatives are incorporated into polymer matrices [59,61,62]. Culbertson et al. investigated three different methods of incorporating a mono-methacrylate POS into dental composites: synthesizing a POS-containing macro-monomer and copolymerizing with dental composite monomers, a one-pot polymerization of POS and monomers, and synthesizing a POS-containing copolymer followed by in situ polymerization with dental monomers. The results showed that simply mixing methacrylate-POS into the dental formulation significantly decreased polymer shrinkage, although a slight reduction in mechanical properties was also seen especially at loadings greater than 10%. This work also indicated that synthesizing a POS-containing macro-monomer allowed the POS to disperse better when mixed with the dental monomers [61,62].

By incorporating methacrylated and octaphenyl POS moieties grafted into Bis-GMA (bis-phenol A-glycidyldimethacrylate), HEMA (2-hydroxyethylmethacrylate) and TEGDMA (tetraethylglycidylmethacrylate)-based dental restorative glass-filled pre-polymers, Dodiuk-Kenig et al. compared the effect of two different POS groups upon thermal, mechanical and physical properties. They demonstrated that the acrylated POS increased the T_g by 5 °C, the composite compressive strength by 7%, the bond shear strength by 36%, and decreased the shrinkage by 28%, while incorporation of octaphenyl POS decreased the compressive strength by 20%, decreased the adhesive shear bond strength by 49% and the shrinkage by 67%. Overall, this group has confirmed that the mechanical properties of dental composites and adhesives are improved by acrylated POS but detrimentally affected by octaphenyl POS, and thus they have concluded that the type of the grafted functional group on the silica cage has a strong influence on dental composite and adhesive behaviour [63].

The main difficulty in developing low shrinkage dental materials is the deficiency in mechanical properties needed to meet clinical requirements. As has been frequently mentioned, POS-modified polymers have good processability, toughness, low cost, mechanical, thermodynamic, and anti-oxidation properties [64,65]. In 2010 Wu et al. evaluated the effect of methacryl POS incorporation into BisGMA and TEGDMA dental composite resins. In this study, POS molecules were incorporated at different weight percents in a range from 0 to 15 wt % and the microstructure was characterized using Fourier-transform infra-red (FTIR) spectroscopy and X-ray diffraction (XRD). Adding 2 wt % POS resulted in an improvement in mechanical properties, including a 15% increase in flexural strength, a 12% increase in compressive strength, and a 15% increase in hardness, as well as the enhancement of the toughness of the resins. It was concluded that incorporation of 2 wt % POS improves the mechanical properties of the dental materials.

The influence of POS on the biocompatibility of methacrylate-based hybrid dental composites was also assessed by Kim el al. [42]. In this study, a reinforced acrylic-based hybrid denture composite resin with POS showed improved biocompatibility. Therefore, taking advantage of the aforementioned properties plus the rigidity, POS could be a potential choice for reducing the shrinkage of dental composites based on multi-methacrylates [61] as well as improving their biocompatibility [42].

Additionally, to prepare a novel dental restorative composite, POS-methacrylate was used to partially (or to completely) replace the commonly used base monomer 2,2′-bis-[4-(methacryloxypropoxy)-phenyl]-propane (BisGMA). The mechanical properties of the composites (i.e., flexural strength, Young's modulus and tensile strength) as well as the volumetric shrinkage, the degree of methacrylate double bond conversion, and the photopolymerization rates were investigated. The results indicated that small percentage substitutions (mass fraction of 10% or less in the resin system) of BisGMA with POS-MA (methacryl-POS cage mixture) improved the flexural strength and Young's modulus of the composites, but large percentage substitutions (mass fraction of 25% or higher in the resin system) led to less desirable mechanical properties, lower conversion of methacrylate double bonds, and slower photopolymerization rates [58].

In general, although using POS derivatives in the dental industry seems promising, the proper dispersion of POS nanocages and their correct synthesis are crucial for the improvement of the mechanical properties of dental materials.

9.4.3 Biosensors

POS has the capability of being combined with a variety of organic compounds in order to define its functionality. The eight vertices of a POS cage can be functionalized with groups such as NH_2, SH, OH, COOH, and in particular ammonium, which leads to a cationic POS that has been widely used in applications such as DNA/gene delivery, drug delivery, and DNA/protein detection.

In particular, POS structures have been used as probes for detecting biomolecules such as DNA and proteins, using a technique known as resonance light scattering (RLS), which was initially considered by Zou et al. [66] (see also Section 7.5.1, Chapter 7). It was found that by adding DNA molecules to a cationic POS aqueous solution the RLS intensity at a wavelength of 360 nm could be significantly improved. Thus, the RLS intensity increased in a positive correlation with the increase in DNA concentration, and was also strongly dependent upon the pH value and ionic strength. Owing to its strikingly sensitive, convenient, rapid, and reproducible features, cationic POS could be potentially useful as a probe to determine DNA concentration via the RLS technique, as the authors suggested.

Many of the POS-containing species such as octa-aminophenyl POS hydrochloride salt (cationic POS) have been employed as new reagents for the RLS determination of DNA, since they have the advantages of good water solubility, good stability across a wide pH range, and high sensitivity. Electrostatic interaction of cationic POS and DNA was shown to enhance the RLS signal [66].

9.4.4 Cardiovascular Implants

Application of biomaterials in cardiovascular system and blood contacting devices necessitates specific characteristics such as blood compatibility and antithrombogenicity. To meet the essential requirements for these applications, we have developed a nanocomposite material by introducing POS moieties into poly(carbonate-urea)urethane (POS-PCU) as a pendent chain group [13,67-70]. Studies on cytocompatibility, antithrombogenicity, and biostability have shown that this new nanocomposite polymer has unique characteristics for these applications [69]. It can be bio-functionalized by modification of the surface in order to attract endothelial progenitor cells from the circulatory blood, and can become endothelialized to enhance bio- and blood-compatibility of the cardiovascular devices made with this material (Fig. 9.5). Currently, development of heart valves [71], percutaneous heart valve prostheses, coronary artery bypass grafts, stent grafts, and nanocomposite-coated stents using POS-PCU are under investigation in our laboratory. For instance, a coronary bypass graft made with this material has already been implanted into a sheep model, and demonstrated a 70% patency rate over two years (unpublished data, see Fig. 9.6) and will soon go into clinical trials. A novel synthetic leaflet heart valve has also been developed based upon the POS-PCU nanocomposite, and is currently undergoing further assessments (Fig. 9.7). This valve could potentially combine the advantages of mechanical and bioprostetic valves while eliminating their drawbacks. In addition, using an electrohydrodynamic spraying approach, we have established the application of POS-PCU for the coating of metallic stents, and demonstrated that this novel polymeric nanocomposite has the potential to be used in the development of a new generation of stents with improved mechanical properties, especially in small diameter stents used in coronary and peripheral arterial applications [72].

POS-PCU has been characterized and assessed for biomedical application in general, and for cardiovascular application in particular, and the results of these studies reveal that POS-PCU nanocomposite confers the enhanced properties needed for biomedical applications, as discussed in the following sections.

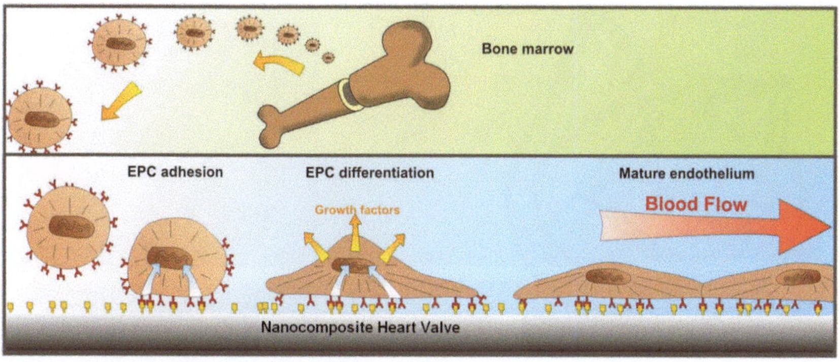

Fig. 9.5 Biofunctionalization of a surface to enhance in situ endothelialization. Biofuntionalized surfaces of heart valve leaflets made from POS-PCU nanocomposite can target several biological processes to promote in situ endothelialization, by promoting the mobilization of EPC from the bone marrow, encouraging cell-specific (circulating EC, EPC, and stem cells) to home to the vascular graft site, providing cell-specific adhesion motifs on the vascular grafts (of a predetermined spatial concentration), and directing the behaviour of the cells post-adhesion to rapidly form a mature fully functioning endothelium with self-repair capability

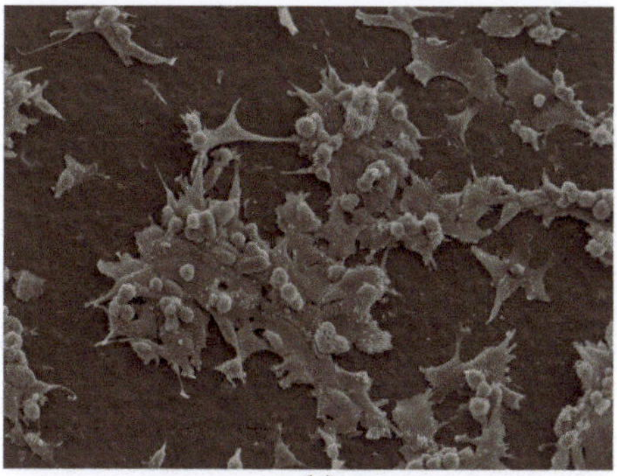

(a)

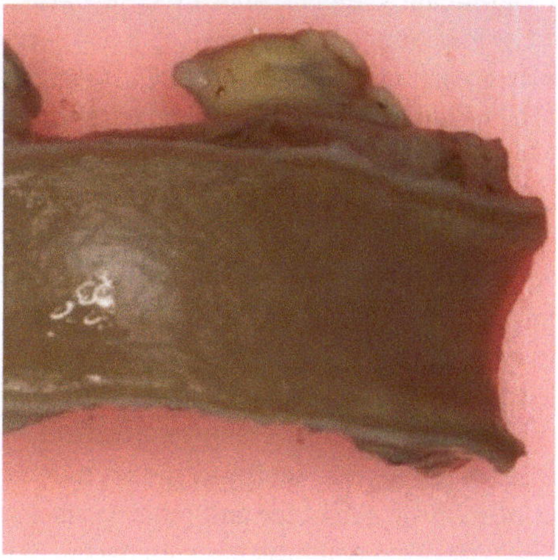

(b)

Fig. 9.6 (a) Scanning electron microscope pictures of endothelial cell adsorption morphology on POS-PCU showing the presence of flat, spindle-shaped cells with numerous filopodia and the absence of cell retraction. This indicates the presence of viable, proliferating cells on POS-PCU at 48 hours (320 x magnification) (b) an explanted POS-PCU bypass graft has been endothelialized after two years implantation in a sheep model

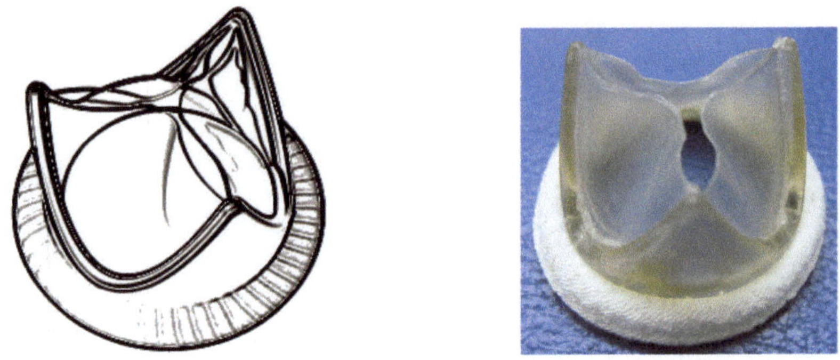

(a) (b)

Fig. 9.7 (a) Trileaflet heart valve design (University College London design) with complex geometry and additional reflection on the leaflets to improve hemodynamic performance and durability (b) a valve prototype fabricated from POS-PCU nanocomposite with a Dacron® suture ring

9.4.4.1 Mechanical Properties

The development of cardiovascular devices such as prosthetic heart valves from polymeric materials requires a broad understanding of the basic mechanical and surface properties of the polymer. In one study [71], the mechanical properties of POS–PCU, including tensile strength, tear strength and hardness, were tested and compared to a control (PCU itself). The surface properties were analyzed by contact angle measurement, and the resistance to platelet adhesion was also investigated. POS–PCU (Shore A hardness 84 ± 0.8) demonstrated significantly higher tensile strength (53.6 ± 3.4 and 55.9 ± 3.9 N mm^{-2} at 25 and 37 °C respectively) than the PCU control (33.8 ± 2.1 and 28.8 ± 3.4 N mm^{-2}). Tensile strength and elongation at break of POS–PCU was significantly higher than PCU at both 25 and 37 °C. POS–PCU showed a relatively low Young's modulus (25.9 ± 1.9 and 26.2 ± 2.0 N mm^{-2}) which was significantly greater than with control PCU (9.1 ± 0.9 and 8.4 ± 0.5 N mm^{-2}) at 25 and 37 °C respectively. There was no significant difference in tear strength between POS–PCU and PCU at 25 °C. However, tear strength increased significantly ($P < 0.001$) (at 37 °C) as the thickness increased from 100 µm (51.0 ± 3.3 N mm^{-1}) to 200 µm (63 ± 1.5 N mm^{-1}). The surface of POS–PCU was significantly less hydrophilic than that of PCU. Thus, the mechanical characterization of the nanocomposite proved it to be an ideal candidate for the use in cardiovascular devices, particularly in synthetic heart valve leaflet applications.

We have also developed and assessed a small porous calibre bypass graft made of POS-PCU nanocomposite via an extrusion-phase-inversion manufacturing technique. Mechanical and elemental analysis (using energy-dispersive X-ray analysis) showed a uniform distribution of POS on the outer and inner surface of the graft, with a favourable mechanical integrity for haemodynamic contacts [73] .

9.4.4.2 Degradative Resistance

Our initial expectation was that the presence of POS nanocores in POS-PCU material would hold together all the components of the polyurethane, and thus prevent degradation, thereby further improving the degradative resistance of the polyurethane. Degradation of these nanocomposites was classified into one of three possible categories namely physiological, oxidative and hydrolytic. FTIR analyses of the nanocomposite after exposure to differing plasma protein fractions (PF I-IV) showed no difference in intensities of Si–O, C–O–C, NHCO (urea) and (NHCO) O (urethane) bands for each bond wavelength, as compared to the control . These results indicate that biological proteins have a negligible degradative effect on these POS-PCU nanocomposites. Hydrolytic enzyme degradation, in particular, phospholipase (PLA), resulted in a significant decrease in the intensity of the Si–O bands of the POS nanocores. In all of these samples, and particularly PLA, we found no evidence of colour change or macroscopic signs of degradation, indicating that although Si-O bands were less prominent in this system, for some reason there was still no significant degradation. Oxidative degradation of these nanocomposites showed a statistically significant reduction in the intensity of Si-O bands of the *t*-butyl $H_2O_2/CoCl_2$ and glutathione/*t*-butyl $H_2O_2/CoCl_2$ degraded-polymers as compared to the control, non-degraded polymer. Detailed analyses of the samples showed a small leftward shift in the peak of the C-O-C bands, indicating a decrease in overall hydrogen bonding.

A scanning electron microscopy (SEM) investigation with FTIR analyses indicated that the weakening of the amide cross-links with the POS segments resulted in increased surface degradation (pit formation), with loss of the POS crystalline peaks (balding) compared to the non-degraded polymer, showing a rough surface with numerous crystalline peaks and intervening flat areas. Stress–strain curves measured for the polymers that were most affected physically and chemically showed that no significant difference in stress–strain behaviour existed between the control, PLA, cholesterol esterase (CE), *t*-butyl-peroxide and PF2-degraded nanocomposites. This data shows that the soft segment of the polyurethane nanocomposite remains intact in spite of accelerated degradation, as illustrated by the preservation of its elasticity. It is important to note, however, that in the case of *t*-butyl-peroxide-degraded nanocomposites, the vessels burst at pressures beyond 220 mmHg, indicating a decrease in mechanical strength as compared to all the other grafts.

As FTIR studies had previously shown, *t*-butyl peroxide and its counterparts affected the Si–O bonds within the graft that are responsible for holding together the nanocomposite by means of a zone of intercalation between the amorphous/soft and the crystalline/hard segments of the polyurethane. The reduction in the strength of the Si–O bond in the *t*-butyl/ $CoCl_2$-degraded POS-PCU sample explains its lower burst strength. The inorganic POS matrix induces a viscoelastic effect in the bulk of the nanocomposite, which is useful in vascular prostheses. Polyurethane degradation begins as a surface phenomenon, and then gradually creeps into the matrix of the polymer. DSC analyses were performed on samples which showed the greatest degradation, namely *t*-butyl peroxide, CE and PLA.

Thermal analysis of the most severely degraded POS-PCU nanocomposites showed that the T_g, an indicator associated with the soft phase of a polymer, did not vary significantly between the non-degraded and degraded samples, thus confirming the stress–strain data for the nanocomposite grafts. We found that the addition of POS-nanocores to PCU increased the final melting temperature of the nanocomposite by approximately 30 °C while decreasing the T_g by a further 10 to12 °C. This paradox may be explained by the fact that the nanocores prevent close packing of polymer chains associated with thermal rearrangement, that in turn results in an increase in dynamic voids within the polymer. This extra space allows for greater segmental mobility of the polymer chain, and therefore lowers the T_g (see also discussions in Chapter 4).

The increase in T_m and the reduced phase separation, on the other hand, are attributable to the increased nanoscale interactions between the constituents, thus binding the POS-PCU nanocomposites together. Furthermore, these nanofilled polyurethanes (unlike PCU which shows a three-phase melt), exhibited a single phase melting process with softening occurring at higher temperatures. In the *t*-butyl-degraded samples, the melt process was two-phase, with lower softening temperatures. These results again signal that both hard and soft phases of the polymer integrate to a far greater extent than in PCU, owing to the nanoscale interaction of the POS-nanocages with the soft phase.

9.4.4.3 Biocompatibility and Biostability

To assess biocompatibility, sheets of POS-PCU were implanted subcutaneously into the back of normal, healthy adult sheep for 36 months. At 36 months, the implants were removed from the animals under general anaesthesia and the surrounding tissue and capsule, if any, was taken for histopathological examination. The explanted nanocomposites exhibited no evidence of an inflammatory layer or capsule formation.

ATR-FTIR analysis of the implanted 2% POS-PCU sheets showed that the Si-O bands at 1109 cm^{-1} for the closed nanocage were intact in all six of the nanocomposite samples, and on both surface and bottom in different regions of the polymer. SEM analyses on the explanted nanocomposite samples showed no significant difference in its surface morphology as compared to the non-degraded control POS-PCU nanocomposite.

The in vitro fibrinogen adsorption tests performed to explain the minimal inflammatory behaviour of this nanocomposite showed a significant increase in fibrinogen adsorption to its surface as compared to a siloxane control. This corresponds to the existing literature, reporting that nanocomposites proportionately adsorb more protein with increasing surface roughness. This could in turn be caused by the effect of an increase in surface area, and therefore it would be logical to assume that the formation of an inflammatory capsule in the case of nanocomposites would be accentuated. Paradoxically however, nanocomposites have been shown to cause minimal inflammation in vivo, in this study as well as in others [74]. The prevalent hypothesis is that proteins such as fibrinogen undergo conformational changes upon adsorption to these surfaces as a result of large contact angle hysteresis.

It can be concluded that POS-PCU nanocomposites are ideal for use at biological interfaces as tissue implants, biomedical devices and even vascular grafts, owing to the very low levels of inflammation elicited by the hosts. We hypothesise that this is due to the intrinsic surface roughness of the polymer, causing increased contact angle hysteresis and conformational change in adsorbed surface fibrinogen, a key initiator for foreign-body reactions.

9.4.4.4 Endothelialization Property

The formation of an endothelial cell layer on the surface of cardiovascular implants enhances the biocompatibility and heamocompatibilty of these devices. In particular, the improvement of the endothelialization of the luminal surface of bypass grafts could potentially improve their patency rates and overall clinical outcomes [75]. Studies have shown the clinical benefit of lining PTFE vascular grafts with endothelial cells, particularly in lower limb arterial bypass grafts [76], in contrast to native uncoated polyethylene terephthalate (PET) grafts. In an in vitro study using Human Umbilical Vein Endothelial Cells (HUVECs), we assessed whether our nanocomposite was safe and compatible with in vitro cell cultures. Apart from indicating its safety as a biomaterial at the cellular level, such information would also serve as a measure of its potential in developing bio-hybrid vascular grafts.

Compared to existing silicone copolymers [76-78], we found that the use of a POS-based nanocomposite improves cell adhesion characteristics. As these nanocages occupy a minimal volume within the polymer, greater surface areas of PU are available, which allows for improved endothelialization. This also confers a greater degree of polarity to the polymer, which could explain its superior hydrophobicity to PCU.

These experiments also indicate that once adherent to its surface, endothelial cells (ECs) are also capable of manifold proliferation in order to form a confluent monolayer. While the PicoGreen® assay showed the excellent proliferating characteristics of ECs on the polymer, microscopic light analysis revealed how this occurred. Prior to achieving cellular confluence, the ECs aligned in a reticular manner. The intervening areas were subsequently filled in, and cellular confluence was achieved [69]. It has also been shown that exposing the EC-seeded POS-PCU conduits to low shear stress before being exposed to physiological shear stress significantly enhanced cell adherence rate. These findings showed the benefits of the pre-seeding method followed by low shear stress, and confirmed the potential of POS-PCU for application in tissue-engineered cardiovascular devices [79].

To enhance endothelialization potential, POS-PCU can be biofunctionlized by modification of its surface using biofunctional peptides such as RGD, a functional part of an extracellular matrix component, fibronectin. In an in vitro study, RGD-modified POS-PCU sheets were assessed to evaluate cell adhesion and proliferation using peripheral blood mononuclear cells (PBMCs) containing EPCs. The experimental data showed that the quality of cell adhesion and proliferation on the polymer was excellent [75]. Qualitative SEM analysis clearly visualized the presence of optimal cell–polymer interactions, with the formation of numerous filopodia at the surface, flattened ECs, and no rounded cells. This suggests that these ECs are capable of morphogenesis and have the ability to proliferate well. In another study, the RGD-modified POS-PCU was used to develop a porous small diameter bypass graft, and tested under static and pulsatile conditions. In this study PBMC containing EPCs isolated from healthy adult volunteers were cultured on nanocomposite graft, and evaluated for cell attachment, growth and endothelialization using Alamar Blue® assay, SEM, and immunostaining for endothelial cell markers including CD34, CD31 and eNOS. This study showed that POS-PCU had bio-functionalization capability, and was able to support cell growth and cell viability, and have a great potential for rapid endothelialization from peripheral blood cells [80].

In a recent study, the effect of surface topography on fibrinogen absorption, and cell attachment and proliferation, was investigated in vitro. In this study, the effect of POS percentage on the surface topography was demonstrated, and AFM studies showed that protein absorption on the soft polymeric component of POS-PCU was higher, compared to the reduced fibrinogen adsorption on the hard nanocage component. The study revealed that alteration of surface topography had a direct effect on fibrinogen molecule adsorption, which could potentially enhance attachment and proliferation of cells. In addition, this study showed that the adsorption of protein on the surface had the potential to improve the biocompatibility of the surface if certain peptide segments are available on the surface [81].

The natural regenerative capacity of the body can be utilized to induce endothelialization of cardiovascular devices. In a very recent study, we assessed the potential of in-situ endothelialization of a heart valve made from POS-PCU nanocomposite. Peripheral blood mono-nuclear cells were separated using Ficoll-Paque centrifugation, with harvested EPCs purified using CD34 microbead labelling and magnetic-activated cell sorting (MACS). Cells were seeded onto 96 well plates coated with either POS\S-PCU, GRGDG/GRGDG-LA modified POS-PCU or PCU polymer controls for a period of 21 days. The cultured cells were studied under light, confocal, and scanning electron microscopy (SEM). In addition, fluorescence-activated cell sorting (FACS) was used to analyze cell surface markers including CD34, CD133, VEGFR2, CD144, CD31 and vWF in order to identify the EPCs' potential proliferation and differentiation toward mature endothelial cells. Cell attachment and growth were observed in all POS-PCU samples, with significantly higher activity than seen within the control polymers ($p<0.05$). Microscopic examination revealed clonal expansion and morphological changes in cells seeded on POS-PCU, GRGDG, and GRGDG-LA modified polymers. The

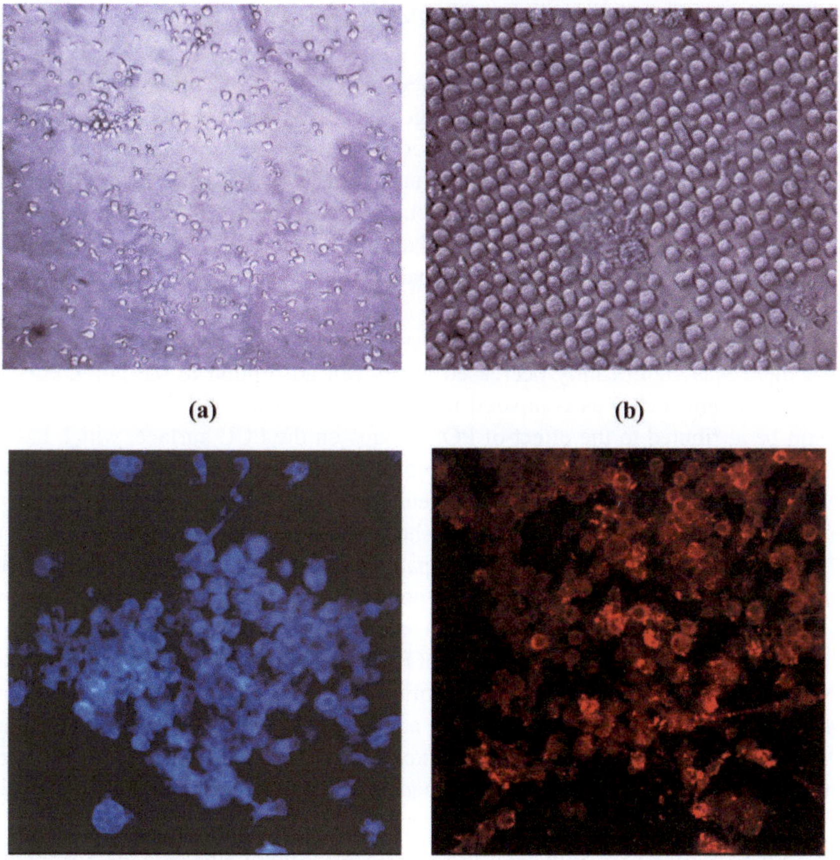

(a) (b) (c) (d)

Fig. 9.8 Morphological changes of isolated cells cultured on POS-PCU nanocomposite polymer. Spindle-shaped morphology of early EPCs at day 7 (a) has been dominated by cobble stone-shaped features at day 21 (b) characteristics for the late EPCs or endothelial cells, immunostaining of the cultured cells at day 14. Cells were stained for vWF (c) and VEGFR2 (d) showing positive expression of these cell surface markers on the cultured cells

frequency of triple positive EPC (cells positive for CD34, CD133 and VEGFR2) increased over time of culture, revealing the proliferation capacity of cultured EPCs. After one week of culture, SEM showed a mixed population of morphologically differentiated endothelial cells and EPCs. The cells expressed increasing levels of mature endothelial cell markers (CD31, vWF, CD144 and VEGFR2) over time with a confluent layer of EC at day 21 (Fig. 9.8). These results support the self-endothelialization potential of cardiovascular implants made with the POS-PCU polymer, and the possibility of an in vivo tissue engineering approach using the natural regenerative capacity of EPC and smart synthetic surfaces based on POS-modified nanocomposite materials.

9.4.4.5 Anti-Thrombogenic Potential

To assess the antithrombogenic properties of POS, we used thromboelastography (TEG), a sensitive indicator of thrombogenicity. This was used as a screening test to ascertain the anti-thrombogenic properties of POS-PCU. It was found that the polymer had significantly lower maximum amplitude (MA) values, indicative of decreased platelet bonding strength, when compared against polystyrene and PCU controls. It was also found that the clots that formed on POS-PCU were significantly unstable, and lyzed in 60 minutes, compared to PCU or to the control polystyrene polymer [68].

Direct ELISA fibrinogen adsorption analyses of the various polymers showed that there was significantly decreased fibrinogen adsorption to the polyurethanes (POS-PCU and PCU) as compared to PTFE. In the case of POS-PCU, this may again be attributed to the effect of POS groups on the PCU surface, which has an unstable surface free energy, and hence reduces both platelet and protein adsorption. Although we found no significant difference between POS-PCU and PCU in terms of fibrinogen adsorption, TEG analyses showed that the strength of the fibrin clot in POS-PCU was much weaker when compared to PCU. This indicated that while the amount of fibrinogen adsorbing to both polymers was similar, the binding strength was weaker in POS-PCU relative to PCU alone.

Platelet adsorption onto cast sheets of POS-PCU was significantly less than both PCU and PTFE sheets in vitro after 120 minutes of contact. After 120 minutes, there was a significant difference in platelet adsorption between POS-PCU, PCU and PTFE. In conjunction with the lower maximum amplitude (MA) values obtained on TEG, this suggested that POS-PCU had an anti-platelet effect, both by repelling platelet surface adsorption, and by reducing the binding strength of the platelets to the polymer, which corresponded to the poor adsorption characteristics exhibited towards fibrinogen, as discussed above [67,68]. The rate of platelet activation was also significantly lower in POS-PCU samples. Direct ELISA analysis quantified the concentration of platelet factor 4 (PF4) released from platelets on the polymer surface. A significantly ($p<0.05$) lower concentration of PF4 was found from the supernatants of platelets incubated on POS-PCU surface compared to the PCU control.

9.4.4.6 Resistance to Calcification and Fatigue

Calcification is a limiting factor in the biomedical application of synthetic materials. In particular, it is considered to be the main restricting factor in the successful application of polymeric heart valves [82]. Calcification also represents a major cause of failure of biological tissue heart valves. It is a complex phenomenon influenced by a number of biochemical and mechanical factors. Recent advances in materials science offer new polymers with improved properties, potentially suitable for the manufacture of synthetic leaflets heart valves. In one study, the calcification-resistance efficacy and fatigue behaviour of POS-PCU were assessed by means of in vitro testing. In particular, thin sheets of nanocomposite, glutaraldehyde fixed bovine pericardium (BP) and polyurethane (PU) were exposed to a calcium solution into a specially designed in vitro accelerated physiological pulsatile pressure system for a period of 31 days and a total of 40×10^6 cycles. The samples were investigated for signs of calcification after exposure to calcium solution by mean of X-ray, microscopic and chemical characterizations. Mechanical and surface properties were also studied by observing stress-strain behaviour, and surface morphology and hydrophobicity. Comparisons showed that under the experimental conditions, the level of calcification for the nanocomposite was considerably lower than for the fixed BP ($p=0.008$) and polyurethane (PU) samples ($p=0.015$) (Fig. 9.9). In addition,

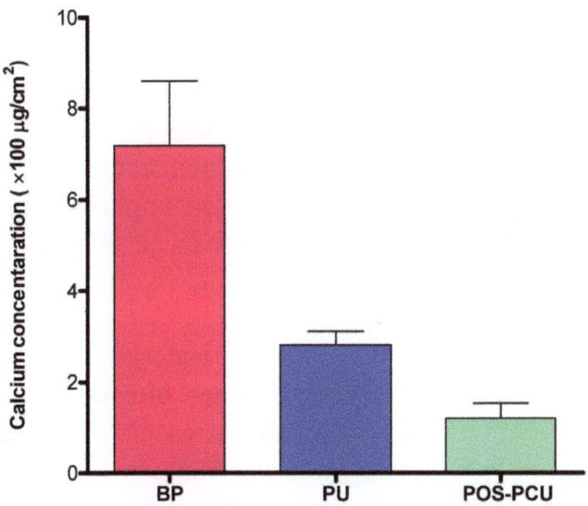

Fig. 9.9 Chemical analysis of calcium deposition. Quantitative analysis of calcium deposition on the samples showed significantly reduced calcium deposition on POS-PCU compared to BP ($p< 0.001$, $n=5$) and to PU ($p< 0.05$, $n=5$)

mechanical properties were unchanged in POS-PCU, while there was a significant deterioration in the PU samples ($p<0.05$). Hydrophobicity was significantly reduced in both the POS-PCU and PU samples ($p<0.0001$). However, the POS-PCU nanocomposite remained more hydrophobic than the PU sample ($p<0.0001$). Less platelet adhered to the POS-PCU compared to the PU ($p<0.0001$). These results indicate that the use of this nanocomposite in synthetic leaflet heart valves may lead to potential advantages in terms of long term performance and durability.

9.4.4.7 Reduced In Vitro Inflammatory Response

The inflammatory response that is mainly triggered by foreign body reaction after transplantation is also a determinant factor in the biocompatibility and the long term durability of biomedical implants. Inflammatory response is a very complex immunologic reaction involving various types of immune cells and immunomediators such as cytokines. All forms of implantation involve some degree of tissue injury, which will initiate two principal reactions, namely inflammation, and the related response of repair. It is still not completely understood which type or extent of inflammation favours optimal biocompatibility, or precisely how the physicochemical properties of the biomaterials induce specific biological responses. However, it is well known that increased immune response and inflammatory reaction adversely affect the implant-host interaction and lead to unfavourable outcomes. In this in vitro setting we investigated potential inflammatory reactions induced by POS-PCU against those of other polymers. Peripheral blood mononuclear cells were exposed to polymeric discs for one week, and the activation of inflammatory cells was assessed by evaluation of cell surface markers and cytokine release. The study revealed that the level of pro-inflammatory cytokines, namely Interleukin-1 (IL-1β/IL-1F2) and Tumor Necrosis Factor (TNFα/TNFSF1A), were significantly lower in the supernatant collected from POS-PCU samples compared to that of the control PCU ($p<0.05$). This was associated with reduced expression of CD86 and CD69, which are leukocyte activation markers ($p<0.05$). The study revealed that incorporation of POS into the PCU polymer could effectively reduce any possible inflammatory reaction resulting from the implantation of the biomedical devices made with these materials. It also enhanced the haemocompatibility of cardiovascular devices by eliminating the well-known effect of leukocyte activation on thrombosis formation.

These characteristics make POS-PCU a material of choice not only for cardiovascular applications, but also for other biomedical applications. This polymer is currently being used in the fabrication of many biomedical devices, including microvascular beds for organ tissue (including liver), muscle, cartilage, breast implants [67], materials for coating quantum dot nanocrystals, small-diameter artificial nasolacrimal ducts, and many others, as discussed in the following sections.

9.4.5 Breast Implants

Silicone implants are extensively used worldwide, especially in breast implants. They were first introduced into breast surgery during the 1960s. Silicone is a relatively inert biomaterial, and was believed at the time to cause minimal inflammation. However, long-term studies are now revealing that silicone delays the healing of wounds [83] and causes capsular contracture, and in Swanson joint prostheses, repetitive movement causes a form of pseudo-inflammation associated with the release of microparticles (silicone debris) [84].

POS derivatives could potentially be an alternative to silicone in breast implant products. In a recent in vivo study [67] it has been shown that the use of POS-PCU tissue implants showed no sign of significant inflammation nor material degradation compared with siloxane controls. Hence, it was concluded that this nanocomposite improved interfacial biocompatibility and had better biological stability compared with conventional silicone biomaterials, making them safer as tissue implants.

9.4.6 Coating Material for Quantum Dot Nanocrystals

Quantum dots (QD) are luminescent nanocrystals that are being intensely studied for use as a new class of optical imaging contrast agents [85]. This next generation of fluorescent probe quantum dots owes its novelty to its large absorption, narrow and discrete multi-colour light emission, bright photoluminescence, high photostability and nano-size. In the next few years, these nanoparticles will bring new and unique capabilities to a variety of biomedical applications [86]. Quantum dots are revolutionizing the processes of tagging molecules, proteins and cells, of imaging, and of cancer guided surgeries in vivo [87,88].

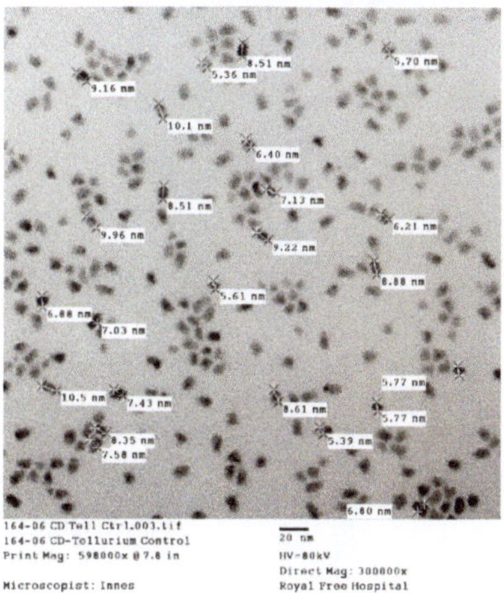

(a)

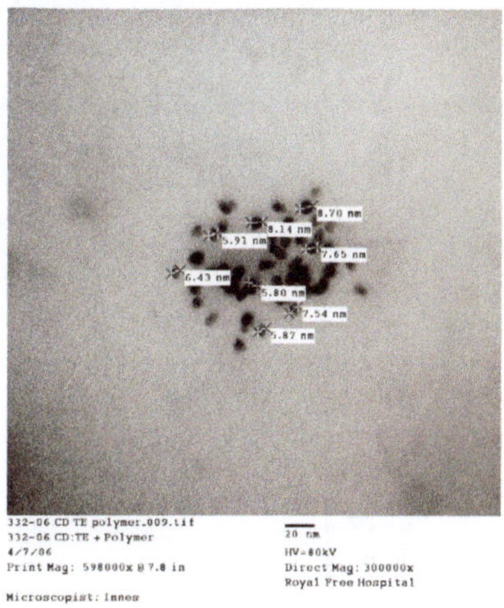

(b)

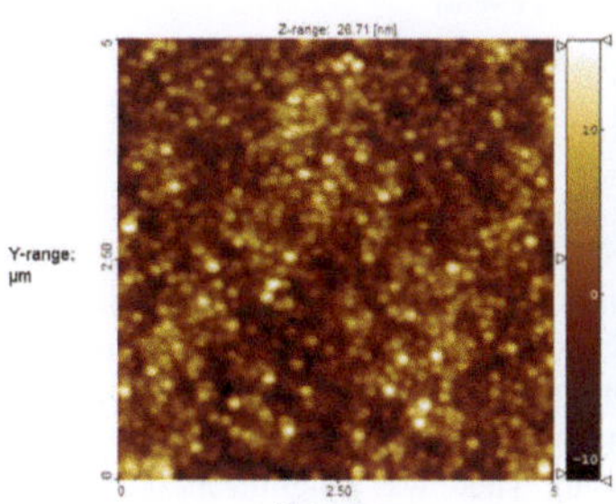

(c)

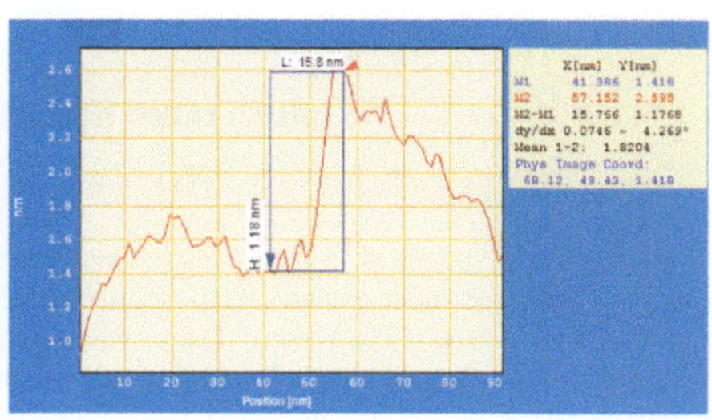

(d)

Fig. 9.10 TEM micrographs and measurements of the sizes of (a) CdTe nanocrystals without NC polymer (b) NCCQD (c) AFM images of NCCQD at AC mode, height map from 5x5 µm scan (d) AFM X-profile measurement of the nanocomposite polymer layer thickness

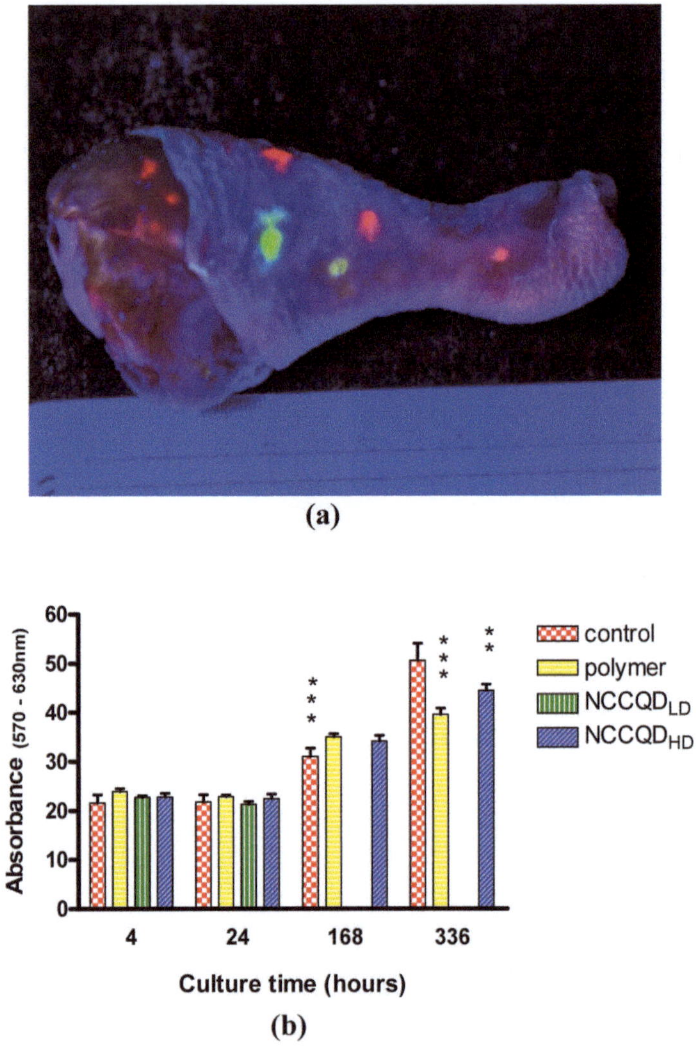

Fig. 9.11 (a) QDs of various sizes have been injected at various depths in a chicken leg as a tissue model to locate their presence under excitation (b) Alamar Blue® assay test on HUVEC exposed to NCCQD-treated CCM for 14 days (336 hours). There was no significant difference ($p > 0.001$) between the control group and the polymer group, neither was there any significant difference ($p > 0.001$) between the control and the NCCQD groups after 4 hours and 24 hours. There was a significant difference ($p < 0.001$) between the control group and the polymer group, as well as between the control group and the CCM treated with NCCQDHD after 7 days (168 hours). There was a significant difference ($p < 0.001$) between the control group and the polymer group, as well as ($p < 0.01$) between the control group and the CCM treated with NCCQDHD after 14 days (336 hours). Absorbance was measured in arbitrary units at 570 nm and the background at 630 nm was subtracted. Data are mean +/- SD (n = 4)

As the engineering of QDs for potential biological applications is being improved, there is concern regarding its degradation in vivo, especially via oxidation [89]. To avoid premature degradation, to enhance biocompatibility, and to reduce potential toxicity, the coating of QDs with biocompatible and biostable polymers has been proposed. POS-PCU is a potential material for coating QDs. POS-PCU has been extensively investigated and it has been shown to be non-toxic [43], biocompatible [69], and biostable [90]. POS-PCU has also been shown to have hydrophilic groups in experiments exploring its putative applications in various biomedical devices. Our group conducted a study to develop, characterize and assess the biocompatibility of QD encapsulated with POS-PCU nanocomposite. In this study the in-vitro cytocompatibility, and any potential cytotoxic effects, were investigated using human endothelial cell culture. The maximum tissue depth at which the nanocomposite-coated quantum dots (NCCQD) could be detected was also investigated. The NCCQD were of narrow size distribution, with a mean hydrodynamic diameter of 10.5 nm, high photostability, excellent monodispersivity and large absorption spectrum with a narrow and discrete emission band at 790 nm (Fig. 9.10). NCCQD were compatible with human umbilical vein endothelial cells (HUVECs), since viable cells were shown to be present after 14 days of growing HUVECs in a cell culture medium exposed to NCCQD. Exposing cells to NCCQD-treated cell culture medium resulted in no apparent damage to cells at concentrations of 2.25×10^{-2} nM (Fig. 9.11). NCCQD were detected at a maximum tissue depth of 10 mm in solid tissue by near infrared camera. POS-PCU provides an opportunity to modify the surfaces of QDs for potential biomedical applications, and abrogates any potential toxic effects of cadmium telluride (CdTe) QD nanocrystals in endothelial cells.

9.4.7 Silver Nanoparticle-Containing Polyhedral Oligosilsesquioxane Polymers

Silver nanoparticles (nanosilver), clusters of silver atoms which range in diameter from 1 to 100 nm, are primarily attracting interest as antibacterial and antimicrobial agents for application in medicine. Recent evidence suggests that nanosilver has a potent anti-inflammatory effect [91,92] and improves wound healing [93]. The antibacterial properties of silver are widely known, as evidenced by its current clinical use in treating burns [94]. The use of this nanoparticle in the structure of synthetic materials can potentially confer antibacterial capacity to the biomaterial. Implantable devices are a major risk factor in the contraction of hospital-acquired infections during medical interventions. Hence modification of the materials with silver nanoparticles could eliminate the risk of infection in devices made from these synthetic materials. Owing to the controversy surrounding the toxicity of silver, research into nanosilver coatings for heart valves and other fully implanted cardiovascular applications is understandably tentative. Since POS-PCU is a potential material of choice for development of biomedical devices, we are currently investigating the antibacterial, mechanical and haemodynamic properties of nanosilver-impregnated POS-PCU. Preliminary results show excellent antibacterial efficacy, with a 99.9% reduction in viable counts of Escherichia coli and Staphylococcus aureus (Fig. 9.12). Therefore, nanosilver-modified POS-PCU

Fig. 9.12 Silver nanoparticle impregnated POS-PCU, as developed by our group, showing antibacterial activity against Escherichia coli, assessed by the disc diffusion method. The clear zone of inhibition around the polymer disc (marked with arrow) shows an area where bacterial growth is inhibited by the nanosilver, and shows silver ions released by our polymer and diffusing through the agar

can be used in the development of implantable devices, in particular implants with a high risk of infection, such as the nasolacrimal duct. Currently, the cytocompatibility and anti-inflammatory properties of nanosilver-modified POS-PCU are under investigation in our lab.

9.4.8 Tissue Engineering

Tissue engineering is a rapidly growing field, combining various aspects of medicine, cell and molecular biology, and materials science and engineering, in order to improve peoples' health and quality of life by regenerating, repairing or replacing diseased human tissues, and by improving organ function [4]. The most widely explored three-dimensional scaffold materials in tissue engineering are categorized as either natural or synthetic materials. The advantage that a synthetic scaffold material has over a natural material is that its production technique is known, its structure is readily controlled on the microscopic and macroscopic scales, it is biodegradable over the time period of genuine tissue formation, and that its biodegradation rate is controllable. In principle, polymer scaffolds are desired to be biocompatible, with the necessary mechanical properties to sustain stresses at the specific site of application [32,95,96]. They should allow good cell adherence, and subsequent cell proliferation and differentiation. Moreover, the non-toxicity of the polymer scaffolds towards the cells is another crucial factor affecting call viability, either by direct contact, or via the products of degradation, and should be taken into consideration [97].

The physical characteristics of porosity, such as pore structure, volume and size, are mainly responsible for the regulation of cell function. Highly porous scaffolds offer significant surface area for cell attachments. The key factor responsible for successful cell adhesion, proliferation and differentiation is pore interconnectivity. Good pore interconnectivity provides a sustainable environment for a uniform cell distribution within the scaffold, and plays an essential role in regulating the diffusion of cell nutrients and waste disposal [32,95,96].

The most common POS-polymer used to date for biomedical and tissue engineering applications is the non-biodegradable nanocomposite based on POS (polyhedral oligomeric silsesquioxane) nanocages and PCU (poly(carbonate-urea) urethane), as discussed in preceding sections. This has been trademarked as NanoBio® by University College London (UCL) and consists of 80% polyhexanolactone and 20% PCU. The incorporation of POS nanocomposites into the poly(caprolactone/carbonate)urethane/urea has resulted in the development of a degradable polymer which preserves its mechanical properties as it undergoes oxidation, hydrolysis, or degradation by plasma protein fractions. The nanocages have a shielding effect upon the soft phase of the polymer [98].

The structure of the POS-PCU scaffold is comprised of three sections including the POS cage, the surrounding urethane creating the hard segment of the polymer, and the softer segment composed of carbonate. The soft segment can influence the characteristics of the scaffold polymer (e.g., viscosity and electrical conductivity) by interaction with solvent, as well as any effect which may occurs as result of soft segment alteration.

Despite the traditional idea that the ideal polymer for tissue engineering should be biodegradable to reduce the possibility of immune system reaction and extrusion, recent research suggests that a non-biodegradable polymer may well be more effective in terms of providing mechanical strength and stability.

In 2007 Gupta et al. [4] studied non-biodegradable POS-PCU and biodegradable POS-PCL-PCU which had been subjected to an electro-hydrodynamic printing technique, for the preparation of biopolymer scaffolds for the tissue engineering applications of small intestine, liver and cartilage. The results of their research demonstrated that the technique offers tremendous potential in the development of organs.

In another study, the cell compatibility of the latest biodegradable formulation of POS-modified poly(hexanolactone/carbonate)urethane/urea containing 80% hexanolactone (caprolactone) with the trade name of UCL-NanoBio® was

a)

b)

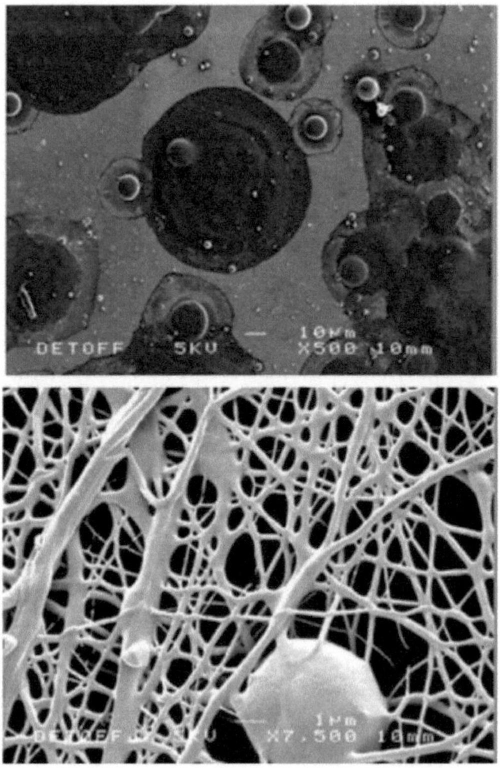

Fig. 9.13 SEM photographs of scaffolds produced using electrospray (a) and electrospinning (b) of 3% (w/w) UCL-NanoBio® polymer solution

investigated. The direct effect of the polymer upon cells was studied by seeding stem cells on to circular discs of the polymer in 24-well plates; these discs were prepared mainly by electro-hydrodynamic jetting. To assess the indirect effect of the polymer, various concentrations of the polymer powder were added to the cell culture medium (CCM) and left on a shaker for 10 days. The precipitate was then removed and the CCM was used for culturing the cells seeded on to the 24-well plates. Cell viability and growth at 48 and 96 hours were assessed using Alamar Blue® and lactate dehydrogenase, and morphology was studied by scanning electron microscopy. Cells were shown to adhere well to the polymer, with cell metabolism being comparable with that found on tissue-culture plastic (TCP). Indirect assessment demonstrated some decrease in cell viability with high concentrations of polymer, but showed no difference in cell death between polymer concentrations. The viability of cells seeded on to the polymer was comparable with that of those seeded on to TCP. Cell viability was comparable on both electro sprayed and electro-spun scaffolds (Fig. 9.13), but infiltration into the scaffold was much more evident on the electro-spun scaffolds. It can be concluded that this new nanocomposite can support the growth and viability of stem cells, and that scaffolds of this polymer nanocomposite fabricated by electro-hydrodynamic jetting routes have potential use for tissue engineering in the future [99].

Application of this biodegradable nanocomposite polymer was also explored for development in the tissue engineering of the small intestine. Rat intestinal epithelial cells were seeded on to the polymer scaffolds for an in vitro study of cell compatibility and proliferation, which was assessed by Alamar Blue® and lactate dehydrogenase assays performed for 21 days post-seeding. The results demonstrated that POS–PCL nanocomposite was produced as a macroporous scaffold with porosity in the range of 40 to 80%, and pore size in the range of 150 to 250 μm. This scaffold was shown to support epithelial cell proliferation and growth. In the present study, a simple and inexpensive method of solvent casting and particulate leaching was employed to fabricate porous scaffolds using a newly developed nanocomposite of silsesquioxane-based PCL (POS–PCL). It was shown that these scaffolds have the physical and chemical properties required for intestinal cells to remain viable within the culture medium exposed to POS–PCL, to adhere to the polymer surface, to remain viable, and to achieve confluence on the scaffold. This in vitro study for tissue engineering of the small intestine provides the option of using an alternative polymer to the conventionally used polyglycolic acid (PGA) or polylactide. The pore size and porosity of the scaffolds should be tailor-made, depending on the type and size of the cell source [100].

As mentioned earlier, the chemical compositions can have a significant influence upon the mechanical properties of tissue engineering scaffolds, and on the adhesion and proliferation of cells inside the scaffolds [101,102]. Therefore, various studies have sought to manipulate the chemical compositions of the synthetic biomaterials in order to achieve better response in tissue engineering applications.

In two recent studies [103,104] the effect of polyhedral oligomeric silsesquioxanes on poly(ester urethane) was analyzed for use in biomedical and tissue engineering applications using various techniques (i.e., microscopy, nuclear magnetic resonance (NMR) spectroscopy, X-ray diffraction, differential scanning calorimetry (DSC), thermogravimetry, and dynamic mechanical analysis DMA). It has been demonstrated that the incorporation of low levels of POS (6 wt %) into the macromolecular structure of poly(ester-urethane) via an in situ homogeneous solution polymerization reaction results in a new hybrid PEU/POS nanocomposite with remarkable improvement in terms of thermal and hydrolytic stability, stiffness, strength and degradation resistance compared to the pure PEU control [104,105].

Further investigation of the structural properties and cell compatibility of the PEU/POS nanocomposite using mouse embryonic stem cells (ESCs) revealed that although incorporation of POS did not have a direct influence on cell adhesion, viability, proliferation or differentiation, it did significantly change the surface architecture of the PEU, into a three-dimensional matrix with regular pores that could thus potentially enhance the biocompatibility of the polymer. The SEM image analysis from this study showed that there were ~950 randomly distributed pores per mm^2 of the matrix, which is about 7.6% of the total matrix area, with a pore diameter size in the range 1–15 μm [103]. Although, these structures are randomly distributed, and the size of the pores is not large enough for cell invasion, the overall matrix appears to be rather uniform. The interconnected grooves simultaneously support the access of the cells to the growth factors, and provide effective nutrients/ waste exchange. Under the cell culture conditions, the growth rate of ESCs was similar to that of gelatin, exhibiting undifferentiated morphology with the expression of pluripotency markers. However, after stimulation for differentiation, the morphology of the ESCs changed dramatically, and the differentiated cells formed a continuous monolayer which was closely embedded in the matrix [103,104]. In summary, these studies suggested that the incorporation of POS into the polymer provided a non-toxic nanocomposite with the potential for fabricating 3D scaffolds, as well as a thin matrix with the desired porosity, mechanical strength and biodegradability, via precise synthetic methods [104,105]. In addition, it provided a conducive environment for cell–cell or cell–matrix interactions, which are essential for tissue formation, and show great promise in ESC-based soft tissue engineering applications [103].

9.5 Other Applications

The application of POS-derived polymers in a variety of other biomedical fields has also been attempted, e.g., the use of POS-modified materials as effective sun protection agents [106] and as films for greenhouse covers [107]. In fact, the versatility of POS moieties offers a wide range of potential applications, with encouraging prospects, especially in the biomedical area.

9.6 Future Prospects

The unique properties of POS nanocomposites will contribute to revolutionary advances in materials science. The combination of POS molecules with polymers holds great promise for the future of biomedical devices. With a suitable nanocomposite building block of a synthetic microvascular network in hand, efforts are now being concentrated on using micro-fabrication technology to recreate the microvascular environment using POS-PCU, and to fuse it into the host using tissue-engineered scaffolds, stem cells and angiogenic factors. Since these materials are safe, resistant to degradation, compliant, biocompatible, anti-thrombogenic, and allow neo-endothelialization, the probability of microvascular occlusion is low. If successful in the long run, the clinical implications would be tremendous, as the ability to 'grow' tissue in vitro would potentially replace tissue transfers and even transplantation. Work continues on developing percutaneous heart valves, breast implants, nanocomposite-coated stents, bypass grafts, and on tissue engineering using these novel POS containing nanocomposites in both non-biodegradable and biodegradable formulations. As an example, our primary results for heart valve prototypes fabricated from this novel nanocomposite were very promising. Besides above-mentioned properties, the resistance to calcification, the enhanced mechanical and surface properties, the capability of attaching bioactive peptides to attract stem cells to the surface of nanocomposite, and the provision of a favourable environment for their proliferation and differentiation into endothelial cells are all among the advantages of using these materials in the development of cardiovascular devices. Hence for tissue engineering approaches, the use of POS-containing scaffold materials seems promising, suggesting its immense potential in this field.

9.7 References

1. Komarneni S (1992) J Mater Chem 2(12):1219-1230.

2. Huang HZ, Qiang Y, Yang XR (2005) J Colloid Interface Sci 282(1):26-31.

3. Roschger P, Grabner BM, Rinnerthaler S, Tesch W, Kneissel M, Berzlanovich A (2001) J Struct Biol; 136(2):126-136.

4. Gupta A, Seifalian AM, Ahmad Z, Edirisinghe MJ, Winslet MC (2007) J Bioactive Compat Polym 22(3):265-280.

5. Alivisatos AP (1996) Science 271(5251):933-937.

6. Gao H, Ji B, Jager IL, Arzt E, Fratzl P (2003) Proc Natl Acad Sci USA 100(10):5597-5600.

7. Lee A, Lichtenhan JD (1998) Macromolecules 31(15):4970-4974.

8. Liu YL, Hsu CY, Su YH, Lai JY (2005) Biomacromolecules 6(1):368-373.

9. Yang HG, Zeng HC (2005) J Am Chem Soc 127(1):270-278.

10. Karajanagi SS, Vertegel AA, Kane RS, Dordick JS (2004) Langmuir 20(26):11594-11599.

11. Meng G, Cao A, Cheng JY, Ajayan PM (2004) J Nanosci Nanotechnol 4(7):712-715.

12. Suhr J, Koratkar N, Keblinski P, Ajayan P (2005) Nat Mater 4(2):134-137.

13. Kannan RY, Salacinski HJ, Edirisinghe MJ, Hamilton G, Seifalian AM (2006) Biomaterials 27(26):4618-4626.

14. Li GZ, Yamamoto T, Nozaki K, Hikosaka M (2001) Polymer 42(20):8435-8441.

15. Maciel GE, Sullivan MJ, Sindorf DW (1981) Macromolecules 14(5):1607-1608.

16. Frye CL, Collins WT (1970) J Am Chem Soc 92(19):5586.

17. Li GZ, Wang LC, Toghiani H, Daulton TL, Koyama K, Pittman CU (2001) Macromolecules 34(25):8686-8693.

18. Gilman JW, Schlitzer DS, Lichtenhan JD (1996) J Appl Polym Sci 60(4):591-596.

19. Brunsvold AL, Minton TK, Gouzman I, Grossman E, Gonzalez R (2004) High Perform Polym 16(2):303-318.

20. Draghi L, Resta S, Pirozzolo MG, Tanzi MC (2005) J Mater Sci Mater Med 16(12):1093-1097.

21. Sanchez C, Soler-Illia GJDA, Ribot F, Lalot T, Mayer CR, Cabuil V (2001) Chem Mater 13(10):3061-3083.

22. Hosseinkhani H, Hosseinkhani M, Kobayashi H (2006) J Bioactive Compat Polym 21(4):277-296.

23. Leu CM, Chang YT, Wei KH (2003) Chem Mater 15(19):3721-3727.

24. Pellice SA, Fasce DP, Williams RJJ (2003) J Polym Sci Part B Polym Phys 41(13):1451-1461.

25. Fu BX, Gelfer MY, Hsiao BS, Phillips S, Viers B, Blanski R (2003) Polymer 44(5):1499-1506.

26. Haddad TS, Lichtenhan JD (1996) Macromolecules 29(22):7302-7304.

27. Huang JC, He CB, Xiao Y, Mya KY, Dai J, Siow YP (2003) Polymer

44(16):4491-4499.

28. Pyun J, Matyjaszewski K, Wu J, Kim GM, Chun SB, Mather PT (2003) Polymer 44(9):2739-2750.

29. Oaten M, Choudhury NR (2005) Macromolecules 38(15):6392-6401.

30. Fu BX, Namani M, Lee A (2003) Polymer 44(25):7739-7747.

31. Jeon GH, Yoon JY, Kim S, Kim SS (2000) Syn Lett 2000;(1):128-130.

32. Langer R, Vacanti JP (1993) Science 260(5110):920-926.

33. Jeoung E, Carroll JB, Rotello VM (2002) Chem Commun (14):1510-1511.

34. He Y, Wang HF, Yan XP (2009) Chemistry 15(22):5436-5440.

35. Frankamp BL, Fischer NO, Hong R, Srivastava S, Rotello VM (2006) Chem Mater 18(4):956-959.

36. Kannan RY, Salacinski HJ, Butler PE, Seifalian AM (2005) Acc Chem Res 38(11):879-884.

37. Naka K, Sato M, Chujo Y (2008) Langmuir 24(6):2719-2726.

38. Naka K, Itoh H, Chujo Y (2002) Nano Lett 2(11):1183-1186.

39. Carroll JB, Frankamp BL, Srivastava S, Rotello VM (2004) J Mater Chem 14:690-694.

40. Silver JH, Lin JC, Lim F, Tegoulia VA, Chaudhury MK, Cooper SL (1999) Biomaterials 20(17):1533-1543.

41. Cui D, Tian F, Ozkan CS, Wang M, Gao H (2005) Toxicol Lett 155(1):73-85.

42. Kim SK, Heo SJ, Koak JY, Lee JH, Lee YM, Chung DJ (2007) J Oral Rehabil 34(5):389-395.

43. Punshon G, Vara DS, Sales KM, Kidane AG, Salacinski HJ, Seifalian AM (2005) Biomaterials 26(32):6271-6279.

44. Li SD, Huang L (2008) Mol Pharmacol 5(4):496-504.

45. Myoung Y, Choi HK (2002) Drug Deliv 9(2):121-126.

46. Carelli V, Coltelli S, Di CG, Nannipieri E, Serafini MF (1999) Int J Pharm 179(1):73-83.

47. McCusker C, Carroll JB, Rotello VM (2005) Chem Commun (8):996-998.

48. Tanaka K, Inafuku K, Naka K, Chujo Y (2008) Org Biomol Chem 6(21):3899-3901.

49. Farokhzad OC, Langer R. (2009) ACS Nano 3(1):16-20.

50. Gu F, Zhang L, Teply BA, Mann N, Wang A, Radovic-Moreno AF (2008) Proc Natl Acad Sci USA 105(7):2586-2591.

51. Kim BS, Park SW, Hammond PT (2008) ACS Nano 2(2):386-392.

52. Sheikh FA, Barakat NA, Kanjwal MA, Aryal S, Khil MS, Kim HY (2009) J Mater Sci Mater Med 20(3):821-831.

53. Dailey LA, Wittmar M, Kissel T (2005) J Control Release 101(1-3):137-149.

54. Carroll JB, Waddon AJ, Nakade H, Rotello VM (2003) Macromolecules 36(17):6289-6291.

55. Zheng L, Farris RJ, Coughlin EB (2001) Macromolecules 34(23):8034-8039.

56. Yuan H, Luo K, Lai Y, Pu Y, He B, Wang G (2010) Mol Pharm 7(4):953-962.

57. Brigger I, Dubernet C, Couvreur P (2002) Adv Drug Deliv Rev 54(5):631-651.

58. Fong H, Dickens SH, Flaim GM (2005) Dental Mater 21(6):520-529.

59. Wheeler PA, Fu BX, Lichtenhan JD, Jia WT, Mathias LJ (2006) J Appl Polym Sci 102(3):2856-2862.

60. Sellinger A, Laine RM (1996) Chem Mater 8(8):1592.

61. Gao F, Tong YH, Schricker SR, Culbertson BM (2001) Polym Adv Technol 12(6):355-360.

62. Schwab JJ, Lichtenhan JD (1998) Appl Organmet Chem 12(10-11):707-713.

63. Dodiuk-Kenig H, Maoz Y, Lizenboim K, Eppelbaum I, Zalsman B,

Kenig S (2006) J Adhesion Sci Technol 20(12):1401-1412.

64. Jang J, Bae J, Kang D (2001) J Appl Polym Sci 82(9):2310-2318.

65. Wu X, Sun Y, Xie W, Liu Y, Song X (2010) Dental Mater 26(5):456-462.

66. Zou QC, Yan QJ, Song GW, Zhang SL, Wu LM (2007) Biosens Bioelectron 22(7):1461-1465.

67. Kannan RY, Salacinski HJ, Ghanavi JE, Narula A, Odlyha M, Peirovi H (2007) Plastic Reconstruct Surgery 119(6):1653-1662.

68. Kannan RY, Salacinski HJ, De Groot J, Clatworthy I, Bozec L, Horton M (2006) Biomacromolecules 7(1):215-223.

69. Kannan RY, Salacinski HJ, Sales KM, Butler PE, Seifalian AM (2006) Cell Biochem Biophys 45(2):129-136.

70. Kannan RY, Salacinski HJ, Odlyha M, Butler PE, Seifalian AM (2006) Biomaterials 27(9):1971-1979.

71. Kidane AG, Burriesci G, Edirisinghe M, Ghanbari H, Bonhoeffer P, Seifalian AM (2009) Acta Biomater 5(7):2409-2417.

72. Bakhshi R, Edirisinghe MJ, Darbyshire A, Ahmad Z, Seifalian AM (2009) J Biomater Appl 23(4):293-309.

73. Sarkar S, Burriesci G, Wojcik A, Aresti N, Hamilton G, Seifalian AM (2009) J Biomech 42(6):722-730.

74. Tanaka K, Tamura J, Kawanabe K, Nawa M, Oka M, Uchida M (2002) J Biomed Mater Res 63(3):262-270.

75. Alobaid N, Salacinski HJ, Sales KM, Ramesh B, Kannan RY, Hamilton G (2006). Eur J Vasc Endovasc Surg 32(1):76-83.

76. Zilla P, Fasol R, Deutsch M, Fischlein T, Minar E, Hammerle A (1987) J Vasc Surg 6(6):535-541.

77. Ai H, Lvov YM, Mills DK, Jennings M, Alexander JS, Jones SA (2003) Cell Biochem Biophys 38(2):103-114.

78. Hesse Y, Kampmeier J, Lang GK, Baldysiak-Figiel A, Lang GE (2003) Graefes Arch Clin Exp Ophthalmol 241(10):823-826.

79. Vara DS, Punshon G, Sales KM, Sarkar S, Hamilton G, Seifalian AM (2008) Artificial Organs 32(12):978-981.

80. de MA, Punshon G, Ramesh B, Sarkar S, Darbyshire A, Hamilton G (2009) Biomed Mater Eng 19(4-5):317-331.

81. Yaseen M, Zhao X, Freund A, Seifalian AM, Lu JR (2010) Biomaterials 31(14):3781-3792.

82. Ghanbari H, Viatge H, Kidane AG, Burriesci G, Tavakoli M, Seifalian AM (2009) Trends Biotechnol 27(6):359-367.

83. Shanklin DR, Smalley DL (1999) Exp Mol Pathol 67(1):26-39.

84. McCarthy DJ, Chapman HL (1988) J Foot Surg 27(5):418-427.

85. Smith AM, Ruan G, Rhyner MN, Nie S (2006) Ann Biomed Eng 34(1):3-14.

86. Jamieson T, Bakhshi R, Petrova D, Pocock R, Imani M, Seifalian AM (2007) Biomaterials 28(31):4717-4732.

87. Pinaud F, Michalet X, Bentolila LA, Tsay JM, Doose S, Li JJ (2006) Biomaterials 27(9):1679-1687.

88. Tan WB, Jiang S, Zhang Y (2007) Biomaterials 28(8):1565-1571.

89. Hardman R (2006) Environ Health Perspect 114(2):165-172.

90. Kannan RY, Salacinski HJ, Odlyha M, Butler PE, Seifalian AM (2006) Biomaterials 27(9):1971-1979.

91. Nadworny PL, Wang J, Tredget EE, Burrell RE (2008) Nanomedicine 4(3):241-251.

92. Sibbald RG, Contreras-Ruiz J, Coutts P, Fierheller M, Rothman A, Woo K (2007) Adv Skin Wound Care 20(10):549-558.

93. Tian J, Wong KK, Ho CM, Lok CN, Yu WY, Che CM (2007) Chem Med Chem 2(1):129-136.

94. Klasen HJ (2000) Burns 26(2):117-130.

95. Mooney DJ, Baldwin DF, Suh NP, Vacanti JP, Langer R (1996) Biomaterials 17(14):1417-1422.

96. Yang S, Leong KF, Du Z, Chua CK (2001) Tissue Eng 7(6):679-689.

97. Kannan RY, Salacinski HJ, Sales K, Butler P, Seifalian AM (2005) Biomaterials 26(14):1857-1875.

98. Raghunath J, Georgiou G, Armitage D, Nazhat SN, Sales KM, Butler

PE (2009) J Biomed Mater Res A 91(3):834-844.

99. Raghunath J, Zhang H, Edirisinghe MJ, Darbyshire A, Butler PE, Seifalian AM (2009) Biotechnol Appl Biochem 52(Pt 1):1-8.

100. Gupta A, Vara DS, Punshon G, Sales KM, Winslet MC, Seifalian AM (2009) Biotechnol Appl Biochem 54(4):221-229.

101. Hwang NS, Varghese S, Elisseeff J (2008) Adv Drug Deliv Rev 60(2):199-214.

102. Yang F, Williams CG, Wang DA, Lee H, Manson PN, Elisseeff J (2005) Biomaterials 26(30):5991-5998.

103. Guo YL, Wang W, Otaigbe JU (2010) J Tissue Eng Regen Med in publication.

104. Wang WS, Guo YL, Otaigbe JU (2009) Polymer 50(24):5749-5757.

105. Nanda AK, Wicks DA, Madbouly SA, Otaigbe JU (2006) Macromolecules 39(20):7037-7043.

106. Chen X, Wang D, Li X, Liu Y, Hu L (2009) Mater Sci Forum 610-613 (2):1039-1043.

107. Hu L, Chen X, Wang D, Liu Y (2009) Polym Mater Sci Eng 100:271-272.

Index

C. Hartmann-Thompson (ed.), *Applications of Polyhedral Oligomeric Silsesquioxanes*, Advances in Silicon Science 3, DOI 10.1007/978-90-481-3787-9,

C

F

G

H

M

Y

Abbreviations

ABS	Acrylonitrile butadiene styrene
ac	Alternating current
Ac	Acetyl
acac	Acetylacetonate
acaF	1-Trifluoroacetyl acetonate
AFM	Atomic force microscopy
AIBN	Azobisisobutyronitrile
AO	Atomic oxygen
Ar	Aryl
AR	Anti-reflection
ATRP	Atom transfer radical polymerization
BCB	Benzocyclobutane
BDGE	1,4-Butanediol diglycidyl ether
BODIPY	Boron dipyrromethene (4,4-difluoro-4-bora-3a,4a-diaza-s-indacene)
BP	Bovine pericardium
^{i}Bu or i-Bu	*iso*-Butyl
^{t}Bu or t-Bu	*tert*-Butyl
c-	*Cyclo*-
C	Crystalline
CCDC	Cambridge Crystallographic Data Centre
CCM	Cell culture medium
CD	γ-Cyclodextrin
CE	Cholesterol esterase
CHP	Cumyl hydroperoxide
CIC	Cover integrated cell
CIE	Commission International de L'Eclairage
COC	Cyclic olefin copolymer
COD	Cyclooctadiene
Cp	Cyclopentyl, η^5-C_5H_5
Cp*	η^5-$C_5(CH_3)_5$
Cp"	η^5-$C_5H_3(SiMe_3)_2$
CP/MAS	Cross polarization magic angle spinning
CTE	Coefficient of thermal expansion
CWA	Chemical warfare agent
Cy	Cyclohexyl

d	Lamellar spacing
D	Silicon atom attached to two oxygen atoms
D	Discotic columnar phase
DBTDL	Dibutyl tindilaurate
DCC	*N,N'*-Dicyclohexylcarbodiimide
dcp	Dicyclopentadienyl
DDM	4,4'-Diaminodiphenylmethane
DDS	4,4'-Diaminodiphenylsulfone
DGEBA	Diglycidol ether of bisphenol A
DLS	Dynamic light scattering
DMA	Dynamic mechanical analysis
DMAc	Dimethylacetamide
DME	1,2-Dimethoxyethane or dimethyl ether
DMFC	Direct methanol fuel cell
DMMP	Dimethyl methylphosphonate
DMSO	Dimethylsulfoxide
DMTA	Dynamic mechanical thermal analysis
2,4-DNT	2,4-Dinitrotoluene
DPE	1,1 Diphenylethylene
DP-PPV	Poly-2,3-diphenyl-1,4-phenylenevinylene
DSC	Differential scanning calorimetry
dvs	Divinyltetramethyldisiloxane
E_a	Activation energy
EBIB	Ethyl 2-bromoisobutyrate
EC	Endothelial cell
ee	Enantiomeric excess
EL	Electroluminescence
ELISA	Enzyme-linked immunosorbent assay
EPC	Endothelial progenitor cell
EPR	Electron paramagnetic resonance
EQE	External quantum efficiency
ESC	Embryonic stem cell
ESR	Electron spin resonance
Et	Ethyl
ETL	Electron transport layer
FAB	Fast atom bombardment
Fc	Ferrocenyl
FEP	Fluorinated ethylene propylene
FL3	Terfluorene
Flu	9-*H*-Fluorene
fppy	Difluorophenylpyridine
FTIR	Fourier transform infrared
G	Glassy
G^0_N	Plateau modulus
GCMS	Gas chromatography mass spectrometry
GEO	Geostationary Earth orbit
GPC	Gel permeation chromatography
GPE	Gelled polymer electrolyte
HDT	Heat distortion temperature
Hex	Hexyl
HFIP	Hexafluoro isopropanol
HHPA	Hexamethyltriethylenetetramine
HIL	Hole injection layer
HMTETA	Hexahydrophthalic anhydride
HOMO	Highest occupied molecular orbital

HPDLC	Holographic polymer dispersed liquid crystal
HPLC	High pressure liquid chromatography
HUVEC	Human umbilical vein endothelial cell
I	Isotropic
ICP-OES	Inductively coupled plasma – optical emission spectroscopy
IEC	Ion exchange capacity
IMS	Ion mobility spectroscopy
IR	Infra red
ITO	Indium tin oxide
IUPAC	International Union of Pure and Applied Chemistry
J	Coupling constant
k	Dielectric constant
LB	Langmuir-Blodgett
LC	Liquid crystal
LCD	Liquid crystal display
LCST	Lower critical solution temperature
LED	Light emitting device
LEO	Low Earth orbit
LER	Line edge roughness
LIDAR	Laser imaging detection and ranging
LUMO	Lowest unoccupied molecular orbital
Lys	Lysine
M	Silicon atom attached to one oxygen atom
MA	Maximum amplitude
M_n	Number average molecular mass
M_w	Weight average molecular mass
MA	Methacrylate
MACS	Magnetic activated cell sorting
MALDI-ToF	Matrix assisted laser desorption ionization time of flight
MAO	Methylaluminoxane
MCPBA	*Meta*-chloroperbenzoic acid
MDI	Methylene diphenyldiisocyanate
Me	Methyl
MEO	Medium Earth orbit
MIBK	Methylisobutyl ketone
MRI	Magnetic resonance imaging
N	Nematic phase
NBB	No bias bend
NBD	Nitrobenzoxadiazole
N_D	Discotic nematic phase
NIPAM	*N*-Isopropyl acrylamide
NMP	Nitroxide-mediated polymerization or *N*-methylpyrrolidinone
NMR	Nuclear magnetic resonance
Oct	Octyl
ODT	Order-disorder transition
OLED	Organic light emitting device
OMent	1*R*,2*S*,5*R*-(–)-menthoxo
MEH-PPV	Poly-2-methoxy-5-[2-ethylhexyloxy]-1,4-phenylenevinylene
MMAO	Modified methylaluminoxane
MS	Mass spectrometry
NLO	Non-linear optic
NMR	Nuclear magnetic resonance
NP	Nanoparticle
OL	Optical limiting

ORMOSIL	Organically modified silica
PAG	Photo acid generator
PAI	Polyamideimide
PAN	Polyacrylonitrile
PANI	Polyaniline
PAS	Polyacetoxystyrene
PBD	Polybutadiene or 2-(4'-*t*-butylphenyl)-5-(4'-diphenyl)-1,3,4-oxadiazole
PBMC	Peripheral blood mononuclear cell
PCL	Poly(ε-caprolactone)
PCU	Polycarbonate urea urethane
PCTFE	Poly(chlorotrifluoroethylene)
PDCPD	Poly(dicyclopentadiene)
PDI	Polydispersity index
PDMS	Poly(dimethylsiloxane)
PE	Polyethylene
PEB	Post exposure bake
PEDOT-PSS	Poly-(3,4-ethylenedioxythiophene)-poly(styrenesulfonate)
PEEK	Poly(ether ether ketone)
PEG	Poly(ethylene glycol)
PEI	Poly(ether imide)
PEM	Polymer electrolyte membrane or proton exchange membrane
PEO	Poly(ethylene oxide)
PES	Poly(phenylene ethersulfone)
PFCB	Perfluorocyclobutyl
PFDOP	Poly-[9,9'-dioctyl-2,7-fluorene-alt-2,5-bis(octyloxy)-1,4-phenylene]
PFE	Poly(fluorenylene-ethynylene)
PFO	Polyfluorene, i.e., poly(9,9'-dihexylfluorenyl-2,7-diyl)
PFT2	Poly-(9,9'-dioctylfluorene-*alt*-bithiophene)
Ph	Phenyl
PHEMA	Poly-(2-hydroxylethylmethacrylate)
PI	Polyimide
PISX	Poly(imide-siloxane)
piq	1-Phenylisoquinoline
PL	Photoluminescence
PLA	Phospholipase
PMMA	Poly(methyl methacrylate)
PMDETA	Pentamethyldiethyltriamine
POM	Polarized optical microscopy
POMS	Polyhedral oligomeric metallasilsesquioxane
POS	Polyhedral oligomeric silsesquioxane
POSS®	Polyhedral oligomeric silsesquioxane product of Hybrid Plastics, Inc.
PP	Polypropylene
ppb	Parts per billion
PPFS	Poly(pentafluorostyrene)
ppm	Parts per million
PPO	Poly(phenylene oxide)
PPS	Poly(phenylene sulfide)
PPSU	Polyphenylsulfone
ppt	Parts per trillion
PPV	Poly(*p*-phenylene vinylene)
ppy	Phenylpyridine
pq	2,2'-Pyridylquinoline
^{i}Pr or i-Pr	*iso*-Propyl
PS	Polystyrene
pt	3-(Pyridin-2'-yl)-1*H*-1,2,4-triazole

PTFE	Poly(tetrafluoroethylene)
PU	Polyurethane
PVA	Poly(vinyl alcohol)
P4VP	Poly(4-vinylpyridine)
PVC	Poly(vinyl chloride)
PVDF	Poly(vinylidene fluoride)
PVK	Poly-(*N*-vinylcarbazole)
Q	Silicon atom attached to four oxygen atoms
QD	Quantum dot
Q-DLTS	Charge-based deep-level transient spectroscopy
R	Alkyl or generic chemical group
RAFT	Radical addition-fragmentation chain transfer
R_g	Radius of gyration
RGD	Peptide with R (arginine) G (glycine) D (aspartic acid) sequence
R_h	Hydrodynamic radius
RH	Relative humidity
RI	Refractive index
RLS	Resonance light scattering
RMS	Root mean square
ROMP	Ring-opening metathesis polymerization
RPM	Revolutions per minute
RT	Room temperature
SAW	Surface acoustic wave
SAXS	Small angle X-ray scattering
SBS	Styrene-butadiene-styrene
SEC	Size exclusion chromatography
SEM	Scanning electron microscopy
SFG	Sum frequency generation
SFIL-R	Step and flash imprint lithography, reverse tone
SIMS	Secondary ion mass spectrometry
SLS	Static light scattering
S_mA	Smectic A phase
S_mC	Smectic C phase
SOFC	Solid oxide fuel cell
T	Silicon atom attached to three oxygen atoms
T_d	Decomposition temperature
TEG	Thromboelastography
T_g	Glass transition temperature
T_8	Si_8O_{12} cage
TBHP	*tert*-Butyl hydroperoxide
TCP	Tissue culture plastic
TEA	Triethylamine
TEM	Transmission electron microscopy
TEOS	Tetraethylorthosilicate
Tf	Triflate
TGA	Thermogravimetric analysis
TGDDM	Tetraglycidyl-4,4'-diaminodiphyenyl methane
THF	Tetrahydrofuran
TIC	Toxic industrial chemical
tmeda	*N,N,N',N'*-tetramethylethylenediamine
TMS	Trimethylsilyl
TNT	Trinitrotoluene
T_{ODT}	Temperature of the order to disorder transition
TOF	Turnover frequency

TON	Turnover number
TSAXS	Transmission small angle X-ray scattering
UHV	Ultra-high vacuum
UV	Ultraviolet
Vi	Vinyl
WAXS	Wide angle X-ray scattering
XPS	X-ray photoelectron spectroscopy
XRD	X-ray diffraction
YAG	Yttrium aluminum garnet

MIX
Papier aus verantwortungsvollen Quellen
Paper from responsible sources
FSC® C105338

If you have any concerns about our products,
you can contact us on
ProductSafety@springernature.com

In case Publisher is established outside the EU,
the EU authorized representative is:
Springer Nature Customer Service Center GmbH
Europaplatz 3, 69115 Heidelberg, Germany

Printed by Libri Plureos GmbH
in Hamburg, Germany